FUNGAL DISEASES OF RICE AND THEIR MANAGEMENT

FUNGAL DISEASES OF RICE AND THEIR MANAGEMENT

Edited by

Deepti Srivastava

Md. Shamim

Malik Mobeen Ahmad

R. S. Upadhyay

AAP | APPLE ACADEMIC PRESS

First edition published 2024

Apple Academic Press Inc.
1265 Goldenrod Circle, NE,
Palm Bay, FL 32905 USA

760 Laurentian Drive, Unit 19,
Burlington, ON L7N 0A4, CANADA

CRC Press
2385 NW Executive Center Drive,
Suite 320, Boca Raton FL 33431

4 Park Square, Milton Park,
Abingdon, Oxon, OX14 4RN UK

Library and Archives Canada Cataloguing in Publication

Title: Fungal diseases of rice and their management / edited by Deepti Srivastava, Md. Shamim, Malik Mobeen Ahmad, R. S. Upadhyay.
Names: Srivastava, Deepti, editor. | Shamim, Md., 1985- editor. | Ahmad, Malik Mobeen, editor. | Upadhyay, R. S., editor.
Description: First edition. | Includes bibliographical references and index.
Identifiers: Canadiana (print) 20230225675 | Canadiana (ebook) 20230225691 | ISBN 9781774912478 (hardcover) | ISBN 9781774912485 (softcover) | ISBN 9781003332169 (ebook)
Subjects: LCSH: Rice—Diseases and pests. | LCSH: Fungal diseases of plants. | LCSH: Rice—Biotechnology.
Classification: LCC SB608.R5 F86 2023 | DDC 633.1/897—dc23

Library of Congress Cataloging-in-Publication Data

Names: Srivastava, Deepti, editor. | Shamim, Md., 1985- editor. | Ahmad, Malik Mobeen, editor. | Upadhyay, R. S., editor.

Title: Fungal diseases of rice and their management / Deepti Srivastava, Md. Shamim, Malik Mobeen Ahmad, R. S. Upadhyay.

Description: First edition | Palm Bay, FL, USA : Apple Academic Press, 2023. | Includes bibliographical references and index. | Summary: "Rice is a widely consumed crop around the world that has tremendous importance and is cultivated almost everywhere except Antarctica. However, various biotic and abiotic stresses have a negative effect on rice cultivation, seriously reducing its yield. This volume examines the bacterial and fungal pathogens that can cause rice diseases and explores how to manage these diseases. The volume covers the economic and environmental impact of rice fungal diseases on global food security and proceeds to delve into diagnostic methods for rice fungal pathogens, discussing both traditional and molecular techniques for detection. The volume discusses the potential of biocontrol agents for the sustainable management of rice fungal diseases and also provides new insights from the bioinformatic tools for rice fungal disease resistance. This book presents a number of key fundamental aspects such as pathogen ecology, epidemiology, and host-pathogen interactions of rice and offers a plethora of control measures to mitigate harm from fungal diseases. It presents an analysis of the current status of resistance genes as well as current strategic and applied methods, such as biotechnology, bioinformatics tools, and biological methods for the effective management of rice fungal diseases. Computational deciphering of blast resistant genes is also discussed. It elaborates on the ecology and research status of diseases and pathogens of rice, including rice sheath blight pathogen, sheath rot disease, rice blast pathogens, rice brown spot disease, rice kernel smut pathogen, rice bakanae/foot rot disease, and others. Fungal Diseases of Rice and Their Management is a valuable resource and guide for research students, rice breeders, rice biotechnologists, and others involved in rice research, production, and cultivation"-- Provided by publisher.

Identifiers: LCCN 2023016595 (print) | LCCN 2023016596 (ebook) | ISBN 9781774912478 (hardcover) | ISBN 9781774912485 (paperback) | ISBN 9781003332169 (ebook)

Subjects: LCSH: Rice--Diseases and pests. | Fungal diseases of plants.

Classification: LCC SB608.R5 F86 2023 (print) | LCC SB608.R5 (ebook) | DDC 633.1/8--dc23/eng/20230614

LC record available at https://lccn.loc.gov/2023016595

LC ebook record available at https://lccn.loc.gov/2023016596

ISBN: 978-1-77491-247-8 (hbk)
ISBN: 978-1-77491-248-5 (pbk)
ISBN: 978-1-00333-216-9 (ebk)

About the Editors

Deepti Srivastava, PhD

Deepti Srivastava, PhD, is Assistant Professor in the Department of Agriculture at Integral University, Lucknow, India. She is the author and co-author of 24 peer-reviewed research articles in reputed national and international journals as well as 25 book chapters. She has one authored book to her credit. Her major research area involves plant biotechnology and molecular breeding in rice. She did gene pyramiding of submergence, blast resistant, and dwarfing gene in one of the finest rice var. Kalanamak. Before joining Integral University, Dr. Srivastava worked at the CSIR-National Botanical Research Institute, Lucknow, where she was engaged in a DBT-funded project that involved development of hybrids of cotton by using a novel male sterility-fertility restorer system and DNA fingerprinting of different bottle gourd germplasms. Her current research work includes development of drought and heat tolerant rice varieties. Dr. Srivastava received a Young Scientist Award from a national and international conference for her research work in the field of molecular breeding.

Dr. Srivastava graduated from Dr. Ram Manohar Lohia Avadh University, Faizabad, India, and MSc (Ag Biotechnology) and PhD (Agricultural Biotechnology) degrees from Narendra Deva University of Agriculture and Technology, Kumarganj, Faizabad, India. During her MSc program, she assessed the diversity of various bottle gourd germplasm with DNA fingerprinting and protein profiling.

Md. Shamim, PhD

Md. Shamim is working as an Assistant Professor and Jr. Scientist in the Department of Molecular Biology and Genetic Engineering at the Bihar Agricultural University, Sabour, India. He is the author or co-author of 50 peer-reviewed journal articles, 30 book chapters, and two conference papers. He has four authored books and three practical books to his credit. He is an editorial board member of several national and international journals.

Dr. Shamim received the Young Scientist Award 2016 for his research work on biotechnology from the Bioved Research Institute of Agriculture, Technology and Sciences, Allahabad, India. He was also awarded Maulana Azad National Fellowship Award from the University Grants Commission, New Delhi, India, during his PhD degree program.

Before joining Bihar Agricultural University, Dr. Shamim worked at the Indian Agricultural Research Institute, New Delhi, where he engaged on heat-responsive gene regulation in wheat. Dr. Shamim also worked at the Indian Institute of Pulses Research, Kanpur, India, on molecular and phylogeny analysis of several Fusarium fungus of pulses as well as at the Biochemistry Department of Dr. Ram Manohar Lohia Institute on plant protease inhibitor isolation and their characterization.

He is a member of the soil microbiology core research group at Bihar Agricultural University (BAU), where he provides appropriate direction and assists with prioritizing the research work on plant growth-promoting rhizobacteria (PGPRs). He has proved himself an active scientist in the area of biotic stress management in rice, especially in yellow stem borer management by isolating protease inhibitor from jackfruit seeds and sheath blight resistance mechanism in wild rice, cultivated rice, and other hosts.

Dr. Shamim acquired his BSc (Biology) degree from Dr. Ram Manohar Lohia Avadh University, Faizabad, India, and MSc (Biotechnology) and PhD (Agricultural Biotechnology) degrees from Narendra Deva University of Agriculture and Technology, Kumarganj, Faizabad, India, with specialization in biotic stress management in rice through molecular and proteomics tools.

Malik Mobeen Ahmad, PhD

Malik Mobeen Ahmad, PhD, is Assistant Professor of Microbial Biotechnology in the Department of Agriculture at Integral University, Lucknow, India, where he has been on the faculty since 2014. With his expertise in the area of microorganisms, he oversees the development of ready-to-use diagnostic kits for mycotoxins or other toxins produced in food crops of importance to developing countries.

Dr. Ahmad received a Young Scientist Award from the Society of Biological Sciences and Rural Development, which recognizes outstanding achievement in biological research leading to the advancement of biotechnology. In 2013, he was given an International Travel Grant by the Department of Science and

Technology, Govt. of India, for attending the 35th Mycotoxin Workshop at Ghent University, Belgium, a program aimed at training eminent microbial biotech scientists to be more efficient in performing science and to discuss ideas with scientists working in the respective area at the international scientific forum. He was also awarded a Senior and Junior Research Fellow of Research Fellowship in Science for Meritorious Students (RFSMS), funded by the University Grants Commission (UGC), Government of India, to carry out his doctoral studies. He has published over 20 scientific papers in national and international journals and five books. His current research focuses on bioprospecting of novel bioactive metabolites from endophytic fungi and their biotransformation.

R. S. Upadhyay, PhD

R. S. Upadhyay is a Professor of Botany since 1997 at the most prestigious University of India, Banaras Hindu University (BHU). He is credited with the discovery of the perfect state of *Fusarium udum*, the causal agent of the wilt of pigeon pea. He is also acclaimed for his contributions to the biology of this pathogen. He has published 143 research papers, 60 book chapters, three edited books, and one monographs of national and international repute. He has produced 27 PhD students ranging on the topics such as biological control of plant pathogens, programmed plant cell death in response to pathogens, bioremediation of toxic effluents, induced resistance in plants and their immunization, plant growth-promoting microbes, chitinase production from actinomycetes, mycorrhizal technology for reclamation of wastelands, role rhizobacteria in detoxifying phytotoxic effect, development of molecular markers in tracking microbes in environment, and effect of biotic and abiotic factors on plants, the molecular basis of plant–microbe interaction. Six PhD students are currently working under his supervision. He has supervised more than 60 MS/MD-level dissertations. His current research interest focuses mainly on biological control of fungal plant pathogens, induced resistance in plants against pathogens, and microbial technology. He is a member of various scientific societies and is on the editorial boards of various journals. Dr. Upadhyay is a recipient of the prestigious Young Scientist Award of the Indian National Science Academy (INSA), Professor L. S. S. Kumar Memorial Award of INSA. He was conferred by the Prime Minister of India with Prof. Hira Lal Chakravorty award of the Indian Science Congress Association in 1986. He has carried out collaborative research at Rothamsted Experimental Station (England; 1988–89) with Dr. David Hornby as a Visiting Fellow of the Royal Society of London; Illinois State University (USA; 1990–91) with Prof. R. K. Jayaswal, Distinguished Professor; and an INSA-JSPS Fellow (Japan; 1994–95) with Prof. Hajimu Komada at Shimane University. Dr. Upadhyay was the Coordinator of the Environmental Science Program at BHU, and currently, he is the Programme Coordinator of the Centre of Advanced Study in Botany, Banaras Hindu University. He was the Convener of the Mycology and Plant Pathology Group of the CAS in Botany at BHU. His biography has been published by various biographical publishing houses including *World Who's Who*.

Contents

Contributors

Abdelrazek S. Abdelrhim
Department of Plant Pathology, Faculty of Agriculture, Minya University, Al-Minya, Egypt

Malik M. Ahmad
Integral Institute of Agricultural Science and Technology (IIAST), Integral University, Dasauli, Lucknow, Uttar Pradesh, India

Salman Ahmad
Department of Agriculture, Integral Institute of Agricultural Science and Technology (IIAST), Integral University, Lucknow, Uttar Pradesh, India

Rajendra Bairwa
Dr. Rajendra Prasad Central Agricultural University, Pusa, Samstipur, Bihar, India

Raina Bajpai
Department of Mycology and Plant Pathology, Institute of Agricultural Sciences, Banaras Hindu University, Varanasi, Uttar Pradesh, India

Prasenjit Barman
Department of Plant Pathology, Uttar Banga Krishi Viswavidyalaya, Pundibari, Cooch Behar, West Bengal, India

Meenakshi Bisht
Department of Agriculture, Integral Institute of Agricultural Science and Technology, Integral University, Lucknow, Uttar Pradesh, India

Rajat Chaudhary
Division of Genetics, I.A.R.I., New Delhi, India

Sameer Akhtar Chaudhary
RASA Life Science Informatics, Pune, Maharashtra, India

Sapana Chaudhary
RASA Life Science Informatics, Pune, Maharashtra, India

Yuvraj Chopra
College of Agriculture, Punjab Agricultural University, Ludhiana, Punjab, India

Sarbani Das
Department of Plant Pathology, Bidhan Chandra Krishi Viswavidyalaya, Mohanpur, Nadia, West Bengal, India

Mona F. A. Dawood
Department of Botany & Microbiology, Faculty of Science, Assiut University, Egypt

Hari Shankar Gaur
Dean, School of Basic Sciences & Research, Sharda University, Greater Noida, Uttar Pradesh, India

Astha Gupta
School of Agricultural Sciences, Sharda University, Greater Noida, Uttar Pradesh, India

Amany A. Hassan
Botany and Microbiology Department, Faculty of Science, New Valley University, Egypt

Raja Husain
Department of Agriculture, Himalayan University, Jullang, Itanagar, Arunachal Pradesh, India

Touseef Hussain
Division of Plant Pathology, ICAR-IARI, Pusa, New Delhi, India

V. B. Jha
Department of Plant Breeding and Genetics, Dr. Kalam Agricultural College, Kishanganj (Bihar Agricultural University, Sabour, Bhagalpur), Bihar, India

Vinny John
Department of Plant Pathology, Sam Higginbottom University of Agriculture, Technology and Sciences, Prayagraj, Uttar Pradesh, India

Balwinder Kaur
Department of Entomology and Nematology, University of Florida, Gainesville, FL, USA

Rajvir Kaur
Department of Plant Breeding and Genetics, Punjab Agricultural University, Ludhiana, Punjab, India

Ravi Kesari
Department of Plant Breeding and Genetics, Bhola Paswan Shastri Agricultural College (Bihar Agricultural University, Sabour, Bhagalpur), Purnia, Bihar, India

N. A. Khan
Department of Plant Molecular Biology and Genetic Engineering, Acharya Narendra Deva University of Agriculture and Technology, Kumarganj, Uttar Pradesh, India

Nadeem Khan
Department of Agriculture, Integral Institute of Agricultural Science and Technology (IIAST), Integral University, Lucknow, Uttar Pradesh, India

Abhishek Kumar
Department of Molecular Biology and Genetic Engineering, Kalam Agricultural College (Bihar Agricultural University, Sabour), Kishanganj, Bihar, India

Deepak Kumar
Divison of Genetics, Indian Agriculture Research Institute, New Delhi, India

Mahesh Kumar
Department of Molecular Biology and Genetic Engineering, Dr. Kalam Agricultural College, Kishanganj (Bihar Agricultural University, Sabour, Bhagalpur), Bihar, India

Sanjeev Kumar
Department of Plant Pathology, Bihar Agricultural University, Sabour, Bhagalpur, Bihar, India

Santosh Kumar
Department of Plant Pathology, Mandan Bharti Agricultural College (Bihar Agricultural University, Sabour, Bhagalpur) Agwanpur Saharsa, Bihar, India

Vinod Kumar
Department of Molecular Biology and Genetic Engineering, Bihar Agricultural University, Sabour, Bhagalpur, Bihar, India

Amit Kumar Maurya
Department of Plant Pathology, Sam Higginbottom University of Agriculture, Technology and Sciences, Prayagraj, Uttar Pradesh, India

Rashmi Maurya
Department of Botany, University of Lucknow, Lucknow, Uttar Pradesh, India

Jhumishree Meher
Department of Mycology and Plant Pathology, Institute of Agricultural Sciences, Banaras Hindu University, Varanasi, Uttar Pradesh, India

Anurag Mishra
Division of Genetics, I.A.R.I., New Delhi, New Delhi, India

Yasser S. Moursi
Department of Botany, Faculty of Science, University of Fayoum, Fayoum, Egypt

Md. Abu Nayyer
Integral Institute of Agricultural Science and Technology, Integral University, Dasauli, Lucknow, Uttar Pradesh, India

Pramila Pandey
Department of Botany, BSNV, PG College, Lucknow, Uttar Pradesh, India

Sonika Pandey
ICAR—Indian Institute of Pulses Research, Kanpur, Uttar Pradesh, India

Hemlata Pant
Assistant Professor, Department of Zoology, C. M. P. PG. College, Prayagraj, Uttar Pradesh, India
ICAR—Indian Institute of Pulses Research, Kanpur, Uttar Pradesh, India
Institute of Molecular Plant Science, University of Edinburgh, Edinburgh, UK

Jayashri Prasanan
RASA Life Science Informatics, Pune, Maharashtra, India

Mohd. Rameez
Department of Agriculture, Integral Institute of Agricultural Science and Technology (IIAST), Integral University, Lucknow, Uttar Pradesh, India

Tushar Ranjan
Department of Molecular Biology and Genetic Engineering, Bihar Agricultural University, Sabour, Bhagalpur, Bihar, India

Md. Mahtab Rashid
Department of Mycology and Plant Pathology, Institute of Agricultural Sciences, Banaras Hindu University, Varanasi, Uttar Pradesh, India

Shivi Rathore
Department of Material Science and Engineering, National Taiwan University of Science and Technology, Taipei, Taiwan

Sakshi Rawat
RASA Life Science Informatics, Pune, Maharashtra, India

Nandini Roy
Department of Plant Pathology, Uttar Banga Krishi Viswavidyalaya, Pundibari, Coochbehar, West Bengal, India

Bholanath Saha
Dr. Kalam Agricultural College, Bihar Agricultural University, Kishanganj, Bihar, India

Sushanta Saha
Directorate of Research, Bidhan Chandra Krishi Viswavidyalaya, Kalyani, Nadia, West Bengal, India

Mohd. Said
Department of Agriculture, Integral Institute of Agricultural Science and Technology (IIAST), Integral University, Lucknow, Uttar Pradesh, India

Dinesh Kumar Saini
Department of Plant Breeding and Genetics, Punjab Agricultural University, Ludhiana, Punjab, India
Division of Plant Pathology, ICAR-IARI, Pusa, New Delhi, India

Karansher Singh Sandhu
Department of Crop and Soil Sciences, Washington State University, Pullman, USA

Md. Shamim
Department of Molecular Biology and Genetic Engineering, Dr. Kalam Agricultural College, Kishanganj (Bihar Agricultural University, Sabour, Bhagalpur), Bihar, India

Vandana Sharma
Division of Genetics, I.A.R.I., New Delhi, India

Mohd. Haris Siddiqui
Department of Agriculture, Integral Institute of Agricultural Science and Technology (IIAST), Integral University, Lucknow, Uttar Pradesh, India

Saba Siddiqui
Department of Agriculture, Integral Institute of Agricultural Science and Technology (IIAST), Integral University, Lucknow, Uttar Pradesh, India

Jagmohan Singh
Indian Agricultural Research Institute, New Delhi, New Delhi, India
Division of Plant Pathology, Indian Agricultural Research Institute, New Delhi, New Delhi, India
Institute of Molecular Plant Science, University of Edinburgh, Edinburgh, UK

Munna Singh
Department of Botany, University of Lucknow, Lucknow, Uttar Pradesh, India

Prakash Singh
Department of Plant Breeding and Genetics, V.K.S. Agricultural College (Bihar Agricultural University, Sabour, Bhagalpur), Dumraon, Bihar, India

D. K. Srivastava
Council of Science and Technology, Lucknow, Uttar Pradesh, India

Deepti Srivastava
Integral Institute of Agricultural Science and Technology, Integral University, Dasauli, Lucknow, Uttar Pradesh, India

Basavaraj Teli
Department of Mycology and Plant Pathology, Institute of Agricultural Sciences, Banaras Hindu University, Varanasi, Uttar Pradesh, India

S. P. Tiwari
Faculty of Agriculture, Mangalaytan University, Jabalpur, India

Nitin Vikram
Department of Agriculture, Zila Parishad Agriculture College, Banda, Uttar Pradesh, India

Garima Yadav
ICAR-Indian Institute of Pulses Research, Kanpur, Uttar Pradesh, India

Prashant Yadav
ICAR—Directorate of Rapeseed Mustard Research, Bharatpur, Rajasthan, India

Abbreviations

ABA	abscisic acid
AFLP	amplified fragment length polymorphism
AG	anatomosis group
AHLs	*N*-acylhomoserine lactones
APX	ascorbate peroxidase
Avr gene	avirulence resistant gene
BCAs	biocontrol agents
Big-PI	GPI-anchored proteins identification
BIL	backcross inbred lines
BioNano	BioNano genomics
BIP	backward inner primer
BjNPR1	*Brassica juncea* nonexpressor of pathogenesis-related genes 1
BLAST	Basic Local Alignment Search Tool
BLB	bacterial leaf blight
BLS	bacterial leaf streak
bp	base pair
BSR2	*broad-spectrum resistance 2*
CAPS	cleaved amplified polymorphic sequences
CAT	catalase
CBIs	chlorine biosynthesis fungicides
CC-NB-LRR	coiled-coil domain, a nucleotide binding domain and leucine rich repeats
CCR	cinnamoyl-CoA reductase
CCTOP	consensus Constrained TOPology prediction
CFU	colony-forming units
CIPDI	cumulative increment in percentage disease index
CK	cytokinin
cM	centimorgan
CMS	cytoplasmic male sterility
CNN	convolutional neural network
CO_2	carbon dioxide
COVID-19	coronavirus disease of 2019.

CRISPER-Cas9	clustered regularly interspaced short palindromic repeats-CRISPER associated protein-9
CRRI	central rice research institute
DArT	diversity array technology
DArT	diversity array technology
DEB	diepoxybutane
DEG	differentially expressed genes
DEP	differentially expressed proteins
DEPGs	differentially expressed protein genes
DAPs	differential accumulation of proteins
DFVF	database of fungal virulence factors
DH	double haploids
DHAR	dehydroascorbate peroxidase
DI	disease incidence
DIBA	dot-immuno-binding assay
DIGA	double immune-diffusion gel assay
DL	deep learning
DNA	deoxynucleotide acid
dNTPs	deoxynucleoside triphosphate
DS	disease severity
EF	elongation factor
ELISA	enzyme-linked immune-sorbent assay
EMS	ethyl methanesulfonate
EPSs	exopolyasaccharides
ERF	ethylene response transcription factor
ET	ethylene
ETI	effector-triggered immunity
ETS	effective-triggered susceptibility
EU	European Union
EWGLI	European Working Group for Legionella Infections
FAO	Food and Agriculture Organization
FAOSTAT	Food and Agriculture Organization of United Nations
FCM	flow cytometry
FIP	forward inner primer
FISH	fluorescence in-situ hybridization
FSB	false smut balls
FSRD	fungal stress response database
FunSecKB	Fungal Secretome Knowledge Base
GA	gibberellic acid

GBS	genotyping by sequencing
GEBV	genomic estimated breeding values
GFP	green fluorescent protein
GLM	general linear model
GMO	genetic modified organism
GO	gene ontology
GOLD	Genomes OnLine Database
GS	genomic selection
GST	glutathione *S*-transferase
GWAS	genome-wide association study
GxE	genetic and environmental risk factors
H_2O_2	hydrogen peroxide
HCN	hydrogen cyanide
HGA	hierarchical genome-assembly process
HMM	hidden Markov model
HPLC	high-performance liquid chromatography
HPLC–MS/MS	high-performance liquid chromatography–mass spectroscopy
HR	hypersensitive response
HSP	heat shock protein
IAA	indoleacetic acid
ICAR	Indian Council of Agricultural Research
IGS	intergenic spacer
InDELS	insertion/deletion
IRGS project	International Rice Genome Sequencing Project
IRP	immune response peptide
IRRI	International Rice Research Institute
ISR	induced systemic resistance
ITS	internal transcribed spacer
JA	jasmonic acid
Kin-LRR	kinase domain and leucine-rich repeats
LAMP	loop-mediated isothermal amplification
LAR	latex agglutination reaction
LC–MS	liquid chromatograph–mass spectrometry
LFD	lateral flow devices
LFIA	lateral flow immunoassays
LMMs	lesion mimic mutants
LOD	logarithm of odds
LPA1	loose plant architecture 1

LPSs	lipopolysaccharides
LRR	lecuine-rich repeats
MAB	marker-assisted breeding
MAGP	marker-assisted gene pyramiding
MALDI-TOF MS	matrix-assisted laser desorption/ionization time-of-flight mass spectrometry
MAMPS	microbe-associated molecular patterns
MAPKs	mitogen-activated protein kinases
MAS	marker-assisted selection
Mb	mega base pair
MBIs	melanin biosynthetic inhibitors
MHAP	MinHash Alignment Process
miRNA	microRNA
MLST	multilocus sequence typing
MNs	meganucleases
mPCR	multiplex PCR
MPs	mating populations
NaCl	sodium chloride
NAD	nicotinamide adenine dinucleotide
NADPH	nicotinamide adenine dinucleotide phosphate
NASBA	nucleic acid sequence-based amplification
NBS	nucleotide-binding site
NBS-LRR	nucleotide-binding site and leucine-rich repeat
NCBI	National Center for Biotechnology Information
NILs	near isogenic lines
NIR	near-infrared
NLR	NOD-like receptor
NPR1	nonexpressor of pathogenesis-related genes-1
O3	ozone
OG	oligogalacturonide
OGAs	oligogalacturonides
OLC	overlap layout consensus
ONT	Oxford Nanopore Technologies
PacBio	Pacific biosciences
PAL1	phenylalanine ammonia-lyase 1
PAMP	pathogen-associated molecular patterns
PAOs	polyamine oxidases
PCR	polymerase chain reaction
PDA	potato dextrose agar

PDAs	plant defense activators
Pfam	protein-domain prediction
PG	polygalactin
PGIPs	polygalacturonase inhibiting proteins
PGPBs	plant growth-promoting bacteria
PGPF	plant growth-promoting fungi
PGPR	plant growth-promoting rhizobacteria
PHA	passive hemagglutination
PHI	pathogen host interaction
PHYLIP	protein distances calculation
PIPDI	periodical increment in percentage disease index
PKS	polyketide synthase
PM	plasma membrane
PMPRRs	plasma membrane (PM)-localized pattern recognition receptors
ppm	parts per million
PR	pathogenesis-related
PRGdb	plant resistance gene database
ProtComp	comparison of known homologous proteins
PRR	pattern recognition receptors
PTI	PAMP-triggered immunity
QoI	quinone outside inhibitor
QS	quorum sensing
QTL	quantitative trait loci
R gene	resistance gene
RAPD	random amplification of polymorphic DNA
RBOHs	respiratory burst oxidase homologs
RCA	rolling circle amplification
RDF	recommended dose
RFLP	restriction fragment length polymorphism
RGSV	rice grassy stunt virus
RIL	recombinant inbred lines
RNA	ribonucleic acid
ROS	reactive oxygen species
RP	resistant protein
RSV	rice stripe virus
RTBV	rice tungro bacilliform virus
RT-PCR	reverse transcriptase (RT)-PCR
RT-PCR	reverse-transcription PCR

RTSV	rice tungro spherical virus
SA	salicylic acid
SAR	systemic acquired resistance
SBR	sheath brown rot
SDS-PAGE	sodium dodoecyl sulphate polyacrylamide gel electrophoresis
Ser/thr-LRR	serine-threonine kinase-like domain and leucine rich repeats
ShB	sheath blight
SignalP3.0 Phobius	signal peptide cleavage site prediction
SLST	single locus sequencing typing
SMRT	single molecule real time sequencing
SMs	secondary metabolites
SNP	single nucleotide polymorphism
SOD	superoxide dismutase
Spp.	species
SSM	site-specific mutagenesis
SSP	small secreted proteins
SSR	simple sequence repeat
SSSLs	single-segment substitution lines
STR	short tandom repeat
SVM	support vector machine
SWEET	sugar will eventually be exported transporter
TAL	transcription activator like
TALEN	transcription activator-like effector nucleases
TALes	transcription activator-like effectors
T-DNA	transfer DNA
TEF	translation elongation factor
TGS	third-generation sequencing
TIR-NB-LRR	protein with Toll-interleukin receptor, a nucleotide binding domain, and a leucine-rich repeat
TLP	thaumatin-like protein
TM	transmembrane
UP-PCR	universally primed-polymerase chain reaction
URP	universal rice primer
VAM	vesicular-arbuscular mycorrhiza
WGS	whole-genome sequencing
WHO	World Health Organization
ZFN	zinc finger nucleases

Preface

Rice is a major food crop that has tremendous importance and is cultivated everywhere in the world except Antarctica. Asia is home to more than half of the world's population and 92% of the rice is cultivated here. Fungal disease of rice poses a serious threat to the world food security because rice is a staple food for more than 60% of the world's population. To overcome the losses due to fungal disease of rice, many management practices such as chemical control, biological control, and conventional breeding are adopted. However, these methods alone are not sufficient to control rice fungal diseases.

Chemical fertilizers are costly and farmers cannot afford them at a large level. In addition, they are not environment-friendly. On the other hand, in conventional breeding, undesired traits are also transferred along with resistance genes. Therefore, there is a need to review all the available methods for control as well as to find possible sustainable methods. Recent advances in biotechnology and bioinformatics helped researchers in finding a new source of resistance. Similarly, biocontrol methods are also used as these are ecofriendly. Any control measures for rice fungal diseases will work only when there is sufficient information on ecology, epidemiology of pathogen, and host–pathogen interactions.

This book presents a number of key fundamental aspects such as pathogen ecology, epidemiology, host–pathogen interactions, and various control measures including analysis of the current status of resistance genes. It also presents current status and perspectives in strategic and applied areas such as biotechnology, bioinformatics tools, biological methods for the effective management of rice fungal diseases.

Here, we tried to provide a comprehensive view to the reader about rice fungal diseases starting from the identification of fungal diseases, and its ecology to the management of bacterial diseases which includes conventional as well as recent methods. This book contains 18 chapters.

Chapter 1 deals with the economical and environmental impact of rice fungal diseases on global food security. Chapter 2 reviews diagnostic methods for rice fungal pathogens. In this, both traditional and molecular techniques for detection are discussed. Chapters 3, 4, and 5 investigate the ecology and host–pathogen interaction of rice blast pathogen along with conventional and molecular tools for its management. Computational deciphering of blast-resistant genes is also discussed.

Chapters 6, 7, and 8 provide insights from the epidemiology of sheath blight pathogen along with its conventional and molecular approaches for management. Chapter 9 and 10 elaborate ecology and research status of sheath rot disease of rice. Chapters 11 and 12 explore the ecology, epidemiology, and identification measures of rice brown-spot pathogen along with comparative analysis of conventional and recent methods for its management. Chapter 13 deals with the biology and ecological status of rice kernel smut disease. Current scenario of management of rice kernel smut disease is also discussed thoroughly in Chapter 14.

Chapter 15 focuses on the identification and pathogenic diversity of rice false-smut pathogen. It also discusses about the scope of resistance sources in the future breeding program. Chapter 16 discusses the role of emerging biotechnological tools for rice bakanae disease. Chapters 17 and 18 elaborate the potential of biocontrol agents for the sustainable management of rice fungal diseases and also provide new insights from the bioinformatic tools for rice fungal disease resistance.

Hence, detailed and updated knowledge related to the rice fungal diseases on a single podium will surely serve as a guide for research students, rice breeders, and rice biotechnologists. In addition to these, it will also help in understanding rice fungal disease genetics and the molecular basis of boosting rice fitness levels thereby enhancing its immunity.

CHAPTER 1

Economical and Environmental Impact of Rice Fungal Diseases on Global Food Security

MD. SHAMIM[1], MAHESH KUMAR[1], SANTOSH KUMAR[2], DEEPTI SRIVASTAVA[3], TUSHAR RANJAN[4], RAVI KESARI [5], RAJA HUSAIN[6], VINOD KUMAR[4], MD. ABU NAYYER[3], and V. B. JHA[7]

[1]*Department of Molecular Biology and Genetic Engineering, Dr. Kalam Agricultural College, Kishanganj, Bihar Agricultural University, Sabour, Bhagalpur, Bihar, India*

[2]*Department of Plant Pathology, Mandan Bharti Agricultural College, Bihar Agricultural University, Sabour, Bhagalpur, Agwanpur Saharsa, Bihar, India*

[3]*Integral Institute of Agricultural Science and Technology, Integral University, Dasauli, Lucknow, Uttar Pradesh, India*

[4]*Department of Molecular Biology and Genetic Engineering, Bihar Agricultural University, Sabour, Bhagalpur, Bihar, India*

[5]*Department of Plant Breeding and Genetics, Bhola Paswan Shastri Agricultural College, Bihar Agricultural University, Sabour, Bhagalpur, Purnia, Bihar, India*

[6]*Department of Agriculture, Himalayan University, Jullang, Itanagar, Arunachal Pradesh, India*

[7]*Department of Plant Breeding and Genetics, Dr. Kalam Agricultural College, Kishanganj, Bihar Agricultural University, Sabour, Bhagalpur, Bihar, India*

Fungal Diseases of Rice and Their Management. Deepti Srivastava, Md. Shamim, Malik Mobeen Ahmad, and R. S. Upadhyay (Eds.)

ABSTRACT

Rice production has a significant position in supplying food for the world's population, as it is used as a staple food in various countries. Climate changes are projected to have an important consequence on food security, agricultural income, production, and cultivation areas around the World. To reduce food insecurity, crop production areas must be increased by 2050. In addition to climate change, rice crop is also subjected to a variety of biotic stresses at different growth stages, with variations in number and intensity. These biotic stresses increase as an outcome of ongoing climate change. Rice-producing area is predicted to fall by 18–51% in the region of the tropics due to global warming over the next century. Biotic stresses not only reduce yield but also harms grain quality. The extensive cultivation of high-yielding, fertilizer-responsive cultivars, and hybrids in recent years has reported an increase in disease incidence and severity during the monsoon season. This chapter emphasizes the economic and environmental impact of rice fugal diseases on global food security.

1.1 INTRODUCTION

Rice is an important staple food crop in India, China, East Asia, Southeast Asia, Africa, and Latin America. Rice feeds up to 70% of the population of the said nations (FAO, 1995). Consumption of rice rises in several developed nations, including North America and the European Union (EU), as a result of dietary habit changes (Faure and Mazaud, 1996). Presently, about 161 million hectares of land are involved in rice production around the world, with an annual output of 678.7 million tons (FAO, 2009). In Asian countries around 90% of the world's rice is produced (about 143 million ha of the area with 612 million tons produced) (FAO, 2009). More than 3 billion Asians eat rice, which provides 30–75% of their total calories (Khush, 2004; Braun and Bos, 2004). For the feeding of the world's population, it is expected that an additional 114 million tons of processed rice will be needed by 2035, representing a 26% increase in production over the next 25 years. The chances of extending the rice-growing region in the immediate future are slim. As a result, the additional rice production required must come from increased productivity (Kumar and Ladha, 2011). The highest yield per unit area of land can only be obtained and maintained if there is a condition for crop safety against the pathogen in addition to high-yielding crop varieties

(Biswal et al., 2017; Srivastava et al., 2010; Srivastava et al., 2017). Among the various biotic stresses, rice disease caused by fungal pathogen is one of the most significant factors disturbing rice production and productivity. Due to biotic factors, there is a loss between 10–15% on a global scale. More than 200 million households around the world rely on rice production as their primary source of income and jobs (Emodi et al., 2016; Darsono et al., 2016).

Rice is the major source of nutrition and provides 500 calories a day to 2.5–3.5 billion people, the majority of them are from developing low-income countries (Smith, 1998; Byerlee et al., 2010; Maclean et al., 2013; Mohanty et al., 2013; Muthayya et al., 2014; FAOSTAT, 2019). It is one of the most important cereals provided by subsistence farmers for food security and income (Lopez, 2008; Kari and Korhonen-Kurki, 2013; McCouch, 2016). Rice was grown on 160 million hectares in 2008, yielding 480–685 million tons (Nguyen, 2002). The global rice production demand in 2020 is expected to be 140 million tons based on the present population growth rate, a 50% rise from 2009 (IIRR, 2005; Chauhan et al., 2000). Although the production of rice has enhanced significantly, however, it is still insufficient to meet the rising global demand (Sasaki and Burr, 2000; Gnanamanickam, 2009). Global rice production has lagged behind demand, and the deficit in rice production has been filled by dipping into surplus stocks (Nguyen, 2000). By 2030, the annual rice shortage is expected to rise from 400,000 tons in 2016 to 800,000 tons (Thirze, 2016; Khush, 2013).

Rice is grown in various agroclimatic conditions all over the world, and its productivity is influenced by biotic stresses (Zhang and Xie, 2014). Climate change-induced biotic stresses will reduce rice blast resistance (Newton and Young, 1996; Goodman and Newton, 2005). Pathogen distribution and development rates may change due to climate change, as well as rice resistance, growth, and metabolism (Bevitori and Ghini, 2014). Climate conditions, such as increase or decrease in temperature and moisture contents in the environments have a major impact on the germination of spores and the growth of lesions and are probable to influence pathogen distribution due to reduced efficiency of precautionary measures (Bevitori and Ghini, 2014; Coakley, 1999).

As a result, climate change would have a significant effect on management strategies that take advantage of host resistance (Chakraborty, 2000). Infield and laboratory studies, as well as modeling-based evaluations, quantitative analyses of the impact of climate change on pathogens are missing (Luck, 2011). As a result, reducing the effect of biotic stresses on rice yield is critical to rising and stabilizing yield. A reduction of 14% in rice yield globally per year is expected by a fungal pathogen (Agrios, 2005). Climate change is predicted to enhance the disease in fresh rice-producing regions due to increasing incidence and severity of the intense condition, as well as an increase in air temperature and concentration of atmospheric CO_2 (Anderson et al., 2004; Rosenzweig, 2005). The conception of options for high-yielding rice varieties adaptation and disease management in future rice-growing situations has been severely restricted due to a lack of investigation of diseases in rice under field environments (Chakraborty and Newton, 2011). Several pathogens are causing disease at their specific timing, humidity, temperature, and are responsible for significant declines in rice yield tolerance (Zeigler and Puckridge, 1995) (Table 1.1). Identification of the pathogen by molecular markers (Datta et al., 2011; Belete and Boyraz, 2019).) is of prime important along with their interaction with the host (Southwood and Ranganathan, 2019) for disease management.

1.2 MAJOR FUNGAL PATHOGENS OF RICE

1.2.1 BLAST DISEASE OF RICE

Blast is the most common fungal disease of rice. The application of fungicide is widely used in temperate or subtropical rice cultivation areas, primarily in Japan, China, South Korea, Taiwan, and Vietnam. Rice blast disease is a polycyclic disease caused by a filamentous ascomycete fungus that infects the aboveground tissues of rice plants with asexual spores (conidia) (Dean et al., 2005; Wilson and Talbot, 2009; Ou, 1985; Couch et al., 2005; Ebbole, 2007; Pennisi, 2010; Hamer, et al., 1988; Bourett and Howard, 1990; Veneault-Fourrey et al., 2006; Dagdas et al., 2012). The extensive phase of leaf dampness, 92–96% relative humidity and air temperatures of 25–280° Celsius are all conducive to sporulation and lesion growth (Agrios, 2005; b et al., 2007; Padmanabhan, 1965). Spore production peaks at night when the relative humidity is 100% and temperature is 22°C (Wilson and Talbot, 2009; Bevitori, 2014).

Fungal growth in rice cells, death of contaminated plant parts tissue, and necrotic injuries, all result within 3–5 days (Ou, 1985; Kingsolver et al., 1984; Wilson and Talbot, 2009). Hundreds of spores per night can be produced by a single lesion for more than 20 days under ideal conditions, resulting in one disease cycle per week (Kato, 2001). Resistance to rice blast pathogen is increased by drought stress and excessive nitrogen application

TABLE 1.1 Major Fungal Pathogen of Rice and Economic Losses.

Sl. No.	Name of the disease	Causative agent	Status	Physiological growth stage of crop	Environmental condition prevails		Per cent losses in rice	Reference
					Temperature range (^{0}C)	Relative humidity (%)		
1	**Blast**	*Pyricularia oryzae*	Major	Milking to maturity	~23–25	71.71–83.98	Approximately 30	Nalley et al. (2016)
2	**Sheath blight**	*Thanatephorus cucumeris*	Major	Seedling to dough	28–32	85–100	Ranged from 10 to 30	Kloepper and Schroth (1978)
3	**Sheath rot**	*Acrocylindrium oryzae*	Major	Booting to dough	20–28	73.85–87.40	Approximately 85	Bigirimana et al. (2015)
4	**Brown spot**	*Helminthosporium oryzae*	Minor	Heading to dough	23.79–32.83	78.91–91.07	Ranged from 4 to 52	Barnwal et al. (2013)
5	**Leaf spot by**	*Nigrospora oryzae*	Minor	Heading to milking	23.51–32.63	73.28–91.93	Approximately 10%	Wang et al. (2017)
6	**Stem rot**	*Sclerotium oryzae*	Minor	Seedling to booting	25.37–32.29	83.32–98.91	Ranged from —5 to 80	Li et al. (1984); Ou (1985); Kumar et al. (2003)
7	**False smut**	*Ustilaginoidea virens*	Minor	Mature	25–35	>95	Approximately 40% in severe infection	Dangi et al. (2020)
8	**Bakanae (Bak)**	*Fusarium fujikuroi*	Minor	Germinated seeds produce ascospores that infect the panicles	27–30	70–80	Approximately 20% in outbreak cases	Suparyono et al. (2009)
9	**Rotting by *Choanephora cucurbitarum***	*Choanephora cucurbitarum*	Reported	Stem elongation to milking	23.85–31.57	78.14–94.48	Minor	Ravat and Basu (2017)

(Bonman and Garrity, 1992; Long et al., 2000; Fukai and Cooper, 1995). Rice blasts could be easily handled if cultural traditions, chemicals, resistant types, and biocontrol agents were all used together. Rice blast is managed by segregating affected grains (Kurahashi et al., 1997).

Furthermore, broadcast disseminating should be avoided because non-uniform seed distribution will result in clusters of high plant densities, generating a favorable microclimate for the growth of rice blast disease (Kato, 1974; Kingsolver et al., 1984). Microbial antagonists (Suprapta, 2012) are now popular as natural products in comparison to chemical fungicides to control blast disease of rice (Martinez et al., 2019; Yoon et al., 2013; Srivastava et al., 2017). Examples include *Streptomyces* bacteria (Law et al., 2017) and *P. fluorescens* Pf7–14 used as a biocontrol agent, which produces antifungal phenazine-1-carboxylic acid (Valasubramanian, 1994). Rice blast can be regulated with fungicides, but caution should be exercised to prevent excessive use of related active ingredients which may create resistance in rice lines (Rossman et al., 1990; Baker, 1997; Kunova et al., 2014). Many rice blast-resistant genes, such as *Piz, Piz-t, Pit, Pik, Pik-m, Pik-p, Pita, Pita-2, and Pib,* are identified and used for generating new resistance rice lines (Chen et al., 200). *Pi21* tends to delay the plant's defensive responses, which could help with defense mechanism optimization (Fukuoka et al., 2009).

1.2.2 SHEATH BLIGHT

Sheath blight is another important disease of rice in the world, coming in second only to blast disease. *Rhizoctonia solani* a fungal disease causes sheath blight in rice (*Oryza sativa* L.). It affects 15–20 million hectares of rice fields in China, resulting in yield losses of 6 million tons per year (Marchetti and Bollich, 1991; Chen et al., 2012; Shamim et al., 2014; Shamim et al., 2016). *R. solani* also influenced 491,932 ha of rice in Japan in 2012 (JPPA, 2013).

Rice sheath blight causes loss up to 50% in 50% in susceptible cultivars in the United States (Prasad and Eizenga, 2008). Depending on the severity of the *R. solani* infection, yield losses were reported from 20–50% (Rao, 1995). About 54.3% of losses by *R. solani* have been estimated to be as high in India (Chahal, 2003). In intensive rice production systems, *R. solani* is particularly problematic. The pathogen may infect plants from over 32 families and 188 genera, indicating that it has a broad host range ((Savary et al., 2000; Srinivasachary et al., 2011).

In India, sheath blight disease was widespread in Srikakulm and Vijayangaram a district of Andhra Pradesh in the year 1993–1994 (Mathur et al., 1999). Further, in Punjab, sheath blight disease caused major losses in the years 1978, 2003, and 2004 (Chahal, 2005). In Bangladesh, up to 40% loss in rice production has been recorded (Shahjahan et al., 1986). *R. solani* infection in rice genotypes was also reported in different severity levels in Karnataka, with disease severity ranging from 6–15%. Some private varieties and hybrids were found to have a high intensity (>50%) of the disease in poorly controlled fields. In the Mandya and Mysuru districts of Karnataka in 2015, a disease severity of more than 60% was reported in hybrid rice (Chetana et al., 2016). Thousands of germplasm samples have been screened by several researchers; however, no real resistance sources are available in cultivated rice genotypes. The explanation for this is that sheath blight resistance is regulated by many minor genes or QTL, each with a small effect. Rice germplasms, such as Jasmine 85, Tetep, Teqing, ARC 10531 have been reported with significant QTLs, whereas few other cultivars showed mild resistance to sheath blight, with around 50 QTLs (Channamallikarjuna et al., 2010; Zou et al., 2000; Zuo et al., 2014; Yadav et al., 2015).

1.2.3 BROWN SPOT

Rice brown spot disease is caused by *Bipolaris oryzae*, a fungal pathogen that belongs to the Deuteromycotina (imperfect fungi) subdivision, class Deuteromycetes, order Moniliales, and family Dematiaceae. Brown spot is one of the most common rice diseases worldwide. The disease was blamed as a major contributor to the 1942 "Great Bengal Famine," which resulted in yield losses of 50–90% and the deaths of 2 million people. In 1900, Breda de Haan named the fungus *Helminthosposium oryzae*, which was later renamed *Drechslera oryzae* (Subramanian and Jain, 1966).

It is named *Bipolaris oryzae* as most conidia germinate from two end cells (Shoemaker, 1959). Ito and Kuribayashi (1927) discovered the teleomorph stage of the fungus, which they called *Ophiobolus miyabeanus*. The fungus was originally classified as belonging to the genus *Cochliobolus* by Drechsler (1934), but it was formally moved to that genus by Dastur (1942). *C. miyabeanus* Drechsler ex Dastur is the current name for this species. The most common species associated with a brown spot in Iran is *B. victoriae* (*C. victoriae*), followed by *B. oryzae (C. miyabeanus*), *B. bicolor*, and *B. indica* (Motlagh and Kaviani, 2008; Ito and Kuribayashi, 1927).

The disease is extremely important in many countries, and it has been confirmed to cause massive grain yield losses (up to 90%) when the leaf spotting process reaches epiphytotic proportions, as it did during the Great Bengal Famine of 1942 (Ghose et al., 1960). Brown spot disease is more common in areas where water is scarce and there is a nutritional imbalance (Kulkarni et al., 1980a; Baranwal et al., 2013). At severely infected locations in West Bengal, India, yield losses in rice cv. Tilakkachery (susceptible) and Bhasamanik (moderately resistant) ranged from 5.6–11.7% and 2.2%, respectively (Chattopadhyay et al., 1975). Severe infection reduced the tillers number as well as grain quality and reduction in weight reported with 30–43% of loss, compared with just 12% for moderate infection and non-significant (negligible) for lower infection ratings (Ou, 1985). The glume blotch process of the disease has been confirmed to cause extra harm and reduce seed germination ability (Kulkarni et al., 1986; Hiremath et al., 1983).

Yield reductions of 18.75–22.50% were observed by Ou (1985); (Singh et al., 1979; Fomba and Singh, 1990; Mia et al., 2001; Chakrabarti, 2001). Ou (1985), found a reduction in yield of 4.6–40%, 3.7–29.1%, 9.28–24.50%, and 26–52% was also reported. The disease's glume blotch mechanism has been shown to cause further harm and reduce seed germinability (Hiremath et al., 1983; Kulkarni et al., 1986). From seedling to milk level, the pathogen attacks the crop (Sunder et al., 2010). The leaf dries up after many spots coalesce. The scorched appearance of the affected nursery can also be recognized from a distance due to the death of the seedlings. The spots on susceptible cultivars are much larger, measuring up to 1 cm or more in length. The symptoms on the leaf sheaths and coleoptiles are the same as on the leaves. *H. rostratum* (*D. rostrata*) causes brown, minute spots, *H. oryzae* (*C. miyabeanus*) causes rectangular brown spot, and *H. halodes* causes leaf tip blight (Ramakrishnan and Subramanian, 1977).

A variety of fungicides have been tested in vitro and in vivo against the brown spot pathogen. The mycelial growth inhibitors Dithane M-45, Kitazin, Hinosan Thiram, Shield, Foltaf, Ridomil, Bitoxazol, PP 296, triadimenol, tridemorph, and edifenphos are highly effective (Gowda and Gowda, 1985; Kumari et al., 1997; Arshad et al., 2013). Hexaconazole and propiconazole were reported to be the most inhibitory to mycelial development, followed by iprobenphos and edifenphos (Sunder et al., 2005). Aside from these, iprodione (Pereira et al., 2002) and strobilurins, such as azoxystrobin, trifloxystrobin, and kresoxim methyl (Thind et al., 2004) are effective inhibitors of the pathogen's mycelial development. Resistance of host plants to this disease has not been reported as other important fungal diseases. Coordinated efforts by different researchers are urgently needed to identify resistance sources and incorporate them into breeding programs.

1.2.4 SHEATH ROT

Sarocladium is a genus of plant pathogens, saprobes, mycoparasites, endophytes, and potential human pathogens that was first described in 1975 (Gams and Hawksworth, 1975). There are currently 16 species in the genus *Sarocladium*, which includes plant pathogens, saprobes, mycoparasites, endophytes, and potential human pathogens (Mew and Gonzales, 2002; Giraldo et al., 2015). *Sarocladium oryzae* was first identified in Taiwan in 1922 as *Acrocylindrium oryzae*, the first organism to be linked to rice sheath rot symptoms. Phylum Ascomycota, the genus belongs to the Hypocreales order. *S. attenuatum* was once thought to be a separate species that caused rice sheath rot, but it is now considered a synonym of *S. oryzae* (Bridge et al., 1989; Bills et al., 2004).

S. oryzae is found in all rice-growing countries around the world (Sakthivel, 2001; Gopalakrishnan et al., 2010). The pathogen grows in size and sometimes coalesces, covering much of the leaf sheath; young panicles remain within the sheath or emerge only partially (Ou, 1985). Within affected sheaths, a thick whitish powdery growth can be observed, and young panicles are rotting. Infection with *S. oryzae* causes chaffy, discolored grains, as well as a reduction in seed viability and nutritional value (Mew and Gonzales, 2002). The following countries have mentioned it so far; Bangladesh, Brunei Darussalam, China, India, Indonesia, Japan, Malaysia, Nepal, Pakistan, Philippines, Saudi Arabia, Sri Lanka, Tajikistan, Thailand, Uzbekistan, Vietnam, Burundi, Cameroon, Côte d'Ivoire, Gambia, Kenya, Madagascar, Niger, Nigeria, Senegal, Tanzania, Mexico, USA, Argentina, Brazil, Venezuela, and Australia (CABI, 2007).

S. oryzae is mainly found in lowland rice-growing areas (Pearce et al., 2001), and the disease thrives in hot, humid weather (Sakthivel, 2001). *S. oryzae* infections were found below 1250 m in Nepal (Sharma et al., 1997). Temperature range of 20–30°C and relative humidity of 65–85% is prevalent for sheath rot disease induction (Sakthivel, 2001). A variety of fungicides have been used to manage sheath rot; however, their effectiveness is limited because they cannot destroy the fungus within the glumes (Sakthivel, 2001). Fungicides and insecticides were used in other control experiments, which yielded better results (Lakshmanan, 1992). Micronutrient foliar spraying is also said to minimize disease occurrence and increase grain yield (Sakthivel, 2001). Various plant extracts, such as neem, pungam oil, and rubber cakes are said to be effective against the disease (Narasimhan et al., 1998; Sakthivel, 2001). Biological control agents are reported to have potential to be beneficial in comparison to chemical fungicides (Sakthivel and Gnanamanickam, 1987; Mew et al., 2004a). Several *Pseudomonas* species effectively regulate *S. oryzae* by promoting antagonism, such as inhibiting fungal growth or inducing systemic resistance response (Saravanakumar et al., 2009). Breeders have initiated work for sheath rot resistance. However, it is restricted by the disease's numerous causes. Sheath rot is a serious problem for high-yielding nitrogen-responsive rice cultivars. Tall rice varieties are resistant to *S. oryzae* (Amin, 1976) in comparison to semi-dwarf rice varieties. A rudimentary toxin preparation-based approach for screening for resistance to *S. oryzae* was also studied for sheath rot control (Hemalatha et al., 1999). The said method was also used by Lakshmanan (1993b) and reported that rice mealy bug a vector of *S. oryzae* pathogen also responsible for the disease. Amin (1976) developed a screening method for resistance to *S. oryzae*, but it does not seem to have been continued.

1.2.5 FALSE SMUT

False smut disease in rice is caused by *Ustilaginoidea virens* (Cooke) Takahashi (teleomorph: *Villosiclava virens*) (Fan et al., 2016a). False smut disease is common in rice-growing part of the World (Brooks et al., 2009; Ladhalakshmi et al., 2012; Jecmen and TeBeest, 2015). It not only reduces grain yield and quality, but it also poses a health risk to humans and animals due to the presence of toxic mycotoxins in rice grains and straws from contaminated plants. Mycotoxins, such as ustiloxins and ustilaginoidins are formed by false smut balls. These false smut balls have fungal colonies in rice florets (Koiso *et al.*, 1994; Nakamura *et al.*, 1994; Zhou *et al.*, 2012; Lu *et al.*, 2015; Wang *et al.*, 2016a; Sun *et al.*, 2017; Wang *et al.*, 2017; Wang *et al.*, 2019).

False smut balls contain chlamydospores and sclerotia, which act as primary inocula to rice crop (Ikegami, 1960; Fan et al., 2010; Yong et al., 2018). A study for understanding different aspects responsible for the development of false smut balls is critical for disease control. False smut has been controlled using cultural, biological, and chemical methods. The effectiveness of various fungicides has been reported by several scientists

(Pannu et al., 2010; Mohiddin et al., 2012). Chemical fungicides are mostly used to control rice false smut disease. Copper fungicides, such as copper oxychloride and copper hydroxide have proven to be extremely powerful (Dodan and Singh, 1997). The most successful treatments were Copper Oxy Chloride (Blitox) 50 WP (0.3%) and Propiconazole (tilt) 25 EC (0.1%). Breeding robust resistance varieties are the most cost-efficient and effective way to treat false smut (Cartwright et al., 2000; Wang et al., 2010). Although there are significant differences in quantitative resistance among rice cultivars, no variety has yet demonstrated complete resistance to *U. virens* (Biswas, 2001; Li et al., 2008; Huang et al., 2015).

1.2.6 STEM ROT

Sclerotium oryzae causes stem rot disease in rice has been an economically significant disease and causes losses of around 35% yield reduction (Singh et al., 2002). The sclerotial type of the causal agent, *S. oryzae* Catt, was first described in Italy (Cattaneo, 1876). Sheath rot has established itself as a major rice disease (Reddy and Gosh, 1985) with 9.6–85% yield losses under favorable weather conditions (Phookan and Hazarika, 1992). Yield losses of up to 85% of rice-growing regions around the world have been recorded by *S. oryzae* (Naeimi et al., 2003). *S. oryzae* causes rice stem rot disease, it is becoming a severe setback in the Indian subcontinent's rice farming (Singh et al., 2002). In the Indian state of Karnataka, this disease became epidemic in 2012–2013, with yield losses of up to 30% recorded (Pramesh and Guruprasad, 2014). In recent years, this disease has also been identified as a principal limitation for rice cultivation in Andra Pradesh a south Indian state (Gopika et al., 2016).

In *S. oryzae,* sclerotia are mainly responsible for the infection, which can be created strewn around in old rice straw on the surface water near rice fields. As soon as the rice leaf sheath comes in contact with a mature rice stem at the waterline, these germinate. When germinated mycelia enter deep into the cavity created at the point of infection, the infection progresses in rice plants (Hussain and Ghaffar, 1993). During the maturation period, new sclerotia from inside the rice tissues are already infected with *S. oryzae*, and they continue in the rice debris and on the soil after harvesting the crops (Keim and Webster, 1975). This disease spreads quickly to other rice crops because of the lightweight nature of sclerotia balls so that these can easily disperse to other rice fields by floating with the help of irrigation water. Due to the rarity of stem rot resistance in rice genotypes (Chand et al., 1985), chemical control of the disease becomes an essential alternative. To find the most effective fungicide against *S. oryzae*, many in vitro studies were conducted. Also at low concentrations, the systemic fungicide Contaf has been shown to minimize mycelial density (Prakash et al., 2012). "Topsin-M," that is, Thiophanate methyl, "Bavistin," that is, Carbendazim (Sumitra et al., 2011), "Thiram," (Kumar et al., 2011), "Hexaconazole," "Propiconazole," and "Butachlor" were among the other systemic fungicides effective against stem rot (Gopika et al., 2011). Several other fungicides, such as tridemorph, tolfos–methyl, carboxin (Vitavax), Ziram, Cuman L-27, and TCM TB, are highly effective at suppressing the germination and development of this pathogen even at low concentrations. Benthiocarb (Saturn), a weedicide, had an inhibitory effect on *S. oryzae* growth and sclerotial formation.

The majority of rice cultivars are susceptible to *S. oryzae* infection (Chand et al., 1985). Figoni et al. (1983) found that three significant wild rice genotypes, *Oryza rufipogon*, *O. spontanea*, and *O. nivara*, are more resistant in comparison to cultivated rice varieties. Furthermore, due to the low heritability of the trait, genetic improvement is tough in cultivated rice varieties (Nayar, 1973). Biological control measures of *S. oryzae* pathogenic activity by bacteria, fungi, or actinomycetes have been also reported by several workers (Lockwood, 1964; Watson and Ford, 1972). Different modes of microbial antagonism are the cause of *S. oryzae* fungistatic (Griffin, 1972). Several fungicides effectively (Benomyl, Edifenphos, Thiophanate Methyl) were used and reported less incidence of stem rot disease in the field condition (Singh et al., 2002; Kumar et al., 2003; Gopika et al., 2016).

1.2.7 FOOT ROT

Fungus *Fusarium moniliformae* Sheldon causes foot rot disease (Bakanae disease). This fungal pathogen is one of the widespread pathogens in every rice-growing area around the world and causes losses in rice yield (Sunder and Satyavir, 1997). Footrot pathogen (*Fusarium moniliformae* Sheldon) is both seed and soil-borne (Ahmad and Raza, 1991). The disease is found in approximately every rice-growing region on the planet, including Africa, Australia, British Guiana, Cameroon, India, Italy, Ivory Coast, Japan, Kenya, Nigeria, Surinam, Tanzania, Thailand, Trinidad, Uganda, USA, Venezuela, Vietnam, Brazil, Ceylon, Indonesia, Philippines, Mexico, China (Singh and Sunder, 2012). Yellowish-green thin leaves, irregular stem elongation, less tillering, and rooting of the root stem joint are the most common symptoms of the disease. This pathogen causes significant economic loss under favorable environmental conditions (Saremi et al., 2008; Yasin et al., 2003). Despite the rice Bakanae's significant economic effects, there are few reliable and successful control methods available, except seed treatment with chemical fungicides.

Bakanae disease has emerged as a serious issue in several rice breeding programs (Li and Luo, 1997; Yang et al., 2003). Treatment of disease seeds with hot water or fungicides is the most common management practice (Hayasaka et al., 2001). Fungicides do not completely control fungal spores, and bakanae strains resistant to fungicides have been identified (Ogawa, 1988; Park et al., 2009; Kim et al., 2010; Lee et al., 2011). The development of resistant varieties may be a fruitful and environmentally sustainable way to combat the disease. However, despite comprehensive screening for bakanae-resistant rice germplasm, very few rice cultivars have been reported (Li et al., 1993; Lv, 1994; Khokhar and Jaffrey, 2002; Kim et al., 2014). There is an urgent need for the identification of new resistance genes from various sources for the development of resistant rice varieties (Lee et al., 2018).

1.2.8 NARROW BROWN SPOT

Narrow brown spot disease/*Cercospora* leaf spot is caused by the fungus *Sphaerulina oryzina* (syn. *Cercospora janseana, Cercospora oryzae*), which can infect leaves, sheaths, and panicles. This pathogen was discovered by Raciborski and named *Napicladium janseanum* Racib in 1900. Metcelf described a similar disease in North America in 1906. *Cercospora oryzae* Miyake was named after a Japanese pathologist who described it in 1906. In 1982, Constantinescu renamed *Cercospora janseana* (Racib.) O. Const. after Raciborski and Miyake (Hollier, 1992). Based on its phylogenetic association with the genus *Passalora* (Braun, 2000). Conidia are 15–60 × 3.6–4.0 μm, hyaline, cylindrical, narrow clavate, 3–10 septate, hyaline, cylindrical, narrow clavate. *Cercospora janseana* is a species of *Cercospora* (Racib). *Passalora janseana Racib*. or O. Const. *Sphaerulina oryzina* K. is an anamorph stage, while U. is an anamorph stage and hara is the fungus's teleomorph stage (Hara, 1918). *Cercospora janseana* produces cercosporin, a photoactivated perylenequinone toxin with a red to maroon pigment, which is a pathogenicity aspect for this pathogen (Daub and Chung, 2007).

The premature drying of the leaves causes a decrease in the yield caused by this disease (Simanjuntak et al., 2019). Since nutrients are prioritized for panicle formation, increased disease infestation on older leaves in comparison to younger leaves can be caused by the movement of divided nutrients (Fensionita, 2006). During prolonged humid conditions, the progression of fungal sporulation is accelerated in the leaves (Groth et al., 2010). In the dark, fungi grow slowly because they require light to activate reactive oxygen, which is toxic to the host plant (Daub et al., 2007). When *C. oryzae* infects the body, it causes elongated lesions with dark edges and a greyish core (Mani, 2015). The appearance of small and elongated lesions with a location parallel to the bone leaf is the first symptom. Narrow brown spot disease has extended across Java's rice-growing regions, causing yield reductions of 30–40% (Simanjuntak et al., 2019).

1.2.9 BAKANAE (BAK)

Bakanae is a rice disease caused by *Fusarium fujikuroi* (Sawada) Wollenw (formerly known as *Fusarium moniliforme* J. Sheld.) and has been known in Japan since 1828 (Ito and Kimura, 1931). Hori (1898) discovered the teleomorph and identified the disease-causing pathogen of Bak as *Fusarium heterosporium* Nees Fujikuro. Sawada (1917) named the fungus *Lisea fujikuroi*, which was later renamed *G. fujikuroi* (Sawada) (Ito and Kimura, 1931) under the genus *Gibberella*, with *Fusarium moniliforme* Sheld as its anamorph. According to Gupta et al. (2015), It is one of the most recent emergent concerns in rice, particularly in aromatic rice (Basmati rice) in India. This disease is becoming a more serious problem to sustainable rice production and in other rice-growing regions. Earlier bakanae was widely studied in Japan in the early 20th century. Before the discovery of fungicides, it was considered a potential threat to rice production. After the discovery of fungicides, the value of bakanae as a constraint in rice production has diminished. However, in recent years, the disease has resurfaced as a problematic disease in seedlings grown in seed boxes for mechanical transplanting or in dry bed nurseries in Japan and other countries. This is partially due to the fungal pathogen's seed-borne existence being overlooked. Other countries, such as Nepal, are still dealing with the problem (Desjardines et al., 2000). Bakanae was successfully managed with benomyl before fungicide resistance was discovered (Ogawa, 1988). As a result, when a single fungicide has been widely and intensively used, chemical seed treatment becomes less efficient. Plant disease control with resistant cultivars is a realistic, cost-effective, and environmentally sustainable strategy. Even though many rice varieties are documented to be resistant or susceptible in the field, there are few studies on host resistance to the Bakanae pathogen but there is no rice variety that has been developed having bakanae disease resistance.

1.2.10 LEAF SPOT BY NIGROSPORA ORYZAE

Nigrospora oryzae (Berk and Br.) Petch, 1924 is the fungus that causes leaf spot disease. The fungus has distinctive septate mycelium that is hyaline when young and darkens with age. Conidiophores develop laterally or terminally on the mycelium's ultimate branches. Since the disease has such a minor impact on the rice crop, it can go unnoticed. Even so, using disease-free seeds is still a good idea. Leaf spot on *O. sativa* has been linked to *N. oryzae* (Wang et al., 2017). The first study of *N. oryzae* causing leaf blight was reported in wild rice *O. rufipogon* in China (Liu et al., 2021).

1.2.11 ROTTING BY CHOANEPHORA CUCURBITARUM

In 2013, sheath rot was discovered in rice cv. Lal swarna samples were collected during surveys in West Bengal, India, and based on cultural, morphological, and pathogenicity investigations, causal organisms were identified as *Choanephora cucurbitarum*. This is the first time *C. cucurbitarum* has been found on rice in West Bengal, India. Ravat et al. (2015) identified *C. cucurbitarum* in rice from West Bengal, India, and stated that as a result of climate change, minor pathogens are replacing major pathogens, causing rigorous destruction under favorable environmental conditions. During the stem elongation to milk level, when temperatures sorts from 23.85°C to 31.57°C and relative humidity ranged from 78.14–94.48%, Rotting by *Choanephora cucurbitarum* was at its peak (Ravat and Basu, 2017).

1.3 STRATEGIES TO MITIGATE RICE DISEASES

Rice output must be increased to counteract the harmful consequences of climate change (Rehmani et al., 2011). Understanding the effects of climate change on air quality, temperature, rainfall, and sea level can help crop managers improve rice yield and production (Nguyen, 2000). Changes in rainfall duration, patterns, intensity, and frequency would have a significant impact on the efficacy of chemical control strategies in rice disease control (Chakraborty et al., 2000). Rainfall measurement followed by fungicide application can increase fungicide coverage on foliage (Shiba and Nagata, 1981), whereas a large amount or intensity of rainfall can lower fungicide coverage on foliage (Shiba and Nagata, 1981). (Phong et al., 2009). Though, after rising global demand for rice as a staple food and the advent of the green revolution, which resulted in the adoption of high-yielding rice varieties, high fertilizers dose, improved irrigation facility and intense cultural techniques have resulted in a significant increase in the prevalence and severity of rice diseases in many countries (Teng, 1990). Among the new emerging fungal diseases, rice blast (*Magnaporthe grisea*) and sheath blight (*Rhizoctonia solani*), the disease is the most common rice disease that inflicts significant economic losses, particularly in South and Southeast Asia (Ling, 1980).

1.4 RICE DISEASES DYNAMICS

Before the development of high-yielding cultivars, rice blast and brown spot were the most common diseases observed during 1940 in India. Bacterial leaf blight and sheath blight have all become serious diseases since the introduction of high-yielding varieties under cultivation. Earlier sheath rot, fake smut, stem rot, and grain discoloration diseases were reported as minor diseases, have recently emerged as major diseases, and are inflicting significant damage causing yield losses. Climate change, crop intensification, and changes in cultural practice are the factors contributing to the emergence of minor disease to major and major to minor. Rice blast disease causes a 35% drop in the overall yield loss in rice, while sheath blight causes a 25% yield reduction in rice production.

1.5 DISEASE RESISTANCE IN RICE: BARRIERS IN TRADITIONAL BREEDING AND THE IMPORTANCE OF NEW MOLECULAR TOOLS

Strategies of disease management are now an extra practical component in contrast to the hasty system of disease control used in an earlier era. Recognizing the economic costs of diseases, researchers have focused their efforts on determining the genetic basis of tolerance and susceptibility. The most effective of the multifaceted management techniques is host plant tolerance, as it removes the need for additional measures to mitigate harm unless other diseases are present. Studies have been conducted for a better understanding of the host–plant relationship in rice resulted in specialized breeding programs for disease resistance. Use of suitable donors for the hybridization purpose of the commercial cultivar, exposure of the segregating population to high levels of disease has been conducted. A further selection of resistant plants with an adequate level of resistance and desirable traits are selected in all steps in the resistance breeding process (Variar and Singh, 2018). Even though many disease resistance genes had been identified, the traditional breeding method had limitations in terms of engineering disease-resistant varieties due to several factors. In a traditional breeding program, long breeding cycles, low selection performance, complexity in hybridization involving distant parents, and lack of additional yield advantage involved for delayed products. Currently, molecular markers are the most advanced methods that are used for the transfer of the desired gene into rice variety (Das et al., 2017). With the advent of developments in molecular markers pyramiding of genes/QTLs that give resistance to biotic pressures and tolerance to different types of abiotic stresses is now achievable. (Das and Rao, 2015; Magar et al., 2014).

1.6 STRATEGIES FOR BETTER YIELD UNDER CLIMATE CHANGE

Crop improvement will aid in the development of a long-term and viable response to the unenthusiastic effects of climate change which needs to develop new strategies to combat the pathogen-causing disease in rice. When the government, non-governmental organizations, and the private sector work together, agricultural technology transfer is more successful (Orford and Beard, 1998; Ojha and Morin, 2001). Technology transmission and implementation of best practices for managing rice fungal disease to boost food security (FAO, 2008). Management along with the researchers from various disciplines to plan vulnerability and develop early warning systems could pave the way for the development of effective and long-term regulation policies to reduce rice disease losses (Bevitori and Ghini, 2014). New tools of molecular breeding can aid in exploiting the intrinsic ability of plants. To classify the underlying QTLs/genes, large-scale screening of tolerance in wild genotypes should be carried out using molecular markers. With advancements in bioinformatics, DNA microarrays, mass spectrometry, RNA sequencing, and other new high-throughput technologies, using genomic approaches, a top-down approach to deciphering the underlying metabolic pathways is now possible. By introducing biotic tolerant characters against important fungal pathogens from wild species, cultivation of gene-conferred resistant rice varieties is a cost-effective, environmentally sustainable way to protect the rice varieties from fungal pathogens and reduce losses.

1.7 CONCLUSIONS

Crop improvement will contribute to the creation of a long-term and effective response to the detrimental effects of climate change; therefore, there is a compelling need to create measures to offset these stresses. Climate change is becoming more visible around the world, and the effects are noticeable in rice production. Extreme biotic factors such as fungal diseases are all consequences of climate change. Damage caused by fungal pathogens poses serious threats to rice production, as well as affecting farmers' livelihoods. To convene the demands of an eternally growing population, rice plants must be protected from biotic stresses that decrease yield and grain quality. Resistance breeding for several biotic stresses in rice is currently progressing quickly thanks to the combination of molecular markers and marker-assisted selection techniques.

KEYWORDS

- **broadcast**
- **climate change**
- **rice yield**
- **wild rice**
- **food security**

REFERENCES

Agrios, G. N. *Introduction to Plant Pathology*, 5th ed.; Elsevier Academic Press Publication: San Diego, CA, USA, 2005; pp 125–170.

Ahmad, R. I.; Raza, T. Association of Fusarium Moniliforme Sheld With the Rice Seeds and Subsequent Infection in Pakistan. *Int. Rice Res. Newslett.* **1991,** *16,* 19–20.

Amin, K. Sources of Resistance to *Acrocylindrium oryzae*–Sheath Rot of Rice. *Plant Dis. Rep.* **1976,** *60,* 72–73.

Anderson, P. K.; Cunningham, A. A.; Patel, N. G.; Morales, F. J.; Epstein, P. R.; Daszak, P. Emerging Infectious Diseases of Plants. Pathogen Pollution, Climate Change and Agrotechnology Drivers. *Trends Ecol. Evol.* **2004,** *19,* 535–544.

Arshad, H. M. I.; Hussain, N.; Ali, S.; Khan, J. A.; Saleem, K.; Babar, M. M. Behaviour of Bipolaris Oryzae at Different Temperatures, Culture Media, Fungicides and Rice Germplasm for Resistance. *Pak. J. Phytopathol.* **2013,** *25,* 84–90.

Baker, B.; Zambryski, P.; Staskawicz, B.; Dinesh-Kumar, S. P. Signaling in Plant-microbe Interactions. *Science* **1997,** *276,* 726–733.

Baranwal, M. K.; Kotasthane, A.; Magculia, N.; Mukherjee, P. K.; Savary, S.; Sharma, A. K.; Singh, H. B.; Singh, U. S.; Sparks, A. H.; Variar, M.; Zaidi, N. A Review on Crop Losses, Epidemiology and Disease Management of Rice Brown Spot to Identify Research Priorities and Knowledge Gaps. *Eur. J. Plant Pathol.* **2013,** *136,* 443–457.

Barnwal, M. K.; Kotasthane, A.; Magculia, N.; Mukherjee P. K.; Savary, S.; Sharma, A. K.; Singh, H. B.; Singh U. S.; Sparks, A. H.; Variar, M.; Zaidi, N. A Review on Crop Losses, Epidemiology and Disease Management of Rice Brown Spot to Identify Research Priorities and Knowledge Gaps. *Eur. J. Plant Pathol.* **2013,** *136,* 443–457.

Belete, T.; Boyraz, N. Biotechnological Tools for Detection, Identification and Management of Plant Diseases. *Afr. J. Biotechnol.* **2019,** *18*(29), 797–807.

Beneault-Fourrey, C.; Barooah, M.; Egan, M.; Wakley, G.; Talbot, N. J. Autophagic Fungal Cell Death is Necessary for Infection by the Rice Blast Fungus. *Science* **2006,** *312,* 580–583.

Bevitori, R.; Ghini, R. Rice Blast Disease in Climate Change Times. Embrapa Arroze Feijao-Artigoem Periodico Indexado (ALICE). [Online] 2014. http://www.alice.cnptia.embrapa.br/alice/handle/doc/1022268 (accessed Aug 8, 2019).

Bigirimana, V. P.; Hua, G. K. H.; Nyamangyoku, O. I.; Höfte, M. Rice Sheath Rot: An Emerging Ubiquitous Destructive Disease Complex. *Front. Plant Sci.* **2015,** *6,* 1066.

Bills, G. F.; Platas, G.; Gams, W. Conspecificity of the Cerulenin and Helvolic Acid Producing "*Cephalosporium Caerulens*", and the Hypocrealean Fungus *Sarocladium Oryzae. Mycol. Res.* **2004,** *108,* 1291–1300.

Biswal, A. K.; Shamim, M.; Cruzado, K.; Soriano, G.; Ghatak, A.; Toleco, M.; Vikram, P. Role of Biotechnology in Rice Production. In *Rice Production Worldwide*; Chauhan, B. S.; et al., Eds.; Springer International Publishing, 2017; pp 787–545.

Biswas, A. False Smut Disease of Rice: A Review. *Environ. Ecol.* **2001,** *19,* 67–83.

Bonman, J.; Garrity, D. Effects of Nitrogen Timing and Split Application on Blast Disease in Upland Rice. *Plant Dis.* **1992,** *76,* 384–389.

Bourett, T. M.; Howard, R. J. In Vitro Development of Penetration Structures in the Rice Blast Fungus (Magnaporthe Grisea). *Can. J. Bot.* **1990,** *68,* 329–342.

Braun, U. Miscellaneous Notes on Some Micromycetes. *Schlechtendalia* **2000,** *5,* 31–56.

Braun, J.; Bos, M. S. The Changing Economics and Politics of Rice: Implications for Food Security, Globalization, and Environmental Sustainability. In "*Rice Is Life: Scientific Perspectives for the 21st Century*"; Toriyama, K., Heong, K. L., Hardy, B., Eds.; pp 7–20. International Rice Research Institute, Los Ban˜ os, Philippines and Japan International Research Center for Agricultural Sciences, Tsukuba, Japan, 2004.

Bridge, P. D.; Hawskworth, D. L.; Kavishe, D. F.; Farnell, P. A. A Revision of the Species Concept in *Sarocladium*, the Causal Agent of Sheath-Rot in Rice and Bamboo Blight, Based on Biochemical and Morphometric Analyses. *Plant Pathol.* **1989,** *38,* 239–245.

Brooks, S. A.; Anders, M. M.; Yeater, K. M. Effect of Cultural Management Practices on the Severity of False Smut and Kernel Smut of Rice. *Plant Dis.* **2009,** *93,* 1202–1208.

Byerlee, D. R.; Dawe, D.; Dobermann, A.; Mohanty, S.; Rozelle, S.; Hardy, B. *Rice in the Global Economy: Strategic Research and Policy Issues for Food Security*; No. 164488; International Rice Research Institute: Los Baños, Philippines, 2010.

CABI. *Crop Protection Compendium.* CAB International: Wallingford, 2007.

Cartwright R. D.; Lee, F. N.; Beaty, T.; Sutton, E. A.; Parsons, C. E. Reaction of Rice Cultivars/ Lines to False Smut, Stem Rot, and Black Sheath Rot Disease. *Univ. Arkansas Agric. Exp. Stn. Res. Ser.* **2000,** *476,* 158–168.

Cattaneo, A. Sullo Sclerotium Oryzae Nouvo Parasite Vegetable Che ha Devasto nel Corrente Anno Molte Risaje di Lombardia e del Novarese. *Rediconti R. Lombard.*; Milano; **1876,** *2ser 9,* 801–807.

Chahal, S. S. Disease Scenario in Rice During Post Revolution Era in Punjab. *Plant Dis. Res.* **2005,** *20*(1), 1–11.

Chakrabarti, N. K. Epidemiological and Disease Management of Brown Spot of Rice in India. In *Major Fungal Diseases of Rice: Recent Advances*; Kluwer Academic Publishers, 2001; pp 293–306.

Chakraborty, S.; Newton, A. C. Climate Change, Plant Diseases and Food Security. An Overview. *Plant Pathol.* **2011,** *60*, 2–14.

Chakraborty, S.; Tiedemann, A. V.; Teng, P. S. Climate Change: Potential Impact on Plant Diseases. *Environ. Pollut.* **2000,** *108*, 317–326.

Chand H.; Singh R.; Panwar D. V. S. Search for Multiple Resistance in Rice, *Indian Phytopathol.* **1985,** *38,* 452–455.

Channamallikarjuna, V.; Sonah, H.; Prasad, M.; Rao, G. J. N.; Chand, S.; Upreti, H. C.; Singh, N. K.; Sharma, T. R. Identification of Major Quantitative Trait Loci qSBRl 1-1 for Sheath Blight Resistance in Rice. *Mol. Breed.* **2010,** *25,* 155–166.

Chattopadhyay, S. B.; Chakrabarti, N. K.; Ghosh, A. K. Estimation of Loss in Yield of Rice Due to Infection of Brown Spot Incited by Helminthosporium Oryzae. *FAO Intern. Rice Comm. Newsl.* **1975,** *24,* 67–70.

Chauhan, B. S.; Jabran, K.; Mahajan, G. *Rice Production Worldwide*; Springer: Berlin/ Heidelberg, Germany, 2017; vol 247.

Chen, D. X.; Chen, X. W.; Wang, Y. P.; Zhu, L. H.; Li, S. G. Genetic Transformation of Rice With Pi-d2 Gene Enhances Resistance to Rice Blast Fungus Magnaporthe Oryzae. *Rice Sci.* **2010,** *17,* 19–27.

Chen, Y.; Zhang, A. F.; Wang, W. X.; Zhang, Y.; Gao, T. C. Baseline Sensitivity and Efficacy of Thifluzamide in *Rhizoctonia Solani*. *Ann. Appl. Biol.* **2012,** *161*, 247–254.

Chethana, B. S.; Deepak, C. A.; Rajanna, M. P.; Ramachandra, C.; Shivakumar, N. Current Scenario of Rice Diseases in Karnataka. *I.J.S.N.* **2016,** *7*(2), 405–412.

Coakley, S. M.; Scherm, H.; Chakraborty, S. Climate Change and Plant Disease Management. *Annu. Rev. Phytopathol.* **1999,** *37*, 399–426.

Couch, B. C.; Fudal, I.; Lebrun, M.-H.; Tharreau, D.; Valent, B.; van Kim, P.; Notteghem, J.-L.; Kohn, L. M. Origins of Host Specific Populations of the Blast Pathogen (Magnaporthe Oryzae) in Crop Domestication With Subsequent Expansion of Pandemic Clones on Rice and Weeds of Rice. *Genetics* **2005,** *170,* 613–630.

Dagdas, Y. F.; Yoshino, K.; Dagdas, G.; Ryder, L. S.; Bielska, E.; Steinberg, G.; Talbot, N. J. Septin-Mediated Plant Cell Invasion by the Rice Blast Fungus (Magnaporthe Oryzae). *Science* **2012,** *336,* 1590–1595.

Dangi B.; Khanal S.; Shah S. A Review on Rice False Smut, it's Distribution, Identification and Management Practices". *Acta Sci. Agric.* **2020,** 48–54.

Darsono, N.; Yoon, D. H.; Raju, K. Effects of the Sintering Conditions on the Structural Phase Evolution and T C of Bi1.6 Pb0.4 Sr2 Ca2 Cu3 O7 Prepared Using the Citrate Sol–gel Method. *J. Supercond. Nov. Magn.* **2016,** *29*, 1491–1497.

Das, G.; Patra, J. K.; Baek, K. H. Insight into MAS: A Molecular Tool for Development of Stress Resistant and Quality of Rice through Gene Stacking. *Front Plant Sci*. **2017,** *8,* 985.

Das, G.; Rao, G. J. N. Molecular Marker Assisted Gene Stacking for Biotic and Abiotic Stress Resistance Genes in an Elite Rice Cultivar. *Front. Plant Sci.* **2015,** *6,* 698.

Dastur, J. F. Notes on Some Fungi Isolated From Black Point Affected Wheat Kernels in the Central Provinces. *Indian J. Agric. Res.* **1942,** *12,* 731–742.

Datta, S.; Choudhry, R. G.; Shamim, M.; Dhar, V. Polymorphism in the Internal Transcribed Spacer (ITS) of the Ribosomal DNA Among Different *Fusarium* Species. *Arch. Phytopathol. Plant Prot.* **2011,** *44* (6), 558–566.

Daub, M. E.; Chung, K. R. Cercosporin: A Phytoactivated Toxin In: Plant Disease. Online APSnet Features, 2007. DOI. 10.1094/APS 2007

Dean, R. A.; Talbot, N. J.; Ebbole, D. J.; Farman, M. L.; Mitchell, T. K.; Orbach, M. J.; Read, N. D. The Genome Sequence of the Rice Blast Fungus. *Magnaporthe Grisea Nat.* **2005,** *434*, 980.

Desjardins, A. E.; Manandhar H. K.; Plattner, R. D.; Manandhar G. G.; Poling S. M.; Maragos C. M. *Fusarium* Species From Nepalese Rice and Production of Mycotoxins and Gibberellic Acid by Selected Species. *Appl. Environ. Microbiol.* **2000,** *66,* 1020–1025.

Dodan, D. S.; Singh, R. Evaluation of Fungitoxicants Against False Smut of Rice. *J. Mycol. Plant Pathol.* **1997,** *27,* 32–34.

Drechsler, C. Phytopathological and Taxonomic Aspects of Ophiobolus, Pyrenophora, Helminthosporium and a New Genus, Cochliobolus. *Phytopathology* **1934,** *24,* 953–985.

Ebbole, D. J. Magnaporthe as a Model for Understanding Host Pathogen Interactions. *Annu. Rev. Phytopathol.* **2007,** *45,* 437–456.

Emodi, A. I.; Madukwe, M. C. Influence of Consumers' Socio-economic Characteristics on Rice Consumption in South East Nigeria. *Libyan Agric. Res. Cent. J. Int.* **2011,** *2*, 105–111.

Fan, J.; Yang, J.; Wang, Y. Q.; Li, G. B.; Li, Y.; Huang, F.; Wang, W.-M. Current Understanding on *Villosiclava Virens*, a Unique Flower-Infecting Fungus Causing Rice False Smut Disease. *Mol. Plant Pathol.* **2016a,** *17,* 1321–1330.

Fan, R.; Wang, Y.; Liu, B.; Zhang, J.; Wang, H.; Hu, D. The Process of Asexual Spore Formation and Examination of Chlamydospore Germination of *Ustilaginoidea Virens* . *Mycosystema* **2010,** *29,* 188–192.

FAO. World Rice Information, issue No.1. FAO, Rome, Italy, 1995.

FAO. Global Review of Good Agricultural Extension and Advisory Service Practices. [Online] 2008. http://faostat.fao.org (accessed on 10 April 2019).

FAO-Food and Agriculture Organization. FAOSTAT Database FAO, Rome, [Online] 2009. www.faostat.fao.org (accessed June, 2010).

Faure, J; Mazaud, F. In *Rice Quality Criteria and the European Market.* Proceedings of the 18th Session of the International Rice Commission, September 5–9, 1996, Rome, Italy, pp 121–131.

Fensionita, A. Perkembangan Hama dan Penyakit Tanaman Padi (Oryza sativa L.) Pada Beberapa Sistem Budidaya (Development of Pest and Diseases of Rice (Oryza sativa L.) in Several Agronomic system), Dissertation Institut Pertanian Bogor, Bogor, 2006.

Figoni, R. A.; Rutger, J. N.; Webster, R. K. Evaluation of Wild Oryza Species for Stem Rot (Sclerotium Oryzae) Resistance. *Plant Dis.* **1983,** *67*, 998–1000.

Fomba, N. N.; Singh, N. Crop Losses Caused by Rice Brown Spot Disease in Mangrove Swamps of Northwestern Sierra Leone. *Trop. Pest Mgm.* **1990,** *36,* 387–393.

Food and Agriculture Organization Corporate Statistical Database (FAOSTAT). Statistical Database. **http://www.fao.org** (accessed Apr 10, 2019).

Fukai, S.; Cooper, M. Development of Drought-Resistant Cultivars Using Physio-Morphological Traits in Rice. *Field Crop. Res.* **1995,** *40,* 67–86.

Fukuoka, S.; Saka, N.; Koga, H.; Ono, K.; Shimizu, T.; Ebana, K.; Yano, M. Loss of Function of a Proline-Containing Protein Confers Durable Disease Resistance in Rice. *Science* **2009,** *325,* 998–1001.

Gams, W.; Hawksworth, D. L. Identity of *Acrocylindrium Oryzae* Sawada and a Similar Fungus Causing Sheath-Rot of Rice. *Kavaka* **1975,** *3,* 57–61.

Ghose, R. L. M.; Ghatge, M. B.; Subramanian, V. Rice in India (revised edn.); ICAR: New Delhi, 1960; p 474.

Giraldo, A.; Gené, J.; Sutton, D. A.; Madrid, H.; De Hoog, G. S.; Cano, J.; et al. Phylogeny of *Sarocladium* (Hypocreales). *Persoonia* **2015,** *34,* 10–24.

Gnanamanickam, S. S. Rice and its Importance to Human Life. In *Biological Control of Rice Diseases*; Springer: Dordrecht, The Netherlands, 2009; pp 1–11.

Goodman, B. A.; Newton, A. C. Effects of Drought Stress and its Sudden Relief on Free Radical Processes in Barley. *J. Sci. Food Agric.* **2005,** *85*, 47–53.

Gopalakrishnan, C.; Kamalakannan, A.; Valluvaparidasan, V. Effect of Seed-Borne *Sarocladium Oryzae*, the Incitant of Rice Sheath Rot on Rice Seed Quality. *J. Plant Prot. Res.* **2010,** *50,* 98–102.

Gopika, K.; Jagadeeshwar, R.; Rao, V. K.; Vijayalakshmi, K. Efficacy of Fungicide Antagonists and Herbicides Against *Sclerotium Oryzae*, Causal Agent of Stem Rot of Rice. *Indian J. Plant Protect.* **2011,** *39* (2), 130–135.

Gopika, K.; Jagadeeshwar, R.; Rao, V. K.; Vijayalakshmi, K. Salient Resarch Findings on Rice Stem Rot Disease (Sclerotium oryzae catt.) and its Management. *Int. J. Plant Animal Environ. Sci.* 2016, *6* (1) 80–82.

Gowda, S. S.; Gowda, K. T. F. In Vitro Studies on the Efficacy of Certain Fungicides Against *Drechslera Oryzae* (Breda de Haan) Subram. and Jain of Brown Disease of Rice. *Pesticides* **1985,** *19,* 41–42.

Griffin, D. M. Ecology of Soil Fungi, Syracuse 0- 10; University Press: Syracuse, New York, 1972; p 193

Groth, D. E.; Hollier, C. Narrow brown leaf spot of rice. In *Louisiana Plant Pathology: Identification and Management Series*; Pub 3105, LSU AgCentre: Baton Rouge, 2010

Gupta, A. K.; Solanki, I. S.; Bashyal, B. M.; Singh, Y.; Srivastava, K. Bakanae of Rice—An Emerging Disease in Asia. *J. Anim. Plant Sci.* **2015,** *25* (6), 1499–1514.

Hamer, J. E.; Howard, R. J.; Chumley, F. G.; Valent, B. A Mechanism for Surface Attachment in Spores of a Plant Pathogenic Fungus. *Science* **1988,** *239,* 288–290.

Hayasaka, T.; Ishiguro, K.; Shibutani, K.; Namai, T. Seed Disinfection Using Hot Water Immersion to Control Several Seed-Borne Diseases of Rice Plants. *Jpn J. Phytopathol.* **2001,** *67,* 26–32.

Hiremath, P. C.; Hegde, R. K.; Kulkarni, B. G. Effect of Seed Infection by Brown Leaf Spot on Rice Seed Germination. *Curr. Res.* **1983,** *12,* 74–76.

Hollier, C. Narrow Brown Leaf Spot. In *Compendium of Rice Diseases*; Webster, p 18.

Hori, S. Researches on "Bakanae" Disease of the Rice Plant. Noji Shikenjyo Seiski **1898,** *12,* 110–119.

Huang, F.; Li, Y.; Shi, J.; Fan, J.; Xu, Y. J.; Wang, W. M. Screening and Polymorphism Analysis of Rice Germplasms for Resistance to False Smut Disease in Sichuan Province. *Acta Phytopathol. Sin.* **2015,** *46* (2), 247–257.

Hussain, S.; Ghaffar, A. Effect of Tillage Practices on the Population and Viability of Sclerotia of Sclerotium Oryzae and Yield of Rice. *Pakistan J. Bot.* **1993,** *25,* 232–234.

Ikegami, H. Studies on the False Smut of Rice, IV. Infection of the False Smut Due to Inoculation With Chlamydospores and Ascospores at the Booting Stage of Rice Plants. *Res. Bull. Fac. Agric. Gifu. Univ.* **1960,** *12,* 45–51.

Indian Institute of Rice Research (IIRR). *Annual Progress Report*; IIRR: Hyderabad, India, 2005; vol 3, pp 133–135.

Ito, S.; Kimura, J. Studies on the 'Bakanae' Disease of the Rice Plant. *Reptr. Hokkaido Nat. Agric. Exp. Sta.* **1931,** *27,* 1–95.

Ito, S.; Kuribayashi, K. Production of the Ascigerous Stage in Culture of Helminthosporium Oryzae. *Ann. Phytopathol. Soc. Japan* **1927,** *2,* 1–8.

Jecmen, A. C.; TeBeest, D. O. First Report of the Occurrence of a White Smut Infecting Rice in Arkansas. *J Phytopathol.* **2015,** *163,* 138–143.

JPPA. Epidemic and Controlling Area in 2012. In Catalogue of agricultural chemicals; Eds.; Japan Plant Protection Association: Tokyo, 2013; pp 572–574.

Kamal, M. M.; Mia, M. A. T. Diversity and Pathogenicity of the Rice Brown Spot Pathogen, Bipolaris Oryzae (Breda de Haan) Shoem. in Bangladesh Assessed by Genetic Fingerprint Analysis. *Bangladesh J. Bot.* **2009,** *38,* 195–200.

Kankanala, P.; Czymmek, K.; Valent, B. Roles for Rice Membrane Dynamics and Plasmodesmata During Biotrophic Invasion by the Blast Fungus. *Plant Cell* **2007,** *19,* 706–724.

Kari, S.; Korhonen-Kurki, K. Framing Local Outcomes of Biodiversity Conservation Through Ecosystem Services: A Case Study From Ranomafana, Madagascar. *Ecosyst. Serv.* **2013,** *3,* 32–39.

Kato, H. Rice Blast Disease. *Pestic. Outlook* **2001,** *12,* 23–25.

Kato, H.; Kozada, T. E_ect of Temperature on Lesion Enlargement and Sporulation of (*Pyricularia oryzae*) in Rice Leaves. *Phytopathology* **1974,** *64,* 828–830.

Keim R.; Webster, R. K. Fungistasis of Sclerotia of *Sclerotium Oryzae. Phytopathology* **1975,** *65,* 283–287.

Khokhar, L. K.; Jaffrey, A. H. Identification of Sources of Resistance Against Bakanae and Foot Rot Disease in Rice. *Pak. J. Agric. Res.* **2002,***17,* 176–177.

Khush, G. S. Harnessing Science and Technology for Sustainable Rice-Based Production Systems. Proceedings of FAO Rice Conference "Rice is life". *Int. Rice Comm. Newsl.* **2004,** *53,* 17–23.

Khush, G. S. Strategies for Increasing the Yield Potential of Cereals: The Case of Rice as an Example. *Plant Breed.* **2013,** *132,* 433–436.

Kim, J. M.; Hong, S. K.; Kim, W.G.; Lee, Y. K.; SH, Y.; Choi, H. W. Fungicide Resistance of *Gibberella fujikuroi* Isolates Causing Rice Bakanae Disease and Their Progeny Isolates. *Kor. J. Mycol.* **2010,** *38,* 75–79.

Kim, M. H.; Hur, Y. J.; Lee, S. B.; Kwon, T. M.; Hwang, U. H.; Park, S. K.; Youn, Y. N.; Lee, J. H.; Cho, J. H.; Shin, D. J.; Kim, T. H.; Han, S. I.; Yeo, U. S.; Song, Y. C.; Nam, M. H.; Park, D. S. Large-Scale Screening of Rice Accessions to Evaluate Resistance to Bakanae Disease. *J. Gen. Plant Pathol.* **2014,** *80,* 408–414.

Kingsolver, C. H.; Barkside, T. H.; Marchetti, M. A. Rice Blast Epidemiology: Bulletin of the Pennsylvania Agricultural Experiment Station; No.853; Pennsylvania State College, Agricultural Experiment Station: State College, PA, USA, 1984; pp 29–40.

Kloepper, J. W.; Schroth, M. N. In *Plant Growth Promoting Rhizobacteria on Radish*, Proceedings of the Fourth Conference Plant Pathogenic Bacteria; INRA: Angers, France, 1978.

Koiso, Y.; Li, Y.; Iwasaki, S.; Hanaoka, K.; Kobayashi, T.; Sonoda, R.; *et al.* Ustiloxins, Antimitotic Cyclic Peptides From False Smut Balls on Rice Panicles Caused by *Ustilaginoidea virens*. *J. Antibiot.* **1994,** *47,* 765–773.

Kulkarni, S.; Ramakrishnan, K.; Hegde, R. K. Incidence of Brown Leaf Spot of Rice Caused by Drechslera Oryzae (Breda de Haan) Subram. & Jain Under Different Agroclimatic Conditions of Karnataka. *Mysore J. Agri. Sci.* **1980a,***14,* 321–322.

Kulkarni, S.; Ramakrishnan, K.; Hegde, R. K. Demonstration on Supervisory Control of Brown Leaf Spot of Rice Caused by Drechslera Oryzae (Breda de Haan) Subram. & Jain ex. M.B. Ellis. *Pl. Pathol. Newsl.* **1986,** *4*, 22.

Kumar, V.; Ladha, J. K.; Direct Seeding of Rice: Recent Developments and Future Research Needs. *Adv. Agron.* **2011,** *111,* 297–413. ISSN 0065-2113, DOI: 10.1016/B978-0-12-387689-8.00001-1.

Kumar, V.; Makkar, H. P. S.; Becker, K. Detoxified Jatropha Curcas Kernel Meal as a Dietary Protein Source: Growth Performance, Nutrient Utilization and Digestive Enzymes in Common Carp (Cyprinus carpio L.) Fingerlings. *Aquacult. Nutr.* **2011,** *17*, 313–326.

Kumar, A.; Singh, R.; Jalali, B. L. Management of Stem Rot of Rice With Resistance Inducing Chemicals and Fungicides. *Indian Phytopathol.* **2003,** *56* (3), 266–69.

Kumari, V.; Jariwala, S.; Rai, B. Effect of Physicochemical Factors on Growth of Drechslera Oryzae. *J. Mycopathol. Res.* **1997,** *35,* 131–136.

Kunova, A.; Pizzatti, C.; Bonaldi, M.; Cortesi, P. Sensitivity of Nonexposed and Exposed Populations of (Magnaporthe Oryzae) From Rice to Tricyclazole and Azoxystrobin. *Plant Dis.* **2014,** *98,* 512–518.

Kurahashi, Y.; Sakawa, S.; Kimboraund, T.; Kagabu, S. Biological Activity of Carpropamid (KTU 3616). A New Fungicide Rice Blast Disease. *J. Pesticide Sci.* **1997,** *22,* 108–112.

Ladhalakshmi, D.; Laha, G. S.; Singh, R.; Karthikeyan, A.; Mangrauthia, S. K.; Sundaram, R. M.; *et al.* Isolation and Characterization of *Ustilaginoidea virens* and Survey of False Smut Disease of Rice in India. *Phytoparasitica* **2012,** *40,* 171–176.

Lakshmanan, P. Sheath Rot of Rice Incited by *Curvularia lunata* in Tamil Nadu, India. *Trop. Pest Manag.* **1992,** *38,* 107–107.

Lakshmanan, P. Efficient Method of Assessing Resistance to Sheath Rot in Rice. *Crop Prot.* **1993b,** *12,* 189–192.

Law, J. W. F.; Ser, H. L.; Khan, T. M.; Chuah, L. H.; Pusparajah, P.; Chan, K. G.; Goh, B. H.; Lee, L. H. The Potential of Streptomyces as Biocontrol Agents Against the Rice Blast Fungus, (*Magnaporthe Oryzae*) (*Pyricularia oryzae*). *Front. Microbiol.* **2017,** *8,* 3.

Lee, S. B.; Hur, Y. J.; Cho, J. H.; Lee, H. H.; Kim, T. H.; Cho, S. M.; Song, Y. C.; Seo, Y. S.; Lee, J.; Kim, T. S.; Park, Y. J.; Oh, M. K.; Park, D. S. Molecular Mapping of *qBK1WD*, A Major QTL for Bakanae Disease Resistance in Rice. *Rice* **2018,** *11,* 3.

Lee, Y. H.; Lee, M. J.; Choi, H. W.; Kim, S. T.; Park, J. W.; Myung, I. S.; Park, K. S.; Lee, S. W. Development of *In Vitro* Seedling Screening Method for Selection of Resistant Rice Against Bakanae Disease. *Res. Plant Dis.* **2011,** *17,* 288–294.

Li, D.; Luo, K. Study of the Relationship Between the Occurrence of Bakanae Disease in Hybrid Rice and the Application of Gibberellin to Seed Reproduction. *J. Hunan Agric. Univ.* **1997,** *23,* 47–49.

Li, D.; Luo, K.; Chen, Z. Studies on Resistance of Rice Varieties to Bakanae Disease and Pathogenicity of Pathogen (*Fusarium Moniliforme*). *Acta Phytopathol. Sin.* **1993,** *23,* 315–319.

Li, Y. S.; Zhu, Z.; Zhang, Y. D.; Zhao, L.; Wang, C. L. Genetic Analysis of Rice False Smut Resistance Using Major Gene Plus Polygene Mixed Genetic Model. Acta Agron. Sin. (in Chinese with English abstract). **2008,** *34* (10), 1728–1733.

Liu, Y. L.; Tang, J. R.; Li, Y.; Zhou, H. K. First Report of Leaf Spots Caused by Nigrospora Oryzae on Wild Rice in China. *Plant Dis.* **2021**. DOI: 10.1094/PDIS-11-20-2522-PDN.

Lockwood, J. L. Soil Fungistasis. *Annu. Rev. Phytopathol.* **1964,** *2,* 341–362.

Long, D. H.; Lee, F. N.; TeBeest, D. O. Effect of Nitrogen Fertilization on Disease Progress of Rice Blast on Susceptible and Resistant Cultivars. *Plant Dis.* **2000,** *84,* 403–409.

Lopez, S. J. TaqMan Based Real Time PCR Method for Quantitative Detection of Basmati Rice Adulteration With Non-basmati Rice. *Eur. Food Res. Technol.* **2008**, *227*, 619–622.

Lu, S.; Sun, W.; Meng, J.; Wang, A.; Wang, X.; Tian, J.; *et al.* Bioactive Bis-Naphtho-Gamma-Pyrones From Rice False Smut Pathogen *Ustilaginoidea virens*. *J. Agric. Food Chem.* **2015,** *63,* 3501–3508.

Luck, J.; Spackman, M.; Freeman, A.; Piotr Treˌ bicki, P.; Griffiths, W.; Finlay, K.; Chakraborty, S. Climate Change and Diseases of Food Crops. *Plant Pathol.* **2011**, *60*, 113–121.

Lv, B. Preliminary Study of Identification for Resistance to Rice Bakanae Disease in Rice Resource. *Plant Prot*. **1994,** *3,* 20–21.

Maclean, J.; Hardy, B.; Hettel, G. *Rice Almanac: Source Book for One of the Most Important Economic Activities on Earth*; International Rice Research Institute: Los Baños, Philippines, 2013.

Magar M. M.; Rani C. V. D.; Anuradha G. Marker Assisted Selection for Bacterial Leaf Blight Resistance in Segregating Populations of Cottondora sannalu. *Int. J. Appl. Sci. Biotechnol.* **2014,** *2,* 229–237.

Mani, K. K. Study of Factors Affecting Growth and Development of Narrow Brown Leaf Spot of Rice Caused by Cercospora Janseana (Racib.) O. Const. LSU Doctoral Dissertations 3568, 2015.

Marchetti, M. A.; Bollich, C. N. Quantification of the Relationship Between Sheath Blight Severity and Yield Loss in Rice. *Plant Dis.* **1991,** *75,* 773–774.

Martinez, J. A.; Dhanasekaran, D. Natural Fungicides Obtained From Plants. In *Fungicides for Plant and Animal Diseases*; InTech: Rijeka, Croatia, 2019. https://www.intechopen.com/books/fungicides-for-plant-and-animal-diseases/natural-fungicidesobtained-from-plants (accessed Apr 8, 2019).

Mathur, K. C.; Reddy, P. R.; Rajamani, S.; Moorthy, B. T. S. Integrated Pest Management in Rice to Improve Productivity and Sustainability. *Oryza* **1999,** *36* (3), 195–207.

McCouch, S. R.; Wright, M. H.; Tung, C. W.; Maron, L. G.; McNally, K. L.; Fitzgerald, M.; Singh, N.; DeClerck, G.; Agosto-Perez, F.; Korniliev, P.; et al. Open Access Resources for Genome-wide Association Mapping in Rice. *Nat. Commun.* **2016**, *7*, 10–532.

Mew, T. W.; Cottyn, B.; Pamplona, R.; Barrios, H.; Xiangmin, L.; Zhiyi, C.; et al. Applying Rice Seed-Associated Antagonistic Bacteria to Manage Rice Sheath Blight in Developing Countries. *Plant Dis.* **2004a,** *88,* 557–564.

Mew, T. W.; Gonzales, P. *A Handbook of Rice Seedborne Fungi.* Science Publishers: Enfield, NH, 2002.

Mia, M. A. T.; Rahman, M.; Pearce, D.; Holderness, M. Effect of Seed Borne Bipolaris Oryzae on Seed Germination and Disease Development in the Field. *Bangladesh J. Pl. Pathol.* **2001,** *17,* 59–62.

Mohanty, S.; Wassmann, R.; Nelson, A.; Moya, P.; Jagadish, S. V. K. *Rice and Climate Change: Significance for Food Security and Vulnerability*; International Rice Research Institute: Los Baños, Philippines, 2013; p 14.

Mohiddin, F. A.; Bhat, F. A.; Gupta, V.; Gupta, D.; Kalha, C. S. Integrated Disease Management of False Smut of Rice Caused by Ustilaginoidea Virens. *Trends Biosci.* **2012,** *5* (4), 301–302.

Motlagh, M. R. S.; Kaviani, B. Characterization of New Bipolaris Spp.: The Causal Agent of Rice Brown Spot Disease in the North of Iran. *Intern. J. Agric. Biol.* **2008,** *10,* 638–642.

Muthayya, S.; Sugimoto, J. D.; Montgomery, S.; Maberly, G. F. An Overview of Global Rice Production, Supply, Trade, and Consumption. *Ann. N. Y. Acad. Sci.* **2014,** *1324,* 7–14.

Naeimi, S.; Okhovvat, S. M.; Hedjaroude, G. A.; Khosravi, V. Sheath Rot of Rice in Iran. *Commun. Agric. Appl. Biol. Sci.* **2003,** *68,* 681–684.

Nakamura, K.; Izumiyama, N.; Ohtsubo, K.; Koiso, Y.; Iwasaki, S.; Sonoda, R.; *et al.* "Lupinosis"-like Lesions in Mice Caused by Ustiloxin, Produced by *Ustilaginoidea Virens*: A Morphological Study. *Nat. Toxins* **1994,** *2,* 22–28.

Nalley, L.; Tsiboe, F.; Durand-Morat, A.; Shew, A.; Thoma, G. Economic and Environmental Impact of Rice Blast Pathogen (Magnaporthe oryzae) Alleviation in the United States. *PLoS One* **2016,** *11* (12), e0167295.

Narasimhan, V.; Rajappan, K.; Ushamalini, C.; Kareem, A. A. Efficacy of New EC Formulations of Neem Oil and Pungam Oil for the Management of Sheath Rot Disease of Rice. *Phytoparasitica* **1998,** *26,* 301–306.

Nayar N. M. Origin and Cytogenetics of Rice. *Adv. Genet.* **1973,** *17,* 153–292.

Newton, A. C.; Young, I. M. Temporary Partial Breakdown of Mloresistance in Spring Barley by the Sudden Relief of Soil Water Stress. *Plant Pathol.* **1996,** *45,* 973–977.

Nigrospora oryzae (Berk. & Broome) Petch, *J. Indian Bot. Soc.* **1924,** *4,* 24.

Ogawa, K. Damage by "Bakanae" Disease and its Chemical Control. *Jpn Pestic Inf.* **1988,** *52,* 13–15.

Ojha, G. P.; Morin, S. R. Partnership in Agricultural Extension: Lessons From Chitwan (Nepal). In *Agricultural Research and Extension Network*; Overseas Development Institute (ODI): London, UK, 2001.

Orford, A.; Beard, J. Making the State Safe for the Market: The World Bank's World Development Report 1997. *Melb. UL Rev.* **1998,** *22,* 195.

Ou, S. H. *Rice Diseases*, 2nd ed.; Common Wealth Mycological Institute: Kew, UK, 1985; p 380.

Padmanabhan, S. Y. Studies on Forecasting Outbreaks of Blast Disease of Rice. *Proc. Indian Acad. Sci. Sect. B* **1965,** *62,* 117–129.

Pannu, P. P. S.; Thind, T. S.; Goswami, S. Standardization Technique for Artificial Creation of False Smut of Rice and its Management. *Indian Phytopathol.* **2010,** *63,* 234–235.

Park, W. S.; Choi, H. W.; Han, S. S.; Shin, D. B.; Shim, H. K.; Jung, E. S.; Lee, S. W.; Lim, C. K.; Lee, Y. H. Control of Bakanae Disease of Rice by Seed Soaking Into the Mixed Solution of Prochloraz and Fludioxnil. *Res. Plant Dis.* **2009,** *15,* 94–100.

Pearce, D. A.; Bridge, P. D.; Hawksworth, D. L. Species Concept in *Sarocladium*, The Causal Agent in Sheath Rot in Rice and Bamboo Blight. In *Major Fungal Diseases of Rice: Recent Advances*;Sreenivasaprasad, S., Johnson, R., Eds.; Springer: Dordrecht, 2001; pp 285–292.

Pennisi, E. Armed and Dangerous. *Science* **2010,** *327,* 804–805.

Pereira, L. A. A.; Coutinho, W. M.; Machado, J. C.; Magalhaes, F. H. L.; Pena, R. C. M. Fungitoxicity In Vitro of Iprodione on Mycelial Growth of Fungi Associated With Rice Seeds. *Revista Brasileira de Sementes* **2002,** *24,* 67–70.

Phong, T. K.; Nhung, D. T. T.; Yamazaki, K.; Takagi, K.; Watanabe, H. Behavior of Sprayed Tricyclazole in Rice Paddy Lysimeters. *Chemosphere* **2009**, *74*, 1085–1089.

Phookan, A. K.; Hazarika, D. K. Distribution of Sheath Rot (ShR) in Six Agroclimatic Zones of Assam, India. *IRRN* **1992,** *17,* 16.

Prakash N.; Puri S. Efficacy of Combination of Systemic and Non-Systemic Fungicides Against Stem Rot of Rice. *Bioscan* **2012,** *7*, 291–294.

Pramesh, D.; Saddamhusen, A.; Muniraju, K. M.; Kirana Kumara, M. A Combination Fungicide for the Management of Sheath Blight, Sheath Rot and Stem Rot Diseases of Paddy. *Int. J. Curr. Microbiol. Appl. Sci.* **2017,** *6* (9), 3500–3509.

Prasad, B. Eizenga, G. C. Rice Sheath Blight Disease Resistance Identified in *Oryza* Species Accessions. *Pl. Dis*. **2008,** *92,* 1503–1509.

Ramakrishnan, G.; Subramanian, C. L. Studies on Helminthosporiose of Rice: Involvement of Different Species in Disease Development. *Madras Agric. J.* **1977,** *64,* 60–61.

Rao, K. M. *Sheath Blight Disease of Rice*. Daya Publishing House: Delhi, 1995; p 146.

Ravat, V. K.; Basu, A. Current Status of Fungal Diseases on Swarna Cultivar (MTU 7029) of Rice in Gangetic Alluvial Zone of West Bengal, India. *Int. J. Curr. Microbiol. App. Sci.* **2017,** *6* (9), 2069–2080.

Ravat, V. K.; Roy S.; Basu A. Rice: A New Host of *Choanephora Cucurbitarum* in West Bengal. *Ann. Plant Protect. Sci.* **2015,** *23* (1), 158–199.

Reddy, C. S.; Ghosh, A. Sheath Rot Incidence and Yield Losses in Rice Due to the Joint Infection of Rice Tungro Virus and Sheath Rot Fungus. *Indian Phytopath.* **1985,** *38* (1), 165.

Rehmani, M. I. A.; Zhang, J.; Li, G.; Ata-Ul-Karim, S. T.; Wang, S.; Kimball, B. A.; Yan, C.; Liu, Z.; Ding, Y. Simulation of Future Global Warming Scenarios in Rice Paddies With an Open-Field Warming Facility. *Plant Methods* **2011,** *7*, 41.

Rosenzweig, C.; Yang, X. B.; Anderson, P.; Epstein, P.; Vicarelli, M. Agriculture: Climate Change, Crop Pests and Diseases. In *Climte Change Futures: Health Ecological Economic Dimensios*; The Center for Health and the Global Environment at Harvard Medical School: Cambridge, MA, USA, 2005; pp 70–77.

Rossman, A. Y.; Howard, R. J.; Valent, B. (Pyricularia Grisea): The Correct Name for the Rice Blast Disease Fungus. *Mycology* **1990,** *82,* 509–512.

Sakthivel, N. "Sheath Rot Disease of Rice: Current Status and Control Strategies." In *Major Fungal Diseases of Rice: Recent Advances*; Sreenivasaprasad, S., Johnson, R., Eds.; Dordrecht: Springer, 2001, pp 271–283.

Sakthivel, N.; Gnanamanickam, S. S. Evaluation of *Pseudomonas Fluorescens* for Suppression of Sheath Rot Disease and for Enhancement of Grain Yields in Rice (*Oryza sativa* L.). *Appl. Environ. Microbiol.* **1987,** *53,* 2056–2059.

Saravanakumar, D.; Lavanya, N.; Muthumeena, K.; Raguchander, T.; Samiyappan, R. Fluorescent Pseudomonad Mixtures Mediate Disease Resistance in Rice Plants Against Sheath Rot (*Sarocladium oryzae*) Disease. *Biocontrol* **2009,** *54,* 273–286.

Saremi, H.; Ammarellou, A.; Marefat, A.; Okhovat, S. M. Binam a Rice Cultivar, Resistant for Root Rot Disease on Rice Caused by *Fusarium Moniliforme* in North-West, Iran. *Intern. J. Bot.* **2008,** *4*, 383–389.

Sasaki, T.; Burr, B. International Rice Genome Sequencing Project: The Effort to Completely Sequence the Rice Genome. *Curr. Opin. Plant Biol.* **2000,** *3*, 138–142.

Savary, S.; Willocquet, L.; Elazegui, F. A.; Csatilla, N.; Teng, P. S. Rice Pest Constraints in Tropical Asia: Quantification and Yield Loss Due to Rice Pests in a Range of Production Situations. *Plant Dis.* **2000,** *84,* 357–369.

Sawada, K. Beitrage Uber Formosas-Pilze no. 14. *Transact. Nat. Hist. Soc. Formosa.* **1917,** *31,* 31–133.

Shahjahan, A. K. M.; Sharma, N. R.; Ahmad, H. U.; Miah, S. A. Yield Loss in Modern Rice Varieties Due to SB in Bangladesh. *J. Agril. Res.* **1986,** *11,* 82–90.

Shamim, M.; Kumar, D.; Srivastava, D.; Hussain, R.; Khan, N. A.; Kumar, M.; Kumar, V.; Kumar, P.; Singh, K. N. Biochemical and Molecular Analysis of Wild Rice and Maize (Non-host) Against *R. solani. J. Pharmacogn. Phytochem.* **2016,** *5* (4), 423–426.

Shamim, M.; Kumar, D.; Srivastava, D.; Pandey, P.; Singh, K. N. Evaluation of Major Cereal Crops for Resistance Against *Rhizoctonia Solani* Under Green House and Field Conditions. *Indian Phytopathol.* **2014,** *67* (1), 42–48.

Sharma, S.; Sthapit, B.; Pradhanang, P.; Joshi, K. "Bacterial Sheath Brown Rot of Rice Caused by *Pseudomonas fuscovaginae* in Nepal," In *Rice Cultivation in Highland Areas*, Proceedings of the CIRAD Conference; Poisson, C., Rakotoarisoa, J., Eds.; Antananarivo, Madagascar, 1997; pp 107–112.

Shiba, Y.; Nagata, T. The Mode of Action of Tricyclazole in Controlling Rice Blast. *Ann. Phytopath. Soc. Jpn.* **1981,** *47*, 662–667.

Shoemaker, R. A. Nomenclature of Drechslera and Bipolaris, Grass Parasites Segregated From 'Helminthosporium'. *Canadian J. Bot.* **1959,** 37, 879–887.

Simanjuntak, F. A. S.; Safni, I.; D Bakti D. Distribution of narrow brown leaf spot disease of rice (Cercospora oryzae Miyake) in North Sumatra. IOP Conf. Ser. Earth Environ. Sci. **2019,** *454* (2020), 012163.

Singh, R.; Kumar, A.; Jalali, B. L. Variability, Predisposing Factors and Management of Stem Rot Caused by Sclerotium Oryzae: An Overview. *Ann. Rev. Plant Pathol.* **2002,** *1,* 275–289.

Singh, R. P.; Singh, P. N.; Shukla, P. Note on Assessment of Losses in Rice Due to Brown Leaf Spot Diseases. *Indian J. Agric. Res.* **1979,** *13,* 57–58.

Smith, B. D. *The Emergence of Agriculture*; Scientific American Library: A Division of HPHLP; Diane Pub Co.: New York, NY, USA, 1998; ISBN 0-7167-6030-4.

Srinivasachary, S.; Willocquet, L.; Savary, S. Resistance to Rice Sheath Blight (*Rhizoctonia Solani* Kuhn) [(Teleomorph: *Thanatephorus Cucumeris* (A. B. Frank,) Donk.] Disease: Current Status and Perspectives. *Euphytica* **2011,** *178,* 1–22.

Srivastava, D.; Shamim, M.; Kumar, M.; Mishra, A.; Pandey, P.; Kumar, D.; Yadav, P.; Sidduqui, H. S.; Singh, K. N. Current Status of Conventional and Molecular Interventions for Blast Resistance in Rice. *Rice Sci.* **2017,** *24*(6), 299–321.

Srivastava, R; Srivastava, K. K.; Sarkar, J. D.; Srivastava, P. Analysis of Problems Faced by the Farmers in Adoption of Control Measures of Diseases of Rice. *J. Interacademicia* **2010,** *14* (2), 260–266.

Subramanian, C. V. Jain, B. L. A Revision of Some Graminicolous Helminthosporia. *Curr. Sci.* **1966,** *35,* 352–355.

Sumitra, P.; Iboton, N.; Tombisana, R. K.; Biocontrol and Fungicidal Management of Stem Rot of Rice. *Plant Dis. Res.* **2011,** *26,* 194.

Sunder, S.; Satyavir. Survival of Fusarium Moniliformein Soil Enriched With Different Nutrients and Their Combinations. *Indian Phytopath.* **1997,** *50,* 474–481.

Sunder, S.; Singh, R.; Agarwak, R. Brown Spot of Rice: An Overview. *Indian Phytopath.* **2014,** *67* (3), 201–215.

Sunder, S.; Singh, R.; Dodan, D. S. Evaluation of Fungicides, Botanicals and Non-Conventional Chemicals Against Brown Spot of Rice. *Indian Phytopath.* **2010,** *63,* 192–194.

Sunder, S.; Singh, R.; Dodan, D. S.; Mehla, D. S. Effect of Different Nitrogen Levels on Brown Spot (Drechslera Oryzae) of Rice and its Management Through Host Resistance and Fungicides. *Pl. Dis. Res.* **2005,** *20,* 111–114.

Suparyono, J. L.; Catindig, A; Castilla, N. P.; Elazegui, F. "Rice Doctor's Bakanae Fact Sheet". *Cereal Knowledge Bank (CKB)* 2009.

Suprapta, D. N. Potential of Microbial Antagonists as Biocontrol Agents Against Plant Fungal Pathogens. *J. ISSAAS* **2012,** *18,* 1–8.

Thind, T. S.; Mohan, C.; Raj, P.; Arora, J. K. Activity Spectrum of Strobilurins, A New Generation of Ecofungicides Against Some Fungal Pathogens. *Indian Phytopath.* **2004,** *57,* 104–106.

Thirze, H. Modelling Grain Surplus and Deficit in Cameroon for 2030. Master's Thesis, Lund University, Lund, Sweden 2016; p 59.

Valasubramanian, R. Biological Control of Rice Blast With (Pseudomonas fluorescens Migula): Role of Antifungal Antibiotic in Disease Suppression; Universidad de Madras: Chennai, India, 1994.

Variar; Singh, U. S. Strategies to Manage Rice Diseases in the Post Genomic Era. *Oryza* **2018,** *55* (Special Issue), 71–81.

Wang, L. P.; Dong, M. Z.; Hao, Z. N.; Tao, R. X. Preliminary Study on Resistance of Rice Varieties to Rice False Smut by Using Method of Natural Infection in Zhejiang Province. *Acta Agric. Jiangxi.* **2010,** *22* (7),73–74.

Wang, X.; Fu, X.; Lin, F.; Sun, W.; Meng, J.; Wang, A.; *et al.* The Contents of Ustiloxins A and B Along With Their Distribution in Rice False Smut Balls. *Toxins* **2016a,** *8,* 262.

Wang, M.; et al. *Persoonia* **2017,** *39,* 118. https://doi.org/10.3767/persoonia.2017.39.06 50 65 Crossref,ISI, Google Scholar

Watson, A. G.; Ford, E. J. Soil Fungistasis- A Reappraisal *Annu. Rev. Phytopathol.* **1972,** *10,* 327–348.

Webster, R. K.; Gunnell, P. K. Eds; APS Press: St. Paul, Minnesota, USA, p 86.

Wilson, R. A.; Talbot, N. J. Under Pressure: Investigating the Biology of Plant Infection by (Magnaporthe Oryzae). *Nat. Rev. Microbiol.* **2009,** *7,* 185.

Yadav, S.; Anuradha, G.; Kumar, R. R.; Vemireddy, L. R.; Sudhakar, R.; Donempudi, K.; Venkata, D.; Jabeen, P.; Narasimhan, Y. K.; Marathi, B.; Siddiq, E. A. Identification of QTLs and Possible Candidate Genes Conferring Sheath Blight Resistance in Rice (*Oryza sativa* L.). *Springer Plus* **2015,** *4,* 175.

Yang, H. F.; Wang, Z. Y.; HZ, W.; Zhu, C. H. Study on the Variation of Rice Bakanae Under the Different Methods of Seedling Raising. *J. Anhui Agric. Sci.* **2003,** *31,*119–124.

Yasin, S. I.; Khan, T. Z.; Akhtar, K. M.; Muhammad, A.; Mushtaq, A. Economic Evaluation of Bakanae Disease of Rice. *Mycopath* **2003,** *1,* 115–117.

Yong, M. L.; Deng, Q. D.; Fan, L. L.; Miao, J. K.; Lai, C. H.; Chen, H. M.; et al. The Role of *Ustilaginoidea Virens* Sclerotia in Increasing Incidence of Rice False Smut Disease in the Subtropical Zone in China. *Eur. J. Plant Pathol.* **2018,** *150,* 669–677.

Yoon, M. Y.; Cha, B.; Kim, J. C. Recent Trends in Studies on Botanical Fungicides in Agriculture. *Plant Pathol. J.* **2013**, *29,* 1.

Zeigler, R. S.; Puckridge, D. W. Improving Sustainable Productivity in Rice-Based Rainfed Lowland Systems of South and Southeast Asia. *GeoJournal* **1995,** *35*, 307–324.

Zhang, F.; Xie, J. In *Genes and QTLs Resistant to Biotic and Abiotic Stresses From Wild Rice and Their Applications in Cultivar Improvements, Rice-Germplasm, Genetics and Improvement*; Yan, W., Bao, J., Eds.; IntechOpen: Rijeka, Croatia, 2014.

Zhou, L.; Lu, S.; Shan, T.; Wang, P.; Sun, W.; Chen, Z.; Wang, S. Chemistry and Biology of Mycotoxins From Rice False Smut Pathogen In *Mycotoxins: Properties, Applications and Hazards*; Melborn, B. J., Greene, J. C., Eds.; Nova Science Publishers: New York, NY, pp 109–130.

Zou, J. H.; Pan, X. B.; Chen, Z. X.; Xu, J. Y.; Lu, J. F.; Zhai, W. X.; Zhu, L. H. Mapping Quantitative Trait Loci Controlling Sheath Blight Resistance in Two Rice Cultivars (*Oryza sativa* L.). *Theor. Appl. Genet.* **2000,** *101,* 569–573.

Zuo, S.; Zhang, Y.; Chen, Z.; Jiang, W.; Feng, M.; Pan, X. Improvement of Rice Resistance to Sheath Blight by Pyramiding QTLs Conditioning Disease Resistance and Tiller Angle. *Rice Sci.***2014,** *21*(6), 318–326.

Southwood, D.; Ranganathan, S. Host-Pathogen Interactions. In *Encyclopedia of Bioinformatics and Computational Biology*, 2019; vol 3, pp 103–112.

CHAPTER 2

A Review of Diagnostic Methods for Rice Fungal Pathogens

BASAVARAJ TELI, RAINA BAJPAI, MD. MAHTAB RASHID, and JHUMISHREE MEHER

Department of Mycology and Plant Pathology, Institute of Agricultural Sciences, Banaras Hindu University, Varanasi, Uttar Pradesh, India

ABSTRACT

Advancement in the diagnosis of rice fungal diseases helps in the rapid detection of disease-causal agents which play a strategic role in managing the disease. Various diagnosis techniques, that is, symptomatic-based, serological, molecular, and biosensor are developed over the period for the diagnosis of fungal pathogens and performing well with appropriate scientific evidences and accuracy to identify rice disease. Adaptation of such techniques in rice crop protection can help in early diagnosis and overcome the crop loss to increase the economics of the farmer. Hence, this chapter summarizes the available techniques best suited for the diagnosis of rice fungal pathogens.

2.1 INTRODUCTION

Plant disease diagnostics is the science of investigating and detecting the source of diseases using a variety of methodologies, which is vital for disease management and control. Disease control efforts can be a waste of time and money if the disease and the disease-causing agent are not properly identified

Fungal Diseases of Rice and Their Management. Deepti Srivastava, Md. Shamim, Malik Mobeen Ahmad, and R. S. Upadhyay (Eds.)

results more losses. As a result, accurate disease diagnosis is critical (Riley et al., 2002). It is not difficult for a plant pathologist to detect a disease when typical symptoms, signs, or indicators of the pathogen are present on the affected plant. Except for an experienced eye, it can be impossible to distinguish between symptoms caused by pathogenic and non-pathogenic microbes in some cases.

Although some nonparasitic causative agencies cause such distinct symptoms that the disease can be clearly identified, this is not the case for the majority of these plant diseases. In such cases, the history of environmental factors is considered for an accurate diagnosis of them (Agrios, 2005). Identification is somewhat easy for almost fungal pathogens because of the production of pathogen structures both vegetative (mycelium, hyphae, appressorium) or reproductive (conidia/spores) can be observed on the host surface (sign of disease), but not always. However, for the majority of fungal infections, such a clear indicator is not present on the host surface, necessitating the search for other symptoms as well as the pathogen within the diseased tissue. In such circumstances, Robert Koch's guidelines, known as "Koch's postulates," are used to make a diagnosis appropriate by isolation and infection on the host and re-isolation from the host tissue (Agrios, 2005).

All of the agricultural crops are attacked by a wide range of diseases and pests lowering the quantity as well as the quality of crop products. The farming community always has to combat more than one disease and/or pest attacking a particular crop often through the application of a broad-spectrum control chemicals which hits the target organisms very less often in normal scenario. The broad-spectrum chemicals used by the farmers for control of disease(s) often go un-targeted due to the lack of proper diagnostics which leads to crop, economic, and environmental losses. These losses can be minimized and a specific treatment can be tailor-made which can contend with specific phytopathogen(s), if the disease(s) are accurately identified and diagnosed in their early stages. The specific treatment of the accurately diagnosed disease would translate to environmental and economic gains. The gains could further magnify if we are able to identify the phytopathogen even before producing visible signs and symptoms.

Rice (*Oryza sativa*) is staple agricultural crop of many countries and is an integral of diet for huge amount of people. The global loss in rice production due to pests inclusive of diseases, animal pests, and weeds is between 20% and 30% (Savary et al., 2012) while, the average loss in rice due to the phytopathogen attack is 12% (Fang and Ramasamy, 2015) of which majority is caused by the fungal phytopathogens. There are as many as 13 economically important rice fungal phytopathogens present in the world (Laha et al., 2017). The symptoms from the attack of fungal phytopathogens are very similar to that of bacterial phytopathogens and their abiotic counterparts often resulting in confusions. The individual thus has to rely on his experiences, guide books, and experts for correct identification. These three methods are though suitable only for identification of similar and typical diseases. However, there are different diseases causing same spots as described earlier and same diseases causing different spots. This increases the chances of misidentification usually leading to incorrect control practices (Yao et al., 2009). With advancement of science, there is increasing methods and techniques of correct and early identification of rice diseases. Various direct and indirect methods have been evolved by the scientists working in the field of plant disease diagnostics for providing accurate identification of diseases/phytopathogens. These advanced methods can also be utilized for rice disease diagnosis in addition to the traditional methods so that correct management practices can be applied. Thus, here in this chapter, we have discussed various methods for early and accurate identification of rice fungal pathogens for better economic and environmental gain.

2.2 IDENTIFY CHARACTERISTIC SYMPTOMS AND SIGN FOR RICE FUNGAL DISEASES

Plant health is affected by various biotic (fungal, bacteria, virus, phytoplasma, nematodes, pest, etc.) and abiotic (Salinity, drought, heat, cold) agents. Various techniques are developed to detect the disease or disease-causing agents either by direct or indirect methods (Fang and Ramasamy, 2015). In direct methods, pathologists detect the causal agents through serological and molecular techniques which results in accurate identification of causal agents accordingly management practices can be followed as described in Section 3 and Section 5. Similarly, in indirect methods, it is necessary to know the infected host plant, infected plant parts, characteristics symptoms, symptoms variability, and sign of causal agents helps in better understanding the cause of disease (Fang and Ramasamy, 2015; Riley et al., 2002). Rice plants are severely attacked by many fungal pathogens and show drastic reduction in quantity and quality of grains. Here, Figure 2.1 shows the various disease symptoms caused by the major fungal pathogens infecting the rice plants. Table 2.1 depicts the disease, causal agents, sign of causal agents, and symptoms variability of the rice fungal diseases.

FIGURE 2.1 Symptoms of major fungal diseases in rice.

2.3 SEROLOGICAL DIAGNOSIS FOR RICE DISEASES

Several serological methods have been invented by researchers, which boost the identification strategy of plant pathogens in a short time interval and also allow the pathogen detection in field condition. Especially in Asia, for the detection of rice pathogens at different pest control stations and agricultural experiment stations, mostly serological methods are being used as these tests are very simple, cheap, and highly sensitive. It is also a feasible technique for disease detection in a larger number of samples quickly, although the first establishment of this unit takes more time because it needs several months to prepare specific antisera against pathogens of rice. The serological methods used in laboratories can be broadly categorized into two major groups; liquid-phase tests and solid-phase tests. The liquid-phase tests included the precipitin test, latex agglutination reaction (LAR), and passive hemagglutination (PHA), however, examples of solid-phase test are the enzyme-linked immune-sorbent assay (ELISA), dot-immunobinding assay (DIBA). Gel-based serological assay, such as double-immunodiffusion gel assay (DIGA) has been also developed for the detection of plant pathogens (Uehara-Ichiki et al., 2013). Most of these techniques are best adapted for bacterial and viral diseases rather than fungal diseases.

In latex agglutination reaction (LAR), latex is used as a carrier particle on which the antigen or the antibody is attached for immunochemical reactions

TABLE 2.1 Economically Important Fungal Diseases of Rice.

Sl. no.	Fungal disease	Causal organism	Yield loss	Signs of biotic causal agents	Symptoms variability	Source of inoculum and epidemiology	Alternate host or collateral host	Plant part affected	References
1	Rice blast	*Magnaporthe oryzae* (Teleomorph) *Pyricularia oryzae* (Anamorph)	10–30% annually, but 100% loss if favorable condition is available. Neck rot and panicle blast can cause 80% loss	Teardrop-shaped hyaline conidia with two septa and melanized mycelium. They are formed on geniculate conidiophore in bunches	Elliptical to spindle-shaped lesion with brown border surrounding the white-gray lesion	Plant debris, air borne conidia, infected weeds Prolonged rain with 95% relative humidity with cool night temperature condition are best suited for disease occurrence and spread	*Oryza sativa, Triticum aestivum, Eleusine coracana* *Pennisetum glaucum* *Panicum repens, Leersia hexandra, Alopecurus carolinianus*, and *Setaria italica*	Leaves, nodes, panicle or spikes	Couch et al. (2005), Jia et al. (2008), Musiime et al. (2005), Rice Knowledge Bank, IRRI (2021), Boddy (2016)
2	Brown spot	*Cochliobolus miyabeanus* (Teleomorph) *Bipolaris oryzae* (Anamorph)	5–45% yield loss	Fungus produces abundant and branching type of melanized mycelium developed by prostrate hyphae, conidia are brown slight curved with 5–6 septa and tapering apex ends	Appearance of brown to gray cylindrical to oval spots consisting of ashy lesions or completely brown lesion. In severe case, the spots coalesce to make large necrosis spots	Infected seed, which give rise to infected seedlings volunteer rice infected rice debris weeds Temperature 16–36°C With 80% relative humidity	*Leersia hexandra*	Leaves, leaf sheath, panicle branches, glumes, spikelets, seeds	Musiime et al. (2005), Rice Knowledge Bank, IRRI (2021), Fahad et al. (2019)
3	False smut	*Ustilaginoidea virens* or *Villosiclava virens* (Anamorph and Teleomorph)	Reduce germination of 20–30% and yield loss from 10–60% under favorable condition	Chlamydospores are spherical to elliptical in nature with radial hyphae. Sclerotia are formed in later stages of disease development	Appearance of yellow-orange spores ball emerge from floral parts. Later bursting of smut ball turns to yellowish-black	Infected seed material and air borne conidia Temperature 25–35°C with heavy rain and high humidity along with wind	*Digitaria marginate* (was assumed) but till now no host is proved	Grains in the panicle part	Rice Knowledge Bank, IRRI (2021), Dangi et al. (2020), Atia (2004)

TABLE 2.1 *(Continued)*

Sl. no.	Fungal disease	Causal organism	Yield loss	Signs of biotic causal agents	Symptoms variability	Source of inoculum and epidemiology	Alternate host or collateral host	Plant part affected	References
4	Sheath blight	*Thanatephorus cucumeris* (Teleomorph) *Rhizoctonia solani* (anamorph)	20–30% yield loss	Septate hyaline mycelium with brown round sclerotia	Elliptical to green gray lesions on leaf sheath. Irregular lesion with white center	Soil borne in nature and spread by irrigation. Temperature 25–30°C with heavy rain or water logging	Wider host range	Leaf sheath	Rice Knowledge Bank, IRRI (2021), Singh and Srivastava (2015)
5	Sheath rot	*Acrocylindrium oryzae* (Teleomorph) *Sarocladium oryzae* (anamorph)	20–85% yield loss and grain discoloration	Conidia are hyaline, smooth, single-celled, and cylindrical in shape	Leaf sheath enveloping young panicles. Flag leaf has oblong or irregular grayish brown lesion and panicle rot	Seed-borne disease and infected plant debris Temperature 25–30°C with high humidity and close planting	Seed-borne, air borne, and also spread by infected plant debris	Leaf sheath, panicle grain	Rice Knowledge Bank, IRRI (2021)
6	Narrow brown leaf spot	*Sphaerulina oryzina* (Teleomorph) *Cercospora janseana = Cercospora oryzae (Anamorph)*	No severe loss	Conidia are hyaline and 3–5 septa with cylindrical and conidiophores are in groups	Small brown light spot on leaf sheath parallel to veins. On lower sheath, net blotch forms. It causes premature death of leaves and leaf sheaths, also premature grain ripening and plant lodging	Temperature 25–32°C with potassium deficit soils, and disease occurs during flowering to maturity stages	Wide host range	Leaves, sheaths, glumes, and panicles	Rice Knowledge Bank, IRRI (2021), Groth and Hollier (2020)
7	**Bakanae disease**	Gibberella fujikuroi (Anamorph) *Fusarium moniliforme* (Teleomorph)	Yield loss of 20%, but in severe case, it may extend to 20–50%	Hyaline mycelium with production of three kinds of spores, such as microconidia, macroconidia, and chlamydospores	Plants are abnormally tall with pale, thin leaves, produce few tillers with partial grain or empty	Infected seeds, plant material, or soil. And also spread by wind and water	Cowpea tomato, banana, proso millet, early water grass, and barnyard grass	Roots or crowns portion	Rice Knowledge Bank, IRRI (2021), Katoch et al. (2019)

to occur. These virus particles or antibodies are attachment to the carrier by the mechanism of adsorption. For detection of rice stripe virus (RSV, a member of the genus *Tenuivirus*, negative sense ssRNA virus), LAR was once the very widely used technique in Japan, due to its high sensitivity than other precipitin assays (Uehara-Ichiki et al., 2013). This method also requires less time and can be performed with minimal facilities. The serological methods used for plant pathogens identification are mostly based upon the enzyme-linked immunosorbent assay (ELISA). In rice also, it is being used for the detection of ustiloxin B in rice false smut balls through indirect competitive enzyme-linked immunosorbent assay (icELISA) Fu et al. (2015).

It uses monoclonal antibody labeled with fluorescent compounds or an enzyme react with a substrate resulting in color change, thereby helps in the plant pathogen detection (Halk and De Boer, 1985; Torrance, 1995). This type of immunoassay method wholly depends on the visualization of the binding pattern of the antigen with its specific antibody (Miller and Martin, 1998). However, the main drawback of ELISA method is, it needs some sophisticated laboratory facilities. To overcome this problem, a new portable instrument is developed, that is, lateral flow device (LFD). The LFD is a very easy method in which the existence of the specific antigen can be detected by simple paper-based dipstick assay without the requirement of any specialized laboratory equipment. Nowadays, the method is becoming more popular and widely used for the rapid detection of plant pathogens in field condition as it enables the disease detection within a few minutes (Boonham et al., 2008; Tomlinson et al., 2010). The dot immunobinding assay (DIBA) or immunoblotting assay techniques are like the ELISA method, but only difference is, here a solid support matrix is required, that is, the membrane onto which sample extracts are spotted. However, due to the interference of chlorophyll, false-positive or false-negative results are often reported (Srinivasan and Tolin, 1992; Chen et al., 2012a).

Before setting up a serological detection and diagnostic systems, some important points must be considered, such as quality of antisera and the type of the epitopes to be recognized. In certain case, antisera produced against a purified virus may retain some contaminants, such as host–plant materials, which may result in nonspecific reactions. Antisera that sense or react with the conformational epitopes should be avoided, and in place of these, antisera recognizing linear epitopes must be chosen for immune detection of denatured proteins. Furthermore, the linear epitopes recognizing antisera could detect the target proteins using Western blot analysis after performing SDS-PAGE for β-mercaptoethanol treated target protein.

However, few drawbacks are there with serological detection techniques. Role of antibody is very crucial and these methods require ample quantity of antibodies to properly detect the target pathogen. In some cases, unwanted nonspecific or cross-reactions may occur leading to false-positive results and finally which gives rise to incorrect diagnosis of disease. Hence, specificity of antibody is the important prerequisite before going for serological detection and due to this reason, it has not been successfully developed against fungal diseases, as they are high variable in nature. Whereas for diagnosing fungal, bacterial or viral pathogen, these methods are well developed, standardized and most widely used (Luchi et al., 2020).

2.4 ADVANCEMENT OF SENSOR TECHNOLOGY IN RICE DISEASE DIAGNOSIS

Use of portable sensors in an advanced technique which has been developed and commercialized for different application in monitoring of environment and diagnostics. These sensors also has application in the field of plant disease diagnostics and can be used for the identification of rice fungal pathogen very efficiently (Table 2.2). The principle behind functioning of sensors for the identification of plant disease is the detection of signals that can be of chemical, electrical, optical, magnetic, electrochemical, or vibrational origin. The detection limit of these sensors can further be improved by the infusing nanoparticle transducers and the specificity could be improved by incorporation of recognition material of biological origin. Those sensors which has a biological recognition element incorporated in itself like enzymes, antibody, DNA, etc., are known as biosensors. These biosensors as explained earlier have advanced recognition capacities and can also be very specific in recognition. Biosensors that are currently present in the market are categorized on the basis of materials they are constructed on. Majority of the biosensor platforms are made of nanomaterials as it increases their detection limit and overall performance (Shipway et al., 2000). Additionally, nanomaterials have other advantages also like they are friendly for the assemblage of biorecognition elements and have higher surface area and electronic conductivity (Fang and Ramasamy, 2015). Alternatively, biosensors are also classified on the basis of biorecognition element used. The biosensors which use antibody and DNA as recognition elements are called affinity biosensors while the ones which uses enzymes are called enzymatic electrochemical biosensors. Additionally, some of the biosensors use bacteriophage as the recognition element and thus are called bacteriophage-based biosensors (Jain et al., 2021).

TABLE 2.2 Development of Sensor Technology for Diagnosis of Rice Fungal Diseases.

Sl. no.	Sensor type	Disease	Procedure	Accuracy	References
1.	Bi-level color image thresholding	Blast disease, brown spot disease, narrow brown spot disease, and sheath blast disease	Captured image recognizes abnormality according to lesion shape and color	Bi-level accuracy of 87% compared to other methods	(Abdullah et al., 2016)
2.	Multispectral imaging sensor	Rice leaf blast	Multispectral leaf blast identification and classification image sensor, which uses three channels of crop leaf and canopy images	Seed blast identification accuracy 95%, and the leaf blast 90%	(Feng et al. 2009)
3.	Graphene based electrochemical DNA biosensor	False smut of rice	On paper electrodes, probe DNA was used as a biological identification element, and oxidized graphene was used to increase the sensor's detection limit and sensitivity. The hybridization of probe single-stranded DNA (ssDNA) and target single-stranded DNA (ssDNA) on the interface surface has been quantified	Enhanced limit of detection, and high selectivity of fungus	Rana et al. (2021)
4.	Ultra-sensitive electrically magnetic-controllable electrochemical biosensor	Rice blast	Detects biochemical *M. oryzae's* chitinases (Mgchi) and cDNA encoding mannose-binding jacalin-related lectin (Osmbl) of rice (*Oryza sativa*) as recognition probe	Low cost, early diagnosis and easy detection of rice blast pathogen	Yang et al. (2014)
5.	Deep convolutional neural network	Rice sheath blight, brown spot	Sensor are trained to detect diseased images, such as pattern, color, and texture for symptomatic detection	Detects rice diseases and pests through video detection	Li et al. (2020)

2.5 MOLECULAR IDENTIFICATION

The principal objective of entire diagnostics ground is to distinguish a quick, precise, and dependable approach to identify disease-causing microbes. Fundamental techniques utilized to identify the microbe generally depend on microscopic, cultural, and morphological methodologies which need a lot of time, labor, and classical taxonomy information (Nilsson et al., 2011). These methodologies, in spite of the fact that the foundation of fungal diagnostics, can prompt the inconsistent outcomes because of the issues in identification. Also, specialists and master with a training in identification of fungi are essential for viable outcomes (Chalupová et al., 2014). There are limitations of conventional methods-initiated utilization of molecular techniques for the pathogen identification and their classification. Numerous molecular methods are progressively turning out to be significant tools in entire procedure of diagnosis of fungus. The techniques used in molecular methods are immunological approaches, nucleic acid-based probe approaches and polymerase chain reaction (PCR) approaches. These approaches rely on genotypic characters imparting quick, exceptionally explicit, compelling, and possibly more precise outcomes. For the quantification and detection of fungi at the point when the morphological characters are not noticeable, enzyme-linked immunosorbent method and an immunological technique are utilized. Nucleic acid-based approaches are used to know the closely related species and to sense pathogen present in minute quantity where a noticeable sign is not present. There are a number of molecular methods like southern hybridization, in situ hybridization, fluorescence in situ hybridization, microarray, and macroarray. Loop-mediated isothermal amplification (LAMP), rolling circle amplification (RCA), and nucleic acid sequence-based amplification (NASBA), multiplex PCR, nested PCR, real-time PCR, reverse transcriptase (RT)-PCR, etc. Every strategy has its own preferences and constraints. Present advanced techniques, such as DNA barcoding and RNA-Seq-based next-generation sequencing assist to overcome the drawbacks of conventional diagnosis methods. The coming progress would enhance the acceptance of these new diagnosis techniques in exploring plant disease (Chang et al., 2012; Tsai et al., 2006). In future, procedures can be additionally evolved dependent on the varied mechanisms of fungicide to quickly identify resistant isolates. Moreover, for managing phytodiseases, detection of resistance levels in phytopathogenic fungi populaces in particular time period might help the cultivators to take an appropriate choice on resistance management plans (Chalupová et al., 2014). Thus, here we discuss some of the important molecular techniques used presently and their advantages and limitations in using commonly for disease diagnosis.

2.5.1 SIMPLE PCR TECHNIQUE

An extraordinary upraised occurred in plant pathology by the invention of PCR. This technique is based on the synthesis of millions of copies of precise parts of DNA via replicating denaturation, annealing, and elongation step using specific primers. It is a specific technique for the detection of plant pathogens. Chen et al. (2014, 2016) utilized this technique for the identification of *Ustilaginoidea virens* causal agent of rice false smut and *Tilletia horrida* which cause rice kernel smut in bulky samples taking specific primers of each disease in seeds of rice. This method relies on the efficiency of DNA extraction and its concentration, whereas the efficiency is also influenced by inhibitors existing in the sample (Fang and Ramasamy, 2015). This technique is also used to identify the effector (AVR-Pita, AVR-Pi54, AVR-Pi9, AVR-Pia, and AVR-Piz/t,) secreted from the fungi, that is, *Magnaporthe oryzae* (Guo et al., 2019).

2.5.2 NESTED PCR

To overcome the issue that is caused by a single pair of primers as precise and correct outcomes are procured by its DNA probes, and nested primers are typically utilized (Compton, 1991). Hence, to attain a very specific and sensitive result, nested PCR is used. For fungal identification, the sensitivity of nested PCR is 1000 times more in comparison to single PCR (Yeo and Wong, 2002). Here, two continuous rounds take place, where one pair of primers is utilized to multiply a bigger DNA region, afterward for the next round, this multiplied DNA sequence behaves as a target using two internal primers. When a sample is included as little as 10 pg of *P. granati* DNA, a nested PCR technique enhanced both sensitivity and detection of *P. granati*, making it possible to determine the causal agent (Yang et al., 2017). This technique was also utilized to detect great yam disease caused by *Colletotrichum gloeosporioides* (Raj et al., 2013), *Eucalyptus* dieback disease caused by *Cylindrocladium scoparium* (Qiao et al., 2016), and false smut of rice caused by *Ustilaginoidea virens* (Tang et al., 2017). The limitations are risk contamination during PCR steps as the different tube is used for two cycles of amplification which could lead to false-positive results (Rahman et al., 2013).

2.5.3 MULTIPLEX PCR

In this method, in single reaction, many pairs of primers are used which allows simultaneous recognition of diverse targeted sequences of DNA that ultimately saves time and money. One major advantage of this technique in plant pathology is its use in multipathogen infection at one time. Simultaneously, amplification of diverse fragments definite to target pathogenic fungi occurs and is followed by detection based on their molecular sizes. Amplicon size influences the efficiency of DNA synthesis. To overcome this problem, the designing of primers should be done cautiously in addition to their relative concentration and annealing temperature (Dasmahapatra and Mallet, 2006). We can use this technique to detect fungicide resistance in rice blast pathogens by observing the presence of scytalone dehydratase (MBI-D) and quinone outside inhibitor (QoI) as melanin biosynthesis inhibitors (Hayashi et al., 2017).

2.5.4 REAL-TIME PCR (Q PCR)

For rapid diagnosis of disease merely amplification data of targeted DNA is not sufficient which we get from conventional PCR methods. So, in the past years, an innovative technique has been introduced, that is, RT-PCR. This method is an improved method of conventional PCR where quantification of DNA could be done in addition to amplification (Mackay, 2004). Here, a fluorescent dye named SYBR Green I or sequence-specific fluorescent-labeled probe is used to observe the reactions throughout the amplification phase (Badali and Nabili, 2012). This fluorescent dye intercalates with DNA thus generating a fluorescent signal which keeps on increasing as the quantity of targeted DNA rises after every amplification cycle (McCartney et al., 2003). Its limitation is the nonspecific nature of the fluorescent dye thus producing incorrect outcomes in the way of primer dimer. Saylor and Yang (2007) performed this technique to detect and quantify *R. solani* AG-1 IA DNA for infected rice plants. Similarly, rice blast was detected by Qi and Yang (2002). Sun et al. (2015) and False smut was detected by Tanaka et al. (2017) by using real-time PCR technique. Franco Ortega et al. (2020) were able to understand the microbiota OUT of *Cladosporium, Alternaria, Myrothecium, or Pyricularia* on the surface of rice leaves.

2.5.5 DNA BARCODING

This is a molecular method, where a minor fragment of DNA is utilized for species identification in eukaryotes. A conserved region, which is universal in target lineages is termed as Barcode. It displays sufficient variation in DNA sequence to differentiate between different species. Here, standardized sequences (500–800 bp) are utilized for the identification of species through the markers effective for the extensive array of taxonomic clusters (Krishnamurthy and Francis, 2012. The efficiency of this technique is based on the hypothesis that intraspecies genetic variations are less comparative to interspecies variations. For fungi, ITS is utilized as a universal barcode region (Schoch et al., 2012; Li et al., 2011). The advantages of the barcoding method are that fungi could be recognized in life-cycle phases which are not appropriate for morphological identification, this method enables species detection relying on phylogenetic analysis of sequences and unveils biodiversity. In a few taxa in which the use of primary barcodes is unproductive. Secondary barcodes are used for the identification of pathogenic fungal species (Aslam et al., 2017). In recent years, DNA barcoding has proven to be highly useful in cases of symptomless plants or the detection of latent infections. *Pyricularia oryzae and Bipolaris oryzae* can be detected by targeting the fungal universal ITS region through metabarcoding technology (Franco Ortega et al. 2020). In the fields of ecology, environment, and medical research, barcodes have been employed to aid in the understanding of culturable and non-culturable microbial populations (Harrison et al., 2020; Hoshino and Inagaki, 2017; Shahraki et al., 2019). Environmental samples studied using barcodes aided population biology and biosustainability.

2.5.6 FISH

Fluorescence in situ hybridization (FISH) was developed after the disadvantage of other radiolabeled probe-based hybridization came into light. This technique is utilized for the fast characterization of microbes like fungi. Specific detection of DNA sequences on chromosomes is done using this cytogenetic method. A fluorescent probe is used in this technique which binds chromosome-specific part illustrating higher homology. This technique is quite versatile as could be utilized for non-dividing cells. A low level of fluorescence intensity is utilized for recognizing by non-dividing cells. Salgado et al. (2018) used a species-specific 24-mer oligonucleotide probe designed to target a region of the rRNA internal transcribed spacer 2 to develop and standardize a FISH approach for imaging *Plasmopara obducens* from *Impatiens walleriana* tissues to study this important pathogen. As a different technique, FISH technique too has a few drawbacks, for example, the preparation of the probe is very complex (Volpi and Bridger, 2008).

2.5.7 LOOP-MEDIATED ISOTHERMAL AMPLIFICATION

LAMP is regarded as an alternative to the PCR technique and is known to be a vigorous and novel method for the amplification of targeted nucleic acid with high specificity in the isothermal circumstances. This is highly useful for the detection and diagnosis of phytopathogens. For the amplification of DNA, a single temperature is needed in LAMP, as it relies on autocycling strand displacement amplification of DNA (Tsui et al., 2011). On the targeted DNA, overall, six exclusive sequences are recognized with the help of a DNA polymerase and two inner and two outer primers making a set of four primers. Forward inner primer (FIP) and backward inner primer (BIP) are the names given to two inner primers, whereas F3 and B3 are the outer primers (Fakruddin, 2011). In comparison to the conventional PCR method, it shows 10 times additional sensitivity and accuracy (Ren et al., 2009) and does not necessitate exclusive equipment for higher precision; moreover, it has lesser preparation steps. Additional lead over the PCR is that less count of inhibition reactions arises in LAMP and there is no time wastage for thermal changes. The primary downside of LAMP innovation is the usage of indirect assessment strategies 1 (Mn2+ dye, gel electrophoresis, SYBR Green I dye, hydroxynaphthol blue dye, complex probe technique, and turbidimetric method) that cannot be distinguished among the desired outcome, and nonspecifically amplified outcome creating incorrect positives. But the use of molecular beacons (MBs) takes care of this issue that binds to target DNA and produces the fluorescence signals.

2.5.8 RNA-SEQ-BASED NEXT-GENERATION SEQUENCING

RNA-Seq is a recently established deep-sequencing technique. SOLiD, Ion Torrent, Illumina's HiSeq are the three utmost commonly utilized next-generation sequencing platforms for the RNA-sequences (Dawei and Peng, 2014). After sequencing, either the alignment of the consequential reads with the reference genome occurs or the assembly of de novo deprived of genomic sequence for the production of a genome-scale transcription map, including both the transcriptional structure and expression of each gene. These methodologies permit the disclosure of new transcripts and impartial detection of the transcripts from several sources, allowing more compelling use of the RNA-Seq for the transcript's detection in addition to transcriptomes classification in the non-model organism (Grabherr et al., 2011). Identification of the fungus *M. oryzae* has been made through this technology (Soanes et al., 2012). Identification of new pathogens and unveiling of comprehensive variations in fungi in a plant is known via RNA-Seq technology (Aslam et al., 2017). Thomas et al. (2019) used RNA sequencing in rice roots to identify the regulation of gene expression during interactions with *Azospirillum brasilense* at two separate time points (1 day and 14 days post-inoculation). At both hours, hundreds of genes were found to be differently expressed in rice roots.

2.6 CONCLUSION

In this chapter, we tried to summarize the various available techniques for the detection and diagnosis of rice fungal pathogens. Apart from the advancement in detection and diagnosis, still the management of the disease is out of control due to shifts in biology over the period. And also, these techniques are highly sensitive, expensive, and not affordable for the farmers. Although, these techniques are helping the plant pathologist or scientific community to study about pathogen biology and understand the behavior of disease spread so that better management practices can be framed to overcome disease and regulate the minimum yield losses to the farming community for sustainable agriculture production.

KEYWORDS

- **biosensor**
- **crop protection**
- **diagnosis**
- **molecular**
- **rice**
- **rice fungal pathogens**
- **serological**
- **symptoms**

REFERENCES

Abdullah, S. N. H. S.; Abdullah, S.; Petrou, M.; Abdullah, S.; Fitriayanshah, A.; Razalan, M. H. A. A Portable Rice Disease Diagnosis Tool Basedon Bi-level Color Image Thresholding. *Appl. Eng. Agric.* **2016,** *32* (4), 295–310.

Atia, M. M. Rice False Smut (*Ustilaginoidea Virens*) in Egypt. *J. Plant Dis. Prot.* **2004,** *111*(1), 71–82.

Badali, H.; Nabili, M. Molecular Tools in Medical Mycology; Where We Are! *Jundishapur J. Microbiol.* **2012,** *6* (1), 1–3.

Bajet, N.; Daquioag, R.; Hibino, H. Enzyme-Linked Immunosorbent Assay to Diagnose Rice Tungro [*Oryza sativa*]. *J. Plant Protect. Trop.* **1985,** *2,* 125–129.

Boddy, L. Pathogens of Autotrophs. In *The Fungi*; Academic Press, 2016; pp 245–292.

Boonham, N.; Glover, R.; Tomlinson, J.; Mumford, R. Exploiting Generic Platform Technologies for the Detection and Identification of Plant Pathogens. *Eur. J. Plant Pathol.* **2008,** *121,* 355–363. https://doi. org/10.1007/s10658-008-9284-3.

Chalupová, J.; Raus, M.; Sedlářová, M.; Šebela, M. Identification of Fungal Microorganisms by MALDI-TOF Mass Spectrometry. *Biotechnol. Adv.* **2014,** *32* (1), 230–241.

Chang, C.-C.; Chen, C.-C.; Wei, S.-C.; Lu, H.-H.; Liang, Y.-H.; Lin, C.-W. Diagnostic Devices for Isothermal Nucleic Acid Amplification. *Sensors* **2012,** *12* (6), 8319–8337.

Chen, Y.; Yang, X.; Yao, J.; Kyaw, E. P.; Zhang, A. F.; Li, Y. F.; Gu, C. Y.; Zang, H. Y.; Gao, T. C. Simple and Rapid Detection of *Tilletia horrida* Causing Rice Kernel Smut in Rice Seeds. *Sci. Rep.* **2016,** *6* (1), 1–8.

Chen, Y.; Yao, J.; Li, Y. F.; Wang, W. X.; Yang, X.; Zhang, A. F. Simple and Rapid Detection of Rice False Smut Pathogen *Ustilaginoidea virens* in Rice Seeds. *Phytoparasitica* **2014,** *42* (3), 371–375.

Compton, J. Nucleic Acid Sequence-Based Amplification. *Nature* **1991,** *350* (6313), 91–92.

Couch, B. C.; Fudal, I.; Lebrun, M. H.; Tharreau, D.; Valent, B.; Van Kim, P.; Nottéghem, J. L.; Kohn, L. M. Origins of Host-Specific Populations of the Blast Pathogen *Magnaporthe Oryzae* in Crop Domestication With Subsequent Expansion of Pandemic Clones on Rice and Weeds of Rice. *Genetics* **2005,** *170* (2), 613–630.

Dangi, B. A Review on Rice False Smut, it's Distribution, Identification and Management Practices. *Acta Sci. Agric.* **2020,** *4*, 48–54.

Dasmahapatra, K.; Mallet, J. DNA Barcodes: Recent Successes and Future Prospects. *Heredity* **2006,** *97* (4), 254–255.

Dawei, W.; Peng, Y. Assessing the Impact of Dominant Sequencing Base Gene Expression Profiling Techniques (SGEPTs) on Phytopathogenic Fungi. *Chiang Mai J. Sci.* **2014,** *41*, 922–9444.

Elazegui, F.; Islam, Z. Diagnosis of Common Diseases of Rice. *Int. Rice Res. Inst*, **2003**.

Fahad, S.; Adnan, M.; Noor, M.; Arif, M.; Alam, M.; Khan, I. A.; Wang, D. Major Constraints for Global Rice Production. In *Advances in Rice Research for Abiotic Stress Tolerance*; Woodhead Publishing, 2019; pp 1–22.

Fakruddin M. Loop Mediated Isothermal Amplification (LAMP) –An Alternative to Polymerase Chain Reaction (PCR). *Bangladesh Res. Pub. J.* **2011,** *5* (4), 425–439.

Fang, Y.; Ramasamy, R. P. Current and Prospective Methods for Plant Disease Detection. *Biosensors* **2015,** *5* (3), 537–561.

Feng, L.; Wu, D.; He, Y. Identification and Classification of Rice Leaf Blast Based on Multi-Spectral Imaging Sensor. *Spectrosc. Spect. Anal.* **2009,** *29* (10), 2730–2733.

Franco Ortega, S.; Ferrocino, I.; Adams, I.; Silvestri, S.; Spadaro, D.; Gullino, M. L.; Boonham, N. Monitoring and Surveillance of Aerial Mycobiota of Rice Paddy Through DNA Metabarcoding and qPCR. *J. Fungi* **2020,** *6* (4), 372.

Groth, D.; Hollier, C. Narrow Brown Leaf Spot of Rice *Cercospora Janseana* (Racib.) O. Const C. oryz" (PDF). *Louisiana State University Agricultural Center*, 2020.

Guo, X.; Zhong, D.; Xie, W.; He, Y.; Zheng, Y.; Lin, Y., Chen, Z.; Han, Y.; Tian, D.;2 Liu, W.; Wang, F.; Wang, Z.; Chen, S. Functional Identification of Novel Cell Death-Inducing Effector Proteins From *Magnaporthe oryzae*. *Rice* **2019,** *12* (1), 1–12.

Halk, E. L.; De Boer, S. H. Monoclonal Antibodies in Plant Disease Research. *Annu. Rev. Phytopathol.* **1985,** *23,* 321–350. https://doi.org/10. 1146/annurev.py.23.090185.001541

Harrison, J.; Calder, W. J.; Shuman, B. N.; Buerkle, C. A. The Quest for Absolute Abundance: The Use of Internal Standards for DNA Barcoding in Microbial Ecology, 2020.

Hayashi, K.; Suzuki, F.; Hayano-Saito, Y. Multiplex PCR Assay for Simultaneous Detection of MBI-D and Q o I Resistance in Rice Blast Fungus. *J. Gen. Plant Pathol.* **2017,** *83* (5), 304–309.

Hoshino, T.; Inagaki, F. Correction: Application of Stochastic Labeling With Random-Sequence Barcodes for Simultaneous Quantification and Sequencing of Environmental 16S rRNA Genes. *PLoS One* **2017,** *12* (1), e0169431.

Iwasaki, M.; Nakano, M.; Shinkai, A. Detection of *Rice Grassy Stunt Virus* in Planthopper Vectors and Rice Plants by ELISA. *Ann. Phytopathol. Soc. Jpn.* **1985,** *51,* 450–458. DOI: 10.3186/jjphytopath.51.450

Jain, U.; Pudake, R. N.; Chauhan, N.; Pareek, S. Advancements in Biosensors for Fungal Pathogen Detection in Plants. In *Biosensors in Agriculture: Recent Trends and Future Perspectives: Concepts and Strategies in Plant Sciences*; Springer, 2021; pp 205–216.

Jia, Y.; Gealy, D.; Lin, M. J.; Wu, L.; Black, H. Carolina Foxtail (*Alopecurus Carolinianus*): Susceptibility and Suitability as an Alternative Host to Rice Blast Disease (*Magnaporthe Oryzae* [Formerly M. Grisea]). *Plant Dis.* **2008,** *92* (4), 504–507.

Katoch, P.; Katoch, A.; Podel, M.; Upreti, S. Bakanae of Rice: A Serious Disease in Punjab. *Int. J. Curr. Microbiol. App. Sci.* **2019,** *8* (5), 129–136.

Krishnamurthy, P. K.; Francis, R. A. A Critical Review on the Utility of DNA Barcoding in Biodiversity Conservation. *Biodivers. Conserv.* **2012,** *21* (8), 1901–1919.

Laha, G. S.; Singh, R.; Ladhalakshmi, D.; Sunder, S.; Prasad, M. S.; Dagar, C. S.; Babu, V. R. Importance and Management of Rice Diseases: A Global Perspective. In *Rice Production Worldwide*; Springer: Cham, 2017; pp 303–360.

Li, D. Z.; Gao, L. M.; Li, H. T.; Wang, H.; Ge, X. J.; Liu, J. Q.; Chen, Z. D.; Zhou, S. L.; Chen, S. L.; Yang, J. B. Comparative Analysis of a Large Dataset Indicates That Internal Transcribed Spacer (ITS) Should be Incorporated Into the Core Barcode for Seed Plants. *Proc. Natl. Acad. Sci.* **2011,** *108* (49), 19641–19646.

Li, H. J.; Hong, S. W.; Kim, H.-J.; Kwak, Y.-S. Development of a Multiplex PCR Method to Detect Fungal Pathogens for Quarantine on Exported Cacti. *Plant Pathol. J.* **2016,** *32* (1), 53–54.

Li, D.; Wang, R.; Xie, C.; Liu, L.; Zhang, J.; Li, R.; Wang, F.; Zhou, M.;. Liu, W. A Recognition Method for Rice Plant Diseases and Pests Video Detection Based on Deep Convolutional Neural Network. *Sensors* **2020,** *20* (3), 578.

Luchi, N.; Ioos, R.; Santini, A. Fast and Reliable Molecular Methods to Detect Fungal Pathogens in Woody Plants. *Appl. Microbiol. Biotechnol.* **2020,** *104,* 2453–2468. https://doi.org/10.1007/s00253-020-10395-4

Luisoni, E.; Milne, R.; Roggero, P. Diagnosis of *Rice Ragged Stunt Virus* by Enzyme-Linked Immunosorbent Assay and Immunosorbent Electron Microscopy. *Plant Dis.* **1982,** 66, 929–932. DOI: 10.1094/PD-66-929

Mackay, I. M. Real-Time PCR in the Microbiology Laboratory. *Clin. Microbiol. Infect.* **2004,** *10* (3), 190–212.

McCartney, H.A.; Foster, S. J.; Fraaije, B. A.; Ward, E. Molecular Diagnostics for Fungal Plant Pathogens. *Pest Manag. Sci.* **2003,** *59* (2), 129–142.

Miller, S. A.; Martin, R. R. Molecular Diagnosis of Plant Diseases. *Annu. Rev. Phytopathol.* **1998,** *26*, 409–432.

Musiime, O., Tenywa, M. M., Majaliwa, M. J. G., Lufafa, A., Nanfumba, D., Wasige, J. E., Woomer, P. L.; Kyondha, M. In Constraints to Rice Production in Bugiri District. African Crop Science Conference Proceedings, 2005; vol. 7, No. pt. 03 of 03, pp 1495–1499.

Nilsson, R. H.; Abarenkov, K.; Larsson, K.-H.; Kõljalg, U. Molecular Identification of Fungi: Rationale, Philosophical Concerns, and the UNITE Database. *Open Appl. Inf. J.* **2011,** 5, 81–86.

Omura, T.; Morinaka, T.; Inoue, H.; Saito, Y. Purification and Some Properties of *Rice Gall Dwarf Virus*, A New Phytoreovirus. *Phytopathology* **1982,** *72,* 1246–1249. DOI: 10.1094/Phyto-72-1246

Omura, T.; Takahashi, Y.; Shohara, K.; Minobe, Y.; Nozu, T. T. Y. Production of Monoclonal Antibodies Against *Rice Stripe Virus* for the Detection of Virus Antigen in Infected Plants and Viruliferous Insects. *Ann. Phytopathol. Soc. Jpn.* **1986,** *52,* 270–277. DOI: 10.3186/jjphytopath.52.270

Qi, M.; Yang, Y. Quantification of *Magnaporthe grisea* During Infection of Rice Plants Using Real-Time Polymerase Chain Reaction and Northern Blot/Phosphoimaging Analyses. *Phytopathology* **2002,** *92* (8), 870–876.

Qiao, T. M.; Zhang, J.; Li, S. J.; Han, S.; Zhu, T. H. Development of Nested PCR, Multiplex PCR, and Loop-Mediated Isothermal Amplification Assays for Rapid Detection of *Cylindrocladium scoparium* on Eucalyptus. *Plant Pathol. J.* **2016,** *32* (5), 414–422. DOI: 10.5423/PPJ.OA.03.2016.0065

Rahman, M.T.; Uddin, M. S.; Sultana, R.; Moue, A.; Setu, M. Polymerase Chain Reaction (PCR): A Short Review. *Anwer. Khan Mod. Med. Coll. J.* **2013,** *4* (1), 30–36.

Raj, M.; Nath, V. S.; Sankar, M. S.; Pravi, V.; Archana, P. V.; Jeeva, M. L.; et al. A Highly Sensitive Nested-PCR Method Using a Single Closed Tube for the Detection of *Colletotrichum Gloeosporioides* Causing Greater Yam Anthracnose. *J. Root Crops* **2013,** *39* (2), 163–167.

Rana, K.; Mittal, J.; Narang, J.; Mishra, A.; Pudake, R. N. Graphene Based Electrochemical DNA Biosensor for Detection of False Smut of Rice (*Ustilaginoidea virens*). *Plant Pathol. J.* **2021,** *37* (3), 291.

Ren, C. H.; Hu, C. Q.; Luo, P.; Wang, Q. B. Sensitive and Rapid Identification of Vibrio Vulnificus by Loop-Mediated Isothermal Amplification. *Microbiol. Res.* **2009,** *164* (5), 514–521.

Rice Knowledge Bank, IRRI [Online] 2021. http://www.knowledgebank.irri.org/training/fact-sheets/pest-management/diseases

Riley, M. B.; Williamson, M. R.; Maloy, O. Plant Disease Diagnosis. *The Plant Health Instructor*, 2002; vol *10*.

Salgado-Salazar, C.; Bauchan, G. R.; Wallace, E. C.; Crouch, J. A. Visualization of the Impatiens Downy Mildew Pathogen Using Fluorescence in Situ Hybridization (FISH). *Plant Methods* **2018,** *14* (1), 1–9.

Savary, S.; Horgan, F.; Willocquet, L.; Heong, K. L. A review of principles for sustainable pest management in rice. *Crop protection*, **2012,** *32*, 54–63.

Sayler, R. J.; Yang, Y. Detection and Quantification of *Rhizoctonia solani* AG-1 IA, the Rice Sheath Blight Pathogen, in Rice Using Real-Time PCR. *Plant Dis.* **2007,** *91* (12), 1663–1668.

Schoch, C. L.; Seifert, K. A.; Huhndorf, S.; Robert, V.; Spouge, J. L.; Levesque, C.A.; Chen, W.; Bolchacova, E.; Voigt, K.; Crous, P. W. Nuclear Ribosomal Internal Transcribed Spacer (ITS) Region as a Universal DNA Barcode Marker for Fungi. *Proc. Natl. Acad. Sci. U S A.* **2012,** *109*(16), 6241–6246.

Shahraki, A. H.; Chaganti, S. R.; Heath, D. Assessing High-Throughput Environmental DNA Extraction Methods for Meta-Barcode Characterization of Aquatic Microbial Communities. *J. Water Health* **2019,** *17* (1), 37–49.

Shipway, A. N.; Katz, E.; Willner, I. Nanoparticle Arrays on Surfaces for Electronic, Optical, and Sensor Applications. *ChemPhysChem* **2000,** *1* (1), 18–52.

Singh, A. K.; Srivastava, J. N. In *Sheath Blight Disease of Paddy and Their Management.* Recent Advances in the Diagnosis and Management of Plant Diseases; Springer: New Delhi, 2015; pp 91–99.

Soanes, D. M.; Chakrabarti, A.; Paszkiewicz, K. H.; Dawe, A. L.; Talbot, N. J. Genome-Wide Transcriptional Profiling of Appressorium Development by the Rice Blast Fungus *Magnaporthe oryzae*. *PLoS Pathogens* **2012,** *8* (2), e1002514.

Srinivasan, I.; Tolin, S. A. Detection of Three Viruses of Clovers by Direct Tissue Immunoblotting. *Phytopathology* **1992,** *82,* 721.

Sun, G.; Liu, J.; Li, G.; Zhang, X.; Chen, T.; Chen, J.; Zhang, H.; Wang, D.; Sun, F.; Pan, H. Quick and Accurate Detection and Quantification of *Magnaporthe oryzae* in Rice Using Real-Time Quantitative Polymerase Chain Reaction. *Plant Dis.* **2015,** *99* (2), 219–224.

Takahashi, Y.; Omura, T.; Kenichiro, S.; Tsuchizaki, T. Rapid and Simplified ELISA for Routine Field Inspection of *Rice Stripe Virus*. *Ann. Phytopathol. Soc. Jpn.* **1987,** *53,* 254–257. DOI: 10.3186/jjphytopath.53.254

Tanaka, E.: Kumagawa, T.; Ito, N.; Nakanishi, A.; Ohta, Y.; Suzuki, E.; Adachi, N.; Hamada, A.; Ashizawa, T.; Ohara, T.; Tsuda, M. Colonization of the Vegetative Stage of Rice Plants by the False Smut Fungus *Villosiclava virens*, as Revealed by a Combination of Species-Specific Detection Methods. *Plant Pathol.* **2017,** 66(1), 56–66.

Tang, J.; Zheng, L.; Jia, Q.; Liu, H.; Hsiang, T.; Huang, J. PCR Markers Derived From Comparative Genomics for Detection and Identification of the Rice Pathogen *Ustilaginoidea virens* in Plant Tissues. *Plant Dis.* **2017,** *101* (8), 1515–1521.

Thomas, J.; Kim, H. R.; Rahmatallah, Y.; Wiggins, G.; Yang, Q.; Singh, R.; Glazko, G.; Mukherjee, A. RNA-seq Reveals Differentially Expressed Genes in Rice (*Oryza sativa*) Roots During Interactions With Plant-Growth Promoting Bacteria, *Azospirillum brasilense*. *PloS One* **2019,** *14* (5), e0217309.

Tomlinson, J. A.; Dickinson, M.; Hobden, E.; Robinson, S.; Giltrap, P. M.; Boonham, N. A Five-Minute DNA Extraction Method for Expedited Detection of *Phytophthora ramorum* Following Prescreening Using *Phytophthora* spp. Lateral Flow Devices. *J. Microbiol. Methods* **2010,** *81,* 116–120. https://doi.org/10.1016/j.mimet.2010.02.006.

Torrance, L. Use of Monoclonal Antibodies in Plant Pathology. *Eur. J. Plant Pathol.* **1995,** *101*, 351–363. https://doi.org/10.1007/BF01874849

Volpi, E. V.; Bridger, J. M. Fish Glossary: An Overview of the Fluorescence in Situ Hybridization Technique. *Biotechniques* **2008,** *45* (4), 385–386.

Wang, Z.-H.; Fang, S.-G.; Zhang, Z.-Y.; Han, C.-G.; Li, D.-W.; Yu, J.-L. Development of an ID-ELISA for the Detection of Rice Black-Streaked Dwarf Virus in Plants. *J. Virol. Methods* **2006,** *134,* 61–65. DOI: 10.1016/j.jviromet.2005.11.019

Wang, Z.; Yu, D.; Li, X.; Zeng, M.; Chen, Z.; Bi, L.; et al. The Development and Application of a Dot- ELISA Assay for Diagnosis of Southern Rice Black-Streaked Dwarf Disease in the Field. *Viruses* **2012,** *4,* 167–183. DOI: 10.3390/v4010167

Wu, J.; Ni, Y.; Liu, H.; Rao, L.; Zhou, Y.; Zhou, X. Development and Use of Three Monoclonal Antibodies for the Detection of Rice Black-Streaked Dwarf Virus in Field Plants and Planthopper Vectors. *Virol. J.* **2013,** *10,* 114. DOI: 10.1186/1743-422X- 10-114

Yang, K.; Lee, I.; Nam, S. Development of a Rapid Detection Method for Peronospora Destructor Using Loop-Mediated Isothermal Amplification (LAMP). *HortScience* **2017,** *52* (9), S413. DOI: 10.21273/HORTSCI.52.9S.S1

Yang, W.; Zhang, H.; Li, M.; Wang, Z.; Zhou, J.; Wang, S., Lu, G.; Fu, F. Early Diagnosis of Blast Fungus, *Magnaporthe oryzae*, in Rice Plant by Using an Ultra-Sensitive Electrically Magnetic-Controllable Electrochemical Biosensor. *Anal. Chim. Acta* **2014,** *850*, 85–91.

Yao, Q.; Guan, Z.; Zhou, Y.; Tang, J.; Hu, Y.; Yang, B. In *Application of Support Vector Machine for Detecting Rice Diseases Using Shape and Color Texture Features*. 2009 International Conference on Engineering Computation, IEEE, 2009; pp 79–83.

Yeo, S. F.; Wong, B. Current Status of Nonculture Methods for Diagnosis of Invasive Fungal Infections. *Clin. Microbiol. Rev.* **2002,** *35* (3), 465–484.

CHAPTER 3

Investigation of Ecology, Molecular, and Host–Pathogen Interaction of Rice Blast Pathogen and Management Approaches

MONA F. A. DAWOOD[1], YASSER S. MOURSI[2], ABDELRAZEK S. ABDELRHIM[3], and AMANY A. HASSAN[4]

[1]*Department of Botany & Microbiology, Faculty of Science, Assiut University, Assiut, Egypt*

[2]*Department of Botany, Faculty of Science, University of Fayoum, Fayoum, Egypt*

[3]*Department of Plant Pathology, Faculty of Agriculture, Minya University, Al-Minya, Egypt*

[4]*Botany and Microbiology Department, Faculty of Science, New Valley University, El-Kharja, Egypt*

ABSTRACT

The rice blast caused by *Magnaporthe oryzae* is a serious threat to global food security. The havoc caused by rice blast depends on the susceptibility of the rice cultivar, the virulence of the pathogen, the environmental conditions, and the application of fungicide. The severity of this pathogen comes from its ability to multiply and survive in ruthless environmental conditions. A good understanding of the causal agent of the disease, its life cycle, occurrence, distribution, epidemiology, symptoms, and management technique will offer a good insight into the disease extent and give appropriate and

Fungal Diseases of Rice and Their Management. Deepti Srivastava, Md. Shamim, Malik Mobeen Ahmad, and R. S. Upadhyay (Eds.)

effective decision-making in its management. So, various strategies have been emerged for controlling the blast disease. These controlling methods are categorized as cultural control, resistant varieties, chemical control, and biological control. Many chemicals were utilized to manage the disease but are found expensive, and less effective over time, so eco-friendly chemical compounds, such as organomercuric, organophosphorus, and copper fungicides were used. *Pseudomonas fluorescens*, *Pyricularia oryzae*, *Bipolaris sorokinianaat*, and *Trichoderma harzianum* are known as the most biological control agents used for hindering the pathogenic fungi. The success and effectiveness of biological control rely on the capability of the biocontrolling microorganism to survive and inhibit itself in the environment.

3.1 INTRODUCTION

Rice (*Oryza sativa* L.) is considered the most staple food among the cereal crops in the world. Food and Agriculture Organization (FAO) considered it as a strategic crop for food security in the world (Zhu et al., 2017; Ahmed et al., 2020). It ranked the second crop after corn with respect to world production. World production of rice paddy reached approximately 782 million tons and 499.1 million tons of milled rice in 2018 (FAO, 2018).

Several biotic and abiotic factors threaten rice plants during the growing season resulting in significant yield loss. One of the major damaging diseases of rice all over the world is blast disease. Rice blast is caused by a pathogenic fungus known as *Magnaporthe oryzae* B. Couch (anamorph. *Pyricularia oryzae* Cavara), belongs to the class filamentous ascomycetes. *M. oryzae* can infect over 50 grasses including economically important crops (i.e., rice, wheat, and barley). On rice, the pathogen can infect all parts of rice plants causing total yield loss under favorable conditions. The symptoms appeared on leaves as white to gray oval or elongated diamond-shaped lesions with dark green to brown borders surrounded by a yellowish halo, collar blast on leaf collars, culm nodes on culms, neck rot on panicle neck nodes, and white panicle and panicle blast on panicles (Miah et al., 2017; Asibi et al., 2019). Although the pathogen attacks all parts of rice plants, the most serious damage occurs when the panicle neck node and/or the panicle are infected. The grain of early infected rice plants may not fill, and the panicle remains erect, but the late infection usually results in weight loss because of partial filling of the grains. The severity of rice blast is relying on the susceptibility of the growing cultivar, the growth stage of the plant, the degree of infection, the environmental conditions, and the history of fungicides application (Asibi et al., 2019).

3.2 EPIDEMIOLOGY AND DAMAGE CAUSED BY *M. oryzae*

M. oryzae is found worldwide, the disease is occurring in more than 85 countries (Pagliaccia et al., 2018). Yield loss due to rice blast has reached up to 50% or more, and for decreasing the outbreaks of the disease by applying fungicides, it costs over $70/ha (Khush and Jena, 2009). Also, it was reported that *M. oryzae* may cause up to 80–100% yield loss when significant epidemics occur and it breaks down whole rice plants within 15–20 days (Prabhu et al., 2009; Agbowuro et al., 2020).

Rice blast epidemics have been observed in Korea, Japan, Vietnam, Brazil, the United States, and China between 2001 and 2005. more than 5 million hectares of rice were devastated (Nunes Maciel, 2011). More than 1700 isolates of *M. oryzae* were collected from 40 countries (Divya et al., 2014). Many workers used pathogenicity and molecular markers traits on diverse rice varieties for investigating the population structure of *M. oryzae*. In the USA, eight lineages were identified from 42 isolates collected from infected USA commercial rice cultivars (Levy et al., 1991). However, in Europe and Africa, a simple population structure was observed which was divided into nine lineages (Takan et al., 2012). Also, a study by Chen et al. (1995) identified 10 lineages were from 1156 *M. oryzae* collected from 38 rice cultivars.

3.3 LIFE CYCLE

The first inoculum is initiated from rice residues, wind spread spores from infected to non-infected neighbor farms (Miah et al., 2017). *M. oryzae* reproduces asexually by forming teardrop-shaped conidia that have three cells, each includes a single nucleus. The spores of *M. oryzae* germinate after adhering tightly to the surface of rice leaves by forming specialized infectious structures called an appressorium, within 24 h of the infection (Talbot, 2003; Skamnioti and Gurr, 2007). The appressorium is released from an apical compartment in the spore tip during hydration (Hamer et al., 1988). The formation of appressorium is regulated by cell cycle progression (Veneault-Fourrey et al., 2006). One nucleus migrates to the newly germinated tube from the three-celled conidia. Four to 6 h after germination, the nucleus undergoes mitosis, and then one of the produced nuclei enters the so formed appressorium, while the rest returns to the conidia. The rest of the spore content is degraded with the three nuclei, while the mature appressorium continues developing with a single nucleus (Wilson and Talbot, 2009). Later, the nucleus in the mature appressorium migrates to the penetration peg, and it undergoes more rounds of mitosis required for developing invasive hyphae. The development of appressorium is regulated by controlling the mitosis process. The stimulation of mitosis enhances the development of appressorium, and vice versa; prevention of mitosis prohibits the degeneration of the conidial nuclei and the conidium. It has been reported that the mitosis process controlled by a temperature-sensitive mutation in a gene termed *MgNIMA*, which is a protein kinase playing a role in mitosis (Osmani et al., 1991) when the temperatures are non-permissive, mutation prevents the formation of the appressoria (Veneault-Fourrey et al., 2006). However, if the temperature is suitable and the appressorium is matured, a high concentration of compatible solutes (glycols) accumulate, which generates a turgor (de Jong et al., 1997). A distinct layer of melanin between chitin rich cell wall of the appressorium and the plasma membrane is fundamental for turgor generation. It also acts as a barrier to the efflux of solute from the appressorium, which accumulates turgor pressure (Chumley and Valent, 1990). Subsequently, the cellular turgor turns into mechanical force, that force is exerted by the emerging penetration peg, supporting penetration of the peg into the cuticle of rice leaves. Once the penetration takes place, the hyphae of *M. oryzae* branch through the leaf tissue and disease lesions develop as a symptom of the blast disease. If the infection occurred as seed-borne, the spores infect the arising tissues of seedlings where the primary leaf and secondary roots were the colonization zones of the produced mycelia. The severity of the disease kills the infected seedlings and those seedlings could be a potential source for the inoculum infecting adjacent plants and for developing blast symptoms (Faivre-Rampant et al., 2013). Under favorable conditions, the symptoms thrive on all parts of the plant, the fungus starts to sporulate, and the disease continues to rapidly spread in the whole field by wind and dewdrop splash (Talbot, 2003). After harvesting of rice plants, the spores will remain on the plant debris of infected rice plants and contaminated seeds, initiating the infection for the next season. *M. oryzae* can survive in the residue of rice plants and can repeat the life cycle (Kingsolver et al., 1984). When the conditions are favorable, the pathogen can complete one cycle a week, and a single lesion can produce each night for more than 20 days (Kato, 2001).

According to the way the fungus obtains its nutrition from plant cells, *M. oryzae* has been classified as hemibiotrophs (Talbot, 2003; Ebbole, 2007). *M. oryzae* forms specialized fungal feeding structures, which are bulbous and branched cells during the early invasion. The feeding structure, analogous to the haustoria in obligate biotrophic fungi, is adapted to deriving nutrition from living plant cells (Ebbole, 2007). In the previous stage, which represents the biotrophic stage of *M. oryzae*, the fungus grows within the host plant cells using a haustoria-like structure for deriving nutrition from the host cells. The fungus moves between cells by means of plasmodesmata. However, when assault hyphae shift to adjacent epidermal cells, the fungus appears to seek out pit field sites. As stomatal guard cells lack plasmodesmata, the fungus cannot enter through stomata guard cells, which is inconsistent with the suggestion of cell-to-cell spread (Kankanala et al., 2007). After prolonged biotrophic stage and as lesions become apparent, the fungus starts to be more necrotrophic which explains why *M. oryzae* has been included under hemibiotrophs (Ebbole, 2007; Kankanala et al., 2007).

3.4 OCCURRENCE AND DISTRIBUTION OF RICE BLAST DISEASES

Southeast Asia is considered to be the origin of rice where records of its cultivation since 148 BCE were convinced. It is planted worldwide in Asia, North and South America, Africa, southern Europe, and many other regions of the globe. It is believed that 23 distinct species of rice are well-known. The *Oryza glaberrima* and *O. sativa have* extensively grown rice species, which have originated from the wild species *Oxymeris fatua* and *Oxymeris breviligulata*, respectively (Muimba-Kankolongo, 2018).

As has been reviewed, the disease is distributed in cool and wet regions, in regions with water scarcity in the nursery, and also in a field with high nitrogen application. The residues as rice straw, reeds, weeds, rushes, and millet-like cereals act as a home for the fungus in spore form which acts as inoculum infecting the cultivated seeds. In a dormant state, conidia and mycelia of *P. oryzae* can survive fertile for 3–5 years (Muimba-Kankolongo, 2018; Shahriar et al., 2020). Mycelial growth is stimulated by low pH from 3.5–6.5, with maximal mycelial growth at pH 6.5 (Takan et al., 2004). It was reported that in moderately infected fields, rice production was reduced by around 50% (Shahriar et al., 2020). Blast disease is likely termed as a rice rotten neck and its causal agent *P. oryzae* has been reported to be the most serious fungus affecting rice.

Rice blast disease is distributed all over the world where the rice is cultivated in 85 countries, in both lowland and upland fields. Blast disease was first recorded in West Africa (Sierra Leone) during 1930, but only light was shed on its severity by 1965 with the use of new varieties in the African continent (Muimba-Kankolongo, 2018).

Blast disease adversely influenced rice production in West Africa, the highest producer of rice in Africa, faced with losses from 3–77% in rice yield (Shahriar et al., 2020). It was first recorded in China as rice fever by Soon Ying-shin in 1637 (Ou and Nuque, 1985), later this disease was reported as Imochi-byo in Japan as documented by Tsuchiya in 1704. In Italy, the first epidemic was recorded in the year of 1828, and it was reported in the year 1913 in Tamil Nadu of India (Padmanabhan, 1965). The disease caused a severe reduction in yield which reached "1–100%" in Japan (Baker and Zambryski, 1997), "21–37%" in Indonesia (Singh and Prasad, 2007), 70% in China (Kato, 2001), "30–100%" in Bangladesh, and "30–50%" in South America and other Southeast Asian countries (Suprapta and Khalimi, 2012). Also, the temperate climate of South Korea caused immense yield reductions every year. During the years 1984–1985, the blast disease caused a 20% loss of the rice cultivated area (Sobrizal and Anggiani, 2007). The epidemics of blast disease have adverse impacts on temperate, tropical, subtropical Asia, Latin America, and Africa (Kapoor and Abhishek, 2014). Nutsugah et al. (2008) stated that rice blast disease is the most damaging epidemic to Ghana's rice yield. Blast pathogen can attack rice in upland and lowland systems at all developmental and growth stages. Nutsugah et al. (2008) reported that rice produced from lowland in Asia's temperate and subtropical climate is immensely susceptible to the disease than tropical upland regions where irrigation was the main factor affecting plants. In Bangladesh, the pathogenicity of blast disease was recorded during the irrigated ecosystem (Boro) and rain-fed ecosystem (Transplanted Aman) in 10 ecological zones (Shahriar et al., 2020). So, rice-*M. oryzae* interaction should be studied extensively to find out the best way to get rid of blast disease which will be included in the following sections.

3.5 DISEASE DEVELOPMENT AND ENVIRONMENTAL CONDITIONS

The development and the spread of rice blast in infected fields are affected by several environmental factors. High severity of rice blast could be expected when the temperature averaged between 18–25°C during the flowering stage with a few days of continuous rain especially when followed by high humidity, sunny, and hot days (Kohli et al., 2011). Cardoso et al. (2008) illustrated that the intensity of the rice blast at 30°C was the highest and increased positively with the period of the wet condition and was low at 25°C with a wet period of less than 10 h. The intensity of blast disease could exceed 85% at 25°C and 40 h of wetting, the penetration of *M. oryzae* spores requires free moisture, and the infection requires high relative humidity between 90% and 92% wettings (Miah et al., 2017). Cloudy weather with high moist and low solar radiation is a perfect condition for severing blast epidemics, especially when considering growing cultivars that are susceptible. Extended leaf dampness with high relative humidity 29–96% at 25–28°C temperature is a favorable condition for sporulation, and the highest rate of sporulation takes place when the temperature is approximately 22°C and the relative humidity is near 100% (Agrios, 2005; Kankanala et al., 2007; Bevitori and Ghini 2014; Wilson and Talbot, 2009).

Rice is cultivated over a wide scope of agroclimatic environments around the globe and its productivity is influenced by biotic anxieties (Zhang et al., 2014). Biotic stresses that emerged from climate change can disable varietal protection from rice blasts (Goodman and Newton, 2005). Climate change may affect pathogen circulation and growth rates, and affect the susceptibility, development, and metabolism of rice (Bevitori and Ghini, 2014). Each phase of the rice blast infection cycle, that is, from spore germination to the formation of the lesion, is altogether impacted by climatic factors, for example, temperature, precipitation, and dew are probably going to influence pathogen distribution because of modified adequacy of preventive methodologies (Coakley et al., 1999; Bevitori and Ghini, 2014). Thus, the management techniques for improving host resistance will be incredibly affected by environmental change (Chakraborty et al., 2000). Fungal diseases alone are assessed to diminish yearly rice production by 14% around the world (Agrios et al., 2005). Expanded recurrence and greatness of outrageous climate occasions joined with high air temperature and CO_2 levels are extended to spread rice infections to new regions (Anderson et al., 2004; Rosenzweig et al., 2005).

Rice blast has gotten harder to control due to the pathogen capacity to endure and facile moving to new fields (Meybeck, 2012). Varietal resistances have lessened because of the presence of new and more destructive strains of the blast pathogen, decreasing the effectiveness of the management, and control tools (Laha et al., 2017). Moreover, fungicides and plant breeding have neglected to give dependable control of rice blast since they are too stable to interpret the dynamic communications between the microbe and rice, which are affected by culture (Briggs, 2000).

Rice blast epidemics are becoming catastrophe when the condition of conidial release is instigated by rain, lack of sunshine, and dew, which extended for long periods (Suzuki, 1975). Rainfall is affecting the conidial dispersion seriously during the beginning of the rainy season and heavy rains (Shahjahan, 1994).

Precipitation (especially low amounts) after fungicide usage may facilitate its presence on leaves (Shiba, 1981); however, excessive precipitation could decrease or wash the fungicide from the leaves (Phong et al., 2009). Further, the growth of rice and its resistance to rice blast is reduced especially under elongated periods of perception cloudy conditions (Suzuki, 1975). The impact of precipitation on the scattering of conidia is generally conspicuous toward the beginning of the season and during heavy rains (Shahjahan, 1994). The rise in sea level is a significant problem for rice production since it could flood low-lying regions and strengthen soil salinity (Dasgupta et al., 2009; Lassa et al., 2016). Also, flooding could transfer the conidia from infected fields to other new areas (Zou and Wei, 2010).

Concerning the CO_2 effect on disease: In a study summarizing the effect of the increased atmospheric CO_2 and O_3 concentrations on yield, Ainsworth (2008) found that a limited increase of CO_2 is predicted to increase the yield, but high O_3 levels and raised air temperature is forecasted to yield reduction. Other workers have revealed that raised CO_2 concentrations probably going to expand the dispersion of rice blast (Kobayashi et al., 2006; Ainsworth, 2008; Gória et al., 2013).

With the increase of CO_2, the canopy and biomass, hence host size increased for a pathogen population (Manning and Tiedemann, 1995; Fuhrer, 2003; Mitchell et al., 2003; Pangga et al., 2004). Furthermore, the presence of a great number of rice residues could increase pathogen survival and inoculum multiplication for subsequent crops and neighboring fields (Coakley et al., 1999). On the other hand, the correlation of plant diseases and high CO_2 levels could be ascribed to alteration in host physiology and anatomy viz., limited nutrients and elevated sugars content of foliage, fiber content, foliar wax, layers of epidermal cells, and mesophyll cells (Plessl et al., 2005; Matros et al., 2006). Moreover, Kobayashi et al. (2006) suggested that a reduction in silicon content could promote leaf blast and sheath blight severity in sensitive rice varieties under high CO_2 levels. The increment of rice foliage wax and the epidermal thickness could be accounted for enhancing physical susceptibility to pathogens, alongside elevated pathogen fecundity and alternation of severity, activity, distribution, and abundance of the pathogen (Plessl et al., 2005; Matros et al., 2006).

In high humidity conditions, rice blast lesions yield plenty of spores, which are transported by wind and act as inoculum for a new infection (Prabhu et al., 2008) and the opposite cut held was recorded under lack of humidity or rainfall (Bevitori and Ghini, 2014). Low humidity or dew triggers the infection of rice blast (Castejón-Muñoz, 1998). Wind affects the host plants via increasing transpiration of rate and silicification of foliage tissue (Kumagaya et al., 1957), hence, supporting the resistance mechanism of the host (Suzuki, 1975). On the other hand, strong winds that move soil particles produce wounds on rice plants, hence, facilitating pathogen invasion (Sakamoto et al., 1940).

High air temperature due to climate change is additionally expected to improve the development and sporulation of the rice blast pathogen (Rajput et al., 2017). The temperature under which rice is developed influences its sensitivity to the blast disease (Asai et al., 1966). For instance, in cold subtropical zones, an elevated temperature stimulates rice blast severity due to an increased risk of invasion (Prabhu et al., 2008). Greer and Webster (2001) stated that rice blast flourishes under temperatures of 17 to 28°C, extended periods of leaf wetness, and high relative humidity. Elevated night temperatures above 20°C result in slow spore dispersal and infestation is diminished, but lesions area increased under fluctuation of daily minimum and maximum air temperatures of 25/32–20/32°C (Espinoza et al., 2003; Bevitori and Ghini, 2014). A mild air temperature of 16 to 24°C may strengthen lesions' sporulation capacity (Bevitori and Ghini, 2014). However, if the predictions of temperature increment occur, the beginning of rice blast may decline in most rice-growing places (Bevitori and Ghini, 2014).

3.6 HOST–PATHOGEN RELATIONSHIP

In response to pathogen infection (early incidence), the redundancy of proteins necessitated in cell signaling is modified in plants, including perception and signal transduction of pathogen. Apoplast receives the early perception of the pathogen via the apoplastic and plasma membrane-localized proteins (Gupta et al., 2015). The apoplast acts as a mediator for nutrients and signals trade-off between plant cells and the neighboring environment. Workers in rice leaves or suspension-cultured cells supplied with compatible and incompatible races of rice blast fungus have addressed the importance of apoplastic proteins in pathogen discrimination (Shenton et al., 2012; Kim et al., 2013). At the PM, microbial invasion influences the cellular transport and membrane characteristics to activate defense signaling at the membrane. Cao et al. (2016) presented in their PM, Proteomic study cumulation of PM transporters, as "vesicle trafficking-related proteins." These proteins share in *M. oryzae* "perception and signal transduction" in rice.

For a confirmed invasion, the pathogen must fight against the plant's fundamental, direct defense strategies, involving physical hindrances, such as cuticle and cell wall, and principally synthesize antimicrobial products to gain nutrients from the host to complete its lifecycle. The microbe also weakens the host's immunity via sending secreted proteins (Vir proteins or Avr effectors) and other metabolites (e.g., lipids, carbohydrates, nucleic acids, etc.) to the host cells to boost pathogenesis (Gupta et al., 2015).

In response, plants have developed an instinctive two-layered immune system, likely coined as a zig-zag model (Jones and Dangl, 2006). The principal line of plant resistance is started from the "perception of extracellular pathogen-associated molecular patterns (PAMPs) by plasma membrane (PM)-localized pattern recognition receptors (PRRs)," in terms of PAMP-triggered immunity, PTI (Boller and Felix, 2009). PTI is an adequately weak immune system; but in many conditions, it is powerful to safeguard against the microbe's assault and shields plants from disease. To defeat the PTI, pathogens emit effector proteins into the host cells, producing "effector-triggered susceptibility (ETS)." However, the "resistance (R)-gene" components of the host can directly or indirectly identify the pathogen's secreted effectors to induce the second immune line termed as "effector-triggered immunity (ETI)," leading to hypersensitive responses, HRs (Cui et al., 2015). The crucial compounds implicated in the interaction of rice and *M. oryzae* are resistance R genes and PRRs from rice, besides effectors and PAMPs from *M. oryzae* (Meng et al., 2019). Owing to the central involvement of R genes in "ETI > 100-R genes" of rice that have been recognized and fight against *M. oryzae* , 27 of which have just been cloned (Liu and Wang, 2016), and an overall of 13 Avr-effector genes have been cloned in *M. oryzae* (Sharma et al., 2012; Liu et al., 2014b). Some investigations found multi-patterns of "R−Avr interaction" in *M. oryzae* (Liu et al., 2014b; Liu and Wang, 2016). Wang et al. (2019a, 2019b) displayed that ZAR1 (CC-type NLR protein ZAR1 protein) form a complex with the plasma membrane, causing cell death initiation and disease resistance. Hence R proteins are effective against pathogen infection via the identification of pathogens' effectors (Wang et al., 2019a, 2019b). These studies shed light on the role of

CC-NLR-mediated ETI in plant resistance. Horsefield et al. (2019) and Wan et al. (2019) demonstrated that "TIR-NBS LRR" induces cell death signaling by breaking down the metabolic cofactor "nicotinamide adenine dinucleotide (NAD^+)" as a reaction to pathogen infection. These investigations confirmed that the R protein involved in *M. oryzae* induces multiple mechanisms to trigger the immune system of rice (Meng et al., 2019).

One of the defense-related proteins called pathogenesis-related (PR) proteins which arise in response to pathogens and abiotic stresses, had antimicrobial activity and is related to signal transduction (van Loon et al., 2006). In this regard, microbial invasion and abiotic stress induce chitinases and glucanases which participated in the breakdown of fungal and oomycete cell walls. Kim et al. (2013) reported that rice blast fungus in rice leaves and suspension-cultured cells induced 10 "PR-5s, 9 chitinases, and 6 PR-1s" in rice plants.

During a pathogen invasion, phytohormones, such as salicylic acid (SA), jasmonic acid (JA), and ethylene (ET) upregulate plant resistance that SA induces the immunity of plants against biotrophic pathogens, while JA and ET adjust plant resistance against necrotrophic and insect pests (Pieterse et al., 2012). SA is responsible for systemic acquired resistance (SAR), which is induced by "a transcriptional cofactor non-expressor of pathogenesis-related genes-1 (NPR1)" in *Arabidopsis* (Dong, 2004). In rice, a similar "NPR1 SA" signaling pathway has been recognized which is mediated by a "WRKY45-dependent "pathway of SA (Shimono et al., 2007). A subsequent study by Jiang et al. (2010) highlighted that SA signaling in rice is divided into "NPR1-dependent and WRKY45-dependent pathways." In dicots, SA is responsible for the resistance against biotrophic pathogens and JA mediates stress responses against herbivores and necrotrophic pathogens (Browse, 2009). Notwithstanding, this model does not completely fit into the monocots. For instance, rice contains high intrinsic SA levels, SA levels are not elevated by compatible and incompatible types of *M. oryzae* (Silverman et al., 1995). Concerning JA, its bioactive form jasmonate isoleucine (JA-Ile) was increased endogenously in rice due to *M. oryzae* infection (Zhang et al., 2018). Yamada et al. (2012) found that JA signaling is induced by biotrophic bacterial pathogen *Xanthomonas oryzae* pv *oryzae* and hemi-biotrophic fungal pathogen *M. oryzae*, demonstrating that the JA signaling pathway may assume a significant part in rice invulnerability against biotrophic and hemi-biotrophic microorganisms, as opposed to dicots (*Arabidopsis*), where it is initiated uniquely in response to necrotrophic pathogens (Meng et al., 2019). Pathogen–host Domain unknown function 26 (DUF26) are proteins related to the plant receptor-like kinase, which are observed in serine/threonine kinases, found to be induced by *M. oryzae* or elicitor treatment. In this regard, LOC_Os04g56430.1 is a DUF26 reported to be mediated by *M. oryzae*, MSP1 treatment, JA application, and wounding, revealing that this protein triggers rice defense by JA pathway (Gupta et al., 2019; Meng et al., 2019).

Although SA and JA are crucial defense-related phytohormones, they work antagonistically to one another in plants for adjusting plant defense signaling (Betsuyaku et al., 2018). Tamaoki et al. (2013) displayed that exogenous application of JA reduced the endogenous level of SA in rice leaves and both JA and SA stimulated some of the common defense marker genes. The co-workers concluded that SA is responsible for the fundamental resistance under the natural environment, and once JA signaling activated, the SA levels decline, thereby SA signaling is prohibited and the defense signaling response of JA is activated. Jiang et al. (2010) discriminated that abscisic acid (ABA) antagonizes SA signaling to deregulate the rice resistance to the pathogen, and A PM proteomics study detected the accumulation of abscisic acid at the early stage of *M. oryzae* infection, implying frustration of host immune response for initial pathogen attack, on the other hand, at a later stage of the infection, cytokinin (CK) was induced (Cao et al., 2016).

The plants induce defense mechanisms via the strengthening of cell walls, the synthesis of phytoalexins, and the production of defense-related proteins. The latter constitute the main ingredients of the signal transduction pathways, hence inducing host defense responses after pathogen perception (Bowles, 1990).

Pathogen attacks severely influenced host cells causing disturbances in cellular metabolism, especially oxidative stress. ROS's inadequate concentration is signaling molecules to manage plant pathways as cell death and pathogen defense. During fungal attack, elevated levels of apoplastic ROS in host cells reinforce the cell wall and local and systemic signal molecules exerted antimicrobial defenses properties, hence is associated with microbial resistance (Waszczak et al., 2018). Such transient and instant synthesis of apoplastic ROS is a distinctive feature of PTI defense signaling and is known as ROS burst (Torres et al., 2006). Waszczak et al. (2018) reported that ROS-inducing enzymes composed of "apoplastic peroxidases, polyamine oxidases (PAOs), and PM-localized NADPH oxidases (respiratory burst oxidase homologs, RBOHs)" are responsible for apoplastic ROS production. In the secretome study of Kim et al. (2013), 15 apoplastic peroxidases and two germins A/oxalate oxidases were highly expressed in rice plants invaded by *Oryzae*. Subsequently, Cao et al. (2016) reported NADPH oxidase encoded by *Osrboh8* in rice leaves experienced *M. oryzae*. Therefore, both apoplastic peroxidases and NADPH oxidases are considered a director for apoplastic ROS, in combination with apoplastic ROS-mediated pathogen resistance.

Plants produce an arsenal of antioxidant molecules and enzymes that limit the content and longevity of ROS, while compartmentalization of ROS synthesis and decactivation determines the role of ROS in the plant (Waszczak et al., 2018; Abdel Latef et al., 2020). Kim et al. (2014) and Bagy et al. (2019) stated exacerbation of ROS-production enzymes such as superoxide dismutase (SOD) and class III peroxidases, as well as enzymes associated with the ROS-detoxifying enzymes viz., "catalase (CAT), glutathione peroxidase, ascorbate peroxidase (APX), glutathione S-transferase (GST), and dehydroascorbate peroxidase (DHAR)" in host plants infected by microbes. In this regard, class III peroxidases have a dual role as H_2O_2–quencher to water and oxygen by peroxidase activity or H_2O_2–producer in the oxidase cycle. Waszczak et al. (2018) and Bindschedler et al. (2006) reported the involvement of class III peroxidases for microbial resistance in *Arabidopsis*. Similarly, in rice, Kim et al. (2013) found an abundance of 20 peroxidases in response to *M. oryzae* infection. A more recent study by Lin et al. (2018) recorded an elevated level of various ROS-scavenging enzymes in rice plants which incompatibly interacted with *M. oryzae*. Moreover, Narula et al. (2018) recorded the upregulation of APX and thioredoxin peroxidase, mixed regulation of GST and thioredoxin, and decrement of DHAR in a resistant cultivar of rice invaded by *M. oryzae*. Thus, Gupta et al. (2019) concluded that a high level of ROS is a defensive tool in response to pathogens, while ROS-synthesizing or -catabolizing enzymes such as class III peroxidases, APX, PM-localized NADPH oxidases (RBOHs), and GST were accumulated to equilibrate redox homeostasis of host cellular. Furthermore, the phenylpropanoid pathway is aided in the production of the antioxidant via participating in the ROS-scavenging system or as signaling molecules and support defense mechanisms in plants (Dixon et al., 2002).

Pathogen infection affects host metabolism and deregulates cellular homeostasis where the risk of protein deterioration due to protease excretion by the pathogens occurs (Wang et al., 2017a). HSPs are the category of molecular chaperones in plant produced in abundance during fungal invasion which prohibits undesirable protein accumulation and promotes refolding of denatured proteins. In response to virulent and avirulent *M. oryzae* strains, 72 hpi rice samples enhanced OsHSP81 production which stimulates callose deposition and ROS production (Lin et al., 2018). Narula et al. (2018) found out that *M. oryzae* infection has caused the enhancement of TPI and protein phosphatase 2C, impairments of FBA, and suppression of

GAPDH in rice tissue which mediated glycolytic flux, cell reproduction, and chromatin alterations as serious efforts to prevent fungal attack. PAMPs, in the interaction of rice and *M. oryzae*, involved chitin, MSP1, and MoHrip1. Overexpression of MSP1 in rice exerts immense resistance against *M. oryzae* and *X. oryzae* pv *oryzae* (Hong et al., 2017).

Meng et al. (2019) reported the identification of more than 100 genes that grant rice blast resistance, such as *Pi-1*, *Pi-5*, and *Pi-z*^{t}, resulting in the production of rice blast resistance cultivars possessing R genes. Ryu et al. (2009) found that *M. oryzae* infection in rice strains carry the *Pi5* R gene with susceptible mutants where 8 proteins were distinctively mediated between resistant and sensitive plants. In the resistance cultivar, three proteins were produced in plenty, such as a 2,3-bisphosphoglycerate-independent phosphoglycerate mutase, triosephosphate isomerase, and an unknown protein, while five proteins were deregulated viz., a fructokinase I, an atpB of chloroplast ATP synthase, a GST an aminopeptidase N, and an unidentified protein. Kinases and phosphatases induce phosphorylation and dephosphorylation of proteins, respectively, influence protein arrangement, causing protein alterations with new enzymatic actions, structural stability, substrate specificity, or intracellular localization (Kline-Jonakin et al., 2011). Phosphoproteomic investigations of resistant and sensitive rice leaves invaded by *M. oryzae* ended to the identification of 53 regulated phosphoproteins. Among them, many ROS-related enzymes were distinguishably managed by phosphorylation in tolerant and sensitive cultivars, involving enzymes associated with the metabolism of ROS rather than accumulation or detoxification (i.e. cysteine synthase and ROS-scavenging enzymes). The same study also presented that ROS-quenching enzymes retarded in the susceptible or resistant strains invaded by *M. oryzae* inoculation, thus phosphorylation triggers an oxidative burst in response to pathogen attack in both cultivars. Further, H_2O_2 was exacerbated higher in the incompatible interaction compared with the compatible interaction during 24 h after inoculation. Susceptible cultivars deregulated phosphorylated cinnamoyl-CoA reductase (CCR) which is responsible for lignin production especially at 8 h after inoculation (Li et al., 2015).

3.7 GENETIC AND MOLECULAR ASPECTS OF BLAST RESISTANCE IN RICE

Several approaches have been proposed to counteract the negative impacts of rice blast, extending from field management to genetic approaches. Rice blast has been intensively studied and investigated throughout the world. Thus, the scientists considered it as a model disease for studying several biological aspects in other diseases, including the genetics, epidemiology, and molecular pathology of the host–parasite interactions. This has been promoted by the availability of the complete genome sequences of both the host (rice) and the parasite *M. oryzae*. The advances in understanding the genetic control of rice blast paved the way to understand the regulatory mechanisms of other plant diseases.

Even though the classical techniques such as conventional breeding have enabled the development of some blast-resistant genotypes, where several major genes were successfully detected and introduced into rice varieties for conferring blast resistant (Korinsak et al., 2011), many mutated resistant varieties have been developed via radiation breeding (Ahloowalia et al., 2004). This remains a marginal success because no durable resistant genotypes have been developed using these techniques. According to Young (1996), blast resistance in rice can be classified into two major types: the complete resistance (qualitative resistance and race-specific) is controlled by major gene(s) (R genes) while the partial-resistance (quantitative character and non-race-specific) is regulated by several genes known as quantitative trait loci (QTLs). About 100 R genes and 500 QTLs were found to be associated with blast resistance in rice (Ashkani et al., 2015b). Out of these, 35 genes have been successfully cloned and investigated (Wang et al., 2017a).

In this section, we will shed light on the different genetic approaches that have been employed to disentangle the genetic architecture of rice blast resistance, including genomic, transcriptomic, and proteomic studies.

3.8 GENOMIC APPROACHES

The rice blast resistance controlled by the major resistance genes (R genes) is not eternal. It decays after 3–5 years because the fungus population is so variable and undergoes adaptation due to mutations or the deletion of the avirulence genes (McDonald and Linde, 2002). The annotated genomes of *M. oryzae* (Dean et al., 2005) and rice (Dean et al., 2005) revolutionized the study of the host–parasite interactions of this disease, making it the model for understanding other plant diseases.

The classical gene identification analysis was the first tool of genomics approaches that has been used to study blast resistance by which dozens of resistance allele's donors have been identified. In Indian landraces, three genes were detected that are conferring complete or partial resistance (Barman et al., 2004; Sharma et al., 2005; Kumar et al., 2010). Additionally, nine resistance genes of blast disease have been identified and cloned by several researchers, reviewed by Srivastava et al. (2017). The distribution of the R genes clearly varied among the rice chromosomes. Chromosome 11 harbors the highest number of those genes followed by chromosome 12, while chromosomes 3 and 7 had the lowest number of the R genes (Ashkani et al., 2016). Interestingly, nine blast resistance genes have been widely used in breeding programs that were co-localized on chromosome 6along with some broad-spectrum resistance genes (Deng et al., 2006; Jiang et al., 2015; Tian et al., 2016). Another tool of the blast resistance genomics is the marker-assisted selection (MAS), it is more simple than conventional breeding as the tolerant genotypes are marked by maintaining few R genes (Young, 1996). MAS is an instrumental approach to select for blast resistance by elucidating the interaction between the pathogen avirulence (Avr) genes and the host R genes (Petit-Houdenot and Fudal, 2017). Several blast-resistant varieties have been developed using MAS (Yan et al., 2017). Several studies and successful development of resistant varieties have been reported (Singh et al., 2013). By combining, MAS and backcross breeding, Jiang et al. (2019) had successfully developed and improved the blast resistance in some lines via the introgression of nine broad-spectrum blast resistance genes. Gene pyramiding is another genomic approach in which R genes from a single donor or various donors are combined in an elite variety (Koide et al., 2009). It has been successfully used to produce blast-resistant genotypes; several blast-resistant genes have been assembled in resistant varieties (Arunakumari et al., 2016). QTL are very important for identifying the partial-resistance genes that show a durable broad-spectrum resistance against many pathogens (Song and Goodman, 2001). It deals with investigating the variation in a biparental population derived from two contrasting parents, that is, a resistant parent is crossed with a susceptible parent, thereafter, progenies are derived from this cross showing a remarkable variation for the trait of interest. Five blast resistance genes have been identified using QTL analysis (Sallaud et al., 2003). About 22 R genes were identified by QTL, and have been successfully cloned and described (Shiping et al., 2003). In 15 diverse populations, about 350 QTLs have been detected for blast resistance (Sato et al., 2006). QTL has been used to identify major and minor genes for blast resistance; it is useful for MAS as well as for cloning the major blast resistance genes (Ashkani et al., 2015a). QTL can be effectively used to improve blast resistance in rice (Guan et al., 2019). The introgression of QTL harbors three dominant blast resistance genes significantly improved the resistance of an elite variety (Hasan et al., 2015a, 2015b). Using QTL mapping, Aglawe et al. (2017) detected two QTL associated with resistance for rice leaf and neck blast. Additionally, they have developed stable recombinant inbred lines that possess both QTLs. Similarly, in a doubled haploid population derived from two contrasting lines, Joshi et al. (2019) mapped a new resistant gene (*Pi-67*) from the well-known broad-spectrum resistant landrace Tetep. This gene has been physically mapped very close to the centromeric region on chromosome 12. In the last two decades, another trait-association technique has been used to identify the causative QTL/genes, namely, Genome-Wide Association Study (GWAS). It does not require biparental populations; alternatively, GWAS deploys the historical linkage disequilibrium that occurred across the evolution of certain species. GWAS for blast resistance was conducted in two populations, Japonica and Indica. The Japonica collection had more quantitative resistance than Indica, two blast resistance loci were identified in the Japonica collection (Raboin et al., 2016). Using GWAS, 25 genomic regions associated with blast resistance were identified in the rice genome (Mgonja et al., 2017). Very recently, many blast-resistance genes have been identified using GWAS by many researchers (Kim and Reinke, 2019; Volante et al., 2020; Wang et al., 2020b). Likewise, the transgenic approach for blast resistance revealed successful results. The transgenic plants showed stable integration as well as inheritance of the transgene. The expression of the rice chitinase gene enhanced the blast resistance (Nishizawa et al., 1999). Similarly, the introduction of *RCC2* gene (*Agrobacterium*-mediated transformation) improved the resistance to blast in Basmati rice under laboratory and field conditions (Asghar et al., 2007). Transgenic rice plants possess two elicitor genes (*MoHrip1* and *MoHrip2*) from *M. oryzae* exhibited high resistance against rice blast (Wang et al., 2017b). Several genes have been successfully transformed into the rice and conferred resistance against blast (Shen et al., 2019; Chandran et al., 2019; Wang et al., 2019b; Singh et al., 2020; Fan et al., 2020; Liu et al., 2020). MicroRNAs (miRNAs) are known as key regulators of growth and defense in eukaryotes. Several reports have demonstrated the role of miRNAs in the immunity of plants against pathogen infections. Expression of miRNAs coding genes has been reported in rice during *M. oryzae* infection (Li et al., 2014). Nine miRNAs were identified in rice leaves after inoculation with *M. oryzae* (Baldrich et al., 2015). Expression of *osa-miR7695* in rice can be used as a biochemical marker to select for blast resistance (Quoc et al., 2019). Similarly, the expression of *osa-miR1873* balanced the rice immunity against *M. oryzae* infection at the expense of yield attributes (Zhou et al., 2019). Contrarily, Zhang et al. (2020) reported that the expression of miR156fhl-3p elevated rice blast resistance without the yield penalty. Genome editing is the most recent approach to improving blast resistance in rice. Among all sequence-specific nucleases (SSNs), the clustered regularly interspaced short palindromic repeats (CRISPR/Cas9) were found to be the most relevant one to edit the rice genome to induce blast resistance (Shan et al., 2015). CRISPR/Cas9-targeted mutagenesis of the ethylene response transcription factor (ERF) gene *OsERF922* improved the blast resistance in rice (Wang et al., 2016). The Knockout of Pi21 (a broad-spectrum resistance gene) by CRISPR/Cas9 has improved rice blast resistance without disturbing the agronomic traits (Nawaz et al., 2020). As a result of the instability of the resistance conferred by R genes, Ning et al. (2020) suggest to developing a core resistant collection with a super performance in the traits of interest and combining both the broad-spectrum and durable resistance genes. This can be achieved using the newly developed approaches, such as GWAS and genome editing.

Collectively, the vast development in the molecular techniques along with the bioinformatics tools effectively improved the identification, transferring, and manipulation of the causative genes of blast resistance in rice. Many approaches have enabled the development of blast-resistant genotypes either independently or in a combination with other approaches.

3.9 THE TRANSCRIPTOMIC APPROACH

However, 100 R genes were identified. Only 35 genes have been cloned and characterized. In this section, we focus on the characterization of the R genes at the transcriptomic levels. Few studies have dealt with the transcriptomic profiling of both the host and pathogen during infection (Kawahara et al., 2012; Dong et al., 2015). This is attributed to some technical issues. The transcriptomic studies on the pathogen *M. oryze* in infected samples is less enriched relative to those of the host rice (Jeon et al., 2019).

Wei et al. (2013) investigated the early response of rice to *M. oryzae*. They found that the signaling and secondary metabolic genes were significantly upregulated, especially WRKYs transcription factors. The transgenic rice plants overexpressing the WRKY gene, *Os*WRKY47 exhibited enhanced resistance to the rice blast fungus. Wang et al. (2020a) studied the alterations in the transcription profile of the gene, namely, *immune response peptide* (IRP). In rice suspension cells, IRP has been upregulated by the *M.*

oryzae chitin treatment. In a cascade manner, the overexpression of IRP triggered the expression of the resistance gene phenylalanine ammonia lyase 1 (PAL1) and enhanced the expression of mitogen-activated protein kinases (MAPKs). The inbred lines of the rice variety "Jao Hom Nin," namely, *Pish*-lines and *Pik*-lines possess one of the blast resistance genes *Pish* or *Pik*. The differentially expressed genes (DEGs) were compared in the *Pish*-lines versus the *Pik*-lines upon treatment with rice blast spore suspension. Ngernmuen et al. (2020) found that some DEGs are specific to the *Pish*-lines (the wall-associated kinase gene) other genes are specific to the *Pik*-lines (zinc-finger-containing protein gene) while some genes are common in both lines. The authors proposed that identifying the specific and common DEGs will be helpful to elucidate the regulation of plant defense. Several studies reported that the overexpression of resistance and defense genes conferred resistance in the transgenic plants upon *M. oryzae* infection. Kanzaki et al. (2002) reported that the rice plants transformed with the wasabi defensin gene showed resistance to *M. oryzae* and the resistance has been inherited over generations. The overexpression of *Os*NPR1 in rice cultivars Gui99 and TP309 effectively enhanced the resistance against *M. oryzae* and triggered the expression of the pathogen resistance (PR) genes (Feng et al., 2011).

The overexpression of the broad-spectrum resistance 1 gene (BSR1) induced the broad-spectrum resistance in rice against two bacterial diseases and two serious fungal diseases (Maeda et al., 2016). Similarly, the overexpression of the calcium-dependent protein kinase *Os*CPK4 enhanced the resistance of rice upon fungal infection (Bundó and Coca, 2016). The overexpression of MoSM1 encodes an immunity-inducing protein from *M. oryzae*, in rice confers broad-spectrum resistance against many fungal and bacterial diseases (Hong et al., 2017). The loss of function of the broad-spectrum resistance Kitaake-1 (BSR-K1) led to the accumulation of *OsPAL1–7* mRNAs in the *bsr-k1* mutant and conferred resistance against two pathogens: *M. oryzae* and *X. oryzae pv oryzae* with nonsignificant effect on the agronomic traits. Additionally, the overexpression of *OsPAL1* in the wild-type rice TP309 has induced the resistance against *M. oryzae*, approving the role of *OsPAL1* (Zhou et al., 2018). The transgenic rice plants transformed with the Cytochrome 450 monooxygenase (CYP71Z18) from maize showed high resistance against *M. oryzae* infection (Shen et al., 2019). Besides inducing the virulence of pathogens, the effector proteins triggering the plant defense machinery, the upregulation of *M. oryzae* systemic defense trigger 1 (MoSDT1) enhanced the blast resistance in rice (Wang et al., 2019a). Not only R genes from rice can confer blast resistance, but genes from other plant species with similar functions can also confer blast resistance in rice. For example, the transformed rice plants overexpressing the nucleotide-binding and leucine-rich repeat genes RPS2 and RPM 1(D505V) from *Arabidopsis* exhibited high broad-spectrum resistance (Li et al., 2019b). Six AAA-ATPase family genes have been induced upon *M. oryzae* infection; the overexpression of them stimulated the pathogenesis-related gene expression and improved the blast resistance in rice; this pathway is mediated by salicylic acid (Liu et al., 2020). As the plasma membrane is the most selective trafficking cellular component, some receptors recognize microbe-associated molecular patterns (MAMPs) and switch on the plant immune system. The induction of BSR1 contingents on the MAMPs-recognition system, supportive of the disturbance of the MAMPs system quenched the broad-spectrum resistance (Kanda et al., 2020).

Taken together, the resistance to fungal and bacterial diseases in rice, in particular, rice blast can be conferred via the overexpression of various sets of genes. More likely, the mechanism of the BSR is widely conserved in different plant species as genes from Arabidopsis and maize conferred BSR in rice.

3.10 THE PROTEOMIC APPROACH

Proteomic analysis is an instrumental tool that grows quickly after the era of genomics and has been widely used to investigate the plant response to various stresses including the plant–microbe interaction. This approach is supported by the progressive development in protein analysis techniques. One of the applications of proteomic analysis is the study of differential expression proteins (DEPs), revealing their potential role in fungal and bacterial disease resistance in rice. The study of blast resistance in rice on the proteomic level got less attention relative to the study on the transcriptomic level. Subsequently, the role of DEPs in blast resistance is poorly understood. Ma et al. (2019) found that 50 DEPs had clearly changed upon *M. oryzae* infection, and conferred race-specific resistance to blast in rice. Using proteomic screening, Wang et al. (2020a) found that small secreted proteins (SSPs) are evolved upon *M. oryzae* infection. Moreover, they found that the gene *immune response peptide* (IRP) encodes the production of the SSPs that regulate the plant immune response. In a comparative study, using the iTRAQ proteome method, the alterations in DEPs have been investigated in durable resistant genotypes versus susceptible ones (Ma et al., 2020). The authors found that the PAMP-induced the immune system in the resistant genotypes at both the transcriptomic and the proteomic levels, while in the susceptible genotypes it has been only induced at the transcriptomic level, and inhibited at the proteomic level. These findings indicate that DEPs are conferring the durable resistance in resistant genotypes. Narula et al. (2018) figured out the role of the nuclear proteome in the signaling and the activation of the immune system to resist the blast disease in rice. They found that 140 immune responsive proteins were associated with nuclear reconstruction, cell division, energy production, and gene regulation. They concluded that the concomitant of the nuclear function with the rearrangement of the host translational machinery regulates the host resistance mechanisms against blast disease. To investigate the role of the R genes *Pi21* in blast resistance, using a combination of CRISPR/Cas9 and iTRAQ, Nawaz et al. (2020) developed mutant lines using CRISPR/Cas9, later, using the iTRAQ-based proteomic analysis, 372 DEPs were identified, out of these, 149 up and 223 were down-regulated, respectively. According to the authors, DEPs involved in photosynthesis, carbohydrate metabolic process, and small molecule metabolic process, as well as the transcriptional levels of genes related to defense signaling were upregulated after the inoculation with *M. oryzae*. Similarly, using mutant lines, Meng et al., 2019 found that a novel R protein coded by the broad-spectrum R gene (*Pita2*) is unique from the protein coded by *Pita* gene, with superiority for the first in conferring the broad-spectrum resistance to blast diseases.

Conclusively, proteomic analysis became an instrumental tool to investigate the variations in blast resistance in rice. Different types of proteins contribute to blast disease resistance in rice, including R proteins and small secreted proteins (SSPs). R proteins confer blast resistance in durable genotypes rather than the susceptible ones.

To sum up, blast resistance in rice is a polygenic trait. BSR is regulated at the transcription and translation levels. Several genes and proteins are associated with the host resistance strategy against blast disease in rice. The complete resistance (endurable and controlled by many R genes) and the partial resistance (durable and affected by several QTLs), combining the genomic regions that harbor the genes of both types is of high importance to ensure better resistance against blast disease. Combining all techniques and tools will be helpful to develop blast-resistant genotypes that are more cost-effective and environmentally friendly than using pesticides.

3.11 MANAGEMENT STRATEGIES FOR RICE BLAST CONTROL

Rice blast disease spread in nearly moat rice fields and the extended host range of *M. oryzae* makes suppression of the disease troublesome. So, several techniques have emerged for higher efficacy of blast disease control. These controlling methods are broadly classified as cultural control, resistant varieties, "chemical control," and "biological control." These ways are functional in producing a less favorable culture for the pathogen in the field and limit the blast disease epidemics (Upadhyay and Bhatta, 2020).

3.11.1 FORECASTING BLAST DISEASE

Disease forecasting in terms of epidemiological disease study assists to find out efficient disease control tools. Thus, this study allows information that helps plant breeders and chemical industries to be in a steady-state to control the disease. For the most destructive disease of rice, some blast forecasting models have been created to predict its circumstance and riskiness to manage the disease operatively. But the disease forecasting model faces problems related to uncertainties and inaccuracies in prediction disease, so they are not widely used. Forecasting methods depend on the host, pathogen, and environmental conditions of disease development (Upadhyay and Bhatta, 2020).

3.11.2 CULTURAL PRACTICES

Intensive use of N-fertilizers, water problems, the availability of infected debris in the field, and sowing infected seeds are the major causes of the distribution and occurrence of disease (Upadhyay and Bhatta, 2020). Thus, selecting healthy and clean seeds minimize seed-borne fungi. Covering or submerging cultivated lands with a large amount of water is advisable to prevent water stress which attenuates disease development because the anaerobic condition is not suitable for the blast pathogen. As high humidity is desirable to blast disease, sowing time should be chosen to prevent flowering at periods of high humidity. Getting rid of the infected crop derbies limit the presence of inoculum in the field for the successive growing season (Upadhyay and Bhatta, 2020). Also, it is recommended that use of an optimum level of nitrogen in split doses could reduce the disease severity. Long et al. (2000) reported that applying overdoses of nitrogen fertilizer and the whole amount in a single time significantly raises disease establishment and severity. Records of leaf blast disease at the panicle primordia stage were 60% in recommended-N dose, 73% in over-N dose, and 43% in split-N treatments. It is realized that Silicon (Si) application lessens the rate of blast fungus appressorial penetration. Hayasaka et al. (2008) proposed that the rate of appressorial penetration after 111 h diminished five-times and the rate of lesions formation reduced 2.5-fold by applying Si from 0–0.8 g per plant.

3.11.3 RESISTANT VARIETIES

Cultivation of blast-resistant variety is a potent practice applied by farmers. Sharma et al. (2012) recorded the identification of 100 quantitative blast Resistance (R) genes and more than 350 "quantitative trait loci (QTLs)" as well as 19 R genes. A single dominant R gene is the potential for avoiding blast disease by *M. oryzae* having the corresponding "*Avirulence* (AVR) gene" (Silue et al., 1992). However, the disease resistance is reduced due to new resistant races of blast fungus, and hence, the varieties which coined to be resistant must be changed periodically. Pyramiding the R genes in rice cultivars is likewise another system created to beat the moderately little impacts of single R genes and further raise the resistance to rice blast. Sustaining high yield is another issue that should be taken into consideration during breeding for disease resistance. Deng et al. (2017) illustrated that adjusting high disease resistance and appropriate yield can be achieved by epigenetic regulation of paired "antagonistic nucleotide-binding leucine-rich (NLR)" receptors. Generally, the selection of resistant varieties differs from one region to another and from one season to another in the same region as well. Several resistant varieties of rice are produced and utilized for breeding programs. Owing to changing in pathogenic races at different seasons and regions, no varieties seem to be resistant to all races. Also, it should be taken into consideration that no microbial race can invade all rice varieties.

3.11.4 CHEMICAL CONTROL

Utilizing chemicals, predominately fungicides, and antibiotics, to manage the rice blast disease is achieved either by pretreatment of grains before sowing or foliar spraying or dusting of rice plants in the field (Upadhyay and Bhatta, 2020). The development of chemical control for rice blast was greatly associated with following the chemical control program of Japan.

Copper fungicides were the dominant blast disease-controlling agent after the Second World War. But, copper-based fungicides had negative impacts on yield production. Later, during the 1950s, the organomercuric compounds rapidly applied instead of copper-based fungicides. During the mid-1960s, the Japanese government prevented the application of organomercuric compounds owing to heavy metal toxicity, although these compounds exerted very low phytotoxicity and were a potent technique in controlling (Upadhyay and Bhatta, 2020). Systemic antibiotic (Blastidin-S) was then progressively applied as a control agent of blast disease. The application of organophosphorus fungicides was developed parallel to antibiotics, but due to their excessive use, the pathogen showed resistance to them. Further, this investigation found that the resistance of rice diminished to zero by stopping antibiotics application. Klittich (2008) reported predominating effective fungicides utilized to control the rice blast disease are coined as plant defense activators (PDAs), chlorine biosynthesis fungicides (CBIs), melanin biosynthetic inhibitors (MBIs), and quinol site mitochondrial respiration inhibitors. The usefulness of these fungicides mainly relies on the applied fungicide, period and way of application, rate of disease during the application, the potency of the forecasting model, and the abundance and/or the development of fungicide-resistant strains. Thus, it is beneficial to apply fungicides with various modes of action in the rotation, or combination secures high performance of fungicide and also decreases the development of fungicide-resistant strains (Upadhyay and Bhatta, 2020).

3.11.5 USING PLANT PRODUCTS

Although chemicals are utilized in a large-scale blast disease control in rice, they cause environmental pollution and impact the health of the biotic community. So, using natural products is an alternative method for nonbiodegradable and persistent chemical compounds. Plant natural products are characterized by their biodegradable, eco-friendly, proportionally inexpensive, target-specific, and easy to use. Currently, ecofriendly chemical compounds such as organomercuric, organophosphorus, copper fungicides, antibiotics like Kasugamycin/Blasticidin, and leaf extracts are applied to control blast disease beneficially.

Among the important plant products, Neem-based commercial botanicals which potentially avoid germination of blast fungal spores. Among these products, neem-based pesticides spray, plant-dried commercial products, such as

Achoole, Neem Gold, Neemzal, and Neem Azal T/S are fruitful botanicals in attenuate rice blast virulence (Pandey, 2018). Several workers recommended the effectiveness of Neem seed kernel extract which represented the second potential method of leaf blast management after *Pseudomonas fluorescence*. Also, the leaf extract of medicinal plants, such as *Calotropis*, *Calendula*, *Eclipta alba*, moringa, etc., showed potential antifungal controlling agents against the severity of blast disease (Jyotsna et al., 2017). Furthermore, the mixed application of bael and tulsi leaf extracts with chemicals is considered a prime technique applied to control leaf blast.

3.11.6 BIOLOGICAL CONTROL

Biological control is considered a feasible approach due to its self-sustaining, effectiveness, and commonly more enduring. Reduction of disease by biological control approach affects disease incidence by minimizing the pathogen inoculum via decrement in production and spread of fertile spores, attenuating their survival and distribution, attenuating pathogen attack, and its severity to the host (Upadhyay and Bhatta, 2020). From the mechanistic effects of microorganisms used for rice, blast management is the use of antagonistic bacteria and stimulation of host resistance. Gnanamanickam and Mew (1992) recorded the potency of *P. fluorescens* (Strain Pf 7–14) in minimizing blast disease significantly in the field experiment. Another controlling microbe against virulent strains of blast pathogen is the use of avirulent strains of *Pyricularia* spp. Arase and Fujita (1992) showed that pre-inoculation with non-pathogenic *Pyricularia* spp. in rice prevents the pathogen to achieve leaf-sheath cells, hence, the plant became highly resistant to blast disease. Manandhar et al. (1998) reported that spraying the avirulent strain of *P. oryzae* and *Bipolaris sorokinianaat* isolates during the fourth leaf stages triggers host systemic resistance and lessens rice blast disease by 20.3% and 30.2%, respectively. Other studies highlighted the capability of plant growths promoting rhizobacteria in activating systemic resistance against rice blast. In conformity, Krishnamurthy and Gnanamanickam (1998) displayed that the potential role of *P. fluorescens* 7–14 and *P. putida* V14i in reducing blast incidence is the activation of Induced Systemic Resistance (ISR) where salicylic acid was elevated which prevents rice blast by 25%. The reduction of disease by *P. fluorescence* was mainly ascribed to antifungal and antibiotic characteristics of the strain which challenged to reduce conidial germination by 70–100. Singh et al. (2012) recorded that seed pretreatment with

Trichoderma harzianum (at 4 g kgG1 seed) immensely suppressed rice blast disease intensity by 10–25%. Subsequent studies showed that several rice rhizobacteria are isolated and identified as inhibitors of rice blast invasion. Spence et al. (2014) isolated and identified *Pseudomonas* bacteria and found that a *Pseudomonas* isolates (EA105), effectually minimized the disease incidence by decreasing the appressoria formation ~90% and slowing down the fungal growth by 76%. Law et al. (2017) reported that the application of *Streptomyces* to the infected rice declines the severity of rice blast disease by 88.3%. Campo et al. (2020) pointed out that using arbuscular mycorrhizal fungi (*Funneliformis mosseae* or *Rhizophagus irregularis*) conferred protection to the rice blast fungus, but the level of mycorrhiza-induced blast resistance varied among host genotypes. The success and effectiveness of biological control rely on the capability of the biocontrol agent to survive and establish itself in the submitted environment.

3.12 CONCLUSION

The productive results are to increase the effective rice blast management to bridge the gap between global rice production and the ever-increasing rice demand. Several attempts have been developed to decrease the damaging impact of blast disease on rice plants caused by *M. oryzae*. In regions of low blast infection areas, control of rice blast is predominately dependent on cultivation-resistant varieties. Chemicals management of blast pathogen is applied to infected plants. Biological control agents, such as *P. fluorescens*, *P. oryzae*, *Bipolaris sorokinianaat*, and *Trichoderma harzianum* were used as eco-friendly tool for rice blast management. Moreover, arbuscular mycorrhizal fungi (*Funneliformis mosseae* or *Rhizophagus irregularis*).

KEYWORDS

- **blast disease**
- **host–pathogen relationship**
- **necrotrophic**
- **pathogen-associated molecular patterns (PAMP)**
- **resistance gene (R gene)**

REFERENCES

Agbowuro, G. O.; Afolabi, M. S.; Olamiriki, E. F.; Awoyemi, S. O. Rice Blast Disease (*Magnaporthe oryzae*): A Menace to Rice Production and Humanity. *Int. J. Pathog. Res.* **2020,** 32–39.

Aglawe, S. B.; Bangale, U. ; Satya, R. D. S. J.; Balija, V.; Pal, B. V.; Kumar, S. S.; Kumar, S. P.; Kumar, S.; Maddamshetty, S.; Maganti, S. M. Identification of Novel QTLs Conferring Field Resistance for Rice Leaf and Neck Blast From an Unique Landrace of India. *Gene Rep.* **2017,** *7,* 35–42.

Agrios, G. N. In *Plant Pathology*, 5th ed.; Elsevier Academic Press: Burlington, MA, USA, 2005; pp 79–103.

Ahloowalia, B. S.; Maluszynski, M.; Nichterlein, K. Global Impact of Mutation-Derived Varieties. *Euphytica* **2004,** 135, 187-204.

Ahmed, T.; Shahid, M.; Noman, M.; Niazi, M. B. K.; Mahmood, F., Manzoor, I.; Zhang, Y.; Li, B.; Yang, Y.; Yan, C.; Chen, J. Silver Nanoparticles Synthesized by Using Bacillus cereus SZT1 Ameliorated the Damage of Bacterial Leaf Blight Pathogen in Rice. *Pathogens* **2020** *9* (3), 160.

Ainsworth, E. A. Rice Production in a Changing Climate: A Meta-analysis of Responses to Elevated Carbon Dioxide and Elevated Ozone Concentration. *Glob. Change Biol.* **2008,** *14* (7), 1642–1650.

Anderson, P. K.; Cunningham, A. A.; Patel, N. G.; Morales, F. J.; Epstein, P. R.; Daszak, P. Emerging Infectious Diseases of Plants: Pathogen Pollution, Climate Change and Agrotechnology Drivers. *Trends Ecol. Evol.* **2004,** *19* (10), 535–544.

Arase, S.; Fujita, K. Induction of Inaccessibility to *Pyricularia oryzae* by Pre-inoculation of *P. grisea*, in Rice Leaf-Sheath Cells. *J. Phytopathol.* **1992,** *134* (2), 97–102.

Arunakumari, K.; Durgarani, C.; Satturu, V.; Sarikonda, K.; Chittoor, P.; Vutukuri, B.; Laha, G.; Nelli, A.; Gattu, S.; Jamal, M. Marker-Assisted Pyramiding of Genes Conferring Resistance Against Bacterial Blight and Blast Diseases Into Indian Rice Variety MTU1010. *Rice Sci.* **2016,** *23,* 306–316.

Asai, G. N.; Jones, M. W.; Rorie, F. G. Influence of Certain Environmental Factors in the Field on Infection of Rice by *Piricularia oryzae*; Army Biological Labs: Frederick, MD, USA, 1966.

Asghar, A.; Rashid, H.; Ashraf, M.; Khan, M. H.; Chaudhry, Z. Improvement of Basmati Rice Against Fungal Infection Through Gene Transfer Technology. *Pak. J Bot.* **2007,** *39* (4), 1277.

Ashkani, S.; Rafii, M.; Shabanimofrad, M.; Ghasemzadeh, A.; Ravanfar, S.; Latif, M. Molecular Progress on the Mapping and Cloning of Functional Genes for Blast Disease in Rice (Oryza Sativa L.): Current Status and Future Considerations. *Crit. Rev. Biotechnol.* **2016,** *36,* 353–367.

Ashkani, S.; Rafii, M. Y.; Shabanimofrad, M.; Miah, G.; Sahebi, M.; Azizi, P.; Tanweer, F. A.; Akhtar, M. S.; Nasehi, A. Molecular Breeding Strategy and Challenges Towards Improvement of Blast Disease Resistance in Rice Crop. *Front. Plant Sci.* **2015a,** *6,* 886.

Ashkani, S.; Yusop, M. R.; Shabanimofrad, M.; Azady, A. R.; Ghasemzadeh, A.; Azizi, P.; Latif, M. A. Allele Mining Strategies: Principles and Utilisation for Blast Resistance Genes in Rice (*Oryza sativa* L.). *Cur. Issues Mol. Biol.* **2015b,** *17,* 57–73.

Asibi, A. E.; Chai, Q.; Coulter, J. A. Rice Blast: A Disease With Implications for Global Food Security. *Agronomy* **2019,** *9* (8), 451.

Bagy, H. M. K.; Hassan, E. A.; Nafady, N. A.; Dawood, M. F. Efficacy of Arbuscular Mycorrhizal Fungi and Endophytic Strain *Epicoccum nigrum* ASU11 as Biocontrol Agents Against Blackleg Disease of Potato Caused by Bacterial Strain *Pectobacterium Carotovora Subsp. Atrosepticum* PHY7. *Biol. Control* **2019,** *134,* 103–113.

Baker, B.; Zambryski, P.; Staskawicz, B.; Dinesh Kumar, S. P. Signaling in Plant Microbe Interactions. *Sci.* **1997,** *276,* 726–733.

Baldrich, P.; Campo, S.; Wu, M. T.; Liu, T. T.; Hsing, Y. I. C.; Segundo, B. S. MicroRNA-Mediated Regulation of Gene Expression in the Response of Rice Plants to Fungal Elicitors. *RNA Biol.* **2015,** *12* (8), 847–863.

Barman, S. R.; Gowda, M.; Venu, R. C.; Chattoo, B. B. Identification of a Major Blast Resistance Gene in the Rice Cultivar 'Tetep'. *Plant Breed.* **2004,** *123* (3), 300–302.

Betsuyaku, S.; Katou, S.; Takebayashi, Y.; Sakakibara, H.; Nomura, N.; Fukuda, H. Salicylic Acid and Jasmonic Acid Pathways are Activated in Spatially Different Domains Around the Infection Site During Effector-Triggered Immunity in Arabidopsis Thaliana. *Plant Cell Physiol.* **2018,** *59* (1), 8–16.

Bevitori, R.; Ghini, R. *Rice Blast Disease in Climate Change Times.* Embrapa Arroze Feijao-Artigo em Periodico Indexado (ALICE); [Online] 2014. http://www.alice.cnptia.embrapa

Bindschedler, L. V.; Dewdney, J.; Blee, K. A.; Stone, J. M.; Asai, T.; Plotnikov, J.; Denoux, C.; Hayes, T.; Gerrish, C.; Davies, D. R; Ausubel, F. M.; Bolwell, G. P. Peroxidase-Dependent Apoplastic Oxidative Burst in Arabidopsis Required for Pathogen Resistance. *Plant J.* **2006,** *47* (6), 851–863.

Boller, T.; Felix, G. A Renaissance of Elicitors: Perception of Microbe-Associated Molecular Patterns and Danger Signals by Pattern-Recognition Receptors. *Annu. Rev. Plant Biol.* **2009,** *60,* 379–406.

Bowles, D. J. Defense-Related Proteins in Higher Plants. *Ann. Rev. Biochem.* **1990,** *59* (1), 873–907.

Briggs, S. In *Functional Genomics and the Development of New Plants*. Presented at the Agriculture Biotechnology International Conference, Toronto, ON, Canada, June 8, 2000; pp 2308–2309.

Browse, J. Jasmonate Passes Muster: A Receptor and Targets for the Defense Hormone. *Annu. Rev. Plant Biol.* **2009,** *60,* 183–205.

Bundó, M.; Coca, M. Enhancing Blast Disease Resistance by Overexpression of the Calcium-Dependent Protein Kinase Os CPK 4 in Rice. *Plant Biotechnol. J.* **2016**, *14* (6), 1357–1367.

Campo, S.; Martín-Cardoso, H.; Olivé, M.; Pla, E.; Catala-Forner, M.; Martínez-Eixarch, M.; San Segundo, B. Effect of Root Colonization by Arbuscular Mycorrhizal Fungi on Growth, Productivity and Blast Resistance in Rice. *Rice* **2020,** *13* (1), 1–14.

Cao, J.; Yang, C.; Li, L.; Jiang, L.; Wu, Y.; Wu, C.; Bu, Q.; Xia, G.; Liu, X.; Luo, Y.; Liu, J. Rice Plasma Membrane Proteomics Reveals *Magnaporthe Oryzae* Promotes Susceptibility by Sequential Activation of Host Hormone Signaling Pathways. *Mol. Plant Microbe Interact.* **2016,** *29* (11), 902–913.

Cardoso, C. A. D. A.; Reis, E. M.; Moreira, E. N. Development of a Warning System for Wheat Blast Caused by *Pyricularia Grisea*. *Summa Phytopathol.* **2008,** *34* (3), 216–221.

Chakraborty, S.; Tiedemann, A. V.; Teng, P. S. Climate Change: Potential Impact on Plant Diseases. *Environ. Pollut.* **2000,** *108* (3), 317–326.

Chandran, V.; Wang, H.; Gao, F.; Cao, X. L.; Chen, Y. P.; Li, G. B.; Zhu, Y.; Yang, X. M.; Zhang, L. L.; Zhao, Z. X.; Zhao, J. H. miR396-OsGRFs Module Balances Growth and Rice Blast Disease-Resistance. *Front. Plant Sci.* **2019,** *9,* 1999.

Chen, D.; Zeigler, R. S.; Leung, H.; Nelson, R. J. Population Structure of *Pyricularia grisea* at Two Screening Sites in the Philippines. *Phytopathology* **1995,** *85* (9), 1011–1020.

Chumley, F. G.; Valent, B. Genetic Analysis of Melanin-Deficient, Nonpathogenic Mutants of *Magnaporthe Grisea. Mol. Plant Microbe Interact.* **1990,** *3* (3), 135–143.

Coakley, S. M.; Scherm, H.; Chakraborty, S. Climate Change and Plant Disease Management. *Annu. Rev. Phytopathol.* **1999,** *37* (1), 399–426.

Dasgupta, S.; Laplante, B.; Murray, S.; Wheeler, D. Sea-Level Rise and Storm Surges: A Comparative Analysis of Impacts in Developing Countries; World Bank Policy ResearchWorking Paper, World Bank: Washington, DC, USA, 2009; p 4901.

de Jong, J. C.; McCormack, B. J.; Smirnoff, N.; Talbot, N. J. Glycerol Generates Turgor in Rice Blast. *Nature* **1997,** *389* (6648), 244–244.

Dean, R. A.; Talbot, N. J.; Ebbole, D. J.; Farman, M. L.; Mitchell, T. K.; Orbach, M. J.; Thon, M.; Kulkarni, R.; Xu, J.-R.; Pan, H. The Genome Sequence of the Rice Blast Fungus *Magnaporthe Grisea. Nature* **2005,** *434,* 980–986.

Deng, Y. Molecular Screening of Blast Resistance Genes in Rice Germplasms Resistant to *Magnaporthe Oryzae. Rice Sci.* **2017,** *24,* 41–47.

Deng, Y.; Zhu, X.; Shen, Y.; He, Z. Genetic Characterization and Fine Mapping of the Blast Resistance Locus Pigm (t) Tightly Linked to Pi2 and Pi9 in a Broad-Spectrum Resistant Chinese Variety. *Theor. Appl. Genet.* **2006,** *113,* 705–713.

Divya, B.; Robin, S.; Rabindran, R.; Senthil, S.; Raveendran, M.; Joel, A. J. Marker Assisted Backcross Breeding Approach to Improve Blast Resistance in Indian Rice (*Oryza Sativa*) Variety ADT43. *Euphytica* **2014,** *200* (1), 61–77.

Dixon, R. A.; Achnine, L.; Kota, P.; Liu, C. J.; Reddy, M. S.; Wang, L. The Phenylpropanoid Pathway and Plant Defence—A Genomics Perspective. *Mol. Plant Pathol.* **2002,** *3* (5), 371–390.

Dong, X. NPR1, all Things Considered. *Cur.Opin. Plant Biol.* **2004,** *7* (5), 547–552.

Dong, Y.; Li, Y.; Zhao, M.; Jing, M.; Liu, X.; Liu, M.; Guo, X.; Zhang, X.; Chen, Y.; Liu, Y.; Liu, Y.; Ye, W.; Zhang, H.; Wang, Y.; Zheng, X.; Wang, P.; Zhang, Z. Global Genome and Transcriptome Analyses of *Magnaporthe Oryzae* Epidemic Isolate 98-06 Uncover Novel Effectors and Pathogenicity-Related Genes, Revealing Gene Gain and Lose Dynamics in Genome Evolution. *PLoS Pathog.* **2015,** *11,* e1004801.

Ebbole, D. J. Magnaporthe as a Model for Understanding Host-Pathogen Interactions. *Annu. Rev. Phytopathol.* **2007,** *45*, 437–456.

Espinoza, I. G.; Shohara, K. Investigación Relativa a la Ocurrencia de Piricularia en Trigo. (v.2); Centro Tecnológico Agropecuário en: Santa Cruz, Bolívia, 2003.

Faivre-Rampant, O.; Geniès, L.; Piffanelli, P.; Tharreau, D. Transmission of Rice Blast From Seeds to Adult Plants in a Non-Systemic Way. *Plant Pathol.* **2013,** *62* (4), 879–887.

Fan, C.; Wang, G.; Wu, L.; Liu, P.; Huang, J.; Jin, X.; Zhang, G.; He, Y.; Peng, L.; Luo, K.; Feng, S. Distinct Cellulose and Callose Accumulation for Enhanced Bioethanol Production and Biotic Stress Resistance in OsSUS3 Transgenic Rice. *Carbohydr. Polym.* **2020,** *232,* 115448.

FAO. FAOSTATE, Food And Agriculture Organization Of The United Nations Statistics Division, 2018.

Feng, J. X.; Cao, L.; Li, J.; Duan, C. J.; Luo, X. M.; Le, N.; Wei, H.; Liang, S.; Chu, C.; Pan, Q.; Tang, J. L. Involvement of OsNPR1/NH1 in Rice Basal Resistance to Blast Fungus Magnaporthe Oryzae. *Eur. J. Plant Pathol.* **2011,** *131* (2), 221.

Fuhrer, J. Agroecosystem Responses to Combinations of Elevated CO2, Ozone, and Global Climate Change. *Agric. Ecosyst. Environ.* **2003,** *97* (1–3), 1–20.

Gnanamanickam, S. S.; Mew, T. W. Biological Control of Blast Disease of Rice (*Oryza Sativa* L.) With Antagonistic Bacteria and its Mediation by a *Pseudomonas* Antibiotic. *Japanese J. Phytopathol.* **1992,** *58* (3), 380–385.

Goodman, B. A.; Newton, A. C. Effects of Drought Stress and its Sudden Relief on Free Radical Processes in Barley. *J. Sci. Food Agri.* **2005,** *85* (1), 47–53.

Gória, M. M.; Ghini, R.; Bettiol, W. Elevated Atmospheric CO2 Concentration Increases Rice Blast Severity. *Tropical Plant Pathol.* **2013,** *38* (3), 253–257.

Greer, C. A.; Webster, R. K. Occurrence, Distribution, Epidemiology, Cultivar Reaction, and Management of Rice Blast Disease in California. *Plant Dis.* **2001,** *85,* 1096–1102.

Guan, H.; Hou, X.; Jiang, Y.; Srivastava, V.; Mao, D.; Pan, R.; Chen, M.; Zhou, Y.; Wang, Z.; Chen, Z. Feature of Blast Resistant Near-Isogenic Lines Using an Elite Maintainer Line II-32B by Marker-Assisted Selection. *J. Plant Pathol.* **2019,** *101,* 491–501.

Gupta, R.; Lee, S. E.; Agrawal, G. K.; Rakwal, R.; Park, S.; Wang, Y.; Kim, S. T. Understanding the Plant-Pathogen Interactions in the Context of Proteomics-Generated Apoplastic Proteins Inventory. *Front. Plant Sci.* **2015,** *6,* 352.

Gupta, R.; Min, C. W.; Kim, Y. J.; Kim, S. T. Identification of Msp1-Induced Signaling Components in Rice Leaves by Integrated Proteomic and Phosphoproteomic Analysis. *Int. J. Mol. Sci.* **2019,** *20* (17), 4135.

Hamer, J. E.; Howard, R. J.; Chumley, F. G.; Valent, B. A Mechanism for Surface Attachment in Spores of a Plant Pathogenic Fungus. *Science* **1988,** *239* (4837), 288–290.

Hasan, M. M.; Rafii, M. Y.; Ismail, M. R.; Mahmood, M.; Rahim, H. A.; Alam, M. A.; Ashkani, S.; Malek, M. A.; Latif, M. A. Marker-Assisted Backcrossing: A Useful Method for Rice Improvement. *Biotechnol. Biotechnol. Equip.* **2015a,** *29,* 237–254.

Hasan, M. M.; Yusop, M. R.; Ismail, M. R.; Mahmood, M.; Rahim, H. A.; Latif, M. A. Performance of Yield and Yield Contributing Characteristics of BC2F3 Population With Addition of Blast Resistant Gene. *Ciênc. Agrotec.* **2015b,** *39* (5), 463–476.

Hayasaka, T.; Fujii, H.; Ishiguro, K. The Role of Silicon in Preventing Appressorial Penetration by the Rice Blast Fungus. *Phytopathology* **2008,** *98* (9), 1038–1044.

Hong, Y.; Yang, Y.; Zhang, H.; Huang, L.; Li, D.; Song, F. Overexpression of MoSM1, Encoding for an Immunity-Inducing Protein From *Magnaporthe Oryzae*, in Rice Confers Broad-Spectrum Resistance Against Fungal and Bacterial Diseases. *Sci. Rep.* **2017**, *7* (1), 1–17.

Horsefield, S.; Burdett, H.; Zhang, X.; Manik, M. K.; Shi, Y.; Chen, J.; Qi, T.; Gilley, J.; Lai, J. S.; Rank, M. X.; Casey, L. W.; Gu, W.; Ericsson, D. J.; Foley, G.; Hughes, R. O.; Bosanac, T.; von Itzstein, M.; Rathjen, J. P.; Nanson, J. D.; Boden, M.; Dry, I. B.; Williams, S. J.; Staskawicz, B. J.; Coleman, M. P.; Ve, T.; Dodds, P. N.; Bostjan Kobe, B. NAD+ cleavage activity by animal and plant TIR domains in cell death pathways. *Sci.* **2019**, *365*(6455), 793-799.

Jeon, J.; Lee, G. W.; Kim, K. T.; Park, S. Y.; Kim, S.; Kwon, S., Huh, A.; Chung, H.; Lee, Y. H. Transcriptome Profiling of the Rice Blast Fungus *Magnaporthe Oryzae* and its Host *Oryza Sativa* During Infection. *Mol. Plant Microbe Interact.* **2020,** *33* (2), 141–144.

Jiang, H.; Li, Z.; Liu, J.; Shen, Z.; Gao, G.; Zhang, Q.; He, Y. Development and Evaluation of Improved Lines With Broad-Spectrum Resistance to Rice Blast Using Nine Resistance Genes. *Rice* **2019,** *12,* 29.

Jiang, J.; Mou, T.; Yu, H.; Zhou, F. Molecular Breeding of Thermo-Sensitive Genic Male Sterile (TGMS) Lines of Rice for Blast Resistance Using Pi2 Gene. *Rice* **2015,** *8,* 11.

Jiang, C. J.; Shimono, M.; Sugano, S.; Kojima, M.; Yazawa, K.; Yoshida, R.; Inoue, H.; Hayashi, N.; Sakakibara, H.; Takatsuji, H. Abscisic Acid Interacts Antagonistically With Salicylic Acid Signaling Pathway in Rice–*Magnaporthe Grisea* Interaction. *Mol. Plant Microbe Interact.* **2010,** *23* (6), 791–798.

Jones, J. D. G.; Dangl, J. L. The Plant Immune System. *Nature* **2006,** *444,* 323–329.

Joshi, S.; Dhatwalia, S.; Kaachra, A.; Sharma, K. D.; Rathour, R. Genetic and Physical Mapping of a New Rice Blast Resistance Specificity Pi-67 From a Broad Spectrum Resistant Genotype Tetep. *Euphytica* **2019,** *215* (1), 9.

Jyotsna, J.; Das, S.; Kumar, B. Efficacy of Aqueous Leaf Extract of Medicinal Plants Against Blast and Brown Spot Disease of Rice. *Int. J. Cur. Microbiol. App. Sci.* **2017,** *6,* 4138–4144.

Kanda, Y.; Nishizawa, Y.; Kamakura, T.; Mori, M. Overexpressed BSR1-Mediated Enhancement of Disease Resistance Depends on the MAMP-Recognition System. *Int. J. Mol. Sci.* **2020,** *21* (15), 5397.

Kankanala, P.; Czymmek, K.; Valent, B. Roles for Rice Membrane Dynamics and Plasmodesmata During Biotrophic Invasion by the Blast Fungus. *Plant Cell* **2007,** *19* (2), 706–724.

Kanzaki, H.; Nirasawa, S.; Saitoh, H.; Ito, M.; Nishihara, M.; Terauchi, R.; Nakamura, I. Overexpression of the Wasabi Defensin Gene Confers Enhanced Resistance to Blast Fungus (*Magnaporthe Grisea*) in Transgenic Rice. *Theor. Appl. Genet.* **2002,** *105* (6–7), 809–814.

Kapoor, P.; Abhishek, K. Past, Present and Future of Rice Blast Management. *Plant Sci. Today* **2014,** *1* (3), 165–173.

Kato, H. Rice Blast Disease. Pesticide Outlook February; 2001, pp 23–25.

Kawahara, Y.; Oono, Y.; Kanamori, H.; Matsumoto, T.; Itoh, T.; Minami, E. Simultaneous RNA-seq Analysis of a Mixed Transcriptome of Rice and Blast Fungus Interaction. *PloS One* **2012,** *7* (11), e49423.

Khush, G. S.; Jena, K. K. Current Status and Future Prospects for Research on Blast Resistance in Rice (Oryza Sativa L.). In *Advances in Genetics, Genomics and Control of Rice Blast Disease*; Springer, Dordrecht, 2009; pp 1–10.

Kim, S. T.; Kim, S. G.; Agrawal, G. K.; Kikuchi, S.; Rakwal, R. Rice Proteomics: a Model System for Crop Improvement and Food Security. *Proteomics* **2014,** *14* (4–5), 593–610.

Kim, S. M.; Reinke, R. F. A Novel Resistance Gene for Bacterial Blight in Rice, Xa43 (t) Identified by GWAS, Confirmed by QTL Mapping Using a Bi-parental Population. *PloS One* **2019,** *14* (2), e0211775.

Kim, S. G.; Wang, Y.; Lee, K. H.; Park, Z. Y.; Park, J.; Wu, J. Kwon, S. J.: Lee, Y. H.; Agrawal, G. K.; Rakwal, R.; Kim, S. T.; Kang, K. Y. In-depth Insight into in Vivo Apoplastic Secretome of Rice-*Magnaporthe Oryzae* Interaction. *J. Proteom.* **2013,** *78*, 58–71.

Kingsolver, C. H.; Barkside, T. H.; Marchetti, M. A. Rice Blast Epidemiology. Bulletin of the Pennsylvania Agricultural Experiment Station, 1984, (853).

Kline-Jonakin, K. G.; Barrett-Wilt, G. A.; Sussman, M. R. Quantitative Plant Phosphoproteomics. *Cur. Opin. Plant Biol.* **2011,** *14* (5), 507–511.

Klittich, C. J. Milestones in Fungicide Discovery: Chemistry That Changed Agriculture. *Plant Health Prog.* **2008,** *9* (1), 31.

Kobayashi, T., Ishiguro, K., Nakajima, T., Kim, H. Y., Okada, M.,Kobayashi, K. Effects of Elevated Atmospheric CO2 Concentration on the Infection of Rice Blast and Sheath Blight. *Phytopathology* 2006, *96* (4), 425–431.

Kohli, M. M.; Mehta, Y. R.; Guzman, E.; De Viedma, L.; Cubilla, L. E. Pyricularia Blast–a Threat to Wheat Cultivation. *Czech J. Genet. Plant Breed.* **2011,** *47* (Special Issue).

Koide, Y.; Kobayashi, N.; Xu, D.; Fukuta, Y. Resistance Genes and Selection DNA Markers for Blast Disease in Rice (*Oryza Sativa* L.). *Japan Agri. Res. Q.* JARQ **2009,** *43,* 255–280.

Korinsak, S.; Sirithunya, P.; Meakwatanakarn, P.; Sarkarung, S.; Vanavichit, A.; Toojinda, T. Changing Allele Frequencies Associated With Specific Resistance Genes to Leaf Blast in Backcross Introgression Lines of Khao Dawk Mali 105 Developed From a Conventional Selection Program. *Field Crops Res.* **2011,** *122,* 32–39.

Krishnamurthy, K.; Gnanamanickam, S. S. Induction of Systemic Resistance and Salicylic Acid Accumulation in Oryza Sativa, L. in the Biological Suppression of Rice Blast Caused by Treatments With *Pseudomonas* spp. *World J. Microbiol. Biotechnol.* **1998,** *14* (6), 935–937.

Kumagaya, S.; Goto, Y.; Hori, C.; Matsuoka, M.; Nakano, R. Annual Change of Silicate Absorption and Effect of Calcium Silicate in the Rice Plant. *Rept. Tokushima Agric. Exp. Stn.* **1957,** *2,* 13–14.

Kumar, P.; Pathania, S.; Katoch, P.; Sharma, T.; Plaha, P.; Rathour, R. Genetic and Physical Mapping of Blast Resistance Gene Pi-42 (t) on the Short Arm of Rice Chromosome 12. *Mol. Breed.* **2010,** *25,* 217–228.

Laha, G. S.; Singh, R.; Ladhalakshmi, D.; Sunder, S.; Prasad, M. S.; Dagar, C. S.; Babu, V. R. Importance and Management of Rice Diseases: A Global Perspective. In *Rice Production Worldwide*; Springer: Cham, Switzerland, 2017; pp 303–360.

Lassa, J. A.; Lai, A. Y. H.; Goh, T. Climate Extremes: An Observation and Projection of its Impacts on Food Production in ASEAN. *Natl. Haz.* **2016,** *84* (1), 19–33.

Latef, A. A. H. A.; Dawood, M. F.; Hassanpour, H.; Rezayian, M.; Younes, N. A. Impact of the Static Magnetic Field on Growth, Pigments, Osmolytes, Nitric Oxide, Hydrogen Sulfide Phenylalanine Ammonia-Lyase Activity, Antioxidant Defense System, and Yield in Lettuce. *Biology* **2020,** *9* (7), 172.

Law, J. W. F.; Ser, H. L.; Khan, T. M.; Chuah, L. H.; Pusparajah, P.; Chan, K. G.; Goh, B. H.; Lee, L. H. The Potential of Streptomyces as Biocontrol Agents Against the Rice Blast Fungus, *Magnaporthe Oryzae* (*Pyricularia Oryzae*). *Front Microbiol.* **2017,** *8,* 3.

Levy, M.; Romao, J.; Marchetti, M. A.; Hamer, J. E. DNA Fingerprinting With a Dispersed Repeated Sequence Resolves Pathotype Diversity in the Rice Blast Fungus. *Plant Cell* **1991,** *3* (1), 95–102.

Li, Z.; Huang, J.; Wang, Z.; Meng, F.; Zhang, S.; Wu, X.; Zhang, Z.; Gao, Z. Overexpression of Arabidopsis Nucleotide-Binding and Leucine-Rich Repeat Genes RPS2 and RPM1 (D505V) Confers Broad-Spectrum Disease Resistance in Rice. *Front Plant Sci.* **2019,** *10,* 417.

Li, Y.; Ye, Z.; Nie, Y.; Zhang, J.; Wang, G. L.; Wang, Z. Comparative Phosphoproteome Analysis of *Magnaporthe Oryzae*-Responsive Proteins in Susceptible and Resistant Rice Cultivars. *J. Proteom.* **2015,** *115,* 66–80.

Lin, S.; Nie, P.; Ding, S.; Zheng, L.; Chen, C.; Feng, R.; Wang, Z.; Wang, L.; Wang, J.; Fang, Z.; Zhou, S.; Ma, H.; Zhao, H. Quantitative Proteomic Analysis Provides Insights Into Rice Defense Mechanisms Against *Magnaporthe Oryzae. Int. J. Mol. Sci.* **2018,** *19* (7), 1950.

Liu, X.; Inoue, H.; Tang, X.; Tan, Y.; Xu, X.; Wang, C.; Jiang, C. J. Rice OsAAA-ATPase1 is Induced During Blast Infection in a Salicylic Acid-Dependent Manner, and Promotes Blast Fungus Resistance. *Int. J. Mol. Sci.* **2020,** *21* (4), 1443.

Liu, W.; Liu, J.; Triplett, L.; Leach, J. E.; Wang, G. L. Novel Insights Into Rice Innate Immunity Against Bacterial and Fungal Pathogens. *Annu. Rev. Phytopathol.* **2014,** *52*, 213–241.

Liu, W.; Wang, G. L. Plant Innate Immunity in Rice: A Defense Against Pathogen Infection. *Natl. Sci. Rev.* **2016,** *3* (3), 295–308.

Long, D. H.; Lee, F. N.; TeBeest, D. O. Effect of Nitrogen Fertilization on Disease Progress of Rice Blast on Susceptible and Resistant Cultivars. *Plant Dis.* **2000,** *84* (4), 403–409.

Ma, S.; Lin, S.; Wang, M.; Zou, Y.; Tao, H.; Liu, W.; Zhang, L.; Liang, K.; Ai, Y.; He, H. Differential Expression Proteins Contribute to Race-Specific Resistant Ability in Rice (*Oryza sativa* L.). *Plants* **2019,** *8* (2), 29.

Ma, Z.; Wang, L.; Zhao, M.; Gu, S.; Wang, C.; Zhao, J.; Tang, Z.; Gao, H.;Zhang, L.; Fu, L.; Yin, Y.; He, N.; Zheng, W.; Xu , Z.. iTRAQ Proteomics Reveals the Regulatory Response to *Magnaporthe Oryzae* in Durable Resistant vs. Susceptible Rice Genotypes. *Plos One* **2020,** *15* (1), e0227470.

Maciel, J. L. N. *Magnaporthe Oryzae*, the Blast Pathogen: Current Status and Options for its Control. *Plant Sci. Rev.* **2011,** *264,* 233–240.

Maeda, S.; Hayashi, N.; Sasaya, T.; Mori, M. Overexpression of BSR1 Confers Broad-Spectrum Resistance Against Two Bacterial Diseases and Two Major Fungal Diseases in Rice. *Breed. Sci.* **2016,** 15157.

Manandhar, H. K.; Jorgensen, H. J. L.; Jorgensen, S. B.; Petersen, V. S. Seed-Borne Infection of Rice by *Magnaporthe Oryzae* and its Transmission to Seedlings. *Plant Dis.* **1998,** *82* (10), 1094–1099.

Manning, W. J.; Tiedemann, A. V. Climate Change: Potential Effects of Increased Atmospheric Carbon Dioxide (CO2), Ozone (O3), and Ultraviolet-B (UV-B) Radiation on Plant Diseases. *Environ. Pollut.* **1995,** *88* (2), 219–245.

Matros, A.; Amme, S.; Kettig, B.; BUCK-SORLIN, G. H.; Sonnewald, U. W. E.; Mock, H. P. Growth at Elevated CO2 Concentrations Leads to Modified Profiles of Secondary Metabolites in Tobacco cv. SamsunNN and to Increased Resistance Against Infection With Potato Virus Y. *Plant Cell Environ.* **2006,** *29* (1), 126–137.

Mcdonald, B. A.; Linde, C. Pathogen Population Genetics, Evolutionary Potential, and Durable Resistance. *Annu. Rev. Phytopathol.* **2002,** *40,* 349–379.

Meng, Q.; Gupta, R.; Min, C. W.; Kwon, S. W.; Wang, Y.; Je, B. I.; Kim, Y. J.;Jeon, J. S.; Agarwal, G. K.; Rakwal, R.; Kim, S. T. Proteomics of Rice—*Magnaporthe Oryzae* Interaction: What Have We Learned so Far? *Front. Plant Sci.* **2019,** 10.

Meybeck, A. In Building Resilience for Adaptation to Climate Change in the Agriculture Sector. Proceedings of the Joint FAO/OECD Workshop; Rome, Italy, April 23–24, 2012.

Mgonja, E. M.; Park, C. H.; Kang, H.; Balimponya, E. G.; Opiyo, S.; Bellizzi, M.; Mutiga, S. K.; Rotich, F.; Ganeshan, V. D.; Mabagala, R.; Sneller, C. Genotyping-by-Sequencing-Based Genetic Analysis of African Rice Cultivars and Association Mapping of Blast Resistance Genes Against *Magnaporthe Oryzae* Populations in Africa. *Phytopathology* **2017,** *107* (9), 1039–1046.

Miah, G.; Rafii, M. Y.; Ismail, M. R.; Sahebi, M.; Hashemi, F. S. G.; Yusuff, O.; Usman, M. G. Blast Disease Intimidation Towards Rice Cultivation: A Review of Pathogen and Strategies to Control. *J. Animal Plant Sci.* **2017,** *27*, 1058–1066.

Mitchell, C. E.; Reich, P. B.; Tilman, D.; Groth, J. V. Effects of Elevated CO2, Nitrogen Deposition, and Decreased Species Diversity on Foliar Fungal Plant Disease. *Glob. Change Biol.* **2003,** *9* (3), 438–451.

Muimba-Kankolongo, A. Cereal Production. Food Crop Production by Smallholder Farmers in Southern Africa, 2018; pp 73–121.

Muñoz, M. C. The Effect of Temperature and Relative Humidity on the Airborne Concentration of "*Pyricularia oryzae*" Spores and the Development of Rice Blast in Southern Spain. *Spanish J. Agri. Res.* **2008,** (1), 61–69.

Narula, K.; Choudhary, P.; Ghosh, S.; Elagamey, E.; Chakraborty, N.; Chakraborty, S. Comparative Nuclear Proteomics Analysis Provides Insight Into the Mechanism of Signaling and Immune Sesponse to Blast Disease Caused by *Magnaporthe Oryzae* in Rice. *Proteomics* **2019,** *19* (3), 1800188.

Nawaz, G.; Usman, B.; Peng, H.; Zhao, N; Yuan, R.; Liu, Y.; Li, R. Knockout of Pi21 by CRISPR/Cas9 and iTRAQ-Based Proteomic Analysis of Mutants Revealed New Insights into *M. oryzae* Resistance in Elite Rice Line. *Genes* **2020,** *11* (7), 735.

Ngernmuen, A.; Suktrakul, W.; Kate-Ngam, S.; Jantasuriyarat, C. Transcriptome Comparison of Defense Responses in the Rice Variety 'Jao Hom Nin'Regarding Two Blast Resistant Genes, Pish and Pik. *Plants* **2020,** *9* (6), 694.

Ning, X. I. A. O.; Yunyu, W.; Aihong, L. Strategy for Use of Rice Blast Resistance Genes in Rice Molecular Breeding. *Rice Sci.* **2020,** *27* (4), 263–277.

Nishizawa, Y.; Nishio, Z.; Nakazono, K.; Soma, M.; Nakajima, E.; Ugaki, M.; Hibi, T. Enhanced Resistance to Blast (*Magnaporthe Grisea*) in Transgenic Japonica Rice by Constitutive Expression of Rice Chitinase. *Theor. Appl. Genet.* **1999,** *99* (3–4), 383–390.

Nutsugah, S. K.; Twumasi, J. K.; Chipili, J.; Sere, Y.; Sreenivasaprasad, S. Diversity of the Rice Blast Pathogen Populations in Ghana and Strategies for Resistance Management. *Plant Pathol. J.* **2008,** *7* (1), 109–113.

Ohyanagi, H.; Tanaka, T.; Sakai, H.; Shigemoto, Y.; Yamaguchi, K.; Habara, T.; Fujii, Y.; Antonio, B. A.; Nagamura, Y.; Imanishi, T. The Rice Annotation Project Database (RAP-DB): Hub for *Oryza Sativa* ssp. Japonica Genome Information. *Nucleic Acids Res.* **2006,** *34,* D741–D744.

Osmani, A. H., O'Donnell, K., Pu, R. T., & Osmani, S. A. Activation of the nimA Protein Kinase Plays a Unique Role During Mitosis That Cannot be Bypassed by Absence of the bimE Checkpoint. *EMBO J.* **1991,** *10* (9), 2669–2679.

Ou, S. H.; Nuque, F. L. Rice Diseases, Seconded. Commonwealth Mycological Institute: Kew and Surrey, UK, 1985.

Padmanabhan, S. Y. Studies on Forecasting Outbreaks of Blast Disease of Rice. Central Rice Research Institute, 1965, pp 117–129.

Pagliaccia, D.; Urak, R. Z.; Wong, F.; Douhan, L. I.; Greer, C. A.; Vidalakis, G.; Douhan, G. W. Genetic Structure of the Rice Blast Pathogen (*Magnaporthe Oryzae*) Over a Decade in North Central California Rice Fields. *Microbial. Ecol.* **2018,** *75* (2), 310–317.

Pandey, S. Effect of Botanicals for Management of Leaf Blast and Enhancing Yield Traits in Rice. *Cur. Bot.* **2018,** 17–21.

Pangga, I. B.; Chakraborty, S.; Yates, D. Canopy Size and Induced Resistance in Stylosanthes Scabra Determine Anthracnose Severity at High CO2. *Phytopathology* **2004,** *94* (3), 221–227.

Petit-Houdenot, Y.; Fudal, I. Complex Interactions Between Fungal Avirulence Genes and Their Corresponding Plant Resistance Genes and Consequences for Disease Resistance Management. *Fron. Plant Sci.* **2017,** *8,* 1072.

Phong, T. K.; Nhung, D. T. T.; Yamazaki, K.; Takagi, K.; Watanabe, H. Behavior of Sprayed Tricyclazole in Rice Paddy Lysimeters. *Chemosphere* **2009,** *74* (8), 1085–1089.

Pieterse, C. M.; Van der Does, D.; Zamioudis, C.; Leon-Reyes, A.; Van Wees, S. C. Hormonal Modulation of Plant Immunity. *Annu. Rev. Cell Develop.Biol.* **2012,** *28*.

Plessl, M.; Heller, W.; Payer, H. D.; Elstner, E. F.; Habermeyer, J.; Heiser, I. Growth Parameters and Resistance Against Drechslera Teres of Spring Barley (*Hordeum Vulgare* L. cv. Scarlett) Grown at Elevated Ozone and Carbon Dioxide Concentrations. *Plant Biol.* **2005,** *7* (6), 694–705.

Prabhu, A. S.; Filippi, M. C.; Silva, G. B.; Lobo, V. L. S.; Morais, O. P. An Unprecedented Outbreak of Rice Blast on a Newly Released Cultivar BRS Colosso in Brazil. In *Advances in Genetics, Genomics and Control of Rice Blast Disease*; Springer: Dordrecht, 2009; pp 257–266.

Prabhu, A. S.; Silva, C. S.; Filippi, M. C. Impacto Potencial das Mudanças Climáticas Sobre as Doenças de Arroz no Brasil. In Mudanças Climáticas: Impactos sobre doenças de plantas no Brasil; Ghini, R., Hamada, E., Eds.; Embrapa Informação Tecnológica: Brasília, DF, Brazil; Embrapa Meio Ambiente: Jaguariúna, Brazil, 2008; pp 140–158.

Quoc, N. B.; Phuong, N. D. N.; Trang, H. T. T.; Phi, N. B.; Chau, N. N. B. Expression of osa-miR7695 Against the Blast Fungus *Magnaporthe Oryzae* in Vietnamese Rice Cultivars. *Eur. J. Plant Pathol.* **2019,** *155* (1), 307–317.

Raboin, L. M.; Ballini, E.; Tharreau, D.; Ramanantsoanirina, A.; Frouin, J.; Courtois, B.; Ahmadi, N. Association Mapping of Resistance to Rice Blast in Upland Field Conditions. *Rice* **2016,** *9* (1), 1–12.

Rajput, L. S.; Sharma, T.; Madhusudhan, P.; Sinha, P. Effect of Temperature on Growth and Sporulation of Rice Leaf Blast Pathogen *Magnaporthe Oryzae*. *Curr. Microbiol. Appl. Sci.* **2017,** *6* (3), 394–401.

Rosenzweig, C.; Yang, X.B.; Anderson, P.; Epstein, P.; Vicarelli, M. Agriculture: Climate Change, Crop Pests and Diseases. In *Climte Change Futures: Health Ecological Economic Dimensios*; The Center for Health and the Global Environment at Harvard Medical School: USA, 2005; pp 70–77.

Ryu, H. S.; Song, M. Y.; Kim, C. Y.; Han, M.; Lee, S. K.; Ryoo, N.; Cho, J.-Il.; Hahn, T.-R. ; Jeon, J. S. Proteomic Analysis of Rice Mutants Susceptible to *Magnaporthe Oryzae*. *Plant Biotechnol. Rep.* **2016,** *3* (2), 167–174.

Sakamoto, M. On the Facilitated Infection of the Rice Blast Fungus, *Piricularia Oryzae* Cav. Due to the Wind. I. *Japanese J. Phytopathol.* **2016,** *10* (2–3), 119–126.

Sallaud, C.; Lorieux, M.; Roumen, E.; Tharreau, D.; Berruyer, R.; Svestasrani, P.; Garsmeur, O.; Ghesqui¨Re, A.; Notteghem, J. -L. Identification of Five New Blast Resistance Genes in the Highly Blast-Resistant Rice Variety IR64 Using a QTL Mapping Strategy. *Theor. Appl. Genet.* **2016,** *106,* 794–803.

Sato, H.; Takeuchi, Y.; Hirabayashi, H.; Nemoto, H.; Hirayama, M.; Kato, H.; Imbe, T.; Ando, I. Mapping QTLs for Field Resistance to Rice Blast in the Japanese Upland Rice Variety Norin 12. *Breed. Sci.* **2006,** *56,* 415–418.

Shafaullah, M. A. K.; Khan, N. A.; Mahmood, Y. Effect of Epidemiological Factors on the Incidence of Paddy Blast (*Pyricularia Oryzae*) Disease. *Pakistan J. Phytopathol.* **2011,** *23* (2), 108–111.

Shahjahan, A. K. M. Practical Approaches to Rice Blast Management in Tropical Monsoon Ecosystems, With Special Reference. In Rice Blast Disease; International Rice Research Institute: Manila, Philippines; Cambridge University Press: Cambridge, MA, USA, 1994; p 465.

Shahriar, S. A.; Imtiaz, A. A.; Hossain, M. B.; Husna, A.; Eaty, M. N. K. Rice Blast Disease. *Annu. Res. Rev. Biol.* **2020,** 50–64.

Shan, Q.; Zhang, Y.; Chen, K.; Zhang, K.; Gao, C. Creation of Fragrant Rice by Targeted Knockout of the Os BADH 2 Gene Using TALEN Technology. *Plant Biotechnol. J.* **2016,** *13* (6), 791–800.

Sharma, T.; Madhav, M.; Singh, B.; Shanker, P.; Jana, T.; Dalal, V.; Pandit, A.; Singh, A.; Gaikwad, K.; Upreti, H. High-Resolution Napping, Cloning and Molecular Characterization of the Pi-k h Gene of Rice, Which Confers Resistance to *Magnaporthe grisea*. *Mol. Genet. Genom.* **2005**, *274,* 569–578.

Sharma, T. R.; Rai, A. K.; Gupta, S. K.; Vijayan, J.; Devanna, B. N.; Ray, S. Rice Blast Management Through Host-Plant Resistance: Retrospect and Prospects. *Agr. Res.* **2012,** *1* (1), 37–52.

Shen, Q.; Pu, Q.; Liang, J.; Mao, H.; Liu, J.; Wang, Q. CYP71Z18 Overexpression Confers Elevated Blast Resistance in Transgenic Rice. *Plant Mol. Bio.* **2019,** *100* (6), 579–589.

Shenton, M. R.; Berberich, T.; Kamo, M.; Yamashita, T.; Taira, H.; Terauchi, R. Use of Intercellular Washing Fluid to Investigate the Secreted Proteome of the Rice–Magnaporthe Interaction. *J. Plant Res.* **2012,** *125* (2), 311–316.

Shiba, Y.; Nagata, T. The Mode of Action of Tricyclazole in Controlling Rice Blast. *Japanese J. Phytopathol.* **1981,** *47* (5), 662–667.

Shimono, M.; Sugano, S.; Nakayama, A.; Jiang, C.-J.; Ono, K.; Toki, S.; et al. Rice WRKY45 Plays a Crucial Role in Benzothiadiazole-Inducible Blast Resistance. *Plant Cell* **2007,** *19,* 2064–2076.

Shiping, L.; Xin, L.; Chaoyang, W.; Xianghua, L.; Yuqing, H. Improvement of Resistance to Rice Blast in Zhenshan 97 by Molecular Marker-Aided Selection. *Acta Botanica Sinica* **2003,** *45,* 1346–1350.

Silué, D.; Notteghem, J. L.; Tharreau, D. Evidence of a Gene-for-Gene Relationship in the *Oryza Sativa-Magnaporthe Grisea* Pathosystem. *Phytopathology* **1992,** *82* (5), 577–580.

Silverman, P.; Seskar, M.; Kanter, D.; Schweizer, P.; Metraux, J. P.; Raskin, I. Salicylic Acid in Rice (Biosynthesis, Conjugation, and Possible Role). *Plant Physiol.* **1995,** *108* (2), 633–639.

Singh, J.; Gupta, S. K.; Devanna, B. N.; Singh, S.; Upadhyay, A.; Sharma, T. R. Blast Resistance Gene Pi54 Over-Expressed in Rice to Understand its Cellular and Sub-Cellular Localization and Response to Different Pathogens. *Sci. Rep.* **2020,** *10* (1), 1–13.

Singh, G.; Prasad, C. S. Evaluation of Fungicides Against Blast in Basmati Rice. *Ann. Plant Pro. Sci.* **2007,** *15* (2), 514–515.

Singh, P. K.; Singh, A. K.; Singh, H. B.; Dhakad, B. K. Biological Control of Rice Blast Disease With *Trichoderma Harzianum* in Direct Seeded Rice Under Medium Low Land Rainfed Conditions. *Environ. Ecol.* **2012,** *30* (3B), 834–837.

Singh, V. K.; Singh, A.; Singh, S. P.; Ellur, R. K.; Singh, D.; Gopala Krishnan, S.; Bhowmick, P. K.; Nagarajan, M.; Vinod, K. K.; Singh, U. D.; Mohapatra, T.; Prabhu, K. V.; Singh, A. K. Marker-Assisted Simultaneous but Stepwise Backcross Breeding for Pyramiding Blast Resistance Genes Piz5 and Pi54 Into an Elite Basmati Rice Restorer Line 'PRR 78'. *Plant Breed.* **2013,** *132* (5), 486–495.

Skamnioti, P.; Gurr, S. J. Magnaporthe Grisea Cutinase2 Mediates Appressorium Differentiation and Host Penetration and is Required for Full Virulence. *Plant Cell* **2007,** *19* (8), 2674–2689.

Sobrizal, S.; Anggiani, S. In Rice Blast Disease in Indonesia. JIRCAS (Japan International Research Center for Agricultural Sciences, Tsukuba, Japan; Working Report, 2007; No. 53, pp 71–79.

Song, F.; Goodman, R. M. Molecular Biology of Disease Resistance in Rice. *Physiol. Mol. Plant Pathol.* **2001,** *59,* 1–11.

Spence, C.; Alff, E.; Johnson, C.; Ramos, C.; Donofrio, N.; Sundaresan, V.;Bais, H. Natural Rice Rhizospheric Microbes Suppress Rice Blast Infections. *BMC Plant Biol.* **2014**, *14* (1), 130.

Srivastava, D.; Shamim, M.; Kumar, M.; Mishra, A.; Pandey, P.; Kumar, D.; Yadav, P.; Siddiqui, M. H.; Singh, K. N. Current Status of Conventional and Molecular Interventions for Blast Resistance in Rice. *Rice Sci.* **2017,** *24,* 299–321.

Suprapta, D. N.; Khalimi, K. Potential of Microbial Antagonists as Biocontrol Agents Against Plant Fungal Pathogens. *J. Int. Society Southeast Asian Agri. Sci.* **2012,** *18* (2), 1–8.

Suzuki, H. Meteorological Factors in the Epidemiology of Rice Blast. *Annu. Rev. Phytopathol.* **1975,** *13* (1), 239–256.

Takan, J. P.; Akello, B.; Esele, P.; Manyasa, E.O.; Obilana, A. B.; Audi, P. O. Finger Millet Blast Pathogen Diversity and Management in East Africa: A Summary of Project Activities and Outputs. *Int. Sorghum Millets Newslett.* **2004,** *45,* 66–69.

Takan, J. P.; Chipili, J.; Muthumeenakshi, S.; Talbot, N. J.; Manyasa, E. O.; Bandyopadhyay, R.; Sere, Y.; Nutsugah, S. K.; Talhinhas, P.; Hossain, M.; Brown, A. E. *Magnaporthe oryzae* Populations Adapted to Finger Millet and Rice Exhibit Distinctive Patterns of Genetic Diversity, Sexuality and Host Interaction. *Mol. Biotechnol.* **2012,** *50* (2), 145–158.

Talbot, N. J. On the Trail of a Cereal Killer: Exploring the Biology of *Magnaporthe Grisea*. *Annu. Rev. Microbiol.* **2003,** *57* (1), 177–202.

Tamaoki, D.; Seo, S.; Yamada, S.; Kano, A.; Miyamoto, A.; Shishido, H.; Miyoshi, S.; Taniguchi, S.; Akimitsu,K.; Gomi, K. Jasmonic Acid and Salicylic Acid Activate a Common Defense System in Rice. *Plant Signal. Beh.* **2013,** *8* (6), e24260.

Tian, D.; Chen, Z.; Chen, Z.; Zhou, Y.; Wang, Z.; Wang, F.; Chen, S. Allele-Specific Marker-Based Assessment Revealed That the Rice Blast Resistance Genes Pi2 and Pi9 Have not Been Widely Deployed in Chinese Indica Rice Cultivars. *Rice* **2016,** *9,* 19.

Torres, M. A.; Jones, J. D. G.; Dangl, J. L. Reactive Oxygen Species Signaling in Response to Pathogens. *Plant Physiol.* **2006,** *141,* 373–378.

Upadhyay, K.; Bhatta B. Rice Blast (*Magnaporthe oryzae*) Management: A Review. *Agri. J.* **2020,** *15* (3), 42–48.

van Loon, L. C.; Rep, M.; Pieterse, C. M. Significance of Inducible Defense-Related Proteins in Infected Plants. *Annu. Rev. Phytopathol.* **2006,** *44,* 135–162.

Veneault-Fourrey, C.; Barooah, M.; Egan, M.; Wakley, G.; Talbot, N. J. Autophagic Fungal Cell Death is Necessary for Infection by the Rice Blast Fungus. *Science* **2006**, *312* (5773), 580–583.

Volante, A.; Tondelli, A.; Desiderio, F.; Abbruscato, P.; Menin, B.; Biselli, C.; Casella, L.; Singh, N.; McCouch, S. R.; Tharreau, D.; Zampieri, E. Genome Wide Association Studies for Japonica Rice Resistance to Blast in Field and Controlled Conditions. *Rice* **2020,** *13,* 71.

Wan, L.; Essuman, K.; Anderson, R. G.; Sasaki, Y.; Monteiro, F.; Chung, E. H.; Nishimura, E. O.; DiAntonio, A.; Milbrandt, J.; Dangl, J. L.; Nishimura, M. T. TIR Domains of Plant Immune Receptors are NAD+-Cleaving Enzymes That Promote Cell Death. *Science* **2019,** *365* (6455), 799–803.

Wang, B.-H.; Ebbole, D. J.; Wang, Z.-H. The Arms Race Between Magnaporthe Oryzae and Rice: Diversity and Interaction of Avr and R Genes. *J. Integr. Agric.* **2017a,** *16,* 2746–2760.

Wang, Z.; Han, Q.; Zi, Q.; Lv, S.; Qiu, D.; Zeng, H. Enhanced Disease Resistance and Drought Tolerance in Transgenic Rice Plants Overexpressing Protein Elicitors From *Magnaporthe oryzae*. *PLoS One* **2017b,** *12* (4), e0175734.

Wang, J.; Hu, M.; Wang, J.; Qi, J.; Han, Z.; Wang, G.; Qi, Y.; Wang, H. W.; Zhou, J. M.; Chai, J. Reconstitution and Structure of a Plant NLR Resistosome Conferring Immunity. *Science* **2019a,** *364* (6435), eaav5870.

Wang, Q.; Tang, J.; Han, B.; Huang, X. Advances in Genome-Wide Association Studies of Complex Traits in Rice. *Theor. Appl. Genet.* **2020b**, *133* (5), 1415–1425.

Wang, F.; Wang, C.; Liu, P.; Lei, C.; Hao, W.; Gao, Y.; Liu, Y. G.; Zhao, K. Enhanced Rice Blast Resistance by CRISPR/Cas9-Targeted Mutagenesis of the ERF Transcription Factor Gene OsERF922. *PloS One* **2016,** *11* (4), e0154027.

Wang, P.; Yao, S.; Kosami, K. I.; Guo, T.; Li, J.; Zhang, Y.; Fukao, Y.; Kaneko-Kawano, T.; Zhang, H.; She, Y. M.; Wang, P.; Xing, W.; Hanada, K.; Liu, R.; Yoji Kawano, Y. Identification of Endogenous Small Peptides Involved in Rice Immunity Through Transcriptomics-and Proteomics-Based Screening. *Plant Biotechnol. J.* **2020a**, *18* (2), 415–428.

Wang, L.; Zhao, L.; Zhang, X.; Zhang, Q.; Jia, Y.; Wang, G.; Li, S.; Tian, D.; Li, W. H.; Yang, S. Large-Scale Identification and Functional Analysis of NLR Genes in Blast Resistance in the Tetep Rice Genome Sequence. Proc. Natl. Acad. Sci. **2019b,** *116* (37), 18479–18487.

Waszczak, C.; Carmody, M.; Kangasja, J. Reactive Oxygen Species in Plant Signaling. *Annu. Rev. Plant Bio.* **2018,** *69,* 209–236.

Wei, T.; Ou, B.; Li, J.; Zhao, Y.; Guo, D.; Zhu, Y.; Chen, Z.; Gu, H.;Li, C.;Qin, G.;

Qu, L. J. Transcriptional Profiling of Rice Early Response to *Magnaporthe oryzae* Identified OsWRKYs as Important Regulators in Rice Blast Resistance. *PloS One* **2013,** *8* (3), e59720.

Wilson, R. A.; Talbot, N. J. Under Pressure: Investigating the Biology of Plant Infection by Magnaporthe Oryzae. *Nature Rev. Microbiol.* **2009,** *7* (3), 185–195.

Yamada, S.; Kano, A.; Tamaoki, D.; Miyamoto, A.; Shishido, H.l Miyoshi, S.; Taniguchi, S.; Akimitsu, K.; Gomi, K. Involvement of OsJAZ8 in Jasmonate-Induced Resistance to Bacterial Blight in Rice. *Plant Cell Physiol.* **2012**, *53* (12), 2060–2072.

Yan, L.; Bai-Yuan, Y.; Yun-Liang, P.; Zhi-Juan, J.; Yu-Xiang, Z.; Han-Lin, W.; Chang-Deng, Y. Molecular Scree Ning of Blast Resistance Genes in Rice Germplasms Resistant to *Magnaporthe Oryzae*. *Rice Sci.* **2017,** 24, 41–47.

Young, N. D. QTL Mapping and Quantitative Disease Resistance in Plants. *Annu. Rev. Phytopathol.* **1996**, *34,* 479–501.

Zhang, S.; Deng, Y. Z.; Zhang, L. H. Phytohormones: The Chemical Language in *Magnaporthe Oryzae*-Rice Pathosystem. *Mycolology* **2018**, *9* (3), 233–237.

Zhang, L.; Fu, L.; Yin, Y.; He, N.; Zheng, W.; Xu, Z. iTRAQ Proteomics Reveals the Regulatory Response to *Magnaporthe Oryzae* in Durable Resistant vs. Susceptible Rice Genotypes. *PloS One* **2020,** *15,* e0227470.

Zhang, F.; Xie, J. Genes and QTLs Resistant to Biotic and Abiotic Stresses From Wild Rice and Their Applications in Cultivar Improvements, Rice-Germplasm, Genetics and Improvement; Yan, W., Bao, J., Eds.; IntechOpen: Rijeka, Croatia, 2014.

Zhou, X.; Liao, H.; Chern, M.; Yin, J.; Chen, Y.; Wang, J.; Zhu, X.; Chen, Z.; Yuan, C.; Zhao, W.; Wang, J.; Li, W.; He, M.; Ma, B.; Wang, J.; Qin, P.; Chen, W.; Wang, Y.; Liu, J.; Qian, Y.; Wang, W.; Wu, X.; Li, P.; Zhu, L.; Li, S.; Ronald, P. C.; Chen, X. Loss of Function of a Rice TPR-Domain RNA-Binding Protein Confers Broad-Spectrum Disease Resistance. *Proc. Natl. Acad. Sci.* **2018,** *115,* 3174–3179.

Zhou, S. X.; Zhu, Y.; Wang, L. F.; Zheng, Y. P.; Chen, J. F.; Li, T. T.; Yang, X. M.; Wang, H.; Li, X. P.; Ma, X. C; Zhao, J. Q.; Pu, M.; Feng, H.; Li, Y.; Fan, J.; Zhang, J. W.; Huang, Y. Y.; Wang, W. M.; Osa_miR1873 Finetunes Rice Immunity Against *Magnaporthe oryzae* and Yield Traits. *J. Integrative Plant Biol.* **2020,** *62* (8), 1213–1226.

Zhu, D.; Wei, H.; Guo, B.; Dai, Q.; Wei, C.; Gao, H.; Hu, Y.; Cui, P.; Li, M.; Huo, Z.; KeXu, K.; Zhang, H. The Effects of Chilling Stress After Anthesis on the Physicochemical Properties of Rice (*Oryza sativa* L.) Starch. *Food Chem.* **2017,** *237,* 936–941.

Zou, L. L.; Wei, Y. M. Driving Factors for Social Vulnerability to Coastal Hazards in Southeast Asia: Results From the Meta-Analysis. *Nat. Hazards* **2010,** *54* (3), 901–929.

CHAPTER 4

A Review of Conventional and Molecular Approaches for the Management of Rice Blast Disease

RASHMI MAURYA[1], MUNNA SINGH[1], and DEEPTI SRIVASTAVA[2]

[1]*Department of Botany, University of Lucknow, Lucknow, Uttar Pradesh, India*

[2]*Integral Institute of Agricultural Science and Technology (IIAST) Integral University, Dashauli, Uttar Pradesh, India*

ABSTRACT

Rice blast disease is caused by *Magnaporthe oryzae* causing huge crop losses worldwide. Recent technological advancement brought new horizons in developing resistance against rice blasts. Conventional breeding approaches have been followed to develop new resistant varieties that took decades to process. The molecular marker-based selection, germplasm screening, Genome-wide association mapping, and quantitative trait locus (QTL) mapping helped in understanding the natural existing sources of resistance. These have been utilized in marker-assisted selection and gene pyramiding to accelerate the development of resistant varieties. Next, allele mining and their utilization in the genetic engineering of rice plants helped in developing cisgenics/intragenics to bring resistant traits from diverse cultivars. A potent approach is to create new sources by artificial mutation/insertion methods will help in providing lost lasting resistance against rice blast.

Fungal Diseases of Rice and Their Management. Deepti Srivastava, Md. Shamim, Malik Mobeen Ahmad, and R. S. Upadhyay (Eds.)

4.1 INTRODUCTION

Rice blast caused by the *Magnaporthe oryzae,* one of the most devastating fungal pathogens of rice marking its presence in 85 countries. The yield losses caused by rice blast were nearly 30–50% under conducive environmental conditions (Yadav et al., 2017). *M. oryzae* is a heterothallic ascomycete, reproduces asexually, or sexually and which establishes hemibiotrophic association with plants. The infection cycle starts with the spreading of conidia on the plant surface and germinates on the leaf surface to form germ tubes and appressoria. In appressoria, internal turgor pressure is created to around 8.0 MPa helps in penetrating inside the cell wall, and vigorously grows leaving lesions on the leaf. Further, it develops conidia to start a new cycle of infection on a new site of infection (Zhang et al., 2016). It can infect rice plants right from seedlings to adult plant stages affecting leaves, nodes, collar, panicles, and even roots (Yadav et al., 2017; Fernandez and Orth, 2018). Another challenge conferred by *M. oryzae,* is via its genome plasticity and variability. Due to its rapid adaptation ability, the resistance machinery of plants are often rendered ineffective leading to severe crop and economic losses. A number of allelic variants have been reported for the rice blast resistance gene in rice (Vasudevan et al., 2016, Lv et al., 2017) Draft genome of *M. oryzae* is 32.8 Mb with seven chromosomes with 11,109 predicted genes. Analysis revealed the presence of a group of secreted proteins, G protein coupled receptors, virulence genes, and enzymes of secondary metabolism. A high rate of active transposable elements has been identified which indicates a highly duplicating nature of the genome. Also, differential regulation of genes has been marked during early infection stages suggesting their probable role in plant-fungal pathogenesis (Dean et al., 2005).

In field, high moisture content/humidity provides conducive environment for sporulation, germination stages for *M. oryzae* infection in rice. High nitrogen fertilizer and drought also render rice crop more susceptible to infection. The high irrigation and excessive water logging could also be an additive reason for infection. The cropping system need to remove any leftovers of previous crop would also an effective way to avoid any inoculum in fresh crop cycle. Next comes the chemical management that requires antifungal treatment which so far is quite effective in managing infection in rice cropping. Ever since it was first discovered, we still do not have any absolute solution for farmers. Emerging technologies have led to various approaches that encourage toward the development of applications for sustainable management of *M. oryzae* infection in rice. An array of strategies/tools has been explored to identify the natural asset that could fit in combating *M. oryzae* while having sustainability to the environment. Genetic breeding and molecular studies have identified DNA markers and mapped *M. oryzae* associated QTLs, genotype resources for the resistance, allele mining, and gene pyramiding, toward cis/transgenic development. Over time, pathogens evade the resistance mechanism of plants via evolving new strategies and transforming into more virulent forms. In the field, the durability of resistance genes could be achieved by understanding its molecular basis. Another emerging approach is interspecies transfer receptor proteins involved in plant defense such as pattern recognition receptors (PRRs). The concept of developing transgenerational plant immunity is another strategy for developing eco-friendly and broad-spectrum-resistant rice varieties (Singh et al., 2018). This chapter includes current conventional and molecular tools in the management of blast resistance in rice.

4.2 GERMPLASM SCREENING, DISEASE SCORING, QTL MAPPING, AND GENOME-WIDE ASSOCIATION MAPPING

Rice plants show varying degree of variability reflected in terms of susceptibility and resistance against blast disease. Rice germplasm from North-Est region of India, Akhanaphou showed field resistance against leaf and neck blast in rice. Resistance was mapped in QTLs qLNBL-5 and qLNBL-7 contributing to 26% and 25% phenotypic variance in blast resistance in rice (Aglawe et al., 2017). So far, more than 100 R genes and 350 QTLs have been reported for blast resistance in rice. In another study, 161 landraces were screened using 28 markers (corresponding to 24 R gene; *Pib, Pb1, Pita, Pi9, Pi2, Pizt, Pid2, Pi33, Pi36, Pi37, Pikm, Pit, Pi5, Pi54, Pish, Pik, Pikp, Pia, Pi25, Pi1, Pi21, Pi56, Pi65*(*t*), and Piz) classified 21 rice landraces as highly resistant, 70 as moderately resistant and 70 as susceptible. More specifically, markers K3957, Pikh, Pi2–i, RM212, and RM302 were found to determine phenotypic blast resistance in the range of 1.4–7.6%. Genetic association studies identified that five markers (*Pik–p, Pikh, Pi2, Pi1, and Pi37*) showed significant association to blast disease of rice with phenotypic variance in the range of 1.4–7.6%. Highest phenotypic variance (7.6%) was observed with the marker, K3957 corresponding to Pik–p gene (Yadav et al., 2019). A large screen of rice germplasm includes 4246 geographically diverse rice accessions belonging to 13 major rice-growing countries identified 289 accessions as broad-spectrum resistance against multiple rice blast isolates. Further, allelic variation of *Pi2* resistance gene was also shown in relation to blast resistance gene (Vasudevan et al., 2014). Another study screened Indian germplasm in a field trial for rice blast resistance and identified Sadanunia, T4M-3-5, and Chakhao Sampark' highly resistant to the blast disease however Kalonunia, Gobindabhog, Konkanijoha were highly susceptible (Mondal et al 2021). Pi54 allele was also reported to deliver blast resistance in rice and identified by screening 885 rice accession of Indian origin. It was recorded a resistant score of 0–3 when infected with field mix of *M. oryzae* strains (Vasudevan et al., 2015)

O. glumaepatula showed field resistance against blast disease and utilized for QTL identification. Two QTLs, qBL3 and qBL7 were mapped on genome having 25 and 34% phenotypic variance toward blast resistance of rice. qBL3 QTL was further targeted for fine mapping, which identified as s Pi68(t); Os03g0281466 (malectin-serine threonine kinase) allele contributing to resistance (Devi et al., 2020). Two other QTLs, qPb11–1(Pb-bd1), and qPb6–1 were mapped in the 40.6 Kb region of the rice genome and Pb-bd1 introgressed line significantly enhanced penicle blast resistance in NJ46 variety (Fang et al., 2019). Another study assessed 311 japonica and indica accessions via genotyping and GWAS which identified 11 accession in the category of high resistance against rice blast disease. A total of 14 marker-trait associations were identified and among them, BRF04, BRF11–1, and BRF06–1 showed blast scores 4.36, 3.45, and 3.08, respectively in the field whereas, BRGC11–2, BRGC11–1, and BRGC06 showed blast score of 1.82, 1.59, and 1.51, respectively in growth chamber conditions (Volante et al., 2020). A screening of 390 *O. sativa* accession with five *M. oryzae* isolates lead to the identification of 60 accessions as highly resistant and 16 were highly susceptible to all five *M. oryzae* isolates. Study revealed 97 loci associated blast resistance were scattered across all 12 chromosomes for resistance against all five *M. oryzae* isolates. Out of these 82 loci associated blast resistance were reported for the first time, nine were located on previously mapped R loci, and six were located in region from which R gene is cloned previously (Kang et al., 2016). Two different backcrossed population (Jin23B/CR071 cross and Jin23B/QingGuAi3 cross population) were also used to identify QTLs responsible for blast resistance in rice. In Jin23B/CR071 population, 3 major, and 13 minor QTLs showed phenotypic variation from 3.50% to 34.08% whereas Jin23B/QingGuAi3 population showed 2 major and 11 minor QTLs with phenotypic variation from 2.42% to 28.95% (Jiang et al., 2020).

4.3 EXPANDING GENETIC RESOURCES VIA ARTIFICIAL MUTATION

In last decades, our focus lies in exploring the naturally occurring genome resources for resistance for pathogenesis. A lot of QTLs and specific alleles have been validated and used for developing blast resistant lines of rice. With the frequent evolution, *M. oryzae* grasp resistant against alleles and rendered the transgenic crops useless. With this pace, we require more genetic diversity that could be created by artificial methods such as physical agent, chemical agents, or biological methods. The physical method includes gamma rays, Ion Beam Radiation, Fast-Neutron Irradiation; chemical agents includes Ethyl Methanesulfonate, N-Methyl-N-Nitrosourea, Sodium Azide, Colchicine for Polyploid Induction and biological methods such as Insertional Mutagenesis, T-DNA Insertion Mutagenesis, Transposon, and Retrotransposon Mutagenesis, Targeted Mutation, genome editing techniques such as zinc-finger nucleases (ZFNs), transcription activator-like (TAL) effector nucleases (TALENs), and clustered regularly interspaced short palindromic repeats (CRISPR) associated with the Cas9 and Cpf1 proteins etc. (Viana et al., 2019). The power of editing and creating novel genomic resources has huge potential that has less explored so far but their targeted utilization can unleash the genomic potential to its fullest.

4.4 BREEDING AND GENE PYRAMIDING APPROACH

Rice genome carry genes that could provide resistance against *M. oryzae* infection in rice but they were present in different germplasm. Also, so far, resistance to rice blast is not absolute with single gene thus, gene marker assisted breeding could be used to bring resistant genes together. Here, contrast germplasm was crossed and progenies were obtained. Various mapping populations such as recombinant inbred lines (RILs), double haploids (DH), Single-segment substitution lines (SSSLs), backcross lines, and F2/F3/F4 lines could be effectively utilized for screening of blast resistant loci and further fine mapping (Ashkani et al., 2015). These progenies could be screened with morphological, biochemical, or DNA markers to select agriculturally important traits. The heterozygous progenies were then self-pollinated to generate multiple progenies having segregating alleles and obtain specific genomic targets contributing to blast resistance. The breeding approaches were utilized to develop resistant lines in rice where sterile lines were used to develop 31 single-gene derived lines and 20 double-gene combination lines. These showed enhanced resistance against *M. oryzae* infection in rice (Jiang et al., 2019). The selected gene should have a strong heritability in genome and the marker should have strong linkage with gene. The markers could be a morphological trait of interest that are visible to the eye or easily detectable morphological features like color, height, grain, aroma etc. Biochemical characteristics could also be any isozyme, protein, or content. Another type of marker is cytological that are represented by the form of chromosome bands, regions of chromatin that become impregnated with specific dyes used in cytology. Any desired trait could be correlated with a chromosomal band pattern indicating probable linkage between them. Several DNA based marker includes microsatellite such as short tendom repeat (STR), simple sequence repeat (SSR), random amplification of polymorphic DNA (RAPD), restriction fragment length polymorphism (RFLP), amplified fragment length polymorphism (AFLP), and single nucleotide polymorphisms (SNPs) acts as majorly utilized in marker assisted selection.

Next, molecular breeding is performed to stack resistant genes from diverse rice variety sources. Gene pyramiding also uses this technique that helps in gene stacking unlinked genes from distant genotype. The genes having no linkage could also be stacked in plant of interest. Here, the known marker was screened at each step to facilitate the overall process. Traditional breeding is accelerated via using markers. For DNA markers, PCR based approaches could be used to select progenies with all the stacked gene. The conventional method of gene pyramiding for developing resistance lines is tedious, labor intensive, time consuming, and less efficient because the pathogen used for screening resistant gene in cultivars might not carry avirulent gene thus making it difficult to screen resistant allele. So, modern screening approaches are key to ease up the breeding process.

4.5 ALLELE MINING: CONVENTIONAL AND ADVANCE APPROACHES

Allele mining is a process of identification of novel allele or any allelic variation driving desired specific phenotypic traits. These phenotypic traits serve as resources in engineering the plants traits. An *in-silico* approach could be utilized to identify resistant allele across large population of germplasm. Blast resistance allele was screened 1883 rice varieties using in-silico approach out of these 15 allelic variants of Pi2/Pi9 locus on chromosome six were identified and validated via transforming it into susceptible germplasm and showing development of resistance in field (Zhou et al., 2020). Another approach following Fluidigm SNP genotyping platform identified 10 loci with race-specific resistance having SNP variation in rice (Kitazawa et al., 2019). A PCR amplification and subsequent Sanger sequencing method was used to establish allelic variation in Pi54 and Pi9 suggesting its broad-spectrum usage in breeding program for rice blast resistance (Thakur et al., 2015; Imam et al., 2016). Ptr gene was mined using a fine mapping strategy using a cross between Amane x Katy germplasm and subsequent screening with microsatellite markers RM3246 and RM1047 which are represented on charomosone12. The resistant phenotype was confirmed using infection assay with blast race (isolate) IB-49 (ML1) (Zhao et al., 2018). Other approach of gene mining includes T-DNA insertion mutant followed by screening which provide array of phenotypes that could be further checked for T-DNA mutated target. This help in identifying new phenotype linked to alleles. Similarly, physical and chemical mutagen treatments develop a mutant library that could be further screened to identify respective allele (Ram et al., 2019). Entrapment tagging is another tool to identify novel targets which could be gene, promotor, or enhancer. In this trap tagged lines are used in which T-DNA or transposable element combined with gene, that help in expression of desired tagged gene. The method has been well established to develop mutant libraries in rice crop (Krishnan et al., 2009)

4.6 GENETIC ENGINEERING, TRANSGENICS, AND MOLECULAR TOOLS

So far, a total of 118 genes are identified and 25 were further characterized as leaf blast resistance gene (Kalia and Rathour, 2019). In order to validate the role of resistant genes in rice, these are transformed amplified from resistant cultivars and transformed into susceptible cultivars. These could be further assessed for developing resistance against rice blast. A *Pi54* gene was transformed into susceptible rice variety; TP309. The proliferating calli was selected on MS-hygromycin and regenerated. The developing plantlet was further induced for rooting and hardening. Developed transgenic lines were screened using genomic PCR, southern blot analysis and expression analysis to confirm presence and expression of transgene. Next, rice transgenic was subjected to infection with *M. oryzae* stain; Mo-ei-79 showed that all the T2 generation showed resistance response as compared to the control rice cultivar (Singh et al., 2020). Several techniques such as functional complementation for pathogenicity, transient expression for subcellular localization, protein expression, and fluorescent marker tagging for spaciotemporal regulation are used for functional validation of resistant genes in rice. Interestingly, R proteins have been extensively characterized for delivering *M. oryzae* resistance in rice. R genes are distributed across all 12 chromosomes except Chro.3. Till today, more than 100 blast resistance genes have been reported. Out of these, 45% are represented from *japonica*, 51% from *indicia* cultivars and only 4% comes from the wild relative of rice cultivars (Sharma, 2012). So far, effort has been made to clone and characterize approximately 25 R genes. The desirable trait of R gene in plant breeding is its dominant and qualitative nature; although a couple of R genes such as pi21 and Pb1 showed quantitative behavior as well (Yadav et al., 2017).

Another approach is to use elicitor genes as transgenes to transform rice which could induce hypersensitive response thus raising the resistance level in transgenic rice plant. *M. oryzae* elicitors such as MoHrip1 and MoHrip2 were transformed in rice variety showing induced expression of SA and ABA related genes. The restricted hyphal growth on these rice transgenic plants were also confirmed by trypan blue staining (Wang et al., 2017). Another group of regulatory proteins such as transcription factors could help in developing blast resistance in rice. For example, OsWRKY53 overexpression in rice enhanced disease resistance to the blast fungus compared control rice plants. Study showed that overexpression of OsWRKY53 is regulated by OsMKK4-OsMPK3/OsMPK6 cascades and regulate multiple defense-related genes including PR genes (Chujo et al., 2014).

With technological advancement, genome editing techniques such as transcription activator-like effector nucleases (TALENs), zinc-finger nucleases (ZFNs), and clustered regularly interspaced short palindromic repeats (CRISPR) associated with the Cas9 and Cpf1 proteins (CRISPR/CAS9) are well established in genome editing in crop plants (Ahmar et al., 2020). A recently developed technique CRISPR/Cas9 has been utilized to create mutant with multiple mutation sites on OsERF922 gene which showed enhanced resistance against rice blast resistance (Wang et al., 2016).

4.7 CONCLUSION AND FUTURE PROSPECT

Developing rice blast resistance is a challenge with existing conventional methods. The chemical methods are effective instead gives an undesirable and non-sustainable environmental hazard. With effects from breeding program to harvest the power of germplasm resources, resistant varieties were developed which could effectively stop crop losses. The stacking of genes from diverse germplasm has been effectively established via using gene pyramiding method. Effects should be made to develop resistant varieties that could provide long lasting and broad-spectrum resistance having wide geographical area. Several techniques such as marker assisted breeding, QTL mapping, gene pyramiding, allele mining, GWAS, genome editing, and genetic engineering transformation have been effectively utilized to study and develop resistance against rice blast resistance. The lost lasting resistance could also be achieved by introgression of two or more resistant genes into susceptible variety from different sources. Thus, the key approach is to combining potential resistant genes in rice germplasm. The R genes having potential for blast resistance should be selected with caution in order to develop rice varieties having long lasting resistance along with wide blast race coverage. The most popular and cost-effective method remain is introgression of potential resistant genes using marker selection that reduces the cost as well as time of procedure. Additionally, it is environment friendly and acceptable for direct consumption as compared to the transgenics, which are still at hold for food crops.

Naturally existing germplasm provide sources for creating resistance against rice blast. New strategies to generate genetic resource via mutagen/insertion have been proven effective in multiple crops. The phenotype created via mutation is result of artificial/novel allelic variation in rice genome. With fast evolving pathogen, asset of novel alleles or mutant library could be of greater importance in creating long lasting resistance. Intriguingly, the molecular and signaling knowledge of blast resistance against multiple blast races will also help in selecting and stacking right gene combination to achieve effective and sustainable resistance varieties in rice.

KEYWORDS

- **rice blast**
- ***Magnaporthe oryzae***
- **allele mining**
- **cisgenics**
- **intragenics**

REFERENCES

Aglawe, S. B.; Bangale, U.; Satya, R. D. S. J.; Balija, V.; Pal, B. V.; Kumar, S. S.; Kumar, S. P.; Kumar, S.; Maddamshettya, S. P.; Maganti S. M. Identification of Novel QTLs Conferring Field Resistance for Rice Leaf and Neck Blast from an Unique Landrace of India. *Gene Rep.* **2017,** *7*, 35–42.

Ahmar, S.; Saeed, S.; Khan, M. H. U.; Khan, S. U.; Mora-Poblete, F.; Kamran, M.; Faheem, A.; Maqsood, A.; Rauf, M.; Saleem, S.; Hong, W.-J.; Jung, K.-H. A Revolution towards Gene-Editing Technology and its Application to Crop Improvement. *Int. J. Mol. Sci.* **2020,** *21* (16), 5665.

Ashkani, S.; Rafii, M. Y.; Shabanimofrad, M.; Miah, G.; Sahebi, M.; Azizi, P.; Tanweer, F. A.; Akhtar, M. S.; Nasehi, A. Molecular Breeding Strategy and Challenges towards Improvement of Blast Disease Resistance in Rice Crop. *Front. Plant Sci.* **2015,** *6*, 886.

Chujo, T.; Miyamoto, K.; Ogawa, S.; Masuda, Y.; Shimizu, T.; Kishi-Kaboshi, M.; Takahashi, A.; Nishizawa, Y.; Miami, E.; Nojiri, H.; Yamane, H.; Okada, K. Overexpression of Phosphomimic Mutated OsWRKY53 Leads to Enhanced Blast Resistance in Rice. *PLoS One* **2014,** *9* (6), e98737.

Dean, R. A.; Talbot, N. J.; Ebbole, D. J.; Farman, M. L.; Mitchell, T. K.; Orbach, M. J.; Thon, M.; Kulkarni, R.; Xu, J.-R.; Pan, H.; Read, N. D.; Lee, Y.-H.; Carbone, I.; Brown, D.; Yo, Y. Y.; Donofrio, N.; Jeong, J. S.; Soanes, D. M.; Djonovic, S.; Kolomiets, E.; Rehmeyer, C.; Li, W.; Harding, M.; Kim, S.; Lebrun, M.-H.; Bohnert, H.; Coughlan, S.; Butler, J.; Calvo, S.; Ma, L.-J.; Nicol, R.; Purcell, S.; Nusbaum, C.; Galagan, J.-E.; Birren, B. W. The Genome Sequence of the Rice Blast Fungus Magnaporthe Grisea. *Nature* **2005,** *434* (7036), 980-986.

Devi, S. R.; Singh, K.; Umakanth, B.; Vishalakshi, B.; Rao, K. V. S.; Suneel, B.; Sharma S. K.; Kadambari, G. K. M.; Prasad, M. S.; Senguttvel, P.; Syamaladevi, D. P.; Madhav M. S. Identification and Characterization of a Large Effect QTL from Oryza Glumaepatula Revealed Pi68 (t) as Putative Candidate Gene for Rice Blast Resistance. *Rice* **2020,** *13* (1), 1–13.

Fang, N.; Wei, X.; Shen, L.; Yu, Y.; Li, M.; Yin, C.; He, W.; Guan, C.; Chen, H.; Zhang, H.; Bao, Y. Fine Mapping of a Panicle Blast Resistance Gene Pb-bd1 in Japonica Landrace Bodao and its Application in Rice Breeding. *Rice* **2019,** *12* (1), 1–12.

Fernandez, J.; Orth, K. Rise of a Cereal Killer: The Biology of *Magnaporthe Oryzae* Biotrophic Growth. *Trends Microbiol.* **2018,** *26* (7), 582–597.

Imam, J.; Mandal, N. P.; Variar, M.; Shukla, P. Allele Mining and Selective Patterns of Pi9 Gene in a Set of Rice Landraces from India. *Front. Plant Sci.* **2016**, *7*, 1846.

Jiang, H.; Feng, Y.; Qiu, L.; Gao, G.; Zhang, Q.; He, Y. Identification of Blast Resistance QTLs based on Two Advanced Backcross Populations in Rice. *Rice* **2020,** *13*, 1–12.

Jiang, H.; Li, Z.; Liu, J.; Shen, Z.; Gao, G.; Zhang, Q.; He, Y. Development and Evaluation of Improved Lines with Broad-Spectrum Resistance to Rice Blast using Nine Resistance Genes. *Rice* **2020,** *12* (1), 1–11.

Kalia, S.; Rathour, R. Current Status on Mapping of Genes for Resistance to Leaf- and Neck-Blast Disease in Rice. *3 Biotech.* **2019,** *9* (6), 1–14.

Kang, H.; Wang, Y.; Peng, S.; Zhang, Y.; Xiao, Y.; Wang, D.; Qu S.; Li Z.; Yan S.; Wang Z.; Liu W.; Ning Y.; Korniliev P.; Leung H.; Mezey J.; McCouch S R.; Wang G.-L. Dissection of the Genetic Architecture of Rice Resistance to the Blast Fungus *Magnaporthe oryzae*. *Mol. Plant Pathol.* **2016,** *17* (6), 959–972.

Kitazawa, N.; Shomura, A.; Mizubayashi, T.; Ando, T.; Nagata, K.; Hayashi, N.; Takahashi, A.; Yamanouchi, U.; Fukuoka, S. Rapid DNA-Genotyping System Targeting Ten Loci for Resistance to Blast Disease in Rice. *Breed. Sci.* **2019,** *69* (1), 68–83.

Krishnan, A.; Guiderdoni, E.; An, G.; Yue-ie, C. H.; Han, C. D.; Lee, M. C.; Yu, S.-M.; Upadhyaya, N.; Ramachandran, S.; Zhang, Q.; Sundaresan, V.; Hirochika, H.; Leung, H.; Pereira, A. Mutant Resources in Rice for Functional Genomics of the Grasses. *Plant physiol.* **2009,** *149* (1), 165–170.

Lv, Q.; Huang, Z.; Xu, X.; Tang, L.; Liu, H.; Wang, C.; Zhou Z.; Xin Y.; Xing J.; Peng Z.; Li X.; Zheng T.; Zhu, L. Allelic Variation of the Rice Blast Resistance Gene Pid3 in Cultivated Rice Worldwide. *Sci. Rep.* **2017,** *7* (1), 1–13.

Mondal, D.; Kantamraju, P.; Jha, S.; Sundarrao, G. S.; Bhowmik, A.; Chakdar, H.; Mandal, S.; Shana, N.; Roy, B.; Bhattacharya, P. M.; Chowdhury, A. K.; Choudhury, A. Evaluation of Indigenous Aromatic Rice Cultivars from Sub-Himalayan Terai Region of India for Nutritional Attributes and Blast Resistance. *Sci. Rep.* **2021,** *11* (1), 1–25.

Ram, H.; Soni, P.; Salvi, P.; Gandass, N.; Sharma, A.; Kaur, A.; Sharma, T. R. Insertional Mutagenesis Approaches and their Use in Rice for Functional Genomics. *Plants.* **2019,** *8* (9), 310.

Sharma, T. R.; Rai, A. K.; Gupta, S. K.; Vijayan, J.; Devanna, B. N.; Ray, S. Rice Blast Management through Host-Plant Resistance: Retrospect and Prospects. *Agric. Res.* **2012,** *1* (1), 37–52.

Singh, J.; Gupta, S. K.; Devanna, B. N.; Singh, S.; Upadhyay, A.; Sharma, T. R. Blast Resistance Gene Pi54 Over-Expressed in Rice to Understand its Cellular and Sub-Cellular Localization and Response to Different Pathogens. *Sci. Rep.* **2020,** *10* (1), 1–13.

Singh, P. K.; Nag, A.; Arya, P.; Kapoor, R.; Singh, A.; Jaswal, R.; Sharma, T. R. Prospects of Understanding the Molecular Biology of Disease Resistance in Rice. *Int. J. Mol. Sci.* **2018,** *19* (4), 1141.

Thakur, S.; Singh, P. K.; Das, A.; Rathour, R.; Variar, M.; Prashanthi, S. K.; Singh A. K.; Singh U. D.; Chand D.; Singh N. K.; Sharma, T. R. Extensive Sequence Variation in Rice Blast Resistance Gene Pi54 makes it Broad Spectrum in Nature. *Front. Plant sci.* **2015,** *6*, 345.

Vasudevan, K.; Gruissem, W.; Bhullar, N. K. Identification of Novel Alleles of the Rice Blast Resistance Gene Pi54. *Sci. Rep.* **2015,** *5* (1), 1–12.

Vasudevan, K.; Vera Cruz, C. M.; Gruissem, W.; Bhullar, N. K. Geographically Distinct and Domain-specific Sequence Variations in the Alleles of Rice Blast Resistance Gene Pib. *Front. Plant Sci.* **2016,** *7*, 915.

Vasudevan, K.; Vera Cruz, C. M.; Gruissem, W.; Bhullar, N. K. Large Scale Germplasm Screening for Identification of Novel Rice Blast Resistance Sources. *Front. Plant Sci.* **2014,** *5*, 505.

Viana, V. E.; Pegoraro, C.; Busanello, C.; Costa De Oliveira, A. Mutagenesis in Rice: The Basis for Breeding a New Super Plant. *Front. Plant Sci.* **2019,** *10*, 1326.

Volante, A.; Tondelli, A.; Desiderio, F.; Abbruscato, P.; Menin, B.; Biselli, C.; Casella, L.; Singh, N.; McCouch, S. R.; Tharreau, D.; Zampieri, E.; Cattivelli, L.; Valè, G. Genome Wide Association Studies for Japonica Rice Resistance to Blast in Field and Controlled Conditions. *Rice* **2020,** *13* (1), 1–17.

Wang, F.; Wang, C.; Liu, P.; Lei, C.; Hao, W.; Gao, Y.; Liu Y.-G.; Zhao, K. Enhanced Rice Blast Resistance by CRISPR/Cas9-targeted Mutagenesis of the ERF Transcription Factor Gene OsERF922. *PloS One.* **2016,** *11* (4), e0154027.

Wang, Z.; Han, Q.; Zi, Q.; Lv, S.; Qiu, D.; Zeng, H. Enhanced Disease Resistance and Drought Tolerance in Transgenic Rice Plants Overexpressing Protein Elicitors from *Magnaporthe Oryzae*. *PLoS One*. **2017,** *12* (4), e0175734.

Yadav, M. K.; Aravindan, S.; Ngangkham, U.; Raghu, S.; Prabhukarthikeyan, S. R.; Keerthana, U.; Marndi, B. C.; Adak, T.; Munda, S.; Deshmukh, R.; Pramesh, D.; Samantaray, P.; Rath, P. C. Blast Resistance in Indian Rice Landraces: Genetic Dissection by Gene Specific Markers. *Plos One*. **2019,** *14* (1), e0211061.

Yadav, M. K.; Aravindan, S.; Ngangkham, U.; Shubudhi, H. N.; Bag, M. K.; Adak, T.; Munda, S.; Samantaray S.; Jena, M. Use of Molecular Markers in Identification and Characterization of Resistance to Rice Blast in India. *PloS One*. **2017,** *12* (4), e0176236.

Zhang, H.; Zheng, X.; Zhang, Z. The Magnaporthe Grisea Species Complex and Plant Pathogenesis. *Mol. Plant Pathol*. **2016,** *17* (6), 796.

Zhao, H.; Wang, X.; Jia, Y.; Minkenberg, B.; Wheatley, M.; Fan, J.; Jia, M. H.; Famoso, A.; Edwards, J. D.; Wamishe, Y.; Valent, B.; Wang, G.-L.; Yang, Y. The Rice Blast Resistance Gene Ptr Encodes an Atypical Protein Required for Broad-spectrum Disease Resistance. *Nat. Commun*. **2018,** *9* (1), 1–12.

Zhou, Y.; Lei, F.; Wang, Q.; He, W.; Yuan, B.; Yuan, W. Identification of Novel Alleles of the Rice Blast-Resistance Gene Pi9 through Sequence-based Allele Mining. *Rice* **2020,** *13* (1), 1–15.

CHAPTER 5

Computational Deciphering of Blast Resistance Genes in Rice

SAMEER AKHTAR CHAUDHARY[1], MALIK M. AHMAD[2], SAPANA CHAUDHARY[1], SAKSHI RAWAT[1], and JAYASHRI PRASANAN[1]

[1]*RASA Life Science Informatics, Pune, Maharashtra, India*

[2]*Department of Agriculture, Integral Institute of Agricultural Science and Technology (IIAST), Integral University, Lucknow, Uttar Pradesh, India*

ABSTRACT

Rice blast disease caused by *Magnaporthe oryzae* has hindered the global production and yield of rice. Hence, it is crucial to study the intricate details of the plant defense mechanism and understand how plants acquire full or partial resistance against the species of *M. oryzae*. Plants encoding nucleotide-binding site leucine-rich repeat (NBS-LRR) like proteins have characteristics of plant-disease resistance genes. This review summarizes the computational analysis trying to decipher and analyzes the underlying role of different blast resistance genes in resistant and susceptible rice cultivars and identifies resistance proteins helpful in developing molecular markers linked to resistance genes.

5.1 INTRODUCTION

Rice is one of the most important crops and staple primary source of food to meet the dietary requirements of people globally (Ngangkham et al., 2010). Therefore, any fungal disease affecting rice plants, be it resistant

Fungal Diseases of Rice and Their Management. Deepti Srivastava, Md. Shamim, Malik Mobeen Ahmad, and R. S. Upadhyay (Eds.)

or susceptible cultivars, is of paramount significance (Chandrakanth et al., 2020), especially the blast disease caused by the fungus *Magnaporthe oryzae* resulting in huge economic losses. The ability of *M. oryzae* to infect rice plants at all stages of growth and across different plant parts (Zhu et al., 2005) has sealed its position as one of the topmost dreaded fungal pathogens of scientific and economic importance (Dean et al., 2012). Though blast disease can be managed through fungicides, resistant cultivars, optimizing fertilizer treatment, and maintaining the planting dates (Bonman, 1992), the fraction of these approaches are quite non-feasible, unprofitable, and could pose experimental risks (Yadav et al., 2017). *Magnaporthe oryzae* B.C. Couch has been well-studied to explore the host–pathogen interactions for many decades (Dean et al., 2005). This has led to many studies focused on understanding host–pathogen interactions and plant resistance (Dean et al., 2005). With the advent of technology and advancements in science, newer tools have become available to explore disease resistance at a genetic level. Therefore, unraveling information regarding the resistance genes underlying host defense mechanisms, including hypersensitive responses (Gururani et al., 2012), would offer invaluable insights (Sharma et al., 2018). The efficient application of functional genomics tools for disease resistance could help us better understand plant defense signaling. Still, it could also reveal novel insights into the interactions between these signaling pathways and other plant processes (Gururani et al., 2012). Resistance (R) genes represent an array of genes crucial to upholding the host defense mechanisms and generally belong to an enormous multigene family, distributed across all the 12 rice chromosomes except for chromosome 3 (Liu et al., 2010). The blast resistance (R) major or minor genes might confer full or partial protection against select or entire species range of *M. oryzae*.

5.2 BLAST RESISTANCE GENES

One of the gene family members, blast R genes, has been the subject of numerous studies at different levels (Chandrakanth et al., 2020). Reported studies using computational analysis have tried to decipher and analyze the underlying role of blast resistance genes in resistant and susceptible rice cultivars (Gururani et al., 2012). Several blast resistance genes, *R* (>100) and Quantitative trait locus (QTLs) (>350) have been identified, and 27 of these genes, *Pib, Pb1, Pita, Pi9, Pi2, Pizt, Pid2, Pid3, Pid3–A4, Pi33, Pii, Pi36, Pi37, Pi50, and Pi65(t)*, *Pit, Pi5, Pik, Pikp, Pikh, Pikm, Pish, Pia,*

PiCO39, Pi25, Pi1, pi21, having been cloned and characterized (Su et al., 2015; Zheng et al., 2016; Zhu et al., 2016). The constitutive expression of blast R genes, being a well-reported phenomenon among different plants, including rice, makes these well-suited for genetic analysis (Li et al., 2019). Most of the rice blast resistance genes encode proteins, resistance proteins (RP) containing a putative central nucleotide-binding site (NBS), and carboxy-terminal leucine-rich repeats (LRR). Primary blast R genes, such as *Pi-ta*, *Pi-d3, Pi-b, Pi-k1,* and *RGA*, encoding Nod-like receptor (NLR) family proteins having a direct or indirect role in interaction with fungal effectors (Zhai et al., 2011) have been cloned and studied (DeYoung and Innes, 2006). Reported literature mentions a few well-characterized structures of blast RP, such as Pita and pi54, while others remain relatively unknown (DeYoung and Innes, 2006; Devanna et al., 2014). However, the mammalian NBS and LRR domain's crystal structures are available to predict similar proteins from plant sources using homology modeling as plant NBS-LRR complete structures are unavailable (McHale et al., 2006). Most of the NBS-LRR proteins exhibited several unidentified domains, represented by 'X'. Though NBS-LRR proteins are involved in plant defense mechanisms, little information is available about their downstream interactors and additional functions, while the mode of action remains a mystery till date (DeYoung and Innes, 2006; Devanna et al., 2014). In certain cases, reported blast R gene-encoded protein sequences await experimental verification of their functions. Predicting structures of uncharacterized proteins would offer additional insights into the structure-function-activity relationship and their evolutionary lineage, thereby promoting their use as biological markers (Doerks et al., 2004).

There are certain challenges and limitations to the continued use of existing blast resistant varieties, though major blast R gene has been quite effective throughout the years (Chen et al., 2006). This has created a constant need to explore minor R genes and new alleles/genes, which has gradually co-evolved in certain rice species as genetic diversity in blast R genes is a major pipeline for sourcing and introgression of major blast R genes into different cultivars for controlling blast (Brar and Kush, 1997). The development of selected cultivars displaying sturdiness and broad-spectrum resistance through different breeding strategies (Table 5.1) involves incorporating R genes in common cultivars *via* marker-assisted selection (Hittalmani et al., 2000; Fukuoka et al., 2014). With the advent of fine mapping of blast R genes, numerous polymerase chain reaction (PCR) based markers (Tables 5.1 and 5.2) have been created to rapidly mine such genes in the unexplored

TABLE 5.1 Chromosomal Locations of Certain Broad-Spectrum R Genes and their Application in Successful Breeding of Rice Plants.

Gene name	Chr.	DNA marker used	Donor	Recipient	Encoding protein	Ref.
Pid3-I1	6	–	MC276	–	CC-NBS-LRR	Inukai et al., 2019
Pi1	11	RM224	C101LAC	BPT5204 GD-7S	NBS-LRR	Kumar et al., 2019; Hua et al., 2012
Pi1/Pi54/Pi-ta	11, 12, 12	Pi54MAS RM224 YL155 YL87 YL155/87	–	MushkBudji	–	Khan et al., 2018
Ptr	12	-	Katy	–	Atypical protein with an armadillo repeat	Zhao et al., 2018
Pi54	11	Pi-54MAS RM224	Tetep	BPT5204 R1 R2	NBS-LRR	Khan et al., 2018; Kumar et al., 2018; Sharma et al., 2012;
Pi9	6	Ins2-3, Clon2-1, Nbs21	75-1-127	E32 R288 Q211S	NBS-LRR	Huang et al., 2018; Zou et al., 2017; Xing et al., 2016; Qu et al., 2006
Pigm	6	M80362	Gumei 4 G8900	Kongyu131 Longjing 26Kenjiandao 6 KT27S	NBS-LRR	Dai et al., 2018; Deng et al., 2017; Tian et al., 2016
Piz-t/Pi54, Pi9/Pi54	6, 11	Z4794, Pikh-1	–	07GY31	–	Xiao et al., 2017
Pi2/Pi54	6, 11	AP5659-5 RM206 AP5930 RM206	–	PB1121 PRR78	–	Ellur et al., 2016; Singh et al., 2012

TABLE 5.1 *(Continued)*

Gene name	Chr.	DNA marker used	Donor	Recipient	Encoding protein	Ref.
Pi40	6	9871.T7E2b	–	Osmancik-97Halilbey	–	Beser et al., 2016
Pi46/Pi-ta	11, 12	RM224 YL155 YL87 YL155/87	–	Hanghui 179	–	Xiao et al., 2016
Pi1/Pi2/Pi33	6, 8, 11	RM224, RM527, RM310	–	Kuboyar	–	Usatov et al., 2016
Pi37	1	–	St. No. 1	RM527	NBS-LRR	Jiang et al., 2015; Lin et al., 2007
Pi64	1	–	Yangmaogu	–	NBS-LRR	Ma et al., 2015
Pi50	6	–	Er-Ba-zhan (EBZ)	–	NBS-LRR	Su et al., 2015
Pike	11	–	Xiangzao 143	–	NBS-LRR	Chen et al., 2015
Pib/Pi54	2, 11	RM208, RM206	–	MR219	–	Tanweer et al., 2015
Pi9/Pi-ta	6, 12	AP5659-5 YL155 YL87, YL155/87	–	Pusa Basmati 1	–	Khanna et al., 2015
***Pi63/ Pikahei-1*(t)**[#]	4	–	Kahei	–	NBS-LRR	Xu et al., 2014
Pi1/Pi2/Pi33	6, 8, 11	M224, RM527, RM25	–	ADT43	–	Divya et al., 2014
Pii	9	–	Hitomebore	–	NBS-LRR	Takagi et al., 2013
Pi1/Pi2	6, 11	RM144, AP22, RM5926, AP5659-5	–	Pusa RH-10	–	Gouda et al., 2013; Liu et al., 2005

TABLE 5.1 *(Continued)*

Gene name	Chr.	DNA marker used	Donor	Recipient	Encoding protein	Ref.
Pi1/Pi2, Pi1/ Pigm	6, 11	RM224, ZJ58.7 AP22	–	GZ63S, 97S, R084, R609	–	Yu et al., 2013
Pi-CO39	11	–	CO39	–	CC-NBS-LRR	Cesari et al., 2013
Pi1/Pi2/D12	6, 11, 12	RM144/RM224 PI2-4/HC28, RM277 RM309	–	Jin 23B	–	Jiang et al., 2012
Pi35#	1	–	Hokkai 188	–	NBS-LRR	Fukuoka et al., 2012
Pia	11	–	Aichi Asahi	–	CC-NBS-LRR	Okuyama et al., 2011
Pik-p	11	–	K60	–	CC-NBS-LRR	Yuan et al., 2011
Pik	11	–	Kusabue	–	CC-NBS-LRR	Zhai et al., 2011
Pi25	6	–	Gumei 2	–	CC-NBS-LRR	Chen et al., 2011
Pb1#	11	–	Modan	–	CC-NBS-LRR	Hayashi et al., 2010
Pish	1	–	Shin-2	–	NBS-LRR	Takahashi et al., 2010
Pi21#	4	–	Owarihatamochi	–	Proline-rich metal-binding protein	Fukuoka et al., 2009
Pi-d3	6	–	Digu	–	CC-NBS-LRR	Shang et al., 2009
Pi5	9	–	Moroberekan	–	CC-NBS-LRR	Lee et al., 2009
Pit	1	–	K59	–	CC-NBS-LRR	Hayashi & Yoshida, 2009
Pikm	11	–	Tsuyuake	–	NBS-LRR	Ashikawa et al., 2008
Pi36	8	–	Q61	–	CC-NBS-LRR	Liu et al., 2007
Pi-d2	6	–	Digu	–	B-lectin receptor kinase	Chen et al., 2006

TABLE 5.1 *(Continued)*

Gene name	Chr.	DNA marker used	Donor	Recipient	Encoding protein	Ref.
Pi2	6	–	C101A51	C815S	NBS-LRR	Zhou et al., 2006
Piz-t	6	–	Toride 1		NBS-LRR	Zhou et al., 2006
Pib/Pik	2, 11	NSb K6415	–	K6415	-	Hayashi et al., 2006
Pi-ta	12	–	Yashiro-mochi	–	NBS-LRR	Bryan et al., 2000
Pib	2	–	Tohoku IL9	–	NBS-LRR	Wang et al., 1994

#Partial resistance observed in Nucleotide-binding site leucine-rich repeat (NBS-LRR); CC-NBS-LRR (Coiled-coil-nucleotide-binding site leucine-rich repeat) R genes.

TABLE 5.2 List of Blast Resistant Genes with their Respective Markers and Selective Primers.

Gene	Marker	Forward (5′-3′)	Reverse (5′-3′)	Marker type	References
Pi65	SNP_3	TGCCACCAGCCATCTTCAACAT	ACCACATCACTCATCGCCATCC	InDel#	Zheng et al., 2016
Pit	tk59-1	ATGATAACCTCATCCTCAATAAGT	GTTGGAGCTACGGTTGTTCAG	FM$	
	tk59-2	ATGATAACCTCATCCTCAATAAGT	CCAAGGGATTAGGTCCTAGTG	FM	Warschefsky et al., 2014
Pikh	Pikh	CAATCTCCAAAGTTTTCAGG	GCTTCAATCACTGCTAGACC	FM	
Pi9	Pi9-i	GCTGTGCTCCAAATGAGGAT	GCGATCTCACATCCTTTGCT	FM	Ingole et al., 2014
Pi2	Pi2-i	CAGCGATGGTATGAGCACAA	CGTTCCTATACTGCCACATCG	FM	
Pi56	CRG4-2	CCTGTCAGTCTTTCCGAGAG	GAATCCGGTAGCTCAAGGTG	Gene specific	Liu et al., 2013
Pi1	RM224	ATCGATCGATCTTCACGAGG	TGCTATAAAAGGCATTCGGG	SSR*	Jiang et al., 2012
Pita/ Pita2	YL155/ YL87	AGCAGGTTATAAGCTAGGCC	CTACCAACAAGTTCATCAAA	FM	Earl et al., 2012; Evanno et al., 2005
Pia	Pia-STS	CTTTTGAGCTTGATTGGTCTGC	CTATTGCACCAGAGGGACCAG	FM	Okuyama et al., 2011

TABLE 5.2 *(Continued)*

Gene	Marker	Forward (5′-3′)	Reverse (5′-3′)	Marker type	References
Pid2	dln2	GCGTCGAAGATGTCCTGAAGCTCA	GGCAGTCGTATTGCTGTGAA	FM	Gao et al., 2010
pi21	pi21_79–3	GATCCTCATCGTCGACGTCTGGC	AGGGTACGGCACCAGCTTG	InDel	Fukuoka et al., 2009
Pikm	k2167	CGTGCTGTCGCCTGAATCTG	CACGAACAAGAGTGTGTCGG	InDel	Ashikawa et al., 2008
Pi1	RM1233	GTGTAAATCATGGGCACGTG	AGATTGGCTCCTGAAGAAGG	SSR	Fuentes et al., 2008
Pi37	RM302	TCATGTCATCTACCATCACACCCA	ATGGAGAAGATGGAATACTTGC	SSR	Lin et al., 2007
	RM212	CTTTCAGCTACTACCAG	CACCCATTTGTCTCTCATTATG	SSR	
Pi25	RM564	CATGGCCTTGTGTATGCATC	ATGCAGAGGATTGGCTTGAG	SSR	Wu et al., 2005
Pi36	RM5647	ACTCCGACTGCAGTTTTTGC	AACTTGGTCGTGGACAGTGC	SSR	Liu et al., 2005
Pi33	RM72	CCGGCGATAAAACAATGAG	GCATCGGTCCTAACTAAGGG	LM	Berruyer et al., 2003
Pi5	40N23R	TGTGAGGCAACAATGCCTATTGCG	CTATGAGTTCACTATGTGGAGGCT	InDel	Jeon et al., 2003
Pib	Pb28	GACTCGGTCGACCAATTCGCC	ATCAGGCCAGGCCAGATTTG	SNP[δ]	Pritchard et al., 2000
Piz	Z56592	GGACCCGCGTTTTCCACGTGTAA	AGGAATCTATTGCTAAGCATGAC	SNP	
Piz-t	Zt56591	TTGCTGAGCCATTGTTAAACA	ATCTCTTCATATATATGAAGGCCAC	SNP	
Pik-p	K3957	ATAGTTGAATGTATGGAATGGAAT	CTGCGCCAAGCAATAAAGTC	SNP	
Pik	K39512	GCCACATCAATGGCTACAACGTT	CCAGAATTTACAGGCTCTGG	SNP	
Pb1	RM2698	ACGCACGCACATCCTCTTCC	CGGTTCTCCATCTGAAATCCCTAGC	LM	Tanksley & McCouch, 1997

[#]InDel = Insertion Deletion; [$]FM = Functional marker; [*]SSR = Simple sequence repeats; [γ]LM = Linked marker; [δ]SNP = Single nucleotide polymorphism.

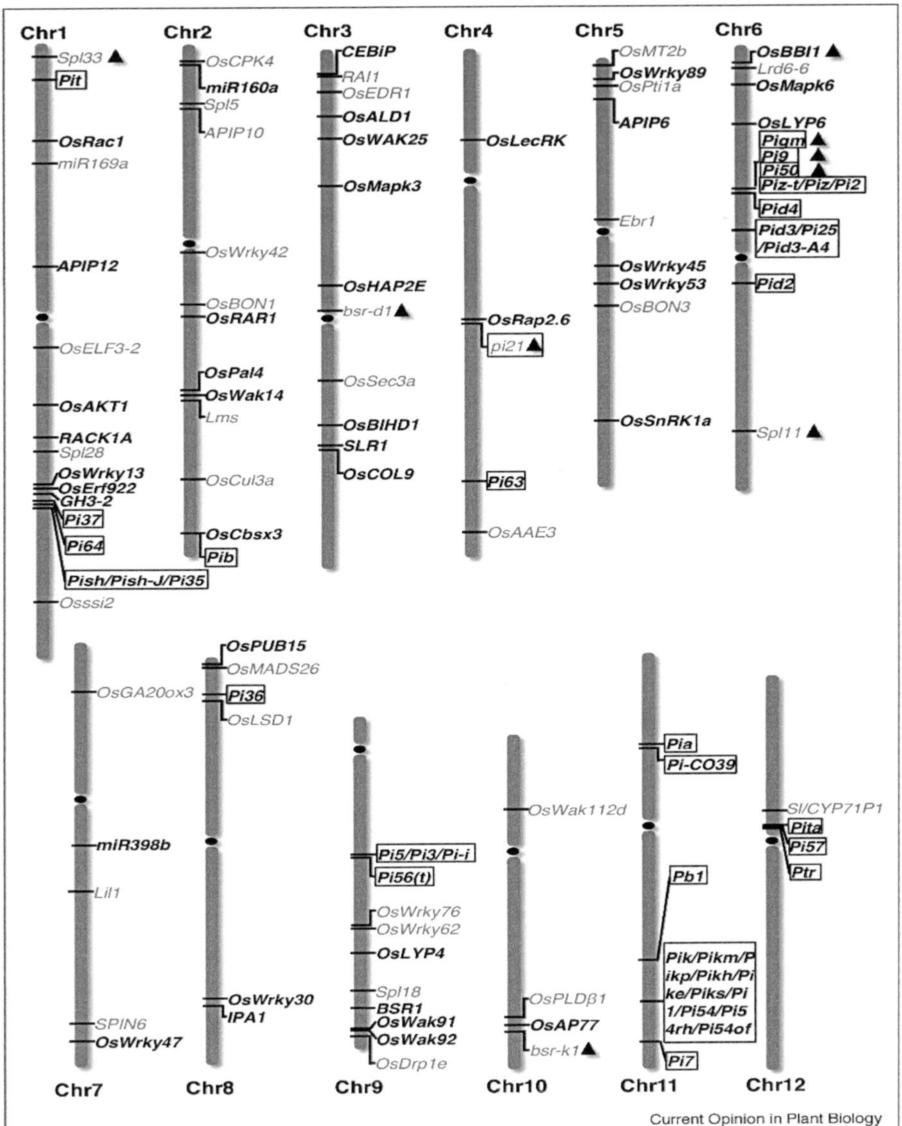

FIGURE 5.1 Association mapping of blast R genes in rice plants.

Source: Reprinted with permission from Li et al., 2019. © 2019 The Authors via Elsevier.

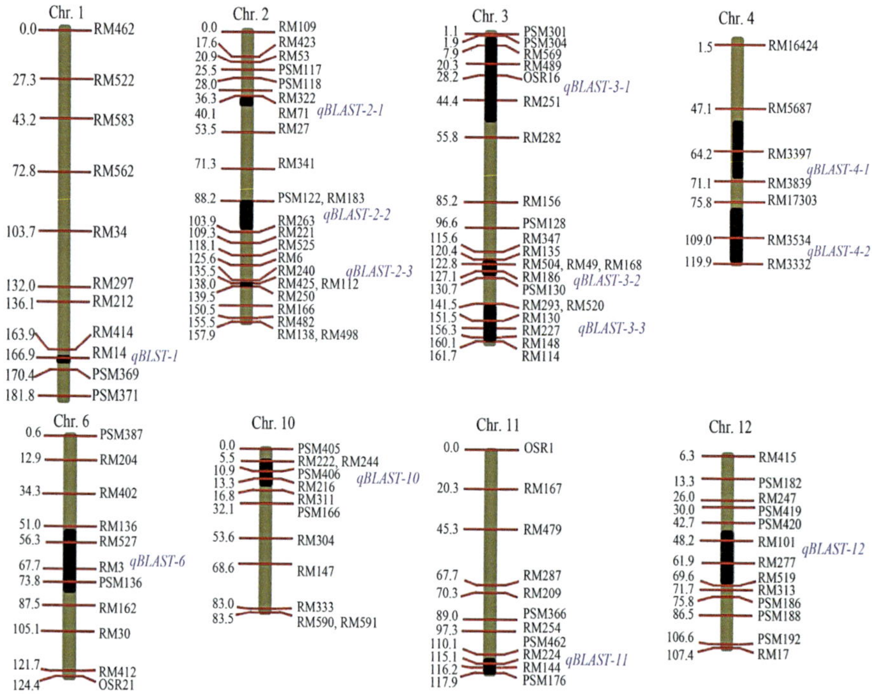

FIGURE 5.2 Linkage mapping of blast R genes in rice plants.

Source: Reprinted from Srivastava et al. (2017) as modified from Zhang et al, 2012. © 2017 Elsevier.

resource of potential discrete alleles, the elite germplasm (Cho et al., 2007; Roy Chowdhury et al., 2012a; Roy Chowdhury et al., 2012b). The blast R genes would assist in the precise controlling of blast disease in rice by the proper use of DNA markers across differential blast species in the identification of specific R genes in diverse elite germplasm (Roy Chowdhury et al., 2012a).

5.3 COMPUTATIONAL ANALYSIS OF BLAST RESISTANCE GENES

Out of the numerous studies that have explored and deciphered the blast R genes and the RPs, a recent study conducted using the seeds of resistant BR2655 and, susceptible HR1 2 rice cultivars performed *in silico* modeling and characterized eight blast RPs (RP1-RP8) using rice reference genome RGAP7, to understand the interaction between the eight RPs and *Avr* effectors (Chandrakanth et al., 2020).

5.3.1 GENOTYPING AND LINKAGE ANALYSIS AND MAPPING

Genotyping was performed for 161 landraces of rice cultivars using twenty-eight functional or linked markers specific to screen the twenty-four blast R genes, *viz., Pib, Pb1, Pi9, Pizt, Pid2, Pi33, Pi36, Pi37, Pikm, Pit, Pita, Pi2, Pi5, Pi54, Pish, Pik, Pikp, Pia, Pi25, Pi1, pi21, Pi56, Pi65(t), and Piz* (Yadav et al., 2019).

5.3.2 RETRIEVAL AND ANALYSIS OF AMINO ACID SEQUENCE

Post amino acid retrieval and analysis (Margelevičius and Venclovas, 2005), the eight RPs, obtained from the resistant BR2655 and susceptible HR1 2 rice cultivars were found to be leucine-rich as demonstrated by the presence of leucine-rich repeats (Chandrakanth et al., 2020).

5.3.3 PHYSICO-CHEMICAL PROPERTIES

The physico-chemical parameters such as molecular weight, pI, amino acid composition, estimated half-life, and instability index can be analyzed using ExPASy Prot. Param (Margelevičius and Venclovas, 2005). The eight RPs obtained from the resistant BR2655 and susceptible HR1 2 rice cultivars were found to exhibit high thermostability based on the high aliphatic index values (Ikai, 1980; Chandrakanth et al., 2020), while the isoelectric points identified their charged potential, such that RP1-RP8 exhibited acidic charges except for RP3 and RP8 (Chandrakanth et al., 2020).

5.3.4 HOMOLOGY MODELING

SWISS-MODEL (Schwede et al., 2003) and modeler 9.22 (Sali and Blundell, 1993) are regularly used to predict the 3D structure of proteins. The eight RPs obtained from the resistant BR2655 and susceptible HR1 2 rice cultivars were modeled using SWISS-MODEL to predict 3D protein structures, which were visualized using UCSF Chimera (Pettersen et al., 2004). SWISS-MODEL servers showed that the eight RPs exhibited a 26–65% query coverage while sharing a low sequence identity of 12.84–18.85% with a RMSD of 1.74 Å–4.50 Å (Chandrakanth et al., 2020). For each RP, the four best templates (sourced from PDB) from the PSI-BLAST (Margelevičius and Venclovas, 2005) output were selected to build the best possible model using modeler 9.22 server, chosen on the basis of Discrete Optimized Protein Energy (DOPE) score (Chandrakanth et al., 2020).

5.3.5 SECONDARY STRUCTURE PREDICTION

The secondary structures predicted based on fold matching, surface cleft analysis, residue conservation, and functional 3D templates using ProFunc servers (Laskowski et al., 2005) indicated the predominant presence of α-helical coils. For the eight RPs obtained from the resistant BR2655 and susceptible HR1 2 rice cultivars, RaptorX (Kalberg et al., 2012) predicted the distribution of structures majorly composed of coils followed by α-helical coils with a small portion of β-strands, and indicating the presence of exposed, medium, and buried residues (Chandrakanth et al., 2020). I-TASSER prediction-based COACH server was quite effective in predicting the active sites (Sharma et al., 2012; Chandrakanth et al., 2020). Based on RMSD, SuperPose successfully provided an estimate of the extent of deviation among different RPs such that the superimposed structures of RP1-4 and RP6-7 showed identical 3D conformations with the exception of RP5 (Chandrakanth et al., 2020).

5.3.6 CONSERVED MOTIF STRUCTURES AND PHYLOGENETIC ANALYSIS OF RP PROTEINS

For eight RPs procured from the resistant BR2655 and susceptible HR1 2 rice cultivars, the sequence similarity of eight RPs was assessed counter to the non-redundant PDB database using PSI-BLAST to identify the protein family. The domains and motifs were then analyzed by CDvist webserver against an Interpro and Pfam 3.0 scan server (Adebali et al., 2014; Kanazin et al., 1996; Chandrakanth et al., 2020). Interpro scan showcased the NB-ARC and LRR domain architectures across eight RPs, P-loop architecture was noticed in RP1-4, RP6-7, while RX-CC domains were seen in RP4 and RP6-8. Domain analysis by CDvist showed the presence of NB-ARC domains across RP1-RP8, while the LRR domain was revealed in RP1 and RP4-7, and a RX-CC domain was found in RP4 and RP6-8 (Chandrakanth et al., 2020).

Post amino acid sequence retrieval from RGAP, sequence diversity was discovered through phylogenetic studies suggesting two clades. The eight RP sequences were evaluated for sequence homology using CLUSTALW, and further phylogenetic studies were conducted to study the evolutionary relationship between reported blast R genes such as *Pita, Pik2, Pib, Pid3, Pi54, Pik1, RGA5, RGA4,* and the eight RPs considered in their study (Chandrakanth et al., 2020).

5.3.7 VALIDATION AND STRUCTURE AND FUNCTION ANALYSIS

ProFunc predicted the biological process, cellular components, and biochemical functions based on 3D structure for all RPs (Brindha et al., 2011). SAVeS Server (https,//services.mbi.ucla.edu/SAVES/) (Mondal et al., 2017) and SuperPose (http,//wishart.biology.ualberta.ca/SuperPose/) are commonly employed to assess the quality of predicted 3D protein structure conformations (Zhang, 2008; Maiti et al., 2004). Ramachandran plot, one of the popular validation scoring statistics, shows the distribution of φ and ψ angles. For the eight RPs acquired from the resistant BR2655 and susceptible HR1 2 rice cultivars, the Ramachandran plot established their amino acid residues that followed dihedral angle conformations and thereby occupying favorable positions (Schwede et al., 2003).

5.3.8 ASSOCIATION ANALYSIS

The general linear model (GLM) was run *via* TASSEL5 software across thousand permutations of available data to test the hypothesis on the genetic association between the blast disease and resistance genes (Bradbury et al., 2007), and substantiated by a $p < 0.05$ for the permutations noted for the most substantial polymorphism in a given region of interest (Fig. 5.1). A linkage map (Fig. 5.2) constructed to identify the location of blast resistance genes and quantitative trait loci from the 'Tianjingyeshengdao' TY/'CO39' recombinant inbred line population has been explored by Wang et al. (2012) and Srivastava et al. (2017). Differences in profiling gene expression of resistance genes in HR12 (S) and BR2655 (R) rice cultivars subjected to *M. oryzae* information gave 69.7 and 75.8 million raw reads, respectively, leading to the identification of 4290 and 7577 differentially expressed genes (DEG), respectively. Keyword search for 'Resistance' gave 38 and 36 transcripts in HR12 and BR2655 rice cultivars, respectively, while "LRR" search resulted in shortlisting of the topmost 17 and 22 transcripts, up regulated in HR 12 and BR2655 cultivars, respectively (Chandrakanth et al., 2020; Yadav et al., 2019). The phenotypic, expression, and structural analysis of the *Pi54* gene and its orthologs phylogeny and alignment of amino acid residues from ORF of *Pi54, Pi54of, and Pi54rh* orthologs, along with the prediction of functional domains and tertiary structures of *Pi54of*, have been well-reviewed by Devanna et al., 2014.

5.4 CONCLUSION

Blast disease, being one of the most destructive diseases in rice plants, conventional measures of controlling blast disease need to be fortified with genetic tools and computation technology to breed resistant cultivars. Thus, the advent and significant advancement in rice genomics has provided valuable scope and tools like gene rotation, association mapping, linkage mapping, allele mining, cloning of R genes, and study of R gene products, gene pyramiding, multiline varieties have offered an opportunity to explore and understand host–pathogen interactions and thereby develop suitable means of controlling blast disease through stacking of multiple R genes, or combinations among others. Though quite an advancement has been made in improving blast resistance in rice cultivars, the quest for developing sustainable blast resistance rice cultivars would be an ongoing battle due to the co-evolving nature of the pathogen.

KEYWORDS

- **rice**
- **blast disease**
- ***Magnaporthe oryzae***
- **NBS-LRR-like proteins**
- **plant disease resistance**
- **genes**

REFERENCES

Adebali, O.; Ortega, D. R.; Zhulin, I. B. CDvist, A Web Server for Identification and Visualization of Conserved Domains in Protein Sequences. *Bioinformatics* **2014**, *31*, 1475–1477

Ashikawa, I.; Hayashi, N.; Yamane, H.; Kanamori, H.; Wu, J.; Matsumoto, T.; Ono, K.; Yano, M. Two Adjacent Nucleotide-binding Site-leucine-rich Repeat Class Genes are Required to Confer Pikm-specific Rice Blast Resistance. *Genetics* **2008,** *180*, 2267–2276.

Berruyer, R.; Adreit, H.; Milazzo, J.; Gaillard, S.; Berger, A.; Dioh, W.; Lebrun, M. H.; Tharreau, D. Identification and Fine Mapping of Pi33, the Rice Resistance Gene Corresponding to the *Magnaporthe grisea* avirulence Gene ACE1. *Theor. Appl. Genet.* **2003,** *107*, 1139–1147.

Beser, N.; Valle, M. M. D.; Kim, S. M.; Vinarao, R. B.; Sürek, H.; Jena, K. K. Marker-assisted Introgression of a Broad-spectrum Resistance Gene, Pi40 Improved Blast Resistance of Two Elite Rice (*Oryza sativa* L.) Cultivars of Turkey. *Mol. Plant Breed.* **2016,** *7*, 1–15.

Bonman, J. Durable Resistance to Rice Blast Disease–Environmental Influences. *Euphytica* **1992,** *63*, 115–123.

Brar, D.; Khush, G. Alien Introgression in Rice. *Plant Mol. Biol.* **1997,** *35*, 35–47.

Brindha, S.; Sailo, S.; Chhakchhuak, L.; Kalita, P.; Gurusubramanian, G.; Kumar, J. N. S. Protein 3D Structure Determination Using Homology Modelling and Structure Analysis. *Sci. Vis.* **2011,** *11*, 125–133.

Bradbury, P. J.; Zhang, Z.; Kroon, D. E.; Casstevens, T. M.; Ramdoss, Y.; Buckler, E. S. TASSEL: Software for Association Mapping of Complex Traits in Diverse Samples. *Bioinformatics* **2007,** *23*, 2633–2635.

Bryan, G. T.; Wu, K. S.; Farrall, L.; Jia, Y.; Hershey, H. P.; McAdams, S. A.; Faulk, K. N.; Donaldson, G. K.; Tarchini, R.; Valent, B. A. Single Amino Acid Difference Distinguishes Resistant and Susceptible Alleles of the Rice Blast Resistance Gene Pi-ta. *Plant Cell.* **2000,** *12*, 2033–2046.

Cesari, S.; Thilliez, G.; Ribot, C.; Chalvon, V.; Michel, C.; Jauneau, A.; Rivas, S.; Alaux, L.; Kanzaki, H.; Okuyama, Y.; Morel, J. B.; Fournier, E.; Tharreau, D.; Terauchi, R.; Kroj, T. The Rice Resistance Protein Pair RGA4/RGA5 Recognizes the *Magnaporthe oryzae* Effectors AVR-Pia and AVR1-CO39 by direct binding. *Plant Cell* **2013,** *25*, 1463–1481.

Chandrakanth, R.; Sunil, L.; Sadashivaiah, L.; Devaki, N. S. In Silico Modelling and Characterization of Eight Blast Resistance Proteins in Resistant and Susceptible Rice Cultivars. *J. Genet. Eng. Biotechnol.* **2020,** *18*, 75.

Chen, J.; Peng, P.; Tian, J. S.; He, Y. G.; Zhang, L. P.; Liu, Z. X.; Yin, D. D.; Zhang, Z. H. Pike. A Rice Blast Resistance Allele Consisting of Two Adjacent NBS-LRR Genes, was Identified as a Novel Allele at the Pik Locus. *Mol. Breeding* **2015,** *35*, 117.

Chen, J.; Shi, Y. F.; Liu, W. Z.; Chai, R. Y.; Fu, Y. P.; Zhuang, J. Y.; Wu, J. L. A Pid3 Allele from Rice Cultivar Gumei 2 Confers Resistance to *Magnaporthe oryzae*. *J. Genet. Genom.* **2011,** *38*, 209–216.

Chen, X.; Shang, J.; Chen, D.; Lei, C.; Zou, Y.; Zhai, W.; Liu, G.; Xu, J.; Ling, Z.; Cao, G.; Ma, B.; Wang, Y.; Zhao, X.; Li, S.; Zhu, L. AB-Lectin Receptor Kinase Gene Conferring Rice Blast Resistance. *Plant J.* **2006,** *46*, 794–804.

Cho, Y. ; Kwon, S.; Choi, I.; Lee, S.; Jeon, J.; Oh, M.; Roh, J.; Hwang, H.; Yang, S.; Kim, Y. Identification of Major Blast Resistance Genes in Korean Rice Varieties (*Oryza sativa* L.) Using Molecular Markers. *J. Crop Sci. Biotech.* **2007,** *10*, 265–276.

Dai, X. J.; He, C.; Zhou, L.; Liang, M. Z.; Fu, X. C.; Qin, P.; Yang, Y. Z.; Chen, L. B. Identification of a Specific Molecular Marker for the Rice Blast-Resistant Gene *Pigm* and Molecular Breeding of Thermo-sensitive Genic Male Sterile Leaf-color Marker Lines. *Mol. Breeding* **2018,** *38*, 72.

Dean, R. A.; Talbot, N. J.; Ebbole, D. J.; Farman, M. L.; Mitchell, T. K.; Orbach, M. J.; Thon, M.; Kulkarni, R.; Xu, J. R.; Pan, H.; Read, N. D.; Lee, Y. H.; Carbone, I.; Brown, D.; Oh, Y. Y.; Donofrio, N.; Jeong, J. S.; Soanes, D. M.; Djonovic, S.; Kolomiets, E.; Rehmeyer, C.; Li, W.; Harding, M.; Kim, S.; Lebrun, M. H.; Bohnert, H.; Coughlan, S.; Butler, J.; Calvo, S.; Ma, L. J.; Nicol, R.; Purcell, S.; Nusbaum, C.; Galagan, J. E.; Birren, B. W. The Genome Sequence of the Rice Blast Fungus *Magnaporthe grisea*. *Nature* **2005,** *434*, 980–986.

Dean, R.; Van Kan, J. A.; Pretorius, Z. A.; Hammond-Kosack, K. E.; Di Pietro, A.; Spanu, P. D.; Rudd, J. J.; Dickman, M.; Kahmann, R.; Ellis, J.; Foster, G. D. The Top 10 Fungal Pathogens in Molecular Plant Pathology. *Mol. Plant Pathol.* **2012,** *13*, 414–430.

Deng, Y. W.; Zhai, K. R.; Xie, Z.; Yang, D. Y.; Zhu, X. D.; Liu, J. Z.; Wang, X.; Qin, P.; Yang, Y. Z.; Zhang, G. M.; Li, Q.; Zhang, J. F.; Wu, S. Q.; Milazzo, J.; Mao, B. Z.; Wang, E. T.; Xie, H. A.; Tharreau, D.; He, Z. H. Epigenetic Regulation of Antagonistic Receptors Confers Rice Blast Resistance with Yield Balance. *Science* **2017,** *355*, 962–965.

Devanna, N. B.; Vijayan, J.; Sharma, T. R. The Blast Resistance Gene Pi54 of Cloned from *Oryza officinalis* Interacts with Avr-Pi54 Through its Novel Non-LRR Domains. *PLoS One* **2014,** *9*, e104840.

DeYoung, B. J.; Innes, R. W. Plant NBS-LRR Proteins in Pathogen Sensing and Host Defense. *Nat. Immunol.* **2006,** *7*, 1243.

Divya, B.; Robin, S.; Rabindran, R.; Senthil, S.; Raveendran, M.; Joel, A. J. Marker-assisted Backcross Breeding Approach to Improve Blast Resistance in Indian Rice (*Oryza sativa*) Variety ADT43. *Euphytica*. **2014,** *200*, 61–77.

Doerks, T.; Mering, C. V.; Bork, P. Functional Clues for Hypothetical Proteins Based on Genomic Context Analysis in Prokaryotes. *Nucleic Acids Res.* **2004,** *32*, 6321–6326.

Earl, D. A.; Von, H. B. M. Structure Harvester, A Website and Program for Visualizing Structure Output and Implementing the Evanno Method. *Conserv. Genet. Resour.* **2012,** *4*, 359–361.

Ellur, R. K.; Khanna, A.; Yadav, A.; Pathania, S.; Rajashekara, H.; Singh, V. K.; Krishnan, G. S.; Bhowmick, P. K.; Nagarajan, M.; Vinod, K. K.; Prakash, G.; Mondal, K. K.; Singh, K.; Prabhu, K. V.; Singh, A. K. Improvement of Basmati Rice Varieties for Resistance to Blast and Bacterial Blight Diseases Using Marker-assisted Backcross Breeding. *Plant Sci.* **2016,** *242*, 330–341.

Evanno, G.; Regnaut, S.; Goudet, J. Detecting the Number of Clusters of Individuals Using the Software STRUCTURE, a Simulation Study. *Mol. Ecol.* **2005,** *14*, 2611–2620.

Fuentes, J. L.; Correa–Victoria, F. J.; Escobar, F.; Prado, G.; Aricapa, G.; Duque, M. C.; Tohme, J. Identification of Microsatellite Markers Linked to the Blast Resistance Gene Pi–1(t) in Rice. *Euphytica* **2008,** *160*, 295– 304.

Fukuoka, S.; Mizobuchi, R.; Saka, N.; Ivan, S.; Matsumoto, T.; Okuno, K.; Yano, M. A Multiple Gene Complex on Rice Chromosome 4 is Involved in Durable Resistance to Rice Blast. *Theor. Appl. Genet.* **2012,** *125*, 551–559.

Fukuoka, S.; Yamamoto, S.; Mizobuchi, R.; Yamanouchi, U.; Ono, K.; Kitazawa, N.; Yasuda, N.; Fujita, Y.; Nguyen, T. T. T.; Koizumi, S.; Sugimoto, K.; Matsumoto, T.; Yano, M. Multiple Functional Polymorphisms in a Single Disease Resistance Gene in Rice Enhance Durable Resistance to Blast. *Sci. Rep.* **2014,** *4*, 4550.

Fukuoka, S.; Saka, N.; Koga, H.; Ono, K.; Shimizu, T.; Ebana, K.; Hayashi, N.; Takahashi, A.; Hirochika, H.; Okuno, K.; Yano M. Loss of Function of a Proline-containing Protein Confers Durable Disease Resistance in Rice. *Science* **2009,** *325*, 998–1001.

Gao, L. J.; Deng, G. F.; Gao, H. L.; Gao, G. Q.; Zhou, M.; Zhou, W. Y.; Yan, Q.; Zhang, J.; Dai, G. X. Establishment and Application of Gene Tagging Linked to Rice Blast Resistance Gene *Pi-d2*. *Southwest China J. Agric. Sci.* **2010,** *23*, 77–82.

Gouda, P. K.; Saikumar, S.; Varma, C. M. K.; Nagesh, K.; Thippeswamy, S.; Shenoy, V.; Ramesha, M. S.; Shashidhar, H. E. Marker-assisted Breeding of Pi-1 and Piz-5 Genes Imparting Resistance to Rice Blast in PRR78, Restorer Line of Pusa RH-10 Basmati Rice Hybrid. *Plant Breeding* **2013,** *132*, 61–69.

Gururani, M.; Venkatesh, J.; Upadhyaya, C. P.; Nookaraju, A.; Pandey, S. K.; Park, S. W. Plant Disease Resistance Genes: Current Status and Future Directions. *Physiol. Mol. Plant Pathol.* **2012,** *78*, 51–65.

Hayashi, K.; Yoshida, H. Refunctionalization of the Ancient Rice Blast Disease Resistance Gene Pit by the Recruitment of a Retrotransposon as a Promoter. *Plant J.* **2009,** *57*, 413–425.

Hayashi, K.; Yoshida, H.; Ashikawa, I. Development of PCR Based Allele-specific and InDel Marker Sets for Nine Rice Blast Resistance Genes. *Theor. Appl. Genet.* **2006,** *113*, 251–260.

Hayashi, N.; Inoue, H.; Kato, T.; Funao, T.; Shirota, M.; Shimizu, T.; Kanamori, H.; Yamane, H.; Hayano-Saito, Y.; Matsumoto, T.; Yano, M.; Takatsuji, H. Durable Panicle Blast-resistance Gene *Pb1* Encodes an Atypical CC-NBS-LRR Protein and was Generated by Acquiring a Promoter Through Local Genome Duplication. *Plant J.* **2010,** *64*, 498–510.

Hittalmani, S.; Parco, A.; Mew, T. V.; Zeigler, R. S.; Huang, N. Fine Mapping and DNA Marker-assisted Pyramiding of the Three Major Genes for Blast Resistance in Rice. *Theor. Appl. Genet.* **2000,** *100*,1121–1128.

Hua, L. X.; Wu, J. Z.; Chen, C. X.; Wu, W. H.; He, X. Y.; Lin, F.; Wang, L.; Ashikawa, I.; Matsumoto, T.; Wang, L.; Pan Q. The Isolation of Pi1, an Allele at thPik Locus which Confers Broad Spectrum Resistance to Rice Blast. *Theor. Appl. Genet.* **2012,** *125*, 1047–1055.

Huang, Y. L.; Yan, Z.; Wang, H.; Shen, G. L.; Zhang, C. H. Directed Improvement of Rice Blast Resistance of Sterile Line Q211S with Molecular Marker-assisted selection. *Chin. Agric. Sci. Bull.* **2018,** *34*, 135–140.

Ikai, A. Thermostability and Aliphatic Index of Globular Proteins. *J. Biochem.* **1980,** *88*, 1895–1898.

Ingole, K. D.; Prashanthi, S. K.; Krishnaraj, P. U. Mining for Major Blast Resistance Genes in Rice Landraces of Karnataka. *Ind. J. Genet. Plant Breeding* **2014,** *74*, 378–383.

Inukai, T.; Nagashima, S.; Kato, M. *Pid3-I1* is a Race-specific Partial-resistance Allele at the *Pid3* Blast Resistance Locus in Rice. *Theor. Appl. Genet.* **2019,** *132*, 395–404.

Jeon, J. S.; Chen, D.; Yi, G. H.; Wang, G. L.; Ronald, P. C. Genetic and Physical Mapping of *Pi5*(t), a Locus Associated with Broad–Spectrum Resistance to Rice Blast. *Mol. Genet. Gen.* **2003,** *269*, 280–289.

Jiang, H.; Feng, Y.; Bao, L.; Li, X.; Gao, G.; Zhang, Q.; Xiao, J.; Xu, C.; He, Y. Improving Blast Resistance of Jin 23B and its Hybrid Rice by Marker-assisted Gene Pyramiding. *Mol. Breeding* **2012,** *30*, 1679–1688.

Jiang, J. F.; Mou, T.; Yu, H. H.; Zhou, F. S. Molecular Breeding of Thermo-sensitive Genic Male Sterile (TGMS) Lines of Rice for Blast Resistance Using *Pi2* Gene. *Rice* **2015,** *8*, 11.

Kallberg, M.; Wang, H.; Wang, S.; Peng, J.; Wang, Z.; Lu, H.; Xu, J. Template-Based Protein Structure Modeling Using the RaptorX Web Server. *Nat. Protoc.* **2012,** *7*, 1511–1522.

Kanazin, V.; Marek, L. F.; Shoemaker, R. C. Resistance Gene Analogues are Conserved and Clustered in Soybean. *Proc. Natl. Acad. Sci. USA* **1996,** *93*, 11746–11750.

Khan, G. H.; Shikari, A. B.; Vaishnavi, R.; Najeeb, S.; Padder, B. A.; Bhat, Z. A.; Parray, G. A.; Bhat, M. A.; Kumar, R.; Singh, N. K. Marker-assisted Introgression of Three Dominant Blast Resistance Genes into an Aromatic Rice Cultivar MushkBudji. *Sci. Rep.* **2018,** *8*, 4091.

Khanna, A.; Sharma, V.; Ellur, R. K.; Shikari, A. B.; Krishnan, G. S.; Singh, U. D.; Prakash, G.; Sharma, T. R.; Rathour, R.; Variar, M.; Prashanthi, S. K.; Nagarajan, M.; Vinod, K K.; Bhowmick, P. K.; Rajashekhara, H.; Singh, N. K.; Prabhu, K. V.; Singh, A. K. Marker-assisted Pyramiding of Major Blast Resistance Genes *Pi9* and *Pita* in the Genetic

Background of an Elite Basmati Rice Variety, Pusa Basmati 1. *Ind. J. Genet.* **2015,** *75*, 417–425.

Kumar, S. V.; Prasad M. S.; Rambabu, R.; Madhavi, K. R.; Bhaskar, B.; Kumar V. A.; Sundaram R. M.; Satya A. K.; Madhav M. S.; Prakasam, V. Marker-assisted Introgression of *Pi-1* Gene Conferring Resistance to Rice Blast Pathogen *Pyricularia oryzae* in the Background of Samba Mahsuri. *Int. J. Curr. Microbiol. App. Sci.* **2019,** *8*, 2133–2146.

Kumar, S. V.; Rambabu, R.; Bhaskar, B.; Madhavi, K. R.; Srikanth, S.; Prakasam, V.; Sundaram, R. M.; Madhav, M. S.; Rao, L. V. S.; Prasad, M S. Introgression of Durable Blast Resistance Gene Pi-54 into Indica Rice cv. Samba Mahsuri, through Marker-assisted Backcross Breeding. *Electr. J. Plant Breeding* **2018,** *9*, 705–715.

Laskowski, R. A.; Watson, J. D.; Thornton, J. M. ProFunc: A Server for Predicting Protein Function from 3D Structure. *Nucleic Acids Res.* **2005,** *33*, W89–W93.

Lee, S. K.; Song, M. Y.; Seo, Y. S.; Kim, H. K.; Ko, S.; Cao, P. J.; Suh, J. P.; Yi, G.; Roh, J. H.; Lee, S.; An, G.; Hahn, T. R.; Wang, G. L.; Ronald, P.; Jeon, J. S. Rice Pi5-mediated Resistance to *Magnaporthe oryzae* Requires the Presence of Two Coiled-coil-nucleotide-binding-leucine-rich Repeat Genes. *Genetics* **2009,** *181*, 1627–1638.

Li, W.; Chern, M.; Yin, J.; Wang, J.; Chen, X. Recent Advances in Broad-spectrum Resistance to the Rice Blast Disease. *Curr. Opin. Plant Biol.* **2019,** *50*, 114–120.

Lin, F.; Chen, S.; Que, Z.; Wang, L.; Liu, X.; Pan, Q. The Blast Resistance Gene Pi37 Encodes a Nucleotide-binding Site Leucine-rich Repeat Protein and is a Member of a Resistance Gene Cluster on Rice Chromosome 1. *Genetics* **2007,** *177*, 1871–1880.

Liu, J.; Wang, X.; Mitchell, T.; Hu, Y.; Liu, X.; Dai, L.; Wang, G. L. Recent Progress and Understanding of the Molecular Mechanisms of the Rice – *Magnaporthe oryzae* Interaction. *Mol. Plant Pathol.* **2010,** *11*, 419–427.

Liu, X. Q.; Lin, F.; Wang, L.; Pan, Q. The in silico Map-based Cloning of Pi36, a Rice Coiled-coil Nucleotide-binding Site Leucine-rich Repeat Gene that Confers Race-specific Resistance to the Blast Fungus. *Genetics* **2007,** *176*, 2541–2549.

Liu, X. Q.; Wang, L.; Chen, S.; Lin, F.; Pan, Q. Genetic and Physical Mapping of Pi36(t), A Novel Rice Blast Resistance Gene Located on Rice Chromosome 8. *Mol. Genet. Genomics* **2005,** *274*, 394–401.

Liu, Y.; Liu, B.; Zhu, X.; Yang, J.; Bordeos, A.; Wang, G.; Leach, J. E.; Leung, H. Fine–mapping and Molecular Marker Development for Pi56(t), A NBS–LRR Gene Conferring Broad Spectrum Resistance to *Magnaporthe oryzae* in Rice. *Theor. Appl. Genet.* **2013,** *126*, 985–998.

Ma, J.; Lei, C.; Xu, X.; Hao, K.; Wang, J.; Cheng, Z.; Ma, X.; Ma, J.; Zhou, K.; Zhang, X.; Guo, X.; Wu, F.; Lin, Q.; Wang, C.; Zhai, H.; Wang, H.; Wan, J. Pi64, Encoding a Novel CC-NBS-LRR Protein, Confers Resistance to Leaf and Neck Blast in Rice. *Mol. Plant-Microbe Interact.* **2015,** *28*, 558–568.

Maiti, R.; Van Domselaar G. H.; Zhang, H.; Wishart, D. S. SuperPose: A Simple Server for Sophisticated Structural Superposition. *Nucleic Acids Res.* **2004,** *32*, 590–594.

Margelevičius, M.; Venclovas Č. PSI-BLAST-ISS, an Intermediate Sequence Search Tool for Estimation of the Position-specific Alignment Reliability. *BMC Bioinform.* **2005,** 6, 185.

McHale, L.; Tan, X.; Koehl, P.; Michelmore, R. W. Plant NBS-LRR Proteins, Adaptable Guards. *Genome Biol.* **2006,** *7*, 212.

Mondal, D.; Ghosh, A.; Roy, D.; Kumar, A.; Shamurailatpam, D.; Bera, S.; Ghosh, R.; Bandopadhyay, P.; Majumder, A. Yield Loss Assessment of Rice (*Oryza sativa* L.) Due to Different Biotic Stresses Under System of Rice Intensification (SRI). *J. Entomol. Zool. Studies* **2017,** *5*, 1974–1980.

Ngangkham, U.; Parida, S. K.; De, S.; Kumar, K. A. R.; Singh, A. K.; Singh, N. K.; Mohapatra, T. Genic Markers for Wild Abortive (WA) Cytoplasm Based Male Sterility and its Fertility Restoration in Rice. *Mol. Breeding* **2010,** *26*, 275–292.

Okuyama, Y.; Kanzaki, H.; Abe, A.; Yoshida, K.; Tamiru, M.; Saitoh, H.; Fujibe, T.; Matsumura, H.; Shenton, M.; Galam, D. C.; Undan, J.; Ito, A.; Sone, T.; Terauchi, R. A. Multifaceted Genomics Approach Allows the Isolation of the Rice Pia-Blast Resistance Gene Consisting of two Adjacent NBS-LRR Protein Genes. *Plant J.* **2011,** *66*, 467–479.

Pettersen, E. F.; Goddard, T. D.; Huang, C. C.; Couch, G. S.; Greenblatt, D. M.; Meng, E. C.; Ferrin, T. E. UCSF Chimera - a Visualization System for Exploratory Research and Analysis. *J Comput Chem.* **2004,** *25*, 1605–1612.

Pritchard, J. K.; Stephens M.; Donnelly P. Inference of Population Structure Using Multilocus Genotype Data. *Genetics* **2000,** *155*,945–959.

Qu, S.; Liu, G.; Zhou, B.; Bellizzi, M.; Zeng, L.; Dai, L.; Han, B.; Wang, G. L. The Broad-Spectrum Blast Resistance Gene *Pi9b* Encodes an NBS–LRR Protein and is a Member of a multigene family in rice. *Genetics* **2006,** *172*, 1901–1914.

Roy Chowdhury, M.; Jia, Y.; Jackson, A.; Jia, M. H.; Fjellstrom, R.; Cartwright, R. D. Analysis of Rice Blast Resistance Gene Pi–z in Rice Germplasm Using Pathogenicity Assays and DNA Markers. *Euphytica* **2012a,** *184*, 35–46.

Roy Chowdhury, M.; Jia, Y.; Jia, M. H.; Fjellstrom, R.; Cartwright, R. D. Identification of the Rice Blast Resistance Gene Pi–b in the National Small Grains Collection. *Phytopathol.* **2012b,** *102*, 700–706.

Sali, A.; Blundell, T. L. Comparative Protein Modelling by Satisfaction of Spatial Restraints. *J. Mol. Biol.* **1993,** *234*, 779–815.

Schwede, T.; Kopp, J.; Guex, N.; Peitsch, M. C. SWISS-MODEL: An Automated Protein Homology-modeling Server. *Nucleic Acids Res.* **2003,** *31*, 3381–3385.

Shang, J. J.; Tao, Y.; Chen, X. W.; Zou, Y.; Lei, C. L.; Wang, J.; Li, X. B.; Zhao, X. F.; Zhang, M. J.; Lu, Z. K.; Xu, J. C.; Cheng, Z. K.; Wan, J. M.; Zhu, L. H. Identification of a New Rice Blast Resistance Gene, Pid3, by Genome-wide Comparison of Paired Nucleotide-binding Site-leucine-rich Repeat Genes and their Pseudogene Alleles, between the Two Sequenced Rice Genomes. *Genetics* **2009,** *182*, 1303–1311.

Sharma, K.; Antunes, I. L.; Rajulapati, V.; Goyal, A. Low-resolution SAXS and Comparative Modeling-based Structure Analysis of Endo-β-1, 4-xylanase a Family 10 Glycosides Hydrolase from *Pseudopedobacter saltans* comb. nov. *Int. J. Biol. Macromol.* **2018,** 112, 1104–1114.

Sharma, T. R.; Rai, A. K.; Gupta, S. K.; Vijayan, J.; Devanna, B. N.; Ray, S. Rice Blast Management through Host–Plant Resistance, Retrospect and Prospects. *Agric. Res.* **2012,** *1*, 37–52.

Singh, V. K.; Singh, A.; Singh, S. P.; Ellur, R. K.; Choudhary, V.; Sarkel, S.; Singh, D.; Krishnan, S. G.; Nagarajan, M.; Vinod, K. K.; Singh, U. D.; Rathore, R.; Prashanthi, S. K.; Agrawal, P. K.; Bhatt, J. C.; Mohapatra, T.; Prabhu, K. V.; Singh, A. K. Incorporation of Blast Resistance into μPRR78, an Elite Basmati Rice Restorer Line, through Marker-Assisted Backcross Breeding. *Field Crops Res.* **2012,** *128*, 8–16.

Srivastava, D.; Shamim, M.; Kumar, M.; Mishra, A.; Pandey, P.; Kumar, D.; Yadav, P.; Siddiqui M. S.; Singh, K. N. Current Status of Conventional and Molecular Interventions for Blast Resistance in Rice. *Rice Science* **2017,** *24*, 299–321.

Su, J.; Wang, W.; Han, J.; Chen, S.; Wang, C.; Zeng, L.; Feng, A.; Yang, J.; Zhou, B.; Zhu, X. Functional Divergence of Duplicated Genes Results in a Novel Blast Resistance Gene Pi50 at the Pi2/9 Locus. *Theor. Appl. Genet.* **2015,** *128*, 2213–2225.

Takagi, H.; Uemura, A.; Yaegashi, H.; Tamiru, M.; Abe, A.; Mitsuoka, C.; Utsushi, H.; Natsume, S.; Kanzaki, H.; Matsumura, H.; Saitoh, H.; Yoshida, K.; Cano, L. M.; Kamoun, S.; Terauchi, R. MutMap-Gap: Whole-Genome Resequencing of Mutant F2 Progeny Bulk Combined with De Novo Assembly of Gap Region Identifies the Rice Blast Resistance Gene Pii. *New Phytol.* **2013,** *200*, 276–283.

Takahashi, A.; Hayashi, N.; Miyao, A.; Hirohiko Hirochika, H. Unique Features of the Rice Blast Resistance *Pish* Locus Revealed by Large Scale Retrotransposon-Tagging. *BMC Plant Biol.* **2010,** *10*, 175.

Tanksley, S. D.; McCouch, S. R. Seed Banks and Molecular Maps, Unlocking Genetic Potential from the Wild. *Science* **1997,** *277*, 1063–1066.

Tanweer, F. A.; Rafii, M. Y.; Sijam, K.; Rahim, H. A.; Ahmed, F.; Ashkani, S.; Latif, M. A. Introgression of Blast Resistance Genes (putative *Pi-b* and *Pi-kh*) into Elite Rice Cultivar MR219 through Marker-Assisted Selection. *Front. Plant Sci.* **2015,** *6*, 1002.

Tian, H. G.; Cheng, H. Q.; Hu, J.; Lei, C. L.; Zhu, X. D.; Qian, Q. Effect of Introgressed Pigm Gene on Rice Blast Resistance and Yield Traits of Japonica Rice in Cold Area. *J. Shenyang Agric. Univ.* **2016,** *47*, 520–526.

Usatov, A. V.; Kostylev, P. I.; Azarin, K. V.; Markin, N. V.; Makarenko, M. S.; Khachumov, V. A.; Bibov, Y. M. Introgression of the Rice Blast Resistance Genes Pi1, Pi2, and Pi33 into Russian Rice Varieties by Marker-Assisted Selection. *Ind. J. Genet.* **2016,** *76*, 18–23.

Wang, G.; Mackill, D. J.; Bonman, J. M.; McCouch, S. R.; Champoux, M. C.; Nelson, R. J. RFLP Mapping of Genes Conferring Complete and Partial Resistance to Blast in a Durably Resistant Rice Cultivar. *Genetics* **1994,** *136*, 1421–1434.

Wang, Y.; Wang, D.; Deng, X.; Liu, J.; Sun, P.; Liu, Y.; Huang, H.; Jiang, N.; Kang, H.; Ning, Y.; Wang, Z.; Xiao, Y.; Liu, X.; Liu, E.; Dai, L.; Wang, G. Molecular Mapping of the Blast Resistance Genes *Pi2-1* and *Pi51(t)* in the Durably Resistant Rice 'Tianjingyeshengdao'. *Phytopathol.* **2012,** *102*, 779–786.

Warschefsky, E.; Penmetsa, R. V.; Cook, D. R.; von Wettberg, E. J. B. Back to the wilds, tapping Evolutionary Adaptations for Resilient Crops through Systematic Hybridization with Crop Wild Relatives. *Am. J. Bot.* **2014,** *101*, 1791–1800.

Wu, J. L.; Fan, Y. Y.; Li, D. B.; Zheng, K. L.; Leung, H.; Zhuang, J. Y. Genetic Control of Rice Blast Resistance in the Durably Resistant Cultivar Gumei 2 against Multiple Isolates. *Theor. Appl. Gffenet.* **2005,** *111*, 50–56.

Xiao, N.; Wu, Y. Y.; Pan, C. H.; Yu, L.; Chen, Y.; Liu, G. Q.; Li, Y. H.; Zhang, X. X.; Wang, Z. P.; Dai, Z. Y.; Liang, C. Z.; Li, A. H. Improving of Rice Blast Resistances in Japonica by Pyramiding Major R Genes. *Front. Plant Sci.* 2**017,** *7*, 1918.

Xiao, W. M.; Luo, L. X.; Wang, H.; Guo, T.; Liu, Y. Z.; Zhou, J. Y.; Zhu, X. Y.; Yang, Q. Y.; Chen, Z. Q. Pyramiding of Pi46 and Pita to Improve Blast Resistance and to Evaluate the Resistance Effect of the Two R Genes. *J. Integr. Agric.* **2016,** *15*, 2290–2298.

Xing, X.; Liu, X. L.; Chen, H. L.; Yang, F. Y.; Li, Y. C.; Liao, H.; You, L.; Liu, J. L.; Dai, L. Y.; Wang, G. L. Improving blast resistance of rice restorer R288 by molecular marker-assisted selection of Pi9 gene. *Crop Res.* **2016,** *30*, 487–491.

Xu, X.; Hayashi, N.; Wang, C. T.; Fukuoka, S.; Kawasaki, S.; Takatsuji, H.; Jiang, C. J. Rice Blast Resistance Gene Pikahei-1(t), a Member of a Resistance Gene Cluster on

Chromosome 4, Encodes a Nucleotide-binding Site and Leucine-rich Repeat Protein. *Mol. Breeding* **2014,** *34*, 691–700.

Yadav, M. K.; Aravindan, S.; Ngangkham, U.; Shubudhi, H. N.; Bag, M. K.; Adak, T.; Munda, S.; Samantaray, S.; Jena, M. Use of Molecular Markers in Identification and Characterization of Resistance to Rice Blast in India. *PloS One* **2017,** *12*, e0176236.

Yadav, M. K.; Aravindan, S.; Ngangkham, U.; Raghu, S.; Prabhukarthikeyan, S. R.; Keerthana, U.; Marndi, B. C.; Adak, T.; Munda, S.; Deshmukh, R.; Pramesh, D.; Samantaray, S.; Rath, P. C. Blast Resistance in Indian Rice Landraces: Genetic Dissection by Gene-Specific Markers. *PLoS One* **2019,** *14*, e0211061.

Yu, M. M.; Dai, Z. Y.; Pan, C. H.; Chen, X. J.; Yu, L.; Zhang, X. X.; Li, Y. H.; Xiao, N.; Gong, H. B.; Sheng, S. L.; Pan, X. B.; Zhang, H. X.; Li, A. H. Resistance Spectrum Difference between Two Broad-Spectrum Blast Resistance Genes *Pigm* and *Pi2*, and their Interaction Effect on Pi1. *Acta Agron. Sin.* **2013,** *39*, 1927–1934.

Yuan, B.; Zhai, C.; Wang, W. J.; Zeng, X. S.; Xu, X. K.; Hu, H. Q.; Lin, F.; Wang, L.; Pan, Q. The Pik-p Resistance to *Magnaporthe oryzae* in Rice is Mediated by a Pair of Closely Linked CC-NBS-LRR Genes. *Theor. Appl. Genet.* **2011,** *122*, 1017–1028.

Zhai, C.; Lin, F.; Dong, Z.; He, X.; Yuan, B.; Zeng, X.; Wang, L.; Pan, Q. The Isolation and Characterization of Pik, A Rice Blast Resistance Gene Which Emerged After Rice Domestication. *New Phytol.* **2011,** *189*, 321–334.

Zhang, Y. Progress and Challenges in Protein Structure Prediction. *Curr. Opin. Struct. Biol.* **2008,** *18*, 342–348.

Zhang Y. X., Yang J. Y,. Shan Z. L., Chen S., Qiao W. H., Zhu X. Y., Xie Q. J., Zhu H. T., Zhang Z. M., Zeng R. Z., Ding X. H., Zhang G. Q. 2012. Substitution mapping of QTLs for blast resistance with SSSLs in rice (Oryza sativa L.). Euphytica, 184(1): 141–150.

Zhao, H. J.; Wang, X. Y.; Jia, Y. L.; Minkenberg, B.; Wheatley, M.; Fan, J. B.; Jia, M. H.; Famoso, A.; Edwards, J. D.; Wamishe, Y.; Valent, B.; Wang, G. L.; Yang, Y. N. The Rice Blast Resistance Gene Ptr Encodes an Atypical Protein Required for Broad-Spectrum Disease Resistance. *Nat. Commun.* **2018,** *9*, 2039.

Zheng, W.; Wang, Y.; Wang, L.; Ma, Z.; Zhao, J.; Wang, P.; Zhang, L.; Liu, Z.; Lu, X. Genetic Mapping and Molecular Marker Development for Pi65(t), a Novel Broad–Spectrum Resistance Gene to Rice Blast Using Next–Generation Sequencing. *Theor. Appl. Genet.* **2016,** *129*, 1035–1044.

Zhu, D.; Kang, H.; Li, Z.; Liu, M.; Zhu, X.; Wang, Y.; Wang, D.; Wang, Z.; Liu, W.; Wang, G. L. A Genome-Wide Association Study of Field Resistance to *Magnaporthe oryzae* in Rice. *Rice (New York, N.Y.)* **2016,** *9*, 44.

Zhu, Y. Y.; Fang, H.; Wang, Y. Y.; Fan, J. X.; Yang, S. S.; Mew, T. W.; Mundt, C. C. Panicle Blast and Canopy Moisture in Rice Cultivar Mixtures. *Phytopathol.* **2005,** *95*,433–438.

Zou, J. F.; Li, Y. C.; Liu, X. L.; Liu, J. L.; Chen, H. L.; Yang, F. Y.; Huang, J.; Liao, H. Improving Blast Resistance of Rice Restorer μE32 and its Hybrid through Molecular Marker-Assisted Selection. *Crop Res.* **2017,** *31*, 11–14.

Zhou, B.; Qu, S.; Liu, G.; Dolan, M.; Sakai, H.; Lu, G.; Bellizzi, M.; Wang, G. L. The Eight Amino-Acid Differences within Three Leucine-Rich Repeats Between Pi2 and Piz-t Resistance Proteins Determine the Resistance Specificity to *Magnaporthe grisea*. *Mol. Plant Microbe Interact.* **2006,** *19*, 1216–1228.

CHAPTER 6

An Update on Epidemiology of Sheath Blight Pathogen of Rice and Its Management

ASTHA GUPTA[1] and HARI SHANKAR GAUR[2]

[1]*School of Agricultural Sciences, Sharda University Greater Noida, India*

[2]*School of Basic Sciences & Research, Sharda University, Greater Noida, India*

ABSTRACT

In agriculture sector, food requirement is rising as world population size is increasing, therefore, food security is an international concern. However, rice, *Oryza sativa* L. is one of the most dominant staple food crops grown over 44 million ha and producing 120 million tons of milled rice in India. There are numerous genetic, biotic, and abiotic factors that influence rice plant and decrease yield and quality. *Rizoctonia solani* is a highly diverse and pathogenic fungus that causes sheath blight disease in rice causing significantly losses yield (~50% or more) under pathogen-favorable climatic conditions (temperature and humidity). Although use of conventional approaches may control the pathogen infection to some extent but application of modern approaches with integrated strategy would be highly effective in overcoming the disease severity. Farmers are using chemical fungicides (mainly Validamycin and Carbendazim), which are effective against major pathogens but these are not environmentally friendly. Combination of

Fungal Diseases of Rice and Their Management. Deepti Srivastava, Md. Shamim, Malik Mobeen Ahmad, and R. S. Upadhyay (Eds.)

appropriate cultural practices, specific macro- and micronutrients amount, superior rice germplasm, and disease-resistant cultivars may play important role to manage to pathogen attack on rice plant under unfavorable climatic conditions. Additionally, plant extracts and ayurvedic formulations have been found to be effective against *R. solani.* These include Garlic (*Allium sativus*) and Turmeric (*Curcuma longa*) followed by Neem (*Azadirachta indica*), Marigold (*Tagetes* spp.), Eucalyptus (*Eucalyptus* spp.), Onion (*Allium cepa*), and Tulsi (*Oscimum sanctum*). Some biological control agents are explained in this chapter, a few of them are species of *Pseudomonas, Bacillus, Trichoderma,* and *Streptomyces*. Plant growth-promoting rhizobacteria (*Bacillus subtilis* and *Bacillus megaterium*), earthworm/vermicompost (*Eudrilus eugeniae and Pontoscolex corethrurus*), and biodecomposable seaweeds and Marine Algal extracts (*Cystoseira myrica* and *Sargassum cinereum*) have also been found useful for sustainable inhibition of *R. solani* pathogen. Climate prediction using smart approaches of detection, machine learning techniques, and their collaboration with resistant breeding and molecular breeding efforts would be novel approach to prevent and minimize sheath blight disease in rice.

6.1 BACKDROP

World food security is major alarm in this time at global level with elevated graph of human population and associated anthropogenic environmental impact. Rice is the most widely grown staple food crop after wheat and maize grown under diverse agro-climates (Fei et al., 2020; Prasad et al., 2020a). In addition, universal rice demand is expected to increase from 496 million tons in 2020 to 555million tons in 2035 (FAO, 2017; Gadal et al., 2019). Moreover, Asian countries rule the inclusive rice production and lead producers of rice are China, India Indonesia, and Bangladesh (Li et al., 2017; Gadal et al., 2019; Fei et al., 2020). Although, India has the largest area, 44 million ha under rice cultivation (Gadal et al., 2019), the productivity is low at 2.7 tons per ha compared to China and USA. Whole plant of rice is useful resource in terms of energy and fodder therefore, extremely important spotlight crop at economical and agricultural level. Rice food grain quality and quantity is core concern of plant breeders for alleviation of poverty and required income source for millions farmers. Fodder is also important because of high content of leaf β-carotene (Kim et al., 2020).

Biotic and abiotic factors affect plant growth as well as the quality and quantity of produce (Mishra et al., 2015, Singh et al., 2020). However, abiotic factors like drought and salinity cause significant yield losses in rice (Razzaq et al., 2020; Mumtaz et al., 2020). Some abiotic stress-tolerant verities have been introduced to accomplish a sustained enhancement in rice production under adverse conditions (Muthu et al., 2020). Nonetheless, biotic factors including bacterial and fungal diseases are the foremost concern of rice crop due to enormous grain loss and productivity (Singh et al., 2020). Biotic factors affect the rice plant through regulation of several metabolic pathways and processes that ultimately reduce the plant yield. Bacterial diseases like bacterial leaf streak, bacterial leaf blight (BLB), and sheath brown rot caused by *Xanthomonas oryzae* and *Pseudomonas fuscovaginae,* respectively, decrease the plant yield worldwide (Jiang et al., 2020; Syed-Ab-Rahman et al., 2020). There are a number of fungal diseases of rice as well which reduce the yield, for example blast (*Magnaparthe grisea*), sheath blight (*Rhizoctonia solani*), sheath rot (*Sacrocladium oryzae* and *Sacrocladium attenuatum*), brown spot (*Cochliobolus miyabeanus*) and scald (*Monographella albescens*) diseases. A recent study claims that among fungal diseases, sheath blight alone causes up to 51.3% potential loss of rice crop in India (Chakrapani et al., 2020a). Sheath blight disease is one of the most destructive infections of rice plant caused by soil-borne fungus, multinucleate, and ubiquitous pathogen *R. solani*. Hence this problem must be addressed in this challenging time. Therefore, development and adoption of appropriate agricultural cultivation practices and agricultural infrastructure are needed to negate the impacts of climate change and pathogen attack (Fei et al., 2020). There are several strategies that can avoid the yield loss through management of fungal infection some of which are explained in this chapter along with detailed description of fungus morphology, mode of action, host–pathogen interactions, etc.

6.2 SHEATH BLIGHT PATHOGEN *R. solani*: A WORLDWIDE THREAT

Sheath blight disease is also known as Oriental sheath/Leaf blight/Snake skin disease. *R. solani* (teleomorph: *Thanatephorus cucumeris*) belongs to phylum Basidiomycota, and family *Ceratobasidiaceae*, is soil saprotrophic and facultative plant parasite. However, *R. solani* is a monocyclic fungus, very complex and diverse in nature that cause sheath blight disease in rice and other crops (Molla et al., 2020). Manifestation of this pathogen was noted all the way through the temperate to tropical rice growing areas in India as well as other countries like China, Netherland, UK, and USA (Singh et al., 2016, Jia et al., 2019). *R. solani* is a soil-waterborne, necrotrophic pathogen, and its pathotypes have a widespread range of host plant species. Interestingly, 14 diverse anastomosis groups, AG1-AG13, and AGBI, of this pathogen have been detected based on several characteristics like distinct physiology, hyphal interactions, and genetic composition (Bhaskar et al., 2020). Sheath blight fungus has its place in anastomosis group AG1-IA of *R. solani* (out of 14 anastomosis groups) that infects rice predominantly in India, Korea, Philippines, and USA (Singh et al., 2016). Recently, it has been analyzed that primary inoculum density (sclerotia) was a major factor in disease outbreak that reduces grain yield (Khoshkdaman et al., 2020).

6.3 SYMPTOMS, INFECTIOUS PROCESS, AND LIFE CYCLE

R. solani shows oval-shaped to globular or unevenly elongated discolored grayish to brownish lesions with brown margins and occasionally similar to skin of a snake, therefore, also known as snake skin disease (Yellareddygari et al., 2014). *R. solani* forms sclerotia or dormant mycelia (basidiospores rare) and causes the disease in basal sheath under favorable condition. Generally, sclerotia can stay in soil and water as dormant stage for long duration that may infect the healthy rice plant in consecutive years. It can spread during field preparation and/or with irrigation water. Dormant sclerotia of *R. solani* and their high variability is main virtue of this epidemic pathogen and also major difficulty in disease management in rice. The life cycle of *R. solani* begins with mycelia growth from sclerotia after attachment onto rice seedlings/plants. Primarily, symptoms exhibited on leaf sheaths and later mycelia move upward along the sheaths and leaves of rice plants. Ultimately, infection cause significant loss of fully matured plants, resulting in damages to the sheaths, leaves, and grains. *R. solani* completes its life cycle with the development of overwintering structures, sclerotia on the different parts of plants (like sheaths, leaves, seeds), in soils and plant water interface (Jia et al., 2019; Molla et al., 2020).

6.4 PLANT–PATHOGEN INTERACTION AND SIGNALING

Relationship between *R. solani* pathogen and rice plant (host–pathogen interactions) establishes various signaling cascades which either develop

FIGURE 6.1 Symptoms of rice sheath blight disease recorded in experimental research field of division of plant pathology.

Source: Indian Agricultural Research Institute, New Delhi.

severe diseases or show tolerance to some extent after immediate pathogen attack depends on their reactions. Development of disease and its intensity depends on several factors like pathogen lodging, stage of plant, climatic conditions, genetic makeup of germplasm, amount of inoculums, and soil. Long duration of pathogen lodging alters canopy design, photosynthetic capability, and total yield of affected plants. Major yield loss has been observed during panicle initiation and flowering time, which limits spikelet filling, seed number, and seed weight (Yellareddygari et al., 2014). However, the pathogen affects a number of physiological mechanisms, biological pathways, and processes, some of them are photosynthesis, transportation of water, nutrients, and carbohydrate assimilates that directly or indirectly regulates plant productivity.

6.5 MODE OF ACTION-REACTION

Recently, it has been reported that there are a number of effector molecules, secreted through fungus that overcome plant defense, which cause infection, disease, and promote fungal colonization on plants. Examples of some effectors are Cytochrome C oxidase assembly protein CtaG/cox11 domain, glycosyl transferase family 2 domain and peptidase inhibitor I9 domain, histone acetyl transferase, histone deacetylase inhibitor, MDR transporter, O-antigen biosynthesis protein, O-methyl sterigmatocystin oxidoreductase, and polygalacturonase pectin lyase (Lo Presti et al., 2015; Ghosh et al., 2019; Molla et al., 2020). Furthermore, it has been observed that *R. solani* AG1-1A secretes oxalate as a virulent factor also (Nagarajkumar et al., 2005; Molla et al., 2020). After attack of pathogen, plants attempt to defend themselves through several signaling pathways to protect and maintain yield biomass and seed grain. This signaling pathway is known as pathogen-responsive signaling pathway, which includes Salicylic acid (SA), jasmonic acid, and ethylene (ET) that directly or indirectly recovers the plant yield through pathogen infection (Molla et al., 2020). Additionally, plant defense system activates and produces some antimicrobial molecules to fight against the pathogen.

6.6 DISEASE SCORING SYSTEMS

Standard Evaluation System (0–9) and SCALE have been developed for disease evaluation in multilocation screening trials where 0 indicated immunity and 9 representing extreme susceptibility (IRRI, 2002; Jia et al., 2019). SCALE is using extensively for disease analysis and disease scoring in rice plant to detect susceptibility and tolerance level. This SCALE has based on relative lesion height (IRRI, 2002) and visual length and area of symptoms (Jia et al., 2019). Table of Standard Evaluation System (0–9) has been described in Table 6.1 with related information of infection visibility in plant system.

Jia et al. (2019) presented symptoms and visibility of sheath blight disease on diseased plant and correlated the phenotype with a SCALE to evaluate diseased plant severity.

TABLE 6.1 Standard Evaluation System (0-9)

Score	SCALE based on relative lesion height (IRRI, 2002)	SCALE based on visual length and area of symptoms (Jia et al., 2019)
0	No infection observed	0%
1	Lesions limited to lower 20% of the plant height	1–10%
3	20–30%	11–30%
5	31–45%	31–50%
7	46–65%	51–70%
9	>65%	70–100%

6.7 THE ENVIRONMENTAL AND ECONOMIC IMPACT OF SHEATH BLIGHT IN RICE

Lindsey et al. (2020) highlighted the Food and Agriculture Organization (FAO) report and discussed that 60% of the global population depends on agriculture and number of disease management strategies adopted and varing globally depending on cultural and traditional practices, including crop rotation, cow's urine, and application of chemical pesticides (streptomycin and copper-mancozeb). Up-to-date reports suggest that *R. solani*, causing rice sheath blight (ShB) disease leads to tremendous yield losses of ~50% at global level (Mishra et al., 2020). Use of susceptible varieties, inappropriate cultivation practices, and unfavorable environment cause significant grain yield losses due to attack and incidence of *R. solani* in rice (Prasad et al., 2020a). There are several factors that affect plant integrity and sensitivity for disease infection that ultimately leads to yield loss. Most promising factors are germplasm susceptibility, crop stages and type, frequency and severity of disease infection, cultivation practices, host–pathogen interaction, plant nutrient levels and climatic conditions (temperature, humidity, and rainfall), anastomosis groups of pathogen, etc. Although, climate is the key factor that regulates physiological and biochemical processes of diseased plant and pathogen through temperature, drought, precipitation, ozone level, and increased CO_2. Association of virulent pathogen, susceptible host plant and favorable environment conditions may establish or initiate the platform for beginning of disease. Usually, temperature, moisture and precipitation control plant fitness, sexual reproduction, and pathogenicity of infectious agent (Jankar et al., 2020). As previous study explained that air temperature (34°C and 26°C) and morning relative humidity (more than 90%) have been investigated more favorable to broaden the sheath blight in field condition (Kaur et al., 2015).

Another important and neglected climatic factor is excess rainfall and duration, intensity, and frequency of rain that may also remove applied fungicides/chemicals from plant surface in field conditions (Jankar et al., 2020).

6.8 CONVENTIONAL APPROACHES TO CONTROL THE SHEATH BLIGHT DISEASE

A number of conventional or advanced approaches have been followed to control the sheath blight disease in rice plants. However, accurate strategy within time duration may more effectively protect crop productivity and pathogen infection intensity and severity that may overcome the yield losses. Some such approaches have been explained sections.

6.8.1 BURNING OF INFECTED TISSUES

Sheath blight affected portion of plant should be separated to inhibit the inoculum spread to save whole plant within specific period of time. Removal of diseased plants can be done through burning or composting of affected plants in the area to control the disease in next season rice crop.

6.8.2 DISEASE-FREE HEALTHY SEEDS

Selection and application of disease-free seeds will led to develop healthier plants, therefore, either treatment with fungicide may resolve fungal infection. In addition to that seed treatment will also control the transfer to healthy seedlings. Therefore, treated seeds should be used for sowing purpose to avoid unwanted fungal infection.

6.8.3 CULTURAL PRACTICES

The cultural practices may generally include broad spacing, stubbles removal, weeds devastation, and incorporation of green manure. Avoiding connections between fields for irrigation can restrict spread of sheath blight pathogen in rice fields. Correct timing of sowing and planting time of susceptible and resistant cultivars of rice may escape the infection. Amount of fertilizers and their types also affect the infection, therefore, phosphorus and nitrogen availability should be standardized. Combination of rice varieties and intercropping system also play important role to improve soil quality (Nattawut et al., 2020), reduce fertilizer consumption (Xu et al., 2020), and to contribute in disease management (Han ct al., 2015; Miah et al., 2017) with improvement of manifold ecosystem services also (Daryanto et al., 2020). Arrangement and grouping of mixed-cropping and rice–duck co-culture with numerous rice cultivars have been found to significantly increase the soil organic matter and total nitrogen contents as compared to mono-cropping systems (Li et al., 2020a). Organic fertilizers produce higher residual nutrients or essential nutrients in soil (except N) subsequent to application of cow manure and compost residue (Surin et al., 2020). The high level of N fertilizer exhibited increased ShB severity and yield loss of rice worldwide (Tang et al., 2007)

6.9 SOURCES OF RESISTANCE

Utilization of disease-resistant cultivars would be sustainable approach to avoid the pesticide exposure in field condition. Disease-resistant cultivars can be selected through screening using disease scoring system from susceptible to resistant scale (Table 6.1). Resistance breeding efforts against sheath blight are quite difficult due to unavailability of appropriate resistant cultivars/parents. Molecular breeding may be right approach for tagging resistant genes for sheath blight in rice. Further, rational replacement of novel resistant rice varieties to pathogens over susceptible varieties with unique alleles or new genes will play central role in the field conditions (Li et al., 2020b).

6.10 SUPERIOR RICE GERMPLASMS AND SPECIAL GENETIC STOCKS

Region-specific improved varieties or germplasm would be precise strategy to avoid infection and maintained plant productivity. However, some sheath blight-resistant rice germplasms have been available namely, LJRIL103 (PI 660982), LJRIL158 (PI 660983), LJRIL186 (PI 660984), and LJRIL220 (PI 660985) that represents agronomic and aromatic traits also (Jia et al., 2019). Rice germplasms or superior varieties with different R resources and specific characteristic (like stem solidity, plant height, and heading date) would be necessary footstep for developing improved rice varieties (Hapsari and Poromarto 2020; Goad et al., 2020). Further, they reported that two local varieties of rice, Padi Hitam and Srikiti (cultivars) exhibited not only moderate resistances to sheath blight but also superior potential of production as compared to control (IR 64 and Inpari 13) due to high content of lignin, cellulose, and hemicelluloses (Hapsari and Poromarto 2020). Out of thirty-one entries/cultivars of rice, eight (R1896-81-2-59-1, R2048-187-2-128-1, ARR-8, FSVT-MS-18, FSVT-MS-20, 213, 1714, 2717), and one variety IR10F36 exhibited moderately resistant reaction (Score-5) under field condition after screening for sheath blight resistant (Prasad et al., 2020a).

6.11 DEVELOPMENT OF MAPPING POPULATIONS

Selection and hybridization of two contrasting parents for sheath blight (resistant/tolerant and susceptible) is the main point for construction of mapping populations in rice. Resistant resources or adaptive germplasms would be good source for development of mapping population (F_2, Backcross inbred lines, Recombinant Inbred Lines, Near Isogenic lines and Double haploids, etc.,) and rice breeding improvement programs through Marker-assisted selection (MAS) (Jia et al., 2019). Identification and accumulation of different resistance genes will be possible through multi lines (combination of NILs and have uniform morphological traits) to overcome diseases (Miah et al., 2017). Goad et al. (2020) constructed advanced-generation recombinant inbred line (RIL) lines resulting from two weed × crop crosses for sheath blight disease in rice.

6.12 CHEMICAL CONTROL/FUNGICIDES

Fungal disease management is extremely dependent on application of chemical fungicides to sustain healthy plant, development, and productivity under field conditions (Molla et al., 2020). Appropriate amount of fungicides at right stage of plants may drastically contribute to minimize sheath blight disease and there are a number of fungicides released in market which are chemically synthesized. Different types of fungicides have been used for a long time and were proven successful in inhibiting the disease, for example, seed treatment using Bavistin and Benlate, soil treatment with Benlate, Kitazin and Hinosan, seed treatment and foliar purpose Topsin-M that reduce seed-borne infection of *R. solani* in rice (Premalatha Dath, 1990; Das and Mishra, 1990). These fungicides are more effective when used in different combinations and particular ratio as some reports suggested: Carbendazim with epoxiconazole and mancozeb (Saha, 2003; Prasad et al., 2006), propiconazole + difenconazole (Kandhari, 2007), and carbendazim 25% + flusilazole 12.5% SE (Goswami et al., 2012). These fungicides are chemically synthesized therefore harmful for human health and environment also, therefore excesses use of fungicides may control disease but will destroy the ecosystem. Recently, six combinations of fungicides were tested against sheath blight disease which did not affect rice yield traits include 1000 grain weight (Parveen et al., 2020). Similarly, five fungicides (tricyclazole, validamycin, propiconazole, hexaconazole, and carbendazim) were selected against *R. solani* as seed treatment alone and in combination with one foliar spray using JK 155 a susceptible cultivar under field conditions (Jain et al., 2020). The results suggested that validamycin seed treatment of 2 g kg^{-1} seed followed by 0.2% foliar spray effectively controlled disease severity 85.3% and increased grain yield by 25.8% over check (Jain et al. 2020). Some cost-effective and efficient fungicides were reported that include Prochloraz 27% + Tricyclazole 23% SE, Tricyclazole 75% WP, Carbendazim 50% WP, and Carbendazim 12% + Mancozeb 63% (Pramesh et al., 2020).

6.13 BIOCONTROL AGENTS

Application of biocontrol agents is a very promising strategy to manage rice sheath blight. Microorganisms such as *Pseudomonas* (Gram-negative bacteria, rod-shaped), *Bacillus* (Gram-positive, rod-shaped bacteria), *Trichoderma* (a fungus helpful like a fungicide), and *Streptomyces* (Gram-positive bacteria, filamentous; useful as antifungal metabolites) are widely spread in the environment like soil, water, and plants. Yang et al. (2017) concluded that crude extract from cowdung was associated with bacteria *Streptomyces cochorusii* strain (NF0919) culture filtrate or fresh cells of *Bacillus amyloliquefaciens* strain (SB177) that provided better disease control than other fungicides against the rice sheath blight pathogen (*R. solani*). Jamali et al. (2020) investigated that *Bacillus subtilis* strain RH5 explored as biocontrol agent that showed noteworthy antagonistic activity (84.41%) against the fungal pathogen *R. solani* through the production of defense-related compounds include indole acetic acid, siderophore, hydrogen cyanide production and phosphate, Zn, K solubility. Further they explained that strain RH5 has antimicrobial peptide biosynthetic genes (bacylisin, surfactin, and fengycin) and some hydrolytic enzymatic enzymes (chitinase, protease, cellulase, xylanase). Recent reports suggest that yeast strains have antagonistic actions against *R. solani* and used as biocontrol agents that could suppress the 60.0–70.3% sheath blight disease in rice (Into et al., 2020). Two cyanobacteria *Nostoc piscinale* and *Anabaena variabilis* have demonstrated very promising control of rice sheath blight disease; their extracts inhibited the mycelial growth of *R. solani* and also increased root length, biomass, and yield of rice plant (Zhou et al., 2020). Another investigation identified potential of biocontrol agent from rhizosphere samples and isolates of Actinobacteria and further antagonism evaluation suggested that six isolates which showed 85% inhibitory activity against *R. solani* (Pralhad et al., 2020). Out of 14 strains of *Trichoderma* (native biocontrol agents), BHU-11, T-4 and BHU-8 acted as plant growth promoter and showed immense disease inhibition 71.48%, 67.28%, and 63.89%, respectively, for *R. solani* with improved economical traits like the number of tillers/hill, plant height, crop yield, and drought avoidance (Mishra et al., 2020). Such kind of strategy not only regulates plant growth and enhances tolerance in rice plant for fungal pathogen *R. solani* but may also help in sustainable development with environment friendly manner.

6.14 PLANT GROWTH-PROMOTING RHIZOBACTERIA (PGPR)

PGPR are rhizosphere-isolated and root-associated free-living soil bacteria that are involved in plant development and resistant against biotic stress including fungal and bacterial diseases also. PGPR enhance tolerance either through direct antagonism to pathogens or induction of systemic resistance to pathogens that regulate resistance via induced systemic resistance (methyl jasmonate and methyl salicylate) in plants and provides plant protection against pathogens (Suryadi et al., 2019). However, PGPR promotes plant growth through several mechanisms and biological processes includes production of iron chelating siderophores, increased N uptake, phytohormone production, nitrogen fixation, and phosphorus solubilization that are associated to pathogen control and exhibiting competition with plant pathogens, synthesis of antibiotics, antifungal metabolites, and defense enzymes (Yellareddygari et al., 2014; Suryadi et al., 2019). Good sources of PGPR are plant tissues and bacteria (*Azotobacter, Azospirillum, Alcaligenes, Arthrobacter, Bacillus, Burkholderia, Enterobacter, Klebsiella, Pseudomonas*, and *Serratia*) that play important role in crop protection and plant productivity (Suryadi et al., 2019). Two strains of PGPR like *B. subtilis* and *Bacillus megaterium* inhibited the growth of *R. solani* in rice (Thakur et al., 2020).

6.15 PATHOGEN AVIRULENCE (AVR) GENES

Pathogen releases effecter molecules that cause infection (pathogenesis) in plants through regulating their innate immunity and recognized by resistance (R) proteins of plants and known as avirulence (AVR) proteins. This interaction leads to hypersensitive response (localized cell death) and resistance in plants. However, R-AVR gene interactions are very complex phenomena in plant-fungi pathosystems and helpful to explore and understand the diseases control process and its management in plant system. As previous a report suggested, (1) AVR genes can be predicted by numerous R genes and vice versa, (2) particular AVR gene may suppress the detection of another AVR gene through its corresponding R gene, and (3) two cooperating R genes are involved to identify an AVR gene (Petit-Houdenot et al., 2017). Detailed study and analysis on antagonistic interactions between AVR genes and corresponding R genes would be promising approach to maintain resistance in rice to sheath blight disease through pyramiding of R genes in agronomically important desired cultivars (Petit-Houdenot et al., 2017). Identified defense response (DR) gene, that is, germin-like proteins (OsGLP), correlated with sheath blight and rice blast disease in rice, further inactivation of these genes enhance the infection (Manosalva et al., 2009). Likewise, Broad-Spectrum DR genes (short sequence motifs in rice promoters) co-expressed in response to *R. solani* and two other rice pathogens (*Magnaporthe oryzae* and *X. oryzae* pv. *oryzae*) that are *cis*-regulatory modules and involved in disease resistance (Tonnessen et al., 2019).

6.16 PLANT EXTRACTS AND AYURVEDIC FORMULATIONS

There are a number of botanicals, or plant extracts, that exhibited antifungal activity and would be useful to control fungus-related violent diseases (Miah et al., 2017). Kora et al., (2020) synthesized silver nanoparticles (Spherical, polydispersed Ag NP of 16.5 nm) from rice leaf extract that showed antifungal activity against *R. solani* mycelium and sclerotia. Further, two treatments of Ag NP, that is, 20 µg/mL and 10 µg/mL entirely repressed disease incidence by *R. solani* isolates and promote seedling vigor index in rice cultivars, respectively (Kora et al., 2020).

The antifungal potentialities of three ayurvedic formulations like bacteridote, fungidote, and viridote were tested against *R. solani* and it was found that fungidote and viridote suppressed the mycelial growth at 0.2 and 0.3% concentrations (Chakrapani et al., 2020b). They concluded, in vitro sensitivity test of ayurvedic formulations will be promising approach toward management of sheath blight disease in rice. Interestingly, fermented mature coconut water had antimicrobial activity and reduced rotting in rice plants sheaths and lesion length significantly ($P < 0.05$), therefore, it would be promising agent to control the disease and increase rice yield (Wahab et al., 2020). In vitro antifungal efficacy of some imperative botanicals has been tested, e.g., garlic, marigold, eucalyptus, turmeric, ginger, onion, and Tulsi, and found that maximum mycelial inhibition was observed due to garlic followed by ginger (80.00%), turmeric (78.51%), marigold (76.29%), and eucalyptus (75.55%) (Chakrapani et al., 2020a).

6.17 R GENE DISCOVERY THROUGH DISEASE EVALUATION METHODS

As per previous research, R gene has not been identified for sheath blight disease in rice (Jia et al., 2019; Molla et al., 2020). Therefore, early identification of R genes is essential to overcome the disease and its infection for speedy revival of crop production. Moreover, the user-friendly sheath blight inoculant can be produced at large scale for R gene discovery under field evaluation using corn chips or rye (2:1 weight ratio) on PDA media and air drying of mycelia and white sclerotia at 24°C followed by treatment (Jia et al., 2019). Further, they explained three evaluation methods for sheath blight disease detection, viz. (1) Detached leaf method, (2) Soft-drink bottle method, and (3) parafilm methods, which are useful for greenhouse experiments to detect and validate the function of R genes. Plant performance against the pathogen can be investigated using disease scoring system for discovery of R gene which would be a good economic strategy due to minimum requirement of resources like time, agricultural field, and fertilizer (Jia et al., 2019).

6.18 NUTRITION MANAGEMENT

Balance concentration of plant nutrients is not only important for normal plant growth and development but also for diseases tolerance. Optimum level of nutrients (N, K, P, Mn, Zn, B, Cl, and Si) has involvement in disease resistance, tolerance, and sustainable agriculture. Macro- and micronutrients (some metals and metalloids), interfere with plant expansion because of their connection in number of metabolic processes but nutrient deficiency and toxicity conditions exhibit some specific symptoms and may be cause of disease also (Ali et al., 2020). There are some essential and nonessential elements which are required for plant development and tolerance against biotic and abiotic stress. Appropriate amount of nutrition makes fundamental contribution for high production system that regulates response of rice against diseases (Miah et al., 2017). Good-quality soils with high organic material and elevated biological activity generally have higher soil fertility that suppresses infection (Chau et al., 2003). However, although Nitrogen is an essential element and its shortage is a limiting factor in achieving high productivity (Miah et al., 2017), additional high and low N levels are associated with obligate and facultative parasites and increase and decrease the severity of infection (Dordas, 2008). There is another important essential nutrient, that is, Phosphorus (P), intensive application of which leads to lesser bioavailability of phosphate (Pi), while higher concentration of P in agriculture soil leads to susceptibility toward fungal diseases (Campos-Soriano et al., 2020). Boron is considered as an essential plant micronutrient and actively involved in several ion, metabolite, and hormone transport reaction but insufficient amount leads to detrimental effect on the yield of agricultural plants because of narrow range between deficiency and toxicity (Brdar-Jokanović, 2020). Further they suggested that deficiency and soil boron toxicity can be resolved by other factors like fertilization, time of application, soil characteristics, temperature, humidity, and leaching that comprise a range of amendments and growing tolerant plant.

Similarly, a nonessential but most important and beneficial element is Silicon (Si) and its high amount in the shoots represented less incidences of rice blast (Prabhu et al., 2001). Recently, one more report suggested that application of carbonized rice husk (CRH), which is a by-product of rice production and cost-effective biochar can be utilized as Si fertilizer and promote carbon sequestration in soil also (Sabes et al., 2020). Further, CRH would be a good agent to the control sheath blight disease in rice as a cultural management.

6.19 ECO-FRIENDLY COELOMIC FLUID OF EARTHWORM/ VERMICOMPOST

Recently, very interesting finding was reported regarding interaction between earthwarm (*Eudrilus eugeniae)* and fungal pathogen *R. solani* tolerance (Nadana et al., 2020). Coelomic fluid (CF: Collected by cold stress method) of earthworm promotes growth parameters (seed germination, lengths of shoot, root and seedlings, and vigor index) of *Oryza sativa* L. subsp. *indica* and resistance against *R. solani* due to presence of plant hormone derivatives (auxin and indole-3-acetyl-L-valine) and ecofriendly insecticidal and other pesticidal compounds. Earthworms are essential soil organisms participating in belowground-aboveground relationships (BAR) and silicon (Si) and play important role in regulating aboveground plant health (*Blanchart* et al., *2020)*. They evaluated amalgamation of earthworms (*Pontoscolex corethrurus*) and Si (crucial element) with NPK (macronutrients) that elevate the biomass gain and decrease the blast disease severity in rice.

6.20 BIO-DECOMPOSABLE SEAWEEDS AND MARINE ALGAL EXTRACTS

Seaweed/sea plants are forms of algae that grow in the sea water and are resources for bioactive compounds useful for human health (Rengasamy et al., 2020) and have antifungal activity too (Uribe et al., 2020; Manam and Subbaiah, 2020; Manam, 2020). However, seaweeds found in coastal areas worldwide and used as bio-fertilizer, seed treatment, foliar spray, and soil application for agricultural farming from prehistoric times in several location like China, Japan, Spain, Britain, and France (Mukherjee and Patel, 2020).

They explained that seaweed extract retained a number of properties that promote nutrient uptake, fruit setting, biotic (disease tolerance), and abiotic resistance (drought, salinity, and temperature). In addition to that seaweeds are eco-friendly due to bio-decomposable, nontoxic nature and therefore, considered as novel strategies better than chemical fertilizer (Mukherjee and Patel, 2020). Recently, colloidal silver nanoparticles (AgNPs) have been synthesized from marine red seaweed *Halymenia porphyroides Boergesen (Crypton)* (Manam and Subbaiah, 2020) and *Colpomenia sinuosa* with good antifungal efficacy (Manam, 2020). Methanolic extract of macroalgae (*Cystoseira myrica* and *Sargassum cinereum*) exhibited antagonistic and anti-oxidant activity against *Fusarium oxysporum* due to increase in the activity of antioxidant enzymes (Computerized axial tomography, POD, and APX) (El-Sheekh et al., 2020). Antioxidant activity of from red seaweed *Pyropia orbicularis* plays key role to enhance the plant resistance (Uribe et al., 2020). Consequently, marine macroalgae or seaweeds would be novel bioagent source for eco-friendly and biological control of soil fungi (El-Sheekh et al., 2020). A good number of bacterial biocontrol agents (30 isolates) from various seaweeds, seawater, and sediments were selected to test fungitoxic effect against *R. solani* and it was concluded that *Bacillus velezensis* could control disease caused by fungal pathogens *R. solani* because of higher level of early accumulation of phenolics and phytoalexins (Raj et al., 2020).

6.21 INTEGRATED DISEASE MANAGEMENT

Use of biocontrol agents with fungicides and antibiotics has been advocated as an important integrated package for sheath blight management. The application of consortium has good potential for disease control (73%) of *R. solani* as compared to single culture application (actinobacteria AUDT502 –70% and validamycin –65%) or chemical control in rice under greenhouse condition (Pralhad et al., 2020). Prasad et al. (2020b) investigated integrated approach of botanicals in various combinations against R. *solani* which are following (1) Neem oil + Teepol (2) Neem powder (3) Karanj oil (4) Karanj powder + Teepol (5) Chili + Garlic +Teepol, and (6) Chli + Teepol + sarf. All the selected treatments appreciably inhibited the mycelial growth and Neem oil resulted in 88.52% growth inhibition (10.33 mm mycelial growth) over untreated treatment. Recently, Sabes et al. (2020) studied four treatments (1) 1.5 t ha^{-1} levels of CRH, (2) 3.0 t ha^{-1} levels of CRH, (3) a spray of fungicide with flutolanil, and (4) control without any application for sheath blight development. They summarized that fungicide application with flutolanil integrated approach and CRH use at 3.0 t ha^{-1} could overcome the disease intensity without affecting rice yield. Raghunandana et al. (2020) tested a novel combination fungicide during Kharif 2015 and Rabi 2015–2016 against rice sheath blight diseases under field condition and found it effective in association to increase in grain yield, this combination included Tricyclazole 22.5% W/V (20.36% W/W) + Azoxystrobin 7.5% W/V (6.79% W/W). In Bangladesh, Azoxystrobin 10% + Tebuconazole 90% fungicides combination was applied and evaluated under diverse seasons and varieties to control sheath blight in rice and suggested to farmers (Parveen et al., 2020). Variable combination of PGPRs (*B. subtilis*) were analyzed under field condition to control the rice sheath blight that is following: T0 (*B. subtilis* ZB87-1/2, 25DAT @1.5gm/lit + *B. subtilis* ZB87-1/2, 50DAT @2.5gm/lit) and T10 (*B. subtilis* ZB87-1/2, 25DAT @2.5gm/lit + *B. subtilis* ZB87-1/2, 50DAT @1.5gm/lit + Hexaconazole-5%EC 25DAT + Hexaconazole-5%EC 50DAT) which decreased the symptoms and controlled the disease upto 60.50% and 68.92%, respectively, with high yield (Thakur et al., 2020).

6.22 MODERN APPROACHES TO CONTROL THE SHEATH BLIGHT DISEASE

6.22.1 *MOLECULAR MARKERS*

Molecular markers have been used in a number of crops for diversity analysis, population structure studies, and genotypic differentiation (Maurya et al., 2013, 2015; Gupta et al., 2013; Gupta, 2020). Deoxyribonucleic acid (DNA)-based molecular markers are Amplified Fragment Length Polymorphism, Simple Sequence Repeats, and Single Nucleotide Polymorphism (SNP), these advanced markers are more authentic as compared to phenotypic markers due to insensitivity to environment (Maurya et al., 2013, 2015; Gupta et al., 2019, 2020). Recently, sequencing based third generation-SNP markers were explored for sheath blight resistance in rice and 333 nonsynonymous SNPs were detected between ShB-susceptible and resistant genotypes of rice (Silva et al., 2012; Goad et al., 2020).

6.22.2 *QUANTITATIVE TRAIT LOCI (QTL) FOR SHEATH BLIGHT*

Quantitative trait loci are chromosomal regions or DNA sequences that shows polygenic inheritance and responsible for phenotypic, agricultural, and economical traits of plants. Early detection of QTLs for desired traits can be execute with the help of polymorphic molecular markers and mapping population, which have been explored in number of plants (Gupta et al., 2020; Goad et al., 2020). As rice varieties or germplasm may not be completely resistant to sheath blight (ShB) disease but they have some chromosomal region/quantitative trait loci that may control or reduced the diseases harshness (Chen et al., 2019). Although, 10 ShB-QTLs were mapped under greenhouse condition (at an early vegetative stage) and confirmed one major ShB-QTL (qShB9-2) in field condition for ShB resistance (Liu et al., 2013). This would be fruitful for rice improvement through introgression of desired candidate genes. Recently, four novel QTLs have been tagged for ShB resistance (Goad et al., 2020). Genome-wide association (GWAS) mapping was performed using natural population for sheath blight disease and it was found that 11 SNP loci were significantly associated with SB resistance in rice (Chen et al., 2019). Further, Chen et al. (2019) reported that two reliable QTLs, including qSB-3 and qSB-6 for sheath blight disease severity, were situated on chromosomes 3 and 6, respectively. Another research group of Oreiro et al. (2020) also investigated GWAS in rice under controlled conditions and identified multiple genomic regions that were located on different chromosomes viz., 1, 4, and 11 that were significantly associated to sheath blight disease.

6.22.3 MAS AND BREEDING

Identification and incorporation of interested candidate gene/loci from donor to recipient individuals with the help of molecular markers and biotechnology considered as MAS or Marker-assisted breeding (MAB). Disease-resistant genes or loci have been introgressed for bacterial blight (*xa13 and Xa21*) and blast (*Pi2 and Pi54*) diseases through marker-assisted backcross breeding in Basmati cultivar and Pusa Basmati 1509 (Sagar et al., 2020). Moreover, resistance breeding programs have been exploring in related species of rice and wild sources for resistance genes/QTLs that may be introgressed or pyramided in several breeding lines (Singh et al., 2020). Since ancient times breeders have been using conventional approachcs to diseases control but there is problem of linkage drag that can be resolved by using molecular marker-based breeding that helps in successful introgression of specific desired disease resistance genes, e.g., BLB (Chukwu et al., 2019). However, pyramiding (engineer multiple genes/loci in one plant) of desired multiple genes through hybridization between transgenic/resistant plants for selected traits can generate multiple biotic stress-resistant cultivars (*Oryza sativa* L. cv. ASD 16) as developed for sheath blight (*tlp*), brown planthopper (*gna*), and bacterial blight resistance (*Xa21*) (Rajesh et al., 2020).

6.22.4 MUTATION APPROACH

Lesion mimic mutants (LMMs) exhibit phenotypes like hypersensitive cell death without pathogen infection or programmed cell death, which is a noteworthy symptom of disease resistance genes mediated defense action (Wang et al., 2011; Jia et al., 2019). Therefore, mutational and molecular investigation may give the basic idea of molecular process and biological pathways of disease infection in plant. Previously, LMM1 has been detected from rice cultivar Katy with fast neutrons treatment and their lesion mimic phenotype (single recessive gene) can be enhanced through blast pathogens and water-related stress that shows resistance to blast and sheath blight disease (Jia, 2005; Jia et al., 2019). Further, identification and evaluation of LMMs in rice would be promising approach to understand the plant innate immunity mechanism also (Jia et al., 2019). Understanding of loss-of-function mutants may reveal some information of wild type loci or alleles (Jia, 2009). Although IR64 rice mutant populations have been constructed with mutagenesis agents including diepoxybutane (D), gamma rays (G), and fast neutrons (F) and developed 21 lesion mimic mutants. These form necrotic lesions in the nonexistence of pathogens that are helpful for categorization of disease resistance pathways and molecular mechanism in rice (Wu et al., 2008).

6.22.5 PLANT BIOTECHNOLOGICAL APPROACHES

Transgenic approaches and RNAi/knockdown of effectors serve a central function to understand the *R. solani* pathogen and sheath blight disease in rice (Molla et al., 2020). Disease resistance gene *BROAD-SPECTRUM RESISTANCE 2*, belongs to CYP78A family and encodes an uncharacterized cytochrome P450 protein that regulates resistance to two *R. solani* anastomosis groups. Its over-expression decreases growth but develops longer seed as compared to control plant of *Arabidopsis* and rice (Maeda et al., 2019). Furthermore, indeterminate domain (IDD3, IDD13) and Loose plant architecture 1 (LPA1) form a transcription factor complex (regulate PIN1a) and over-expression of IDD13 and RNAi plants (lpa1 and IDD13) was associated with sheath blight resistance in rice (Sun et al., 2020). However, rice small-interfering RNA (siR109944) expression was related to susceptibility against *R. solani* that diminishes plant immunity and pretreatment with auxin-induced sheath blight resistance through regulating endogenous auxin homeostasis in rice (Qiao et al., 2020). In addition, Glutathione-S-transferase (tau class defense gene: *OsGSTU5*) was investigated in rice and results suggested that pathogen infection and its symptoms (lesion cover area and hyphal penetration) were higher in knockdown line as compared to over-expression line (more tolerant for *R. solani*) (Tiwari et al., 2020). R. *solani* infection significantly regulates LPA1 expression in the leaves and over-expression of LPA1 considerably promotes the resistance of rice to sheath blight by the activation of PIN1a (Sun et al., 2019, 2020).

6.22.6 PREDICTION MODEL SYSTEMS/MACHINE LEARNING TECHNIQUE: SMART APPROACH

Disease predication and its targeted management are possible through advanced approaches and early prediction model of computer. Machine learning and deep learning would be right choice of technology to control the severity and timely recovery of plant diseases without economic loss to farmers. There are several models and programs that can establish the relation between climatic conditions (temperature, relative humidity, and soil moisture) and disease severity that would be useful for pathogen infection likelihood and forecast out breaks within time for speedy recovery. Deep learning-based video detection architecture with a custom backbone was projected for detecting plant diseases in videos to build a real-time crop diseases including sheath blight disease of rice (Li et al., 2020c). Similarly, deep learning-based convolutional neural networks (CNN) was helpful to detect rice sheath blight disease using images captured in real-life scenario with 99.53% accuracy using VGG16 (CNN architecture) including 138 million number of parameters (Rahman et al., 2020). In future, this strategy will be very human friendly due to memory efficient architectures and mobile application. Integrated approach of machine learning along with near-infrared (NIR) and Raman spectroscopy has helped to detect rice sheath blight (ShB) at early stage through evaluation of NIR and Raman spectra from leaves of the ShB-susceptible rice cultivar (Conrad et al., 2019).

6.22.7 OMICS STUDIES TO UNDERSTAND HOST DEFENSE

Genomics, transcriptomics, proteomics, and metabolomics technology collectively known as "OMICS" that deal with DNA, RNA, Protein, and related metabolites, respectively, established the relationship among molecular pathways, biological processes, and disease interactions. Application of these molecular biology technology helped to develop first draft genome sequence of *R. solani* AG1-1A (36.94-Mb) and annotated 6156 genes, 257 genes specific for host–pathogen interaction and 25 candidate effector genes (Zheng et al., 2013). Later Malaysian isolate of *R. solani* published with small genome size of 28.93 Mb approximately (Nadarajah et al., 2017). Advance genomic and transcriptome sequencing approaches facilitate to detect AG1-1A specific genes, putative pathogenesis genes, virulence factor/effector genes, and candidate genes for sheath blight resistance (Silva et al., 2012; Hane et al., 2014; Ghosh et al., 2014, 2018) and candidate pathogenic novel miRNA-like small RNAs "miRNAs" also (Lin et al., 2016). Further, alternative splicing of significant pathogenic genes of *R. solani* AG1-1A has been recognized in rice infection (Xia et al., 2017). In future, application of OMICS tools and techniques on *R. solani* may yield information about infections genes/DNA region or their corresponding proteins, which regulate the sheath blight disease in rice and other plants (Molla et al., 2020). In rice, two contrasting cultivars (resistant-Teqing and susceptible-Lemont) were used to investigate differentially abundant proteins against *R. solani* through proteomic analysis (Ma et al., 2020). They detected and quantified total 755 proteins, which participated in rice immunity for *R. solani* pathogen and were involved in glyoxylate and dicarboxylate metabolism; glycine, serine, and threonine metabolism; unsaturated fatty acid biosynthesis; and glycolysis/gluconeogenesis regulation pathways.

6.23 FUTURE THRUSTS

Sheath blight is very severe disease that drastically decreases crop productivity and quality of rice grain. Conventional approaches are useful for disease control but integrated management would be more powerful, for example: Combination of plant extract with PGPR, seaweeds, earthworms, or mixture of fungicides in specific amounts. Fungicides are often not recommended due to environmental and human health hazards. Identification of R gene will be great breakthrough for management of this disease; It has not been adequately explored yet and needs further investigation. Tagging of quantitative trait loci or R gene for sheath blight resistance and introgression in high yielding rice cultivar will be effective. Screening of large number of resistant germplam should be undertaken, to find lines that may be used in further conventional and molecular breeding programs at desired location. Molecular analysis of *R. solani* and resistant germplams of rice will be informative to detect the disease causing regions or tolerant/resistant loci, respectively. However, plant biotechnological and OMICS approaches may give the basic idea about plant-pathogen interaction, regulating molecular processes and biological pathways. Development of nanoparticles with good antifungal activity against *R. solani* for mycelium and sclerotia would also be a promising approach. In addition, establishment of prediction model (associated with *R. solani* infection and climatic conditions) will be the future strategy for early/pre-detection of sheath blight disease and pathogen incidence for sustainable management.

ACKNOWLEDGMENTS

I am thankful to my family members, especially my husband Mr. Pankaj Gupta, for consistent support. I am grateful to Dr. Hari Shankar Gaur, Dean, School of Basic Sciences & Research, Sharda University for critical review and valuable suggestions. We owe special thanks to Dr. Rashmi Aggarwal (Principal Scientist & Head) and Dr. Bishnu Maya Bashyal (Senior Scientist), for evaluating rice sheath blight disease symptoms in division of plant pathology, ICAR - Indian Agricultural Research Institute, New Delhi.

KEYWORDS

- ***Rizoctonia solani***
- **rice**
- **sheath blight disease**
- **conventional**
- **modern approaches**
- **biological control**

REFERENCES

Ali, A.; Bhat, B. A.; Rather, G. A.; Malla, B. A.; Ganie, S. A. *Plant Micronutrients*; Springer: Cham. Switzerland; 2020, pp 257–284; Proteomic Studies of Micronutrient Deficiency and Toxicity.

Bhaskar, R. T.; Chopperla, R.; Prathi N. B.; Balakrishnan, M.; Prakasam, V.; Laha, G. S.; Balachandran, S. M.; Mangrauthia, S. K. A Comprehensive Gene Expression Profile of

Pectin Degradation Enzymes Reveals the Molecular Events during Cell Wall Degradation and Pathogenesis of Rice Sheath Blight Pathogen *Rhizoctonia solani* AG1-IA. *J. Fungi.* **2020,** *6* (2), 71.

Blanchart, E.; Ratsiatosika, O.; Raveloson, H.; Razafimbelo, T.; Razafindrakoto, M.; Sester, M.; Becquer, T.; Bernard, L.; Trap, J. Nitrogen Supply Reduces the Earthworm-Silicon Control on Rice Blast Disease in a Ferralsol. Appl. Soil Ecol. **2020,** *145, 103341.*

Brdar-Jokanović, M. Boron Toxicity and Deficiency in Agricultural Plants. *Int. J. Mol. Sci.* **2020,** *21* (4), 1424.

Campos-Soriano, L.; Bundó, M.; Bach-Pages, M.; Chiang, S. F.; Chiou, T. J. San Segundo B. Phosphate Excess Increases Susceptibility to Pathogen Infection in Rice. *Mol. Plant Pathol.* **2020,** *21* (4), 555–570.

Chakrapani, K.; Sinha, B.; Chanu, W. T.; Chakma, T.; Siram, T. Assessing in Vitro Antifungal Activity of Plant Extracts against *Rhizoctonia solani* causing Sheath Blight of Rice (*Oryza sativa*. L). *J. Pharmacogn. Phytochem.* **2020a,** *9* (1), 1497–1501.

Chakrapani, K.; Sinha, B.; Siram, T.; Chanu, W. T.; Chakma, T.; Thangjam, B. In Vitro Sensitivity Test of Ayurvedic Formulations against Growth of *Rhizoctonia solani* causing Sheath Blight of Rice in Manipur. *Int. J. Agric. Invent.* **2020b,** *5* (1), 72–74.

Chau, L. M.; Cat, H. D.; Ben, P. T.; Phuong, L. T.; Cheng, J.; Heong, K. L. Impacts of Nutrition Management on Insect Pests and Diseases of Rice. *Omonrice.* **2003,** *11*, 93–102.

Chen, Z.; Feng, Z.; Kang, H.; Zhao, J.; Chen, T.; Li, Q.; Gong, H.; Zhang, Y.; Chen, X.; Pan, X.; Liu, W. Identification of New Resistance Loci against Sheath Blight Disease in Rice through Genome-wide Association Study. *Rice Sci.* **2019,** *26* (1), 21–31.

Chukwu, S. C.; Rafii, M. Y.; Ramlee, S. I.; Ismail, S. I.; Hasan, M. M.; Oladosu, Y. A.; Magaji, U. G.; Akos, I.; Olalekan, K. K. Bacterial Leaf Blight Resistance in Rice: A Review of Conventional Breeding to Molecular Approach. *Mol. Biol. Rep.* **2019,** *46* (1), 1519–1532.

Conrad, A. O.; Lee, D. Y.; Li, W.; Wang, G. L.; Bonello, P. E. Machine Learning-based Early Rice Disease Detection using Spectral Profiles. *Plant Health.* **2019**.

Daryanto, S.; Fu, B.; Zhao, W.; Wang, S.; Jacinthe, P. A.; Wang, L. Ecosystem Service Provision of Grain Legume and Cereal Intercropping in Africa. *Agric. Syst.* **2020,** *178*, 102761.

Das, S. R.; Mishra, B. Field Evaluation of Fungicides for Control of Sheath Blight of Rice. *Indian Phytopath.* **1990,** *43*, 94–96.

Dordas, C. Role of Nutrients in Controlling Plant Diseases in Sustainable Agriculture. A review. *Agron. Sustain. Dev.* **2008,** *28* (1), 33-46.

El-Sheekh, M. M.; Mousa, A. S.; Farghl, A. A. Biological Control of Fusarium Wilt Disease of Tomato Plants Using Seaweed Extracts. *Arab. J. Sci. Eng.* **2020,** *24*, 1–4.

FAO. *The future of food and agriculture - Trends and challenges*; Rome, 2017. http://www.fao.org/3/a-i6583e.pdf

Fei, L.; Meijun, Z.; Jiaqi, S.; Zehui, C.; Xiaoli, W.; Jiuchun, Y. Maize, Wheat and Rice Production Potential Changes in China under the Background of Climate Change. *Agric. Syst.* **2020,** 102853.

Gadal, N.; Shrestha, J.; Poudel, M. N.; Pokharel, B. A Review on Production Status and Growing Environments of Rice in Nepal and in the World. *Arch. Agric. Environ. Sci.* **2019,** *4* (1), 83–87.

Ghosh, S.; Gupta, S. K.; Jha, G. Identification and Functional Analysis of AG1-IA Specific Genes of *Rhizoctonia solani*. *Curr. Genet.* **2014,** *60*, 327–341.

Ghosh, S.; Kanwar, P.; Jha, G. Identification of Candidate Pathogenicity Determinants of *Rhizoctonia solani* AG1-IA, which causes Sheath Blight Disease in Rice. *Curr. Genet.* **2018,** *64*, 729–740.

Ghosh, S.; Mirza, N.; Kanwar, P.; Tyagi, K; Jha, G. Genome Analysis Provides Insight about Pathogenesis of Indian Strains of *Rhizoctonia solani* in Rice. *Funct. Integr. Genomics,* **2019,** *19*, 799–810.

Goad, D. M.; Jia, Y.; Gibbons, A.; Liu, Y.; Gealy, D.; Caicedo, A. L.; Olsen, K. M. Identification of Novel QTL Conferring Sheath Blight Resistance in Two Weedy Rice Mapping Populations. *Rice.* **2020,** *13*, 1–10.

Goswami, S.; Thind, T. S.; Kaur, R.; Kaur, M. Management of Sheath Blight of Rice with Novel Action Fungicides. *Indian Phytopath.* **2012,** *65*, 92–93.

Gupta, A. Evaluation of Morphological, Ecological and Molecular Diversity in *Arabidopsis* and Future Prospective: A Review. *J. Biol. Nat.* **2020,** *12*, 32–40.

Gupta, A.; Bhardwaj, A.; Sawant, S. V.; Yadav, H. K. Utilization and Characterization of Genome-wide SNP Markers for Assessment of Ecotypic Differentiation in *Arabidopsis thaliana*. *Int. J. Curr. Microbiol. App. Sci.* **2019,** *8* (6), 158–173.

Gupta, A.; Jaiswal, V.; Sawant, S. V.; Yadav, H. K. Mapping QTLs for 15 Morpho-metric Traits in *Arabidopsis thaliana* using Col-0× Don-0 Population. *Physiol. Mol. Biol. Plants* **2020,** *1*, 1–4.

Gupta, A.; Maurya, R.; Roy, R. K.; Sawant, S. V.; Yadav, H. K. AFLP based Genetic Relationship and Population Structure Analysis of Canna—An Ornamental Plant. *Sci. Hortic.* **2013,** *154*, 1–7.

Hane, J. K.; Anderson, J. P.; Williams, A. H.; Sperschneider, J.; Singh, K. B. Genome Sequencing and Comparative Genomics of the Broad Host Range Pathogen R. *solani* AG8. *PLOS Genet.* **2014,** *10*, e1004281.

Hapsari, L. L.; Poromarto, S. H. Characters of Local Varieties cv. Padi Hitam and Srikiti as the Moderate Resistant Rice to Sheath Blight caused by *Rhizoctonia solani*. *Ann. Agri Bio Res.* **2020,** *25* (1), 54–60.

Into, P.; Khunnamwong, P.; Jindamoragot, S.; Am-in, S.; Intanoo, W.; Limtong, S. Yeast Associated with Rice Phylloplane and their Contribution to Control of Rice Sheath Blight Disease. *Microorganisms*. **2020,** *8* (3), 362.

IRRI. *Standard Evaluation System for Rice*, 5th edn., Nov. 2002, INGER, Genetic Resources Center, IRRI, P.O. Box 933, 1099, Manila, Philippines, 2002, pp 56.

Jain, A. K.; Joshi, R. P. Efficacy of Fungicides for Management of Banded Leaf and Sheath Blight of Kodo Millet caused by *Rhizoctonia solani* Kühn. *Ann. Plant Prote. Sci.* **2020,** *28* (1), 60–62.

Jamali, H.; Harma, A.; Srivastava, A. K. Biocontrol Potential of *Bacillus Subtilis* RH5 against Sheath Blight of Rice caused by *Rhizoctonia solani*. *J. Basic Microbiol.* **2020,** *60* (3), 268–280.

Jankar, K.; Mane, S. S.; Tatte, R.; Kose, P. Plant Disease Management Strategies under Changing Climate Scenario. *J. Pharmacogn. Phytochem.* **2020,** *9* (4), 950–954.

Jia, Y. Registration of Lesion Mimic Mutant of Katy Rice. *Crop Science.* **2005,** *45,* 1675.

Jia, Y. Understanding the Molecular Mechanisms of Rice Blast Resistance using Rice Mutants. In Shu QY, editor.; *Induced Plant Mutations in the Genomics Era*. Food and Agriculture Organization of the United Nations: Rome, 2009, 375–378.

Jia, Y.; Jia, M. H.; Wang, X.; Zhao, H. A. Toolbox for Managing Blast and Sheath Blight Diseases of Rice in the United States of America. In *Protecting Rice Grains in the Post-Genomic Era. IntechOpen.* **2019,** 1–18.

Jiang, N.; Yan, J.; Liang, Y; Shi, Y.; He, Z.; Wu, Y.; Zeng, Q.; Liu, X.; Peng J. Resistance Genes and their Interactions with Bacterial Blight/Leaf Streak Pathogens (*Xanthomonas oryzae*) in Rice (*Oryza sativa* L.)—an Updated Review. *Rice.* **2020,** *13* (1), 1–2.

Kandhari; Janki. Management of Sheath Blight of Rice through Fungicides and Botanicals. *Indian Phytopath.* **2007,** *60*, 214–217.

Kaur, A.; Dhaliwal, L. K.; Pannu, P. P. Role of Meteorological Parameters on Sheath Blight of Rice under Different Planting Methods. *Int. J. Bio.resour. Stress Manag.* **2015,** *6* (2), 214–219.

Khoshkdaman, M.; Mousanejad ,S.; Elahinia, S. A.; Ebadi, A. A.; Padasht-Dehkaei, F. Impact of Soil-borne Inoculum on Sheath Blight Disease Development in Rice. *J. Crop Prot.* **2020,** *9* (4), 625–635.

Kim, J. G.; Park, B. C.; Liu, C.; Zhao, G. Q.; Kim, H. J.; Ahn, E. K. Productivity and Quality of Whole Crop Rice Varieties in Relation to Plant Components. *Grassland Sci.* **2020,** *66* (1), 40–47.

Kora, A. J.; Mounika, J.; Jagadeeshwar, R. Rice Leaf Extract Synthesized Silver Nanoparticles: An in Vitro Fungicidal Evaluation against *Rhizoctonia solani*, the Causative Agent of Sheath Blight Disease in Rice. *Fungal Biol.* **2020,** *124* (7), 671–681.

Li, C.; Yang, J.; He, X.; Zhu, S.; Zhu, Y. Agricultural Biodiversity for Crop Disease and Pest Management. Springer, Cham. *Integr. Biol. Control.* **2020b,** 117–132.

Li, D; Wang, R.; Xie, C.; Liu, L.; Zhang, J.; Li, R.; Wang, F.; Zhou, M.; Liu, W. A Recognition Method for Rice Plant Diseases and Pests Video Detection based on Deep Convolutional Neural Network. *Sensors.* **2020c,** *20* (3), 578.

Li, M.; Li, R.; Zhang, J.; Wu, T.; Liu, S.; Hei, Z.; Qiu, S. Effects of the Integration of Mixed-cropping and Rice–duck Co-culture on Rice Yield and Soil Nutrients in Southern China. *J. Sci. Food Agric.* **2020a,** *100* (1), 277–286.

Li, W. T.; Zhu, Z. W.; Chern, M. S.; Yin, J. J.; Yang, C; Ran, L; Cheng, M.; He, M.; Wang, K.; Wang, J. and Zhou, X. A Nature Allele of a Transcription Factor in Rice Confers Broad-Spectrum Blast Resistance. *Cell.* **2017,** *170*, 114–126.

Lin, R.; He, L.; He, J.; Qin, P.; Wang, Y.; Deng, Q.; Yang, X.; Li, S.; Wang, S., Wang, W.; Liu, H. Comprehensive Analysis of MicroRNA-Seq and Target mRNAs of Rice Sheath Blight Pathogen Provides New Insights into Pathogenic Regulatory Mechanisms. *DNA Res.* **2016,** *23*, 415–425.

Lindsey, A. P.; Murugan, S.; Renitta, R. E. Microbial Disease Management in Agriculture: Current Status and Future Prospects. *Biocatal. Agric. Biotechnol.* **2020,** 23, 101468.

Liu, G.; Jia, Y.; McClung, A.; Oard, J. H.; Lee, F. N.; Correll, J. C. Confirming QTLs and Finding Additional Loci Responsible for Resistance to Rice Sheath Blight Disease. *Plant disease.* **2013,** *97* (1), 113–117.

Lo Presti, L.; Lanver, D.; Schweizer, G.; Tanaka, S.; Liang, L.; Tollot, M.; Zuccaro, A.; Reissmann, S.; Kahmann, R. Fungal Effectors and Plant Susceptibility. *Annu. Rev. Plant Biol.* **2015,** *66*, 513–545.

Ma, H.; Sheng, C.; Qiao, L.; Zhao, H.; Niu, D. A Comparative Proteomic Approach to Identify Defence-related Proteins between Resistant and Susceptible Rice Cultivars Challenged with the Fungal Pathogen *Rhizoctonia solani*. *Plant Growth Regul.* **2020,** *90* (1), 73–88.

Maeda, S.; Dubouzet, J. G.; Kondou, Y.; Jikumaru, Y.; Seo, S.; Oda, K.; Matsui, M.; Hirochika, H.; Mori, M. The Rice CYP78A Gene BSR2 Confers Resistance to *Rhizoctoniasolani* and Affects Seed Size and Growth in *Arabidopsis* and Rice. *Sci rep.* **2019,** *9* (1), 1–4.

Manam, V. K.; Subbaiah, M. Biosynthesis and Characterization of Silver Nanoparticles from Marine Macroscopic Red Seaweed *Halymenia Porphyroides Boergesen* (Crypton) and Its Antifungal Efficacy against Dermatophytic and Non-dermatophytic Fungi. *Asian J. Pharm. Clin. Res.* **2020,** *13* (8), 174–181.

Manam. D. V. Biosynthesis and Characterization of Silver Nanoparticles from Marine Seaweed *Colpomenia Sinuosa* and its Antifungal Efficacy. *Int. J. Pharma. Bio. Sci.* **2020.**

Manosalva, P. M.; Davidson, R. M.; Liu, B.; Zhu, X.; Hulbert, S. H.; Leung, H.; Leach, J. E. A Germin-like Protein Gene Family Functions as a Complex Quantitative Trait Locus Conferring Broad-spectrum Disease Resistance in Rice. *Plant Physiol.* **2009,** *149*, 286–296.

Maurya, R.; Gupta, A.; Singh, S. K.; Rai, K. M.; Katiyar, R.; Sawant, S. V.; Yadav, H. K. Genomic-derived Microsatellite Markers for Diversity Analysis in *Jatropha curcas*. *Trees.* **2015,** *29* (3), 849–858.

Maurya, R.; Gupta, A.; Singh, S. K.; Rai, K. M.; Sawant, S. V.; Yadav, H. K. Microsatellite Polymorphism in *Jatropha curcas* L.—A Biodiesel Plant. *Ind. Crops Prod.* **2013,** *49*, 136–142.

Miah, G.; Rafii, M. Y.; Ismail, M. R.; Sahebi, M.; Hashemi, F. S.; Yusuff, O.; Usman, M. G. Blast Disease Intimidation Towards Rice Cultivation: A Review of Pathogen and Strategies to Control. *J. Anim. Plant. Sci.* **2017,** *27*, 1058–1066.

Mishra, A. K.; Mottaleb, K. A.; Khanal, A. R.; Mohanty, S. Abiotic Stress and Its Impact on Production Efficiency: The Case of Rice Farming in Bangladesh. *Agric. Ecosyst. Environ.* **2015,** *1* (199), 146–53.

Mishra, D.; Rajput, R. S.; Zaidi, N. W.; Singh, H. B. Sheath Blight and Drought Stress Management in Rice (*Oryza sativa*) through Trichoderma spp. *Indian Phytopathol.* **2020,** *73* (1), 71–77.

Molla, K. A.; Karmakar, S.; Molla, J.; Bajaj, P.; Varshney, R. K.; Datta, S. K.; Datta, K. Understanding Sheath Blight Resistance in Rice: The Road Behind and the Road Ahead. *Plant Biotechnol. J.* **2020,** *18* (4), 895–915.

Mukherjee, A.; Patel, J. S. Seaweed Extract: Biostimulator of Plant Defense and Plant Productivity. *Int. J. Environ. Sci. Technol.* **2020,** *17* (1), 553–558.

Mumtaz, M. Z.; Saqib, M.; Abbas, G.; Akhtar, J.; Ul-Qamar, Z. Drought Stress Impairs Grain Yield and Quality of Rice Genotypes by Impaired Photosynthetic Attributes and K Nutrition. 水稻科学. **2020,** *27* (1):5–9.

Muthu, V.; Abbai, R.; Nallathambi, J.; Rahman, H.; Ramasamy, S.; Kambale, R.; Thulasinathan, T.; Ayyenar, B.; Muthurajan, R. Pyramiding QTLs Controlling Tolerance against Drought, Salinity, and Submergence in Rice through Marker Assisted Breeding. *PloS One.* **2020,** *15* (1), e0227421.

Nadana, G. R. V.; Rajesh, C.; Kavitha, A.; Sivakumar, P.; Sridevi, G.; Palanichelvam, K. Induction of Growth and Defense Mechanism in Rice Plants towards Fungal Pathogen by Eco-friendly Coelomic Fluid of Earthworm. *Environ. Technol. Innov.* **2020,** 101011.

Nadarajah, K.; Mat Razali, N.; Cheah, B. H.; Sahruna, N. S.; Ismail, I.; Tathode, M.; Bankar, K. Draft Genome Sequence of *Rhizoctonia solani* Anastomosis Group 1 Subgroup 1A Strain 1802/KB Isolated from Rice. *Genome Announc.* **2017,** *5*, e01188–e01117.

Nagarajkumar, M.; Jayaraj, J.; Muthukrishnan, S.; Bhaskaran, R.; Velazhahan, R. Detoxification of Oxalic Acid by Pseudomonas Fluorescens Strain PfMDU2: Implications for the Biological Control of Rice Sheath Blight Caused by *Rhizoctonia. solani*. *Microbiol. Res.* **2005,** *160*, 291–298.

Nattawut, S.; Lampan, K.; Sukunya, Y.; Sudteerak, S.; Patcharaporn, S.; Potjana, S.; Nittaya, P. Residual Effects of Intercrop System and Rice as Mono Crop on Growth and Yield of Maize in Rice-based Production System. *Int. J Agric. Technol.* **2020,** *16* (2), 309–318.

Oreiro, E. G.; Grimares, E. K.; Atienza-Grande, G.; Quibod, I. L.; Roman-Reyna, V.; Oliva R. Genome-Wide Associations and Transcriptional Profiling Reveal ROS Regulation as One Underlying Mechanism of Sheath Blight Resistance in Rice. *Mol. Plant Microbe. Interact.* **2020,** *33* (2), 212–222.

Parveen, S.; Ali, M. A.; Ali, M. A. Evaluation of Fungicides Combinations against Sheath Blight Disease of Rice in Bangladesh. *Eco-friendly Agril. J.* **2020,** *13* (01), 01–06.

Petit-Houdenot, Y.; Fudal, I. Complex Interactions between Fungal Avirulence Genes and their Corresponding Plant Resistance Genes and Consequences for Disease Resistance Management. *Front. Plant Sci.* **2017,** *8*, 1072.

Prabhu, A. S.; Barbosa Filho, M. P.; Filippi, M. C.; Datnoff, L. E.; Snyder, G. H. Silicon from Rice Disease Control Perspective in Brazil. Elsevier. *Studies plant sci.* **2001,** 8, 293–311.

Pralhad, S. P. Evaluation of Actinobacteria for Biocontrol of Sheath Blight in Rice. *J. Pharmacog. Phytochem.* **2020,** *9* (3), 371–386.

Pramesh, D.; Chittaragi, A.; Nagaraj, T.; Saddamhusen, A.; Chidanandappa, E.; Sharanabasav, H.; Raghunandana, A.; Mahesh, M. H.; Kumar, S. H.; Mahantashivayogayya, K.; Reddy, B. G. Bio-efficacy of Combination Fungicide Prochloraz 27%+ tricyclazole 23% SE for the Management of Blast Disease of Rice. *J. Pharmacog. Phytochem.* **2020,** *9* (4), 3027–3032.

Prasad, N.; Singh, N.; Avinash, P.; Tiwari, P. K. Efficacy of Botanical Plant Product and Extracts against *Rhizoctonia solani* Kuhn causing Sheath Blight Disease of Rice under In-Vitro Condition. *J. Pharmacogn. Phytochem*. **2020b,** *9* (3), 312–315.

Prasad, N.; Singh, N.; Avinash, P.; Tiwari, P. K. Screening/Rescreening of Rice Entries for Sheath Blight Resistant under Field Condition. *J. Pharmacogn. Phytochem*. **2020a,** *9* (3), 444–446.

Prasad, P.S.; Naik, M. K.; Nagaraju, P. Screening of Genotypes, Fungicides, Botanicals and Bio-agents against *Rhizoctonia solani*, the Incitant of Sheath Blight of Rice. *Proc. Natnl. Seminar New Front. Pl. Pathol.* **2006,** 139.

Premalatha Dath, A. *Sheath Blight of Rice and its Management*. Associated Publishing Co., Shidipura, Karol Bagh, New Delhi, 1990, pp129.

Qiao, L.; Zheng, L.; Sheng, C.; Zhao, H.; Jin, H.; Niu, D. Rice siR109944 Suppresses Plant Immunity to Sheath Blight and Impacts Multiple Agronomic Traits by Affecting Auxin Homeostasis. *The Plant J.* **2020,** *102* (5), 948–964.

Raghunandana, M. H.; Maruti, B. T.; BT, R.; Sujay Huruli, B. G. Bio-efficacy of Tricyclazole 22.5% W/V+ azoxystrobin 7.5% W/V against Sheath Blight Disease of Paddy. *IJCS*. **2020,** *8* (4), 2928–2931.

Rahman, C. R.; Arko, P. S.; Ali, M. E.; Khan, M. A.; Apon, S. H.; Nowrin, F.; Wasif, A. Identification and Recognition of Rice Diseases and Pests using Convolutional Neural Networks. *Biosyst Eng.* **2020**, *194*, 112–120.

Raj, T. S.; Muthukumar, A.; Renganathan, P.; Suji, H. A. Efficacy of Seaweed and Seawater Associated *Bacillus velezensis* against Sheath Blight of Rice caused by *Rhizoctonia solani kuhn*. *Plant Arch.* **2020,** *20* (1), 3727–3731.

Rajesh, T.; Maruthasalam, S.; Kalpana, K.; Poovannan, K.; Kumar, K. K.; Kokiladevi, E.; Sudhakar, D.; Velazhahan, R.; Balasubramanian, P. Pyramiding Insect and Disease Resistance in an Elite Indica Rice Cultivar asd16. *Biologia Plantarum.* **2020,** *64*, 77–86.

Razzaq, A.; Ali, A.; Safdar, L. B.; Zafar, M. M.; Rui, Y.; Shakeel, A.; Shaukat, A; Ashraf, M.; Gong, W.; Yuan, Y. Salt Stress Induces Physiochemical Alterations in Rice Grain Composition and Quality. *J. Food Sci.* **2020,** *85* (1), 14–20.

Rengasamy, K. R.; Mahomoodally, M. F.; Aumeeruddy, M. Z.; Zengin, G.; Xiao, J.; Kim, D. H. Bioactive Compounds in Seaweeds: An Overview of their Biological Properties and Safety. *Food chem. Toxicol.* **2020,** *135*, 111013.

Sabes, P. L.; Lon, M. M.; Peter, M. A.; Maruyama, J.; Koyama, S.; Watanabe, T.; Koizumi, S. Effect of Increased Silicon Content of Paddy Rice on Sheath Blight Development through Carbonized Rice Husk Application. *Jpn. Agric. Res. Q.* **2020,** *54* (2), 145–151.

Sagar, V.; Dhawan, G.; Krishnan, S. G.; Vinod, K. K.; Ellur, R. K.; Mondal, K. K.; Rathour, R.; Prakash, G.; Nagarajan, M.; Bhowmick, P. K.; Bollinedi, H. Marker Assisted Introgression of Genes Governing Resistance to Bacterial Blight and Blast Diseases into an Elite Basmati Rice variety, 'Pusa Basmati 1509'. *Euphytica.* **2020,** *216* (1), 1–8.

Saha, S. Evaluation of New Fungicidal Formulations against Sheath Blight of Rice in West Bengal. *Environ. Ecol.* **2003,** *21*, 237–239.

Silva, J.; Scheffler, B.; Sanabria, Y.; De Guzman, C.; Galam, D.; Farmer, A.; Woodward, J.; May, G.; Oard, J. Identification of Candidate Genes in Rice for Resistance to Sheath Blight Disease by Whole Genome Sequencing. *Theor. Appl. Genet.* **2012,** *124*, 63–74.

Singh, P.; Verma, R. L.; Singh, R. S.; Singh, R. P.; Singh, H. B.; Arsode, P.; Kumar, M.; Singh, P. K. Biotic Stress Management in Rice (*Oryza sativa* L.) through Conventional and Molecular Approaches. In *New Frontiers in Stress Management for Durable Agriculture*, Springer, Singapore, **2020,** 609–644.

Singh, R.; Sunder, S.; Kumar, P. Sheath Blight of Rice: Current Status and Perspectives. *Indian Phytopathol.* **2016,** *69* (4), 340–351.

Sun, Q.; Li, D. D.; Chu, J.; Li, S.; Zhong, L. J.; Han, X.; Xuan, Y. H. Indeterminate Domain Proteins Regulate Rice Defense to Sheath Blight Disease. *Rice.* **2020,** *13* (1), 1–2.

Sun, Q.; Li, T. Y.; Li, D. D.; Wang, Z. Y.; Li. S.; Li, D. P.; Han, X.; Liu, J. M.; Xuan, Y. H. Overexpression of Loose Plant Architecture 1 Increases Planting Density and Resistance to Sheath Blight Disease via Activation of PIN-FORMED 1a in Rice. *Plant Biotechnol. J.* **2019,** *17* (5), 855.

Surin, P.; Chomngam, K.; Yampracha, S. Effects of Different Organic Fertilizers on Residual Nutrients, Rice Growth and Yield. *Int. J. Agric. Technol.* **2020,** *16* (2), 443–456.

Suryadi, Y.; Susilowati, D. N.; Fauziah, F. Management of Plant Diseases by PGPR-Mediated Induced Resistance with Special Reference to Tea and Rice Crops. In *Plant Growth Promoting Rhizobacteria for Sustainable Stress Management,* Springer, Singapore, **2019,** 65–110.

Syed-Ab-Rahman, S. F.; Carvalhais, L. C.; Omar, D. Development of Plant-based Emulsion Formulations to Control Bacterial Leaf Blight and Sheath Brown Rot of Rice. *Heliyon.* **2020,** *6* (1), e03151.

Tang, Q.; Peng, S.; Buresh, R. J.; Zou, Y.; Castilla, N. P.; Mew, T. W.; Zhong, X. Rice Varietal Difference in Sheath Blight Development and its Association with Yield Loss at Different Levels of N Fertilization. *Field Crops Res.* **2007,** *102*, 219–227.

Thakur, M.; Tamin, V. U.; Tiwari, P. K. Biological control of sheath blight of rice. *J. Pharmacog. Phytochem.* **2020,** *9* (2), 2264–2267.

Tiwari, M.; Srivastava, S.; Singh, P. C.; Mishra, A. K.; Chakrabarty, D. Functional Characterization of tau class glutathione-S-transferase in Rice to provide Tolerance against Sheath Blight Disease. *3 Biotech.* **2020,** *10* (3), 84.

Tonnessen, B. W.; Bossa-Castro, A. M.; Mauleon, R.; Alexandrov, N.; Leach, J. E. Shared Cis-regulatory Architecture Identified across Defense Response Genes is Associated with Broad-Spectrum Quantitative Resistance in Rice. *Sci. rep.* **2019,** *9* (1), 1–3.

Uribe, E.; Vega-Gálvez, A.; García, V.; Pastén, A.; Rodríguez, K.; López, J.; Scala, K. D. Evaluation of Physicochemical Composition and Bioactivity of a Red Seaweed (*Pyropia orbicularis*) as Affected by Different Drying Technologies. *Drying Technol.* **2020,** *38* (9), 1218–1230.

Wahab, M. Z.; Hasan, N. B.; Bakhari, F. A.; Anuar, I. S.; Sapak, Z. Effect of Antimicrobial Activity from Fermented Mature Coconut Water to Sheath Brown Rot of Rice Disease. *Int. J. Agric. Resour. Governance Ecol.* **2020,** *16* (2), 110–122.

Wang, X.; Wu, D.; Jia, Y.; Zhang, M. Characterization of Rice Lesion Mimic Mutants of 93-11 for a Better Understanding of General Host Defense Response to both Rice Blast and Sheath Blight Diseases [abstract]. In *Fifth International Rice Blast Conference*, August 12-14, Little Rock, Arkansas, 2011, p.23.

Wu, C.; Bordeos, A.; Madamba, M. R.; Baraoidan, M.; Ramos, M.; Wang, G. L.; Leach, J. E.; Leung, H. Rice Lesion Mimic Mutants with Enhanced Resistance to Diseases. *Mol. Genet. Genomics.* **2008,** *279* (6), 605–619.

Xia, Y.; Fei, B.; He, J.; Zhou, M.; Zhang, D.; Pan, L.; Li, S.; Liang, Y.; Wang, L.; Zhu, J.; Li, P. Transcriptome Analysis Reveals the Host Selection Fitness Mechanisms of the *Rhizoctonia solani* AG1IA Pathogen. *Sci. Rep.* **2017,** *7* (1), 1–6.

Xu, Z.; Li, C.; Zhang, C.; Yu, Y.; Van der Werf, W.; Zhang, F. Intercropping Maize and Soybean Increases Efficiency of Land and Fertilizer Nitrogen Use; A Meta-analysis. *Field Crops Res.* **2020,** *246*, 107661.

Yang, J. H.; Zhang, W.W.; Zhuang, Y. Q.; Xiao, T. Biocontrol Activities of Bacteria from Cowdung against the Rice Sheath Blight Pathogen. *J. Plant Dis. Protect.* **2017,** *124* (2), 131–141.

Yellareddygari, S. K. R.; Reddy, M.S.; Kloepper, J. W.; Lawrence, K. S.; Fadamiro, H. Rice Sheath Blight: A Review of Disease and Pathogen Management Approaches. *J. Plant Pathol. Microbiol.* **2014,** *5* (4), p1.

Zhou, Y.; Bao, J.; Zhang, D.; Li, Y.; Li, H.; He, H. Effect of Heterocystous Nitrogen-fixing Cyanobacteria Against Rice Sheath Blight and the Underlying Mechanism. *Appl. Soil Ecol.* **2020,** *153*, 103580.

CHAPTER 7

Insights from Genetics, Breeding, and Molecular Approaches for Rice Sheath Blight Disease Resistance

RAJA HUSAIN[1,3,4], NITIN VIKRAM[2], SONIKA PANDEY[3], GARIMA YADAV[3], N. A. KHAN[4], MD. SHAMIM[5], DEEPTI SRIVASTAVA[6], TOUSEEF HUSSAIN[7], and S. P. TIWARI[8]

[1]*Department of Agriculture, Himalayan University, Jullang, Itanagar, Arunachal Pradesh, India*

[2]*Department of Agriculture, Zila Parishad Krishi Mahavidyalaya, Banda, Uttar Pradesh, India*

[3]*ICAR-Indian Institute of Pulses Research, Kanpur, Uttar Pradesh, India*

[4]*Department of Plant Molecular Biology and Genetic Engineering, Narendra Deva University of Agriculture and Technology, Kumarganj, Ayodhya, Uttar Pradesh, India*

[5]*Department of Molecular Biology and Genetic Engineering, Dr. Kalam Agricultural College (BAU, Sabour), Kishanganj, Bihar, India*

[6]*Integral Institute of Agriculture Sciences and Technology, Integral University, Lucknow, Uttar Pradesh, India*

[7]*Division of Plant Pathology, ICAR-IARI, Pusa, New Delhi, India*

[8]*Faculty of Agriculture, Mangalaytan University, Jabalpur, India*

Fungal Diseases of Rice and Their Management. Deepti Srivastava, Md. Shamim, Malik Mobeen Ahmad, and R. S. Upadhyay (Eds.)

ABSTRACT

Rhizoctonia solani causes sheath blight, which is a widespread disease. ShB reduced severe yield loss by 30–50% in rice-growing regions. No effective management has been developed until now against this ailment. Many cultivation practices are being adopted to block the growth of *Rhizoctonia solani*. Fertilizers including fungicides could be used in paddy crops but chemicals are being proven to be health hazardous. Apart from this, several breeding approaches like Marker-assisted selection, gene pyramiding, and mutation breeding are commonly used for the development of the resistant cultivar against sheath blight. There are so many QTLs/genes related to SHB that have been identified with the help of proteomics, genomics, and transcriptomics. Sheath blight is governed by quantitative traits so the development of sheath blight resistant rice varieties is much more difficult through the traditional breeding approach. Hence, the pyramiding of QTLs may result in stable and potential rice cultivars. *Japonica* is one of the resistant cultivar sheath blight which is caused by the pyramiding of the genes qSB-9TQ and qSB-7TQ on chromosomes 9 and 7, respectively. NILs with both TAC1TQ and qSB-9TQ displayed higher resistance than those with just one of them. Improved Pusa Basmati was created by adding the blast resistance gene Pi54 and sheath blight resistant quantitative trait loci (QTL) from Tetep, qSBR11-1. For QTL and association mapping, SNPs are widely used. Rice plants have a huge number of SNP markers and a highly dense rice genome has been built by utilizing 3.6 million SNPs. A study performed the identification of QTLs position, six QTLs were found which were related to sheath blight by using 128 lines of a hybrid between Jasmin 85 (resistant) and Lemont produced an F2 clonal population (susceptible). In another study, 15 more QTLS were derived using a inbreed line population of Tequing and Lemont. Resistant genes like PR-3 chitinase and PR-5 (thaumatin-like protein) like Plant pathogenesis-related (PR) genes provided resistance against sheath blight disease. Instead of a single PR gene, a combination of two PR genes results in a more efficient and greater level of sheath blight resistance. When the Endo-chitinase gene (cht42) was expressed in transgenic rice, sheath blight disease was reduced by up to 62%.

7.1 INTRODUCTION

Rhizoctonia solani Kuhn causes rice sheath blight, which is one of the most frequent and damaging diseases of rice (Rao et al., 2020). SBH is also known as Snake-skin disease or rotten foot stalk or mosaic foot stalk because of its unique symptoms (Molla et al., 2020). In Japan, sheath blight was reported in 1910 and it is recognized as the second most important rice disease which reduced the significant amount of yield losses approximately 30–50% in rice growing region (Zhu et al., 2019). As against SHB, no completely resistant cultivar has been developed.

In the present scenario, fungicides, and many cultivation practices are being adopted to block the growth of *Rhizoctonia solani* (Singh et al., 2019). Apart from several breeding approaches like marker-assisted selection, gene pyramiding, and mutation breeding have been used for the development of the resistant cultivar against SBH. Advances in omics such as proteomics, genomics, and transcriptomics, QTLs/genes related to SBH resistance have been identified. Integrative management strategies have been recently researched closely (Molla et al., 2020). During the last decades, the wetness in the field increased the severity of SHB because of the application of high dose of nitrogen fertilizer (Yellareddyrari et al., 2014). Higher crop density and high humidity encourage SHB to become more severe (Kannaiyan and Prasad, 1983). The chapter focused on the breeding and molecular approaches for the management of SHB in rice. *Rizoctonia solani* is a soil-water-borne, highly infectious fungus with a total of 14 distinct anastomosis groups (AG) for different *Rhizoctonia solani* isolates (Carling et al., 2002a). Among 14 AG groups, 13 groups were denoted as AG1 to AG13, and 14th group was denoted as AGB1 this is known as bridging isolation. On the basis of DNA sequencing study the morphology and homology of sclerotia, *Rhizoctonia solani* AG1 isolates have subdivided into three subgroups IA, IB, and IC (Sneh et al., 1991) the ID, IE, and IF were proposed and argued based on isozyme studies and rDNA-ITS sequences, respectively. (Liu and Sinclair, 1993; Priatmojo et al., 2001). The most accepted isolates or *Rhizoctonia solani* that is AG1 - IA is the causal organism of SBH in rice.

7.2 OVERVIEW OF THE PATHOGENS AND SHEATH BLIGHT DISEASES

Rhizoctonia solani is a necrotrophic soil-borne fungus that is a widespread saprotrophic and facultative plant parasite that survives in soil (Ogoshi, 1996; Anees et al., 2010). It infects paddy plants and promotes sheath blight disease symptoms. *R. solani* has a very wide range of host plants about 250 different plant species (Chahal et al., 2003; Dath, 1990). *R. solani* is a species complex of several anastomosis groups based on the hyphal fusion condition of isolates (Table 7.1) which differ in their both phenotypic and genotypic characters (Carling et al., 2002; Anees et al., 2010; Summer, 1996).

Rhizoctonia solani sclerotia has the ability to survive for two years and under unfavorable survival conditions. During the field preparation by flooding irrigation, *R. solani* spores spread (Webster and Gunnell, 1992; Brooks, 2007). Sclerotia/hyphae binds the plant, or any plant part and if the conditions are favorable then it infects plant, and the symptoms of sheath blight disease appear. Primarily the initial symptoms appear on the leaf sheath near the water level line as water-soaked lesions and the secondary infections and symptoms are due to hyphae growing upward toward the upper uninfected plant, producing so many lesions and sclerotia on leaf sheaths for the disease cycle (Webster and Gunnell, 1992; Brooks, 2007).

Rice sheath blight disease, caused by *Rhizocotnia solani*, is a worldwide devastating disease that can cause severe yield loss as well as quality degradation (Brooks, 2007; Lee and Rush, 1983; Nagarajkumar et al., 2004).

The severity of sheath blight can be increased by rice cultivar morphological features and their relationship with SBH development indicates that the plant modifies the microclimate and light transmission inside the dense canopy, favoring disease development (Han et al., 2003; Tang et al., 2007; Brooks, 2007).

Low sunshine, high humidity, and high temperatures are all favorable for SHB formation (28–32°C).

R. solani spreads rapidly through runner hyphae to upper plant sections as well as neighboring plants in favorable conditions.

In most cases, disease growth occurs at a faster rate during the early phases of heading and grain filling.

Symptoms on rice typically begin above water level or above soil level in upland conditions with pale, greenish-gray, ellipsoidal lesions that are about 2–3 cm long at first and have a dark brown margin, and then enlarge and coalesce, changing color to fawn or off-white with a brown or purplish-brown margin.

Lesions can form anywhere on the leaf sheath, but they usually start on the outer part of the sheath and then spread to the inner sheath. They can even be found on the leaf blade in humid warm regions and in the critical stage, a strong attack can lead to plant death (Gangopadhyay and Chakrabarti, 1982).

Sclerotia can infect a variety of plants, including barley, lettuce, tomato, sorghum, and maize, in addition to rice (Zhang et al., 2009).

Without SBH protection techniques, yield losses range from 10 to 30% (Xie et al., 2008) and can reach 50% during peak years (Meng et al., 2001; Ou, 1985; Damicone et al., 1993).

Rice seedlings may become infected with *R. solani* if they are transferred into diseased soil.

They germinate when they come in contact with a rice plant and the emerging hyphae form infection cushions on the exposed leaf sheath (Lee and Rush, 1983; Gangopadhyay and Chakrabarti, 1982) and now infected cushions, haustoria grow, and penetrate to the host tissue and the typical lesions will develop. Sclerotia of SHB are generally formed and developed after the death of invaded cells and the fungal growth has been found most prolific. However, the exact elements that cause sclerotia to emerge from the soil differ greatly in the literature.

In fields with a history of high sheath blight infection, Lee (1980) found 216 to 701 sclerotia per liter of soil in the top 0–0.6 cm layer, with viability ranging from 42.3 to 51.4%.

At depths ranging from 0.6 to 7.6 cm, this was reduced to 27 87 sclerotia with a viability range of 14.2–30.3%.

It has been thoroughly documented and studied that sclerotia can live for up to two years.

Because of the importance of SHB in rice production, substantial research, and extension operations have been conducted around the world in order to establish integrated management techniques based on cultural practices, resistant cultivars, and chemical control and as well as biological control for the effective control of SBH.

Management of sheath blight disease in rice is difficult due the broad host range and high genetic variability of its pathogen. From the existing rice germplasm, finding any suitable stage of natural resistance is quite challenging.

It is therefore imperative to discover any long-term solutions to tackle the *R. solani* infection in order to reduce rice yield losses while also lessening the threat to global food security.

One of the alternate methods for using organic fungicides and avoiding the use of harmful fungicides is to create resistance at the genetic level.

This chapter is mainly based on the efforts to understand the relationship between host and pathogen and also to find the gene loci/marker resistance and modify the host genome by transgenic development. Signaling pathways play an important role during pathogen attack; plants have been able to protect themselves by activating the signaling complex pathways. There are so many complexes such as salicyclic acid, jasmonic acid, and ethylene that plays important role in pathogen-responsive signaling pathways (Glazebrook, 2005; Kunkel and Brooks, 2002; Nagarajkumar et al., 2005; Garc et al., 2006). Generally, salicyclic acid mediated signaling induces resistance against the biotrophic pathogens while jasmonic acid mediated signaling induces resistance against the nectrotrophic fungi (Browse, 2009; Glazerbrook, 2005; Nagarajkumar et al., 2005; Oka et al., 2013). The development of sheath blight disease with high rate during flowering while rice has dense canopy, forming a favorable microclimate for pathogen growth and spread (Brooks, 2007). *Rhizoctonia solani* has the potential to infect the seed during mature plants also, causing moderate to significant yield losses depending on the plant part that is affected. The appearance of so many symptoms such as formation of lesions, plant lodging, and empty grains at rice plants. Softness of stem in lower leaves appears due to large lesions on the infected sheath and this initiates stem lodging. Lodging alters the normal rice canopy design, affecting photosynthetic ability, and total biomass production (Wu et al., 2012; Hitaka, 1969). During paddy sheath blight epidemics, this interrupts the translocation of water, nutrients, and carbohydrate assimilates through the vascular bundle channels and affects the grain filling. It also affects during the panicle initiation/flowering that cause the reduction in total seed weight; reason behind is lower percentage of filled spikelet's and it may result in significant yield losses (Wu et al., 2012; Nagarajkumar et al., 2004). Factors for the sheath blight disease severity, spread, and intensity are paddy developmental stages as infection, ecological surroundings, cultivar resistance, cultural, and seasonal crop practices, previous crop residues, planting material (Groth et al., 1992; Gangopadhyay and Chakrabarti, 1982). The application of high rates of nitrogen containing fertilizers, double cropping system, high plant densities, early maturing, high tiller, and susceptible varieties have been shown to increase the sheath blight severity in most of world's paddy growing areas (Lee and Rush, 1983; Nagarajkumar et al., 2004). Paddy plants respond to all type of stresses in their surroundings, attack by all insects and pests as well as physical stress (Gullino et al., 2000; Glazerbrook, 2001; Ramonell and Somerville, 2002). A virulent *R. solani* pathogen overcomes a plant defense mechanism by evading the effects of activated defenses, eschewing triggering plant guarding or suppressing the plant resistance interactions. There are so many control measures reported in literature such as cultural control, chemical control (Van et al., 1998; Shamim et al., 2016; Shamim et al., 2018; Shamim et al., 2020). Most of the fungicides are derived from β-methoxyacrylate and Azoxystrobin is the most suitable fungicide to control sheath blight (Grichar et al., 2004).

In this chapter basically organic fungicides are the safe to the human health but preparation method, storing, and application of these fungicides are so difficult. So, if chemical control is an alternative effective method to manage the disease; application of these chemicals with the perfect quantity and best method of application or spearing it may be safe for human health as well as effective in managing the disease. Often it has a significant impact on human health and the typical environment through the pollution of soils, above and below ground water resources, and the food supply chain. One another method is biological control using PGPR for controlling the sheath blight and buffer zone between the soil microorganism and plant pathogens is important for natural interference to inhibit and reduce disease development (Knight et al., 1997; Kohl et al., 2011). PGPR is also used commercially for enhancing plant health and growth through seed treatment resulting in increased shoot length of seedlings. These PGPR are ecofriendly in applying and they are typically available in the nature and provide protection against pathogens (Kohl, 2011). There is advancement in PGPR for controlling sheath blight disease which is the development of antagonistic bacteria with plant and root stimulating ability. Local accessions, cultivars, landraces, and advanced breeding lines have attempted to identify the sources of SHB resistance. Chen et al. (2000), Eizenga et al. (2000), Jia et al. (2012) and Dey et al. (2016) found so many genotypes for sheath blight resistance but there is no absolute protection (Shamim et al., 2014). An advanced understanding of microbial biomass dynamics is required before amending the farming practices to increase plant growth and yield. PGPR might be more productive when combined with other sheath blight disease-controlling methods used through an integrated approach.

Fig. a **Fig. b**

FIGURE 7.1 Sheath blight disease in rice. (a) *R. solani* infection in rice; (b) Sclerotia of *R. solani* on PDA (potato dextrose agar) media.

7.3 MARKER ASSISTED SELECTION (MAS) IN RICE

Marker-assisted selection occurs when a trait is selected by utilizing an associated marker rather than phenotypic traits. Cultivar identification, selection of genetically varied parents, identification of the genomic region for the desired trait, and analysis of heterosis are the breeding materials used for marker-assisted selection (Collard and Mackill, 2008). There are so many traits that are controlled by a single gene are called quantitative traits and they are present on chromosome, known as quantitative trait loci (QTL), but some traits of economic importance are controlled by more than one gene (Miah and Ismail, 2013; Vikal and Bhatia, 2017). Those genes that have the larger effect than other genes in QTL are called major gene.

In marker-assisted selection, the following properties are necessary for a DNA-based marker (Collard and Mackill, 2008): (1) Reliability: A molecular marker should be tightly linked to the target traits, less than 5 cm distance is preferable. (2) Quantity and quality of DNA: Some DNA-based marker technique (e.g. RFLP) requires a large amount of DNA with good quality. (3) Reproducible: A DNA-based marker should be highly reproducible and transferable among researchers and laboratories. (4) Polymorphism: A molecular marker should have a high level of polymorphism. It should be able to differentiate the homozygocity and heterozygocity in progeny. (5) Cost effective: The cost of the DNA marker should be cost-effective, cheap, user-friendly, and easy to apply on a large sample/population. The aim of using DNA based marker are basically for following purposes: (1) To find out the desired allele(s) (dominant or recessive) across generations; in order to accumulate desired alleles, (2) Among segregating populations to find out the most suitable individuals, based on the allelic composition of a part or of the entire genome, and (3) The elimination of putative connections between advantageous alleles and unfavorable loci. The molecular marker system enables the high-density DNA marker map to be constructed for a large number of important agriculture crop species providing the need for MAS.

Breeding programs may help to increase the efficiency and effectiveness through the marker assisted selection. Breeders can select the desired plant at the seedling stage; it increases the reliability to select a single plant (Jena and Mackill, 2008). This saves the time as well as cost. In foreground selection breeders are able to discard those plants that are not carrying the target gene; the molecular marker-based selection of plants is more reliable due to the variability in the environmental factors which affect field trials (Bishwas et al., 2016).

Equipment and consumable materials are required to develop and maintain a laboratory for molecular breeding. Development of marker also requires a high initial cost. Low reliability of marker to determine phenotype is another limitation. QTL can be affected by sampling bias in a small population which is detected by high LOD. A high LOD score value is not suitable for marker-assisted selection. The genetic organization of the plant may influence the effect of QTL.

7.4 GENE PYRAMIDING IN RICE BREEDING PROGRAM

Gene pyramiding is the process of combining sensible features from numerous parents into a single genotype in order to generate elite lines and variations. Watson and Singh were the first to propose the concept of gene pyramiding (1953). Gene pyramiding's major goal is to improve trait performance by combining several complimentary genes, compensating for deficits with introgression genes from other sources and boosting durability (Malav et al., 2016). The main goals of gene pyramiding are to improve disease management and increase crop output in order to advance agriculture.

Gene pyramiding may be described in two parts: (1) Root genotype: This aim at cumulating the entire target gene in a single genotype (2) The target gene is fixed into the homozygous strategies during the fixation stage, which determines the ideal genotype from a single genotype. Each node of the tree is known as an intermediate genotype and has two parents. (1) The target trait's or gene's characteristics Reproductive traits are the two key elements that influence gene pyramiding.

Gene pyramiding is a method of integrating numerous disease resistance genes for a given pathogen race.

This is harder to achieve conventional method. Main use of gene pyramiding is to improve exiting elite cultivar and eliminate extensive phenotype. Gene pyramiding controls the linkage drag and eliminates breeding duration.

7.5 BREEDING STRATEGIES AND ROLE OF PLANT BREEDING IN SHEATH BLIGHT MANAGEMENT

The use of resistant cultivars, chemical, cultural, and biological approaches are the options for managing sheath blight disease. Host resistance is the most appropriate approach out of all the ones accessible. The sources of sheath blight resistance have been identified using local accessions, cultivars, landraces, and/or advanced breeding lines. Several genotypes for sheath blight resistance have been reported by Chen et al. (2000), Eizenga et al. (2002), Jia et al. (2012), Dey et al. (2016). However, no absolute resistant genotypes have been established. Disease escape and physiological resistance are the two pathways for sheath blight resistance, according to Sattari et al. (2014). According to Sharma et al. (2009), short stature at the sd-1 semi-dwarfing gene was highly associated with increased sheath blight infection.

Traditional breeding is problematic for developing sheath blight-resistant rice varieties because sheath blight resistance is driven by quantitative features. Pyramiding QTLs may result in cultivars that are both stable and promising. Chen et al. (2014) produced Japonica rice resistance to sheath blight by pyramiding qSB-9TQ and qSB-7TQ on chromosomes 9 and 7, respectively. According to Zuo et al. (2014), NILs carrying both TAC1TQ and qSB-9TQ demonstrated higher resistance than NILs carrying only one of them. To improve Pusa Basmati, Singh et al. (2012) transplanted the blast resistance gene Pi 54 and SBH resistance quantitative trait loci from Tetep and qSBR11-1 to create multiple disease resistance loci for basmati rice.

7.5.1 MARKERS USED IN RICE BREEDING

Rice is a good source of vitamins and minerals and has nutritional value. On the world level production, rice is very high, but it is not sufficient to meet the requirements of growing populations (Leegood et al. 2010). So, for finding the solution to this problem various genomic and proteomic-based techniques are being employed. These molecular approaches are very effective but the cost of these techniques is very high. Conventional techniques which are normally used for improving rice quantity and quality are very much time consuming.

Markers used in rice breeding is of three types:

1. **Morphological:** They are used for the characterization of phenotypic traits like color, size, shape etc.
2. **Biochemical:** The basic principle on which biochemical markers work are enzymatic variations (Agarwal et al., 2008). These markers are still chosen by conventional rice breeders.
3. **Molecular/DNA markers:** These are used for the polymorphism determination. Hybridization markers-restriction fragment length polymorphism markers are the best example of this category (Collard et al., 2005). The first rice molecular map was constructed using 135 RFLP markers by crossing indicia and Japonica cultivars. PCR based markers- In this category RAPD, AFLP, and SSRS are included (Shah et al., 2015; McCouch et al., 2002).

In rice, most commonly SSR markers are used, and rice full genome has a large amount SSR markers. In rice research, various SSR markers have been detected with high polymorphism. SSR markers are frequently used for germplasm characterization but there are some drawbacks like they are difficult to develop. There is trouble in merging SSR data among different working groups. To overcome these problems SNPs are currently used.

Markers based on DNA sequence- Single nucleotide polymorphism-based markers are included in this category. These markers are used for the determination of difference between two DNA sequences based on nucleotide substitution. Both non-coding and coding regions of DNA have SNPs. SNPs can be used for molecular breeding. SNPs are commonly used in QTL and association mapping. These markers are used to study complex genetic traits (Jain et al., 2010). Rice plants have many SNP markers (Zhou et al., 2012, Huang et al., 2010). By using 3.6 million SNPs a highly dense rice genome has been constructed.

7.5.2 QTLS FOR DISEASE RESISTANCE TO RICE SHEATH BLIGHT

Rhizoctonia solani Kuhn causes sheath blight, a serious rice disease. This sickness was originally documented in Japan in 1910 (Miyaki et al., 1910). A susceptibility index for rice sheath blight has been discovered in the field. This disease is resistant to wild accessions. There are only a few resistance loci accessible for this disease (Eizenga and colleagues, 2002). This illness is resistant to only a few kinds (Srinivasachary et al., 2011; Jia et al., 2009).

Before the cultivation of semi-dwarf varieties this disease was not a limiting disease but after the cultivation of semi-dwarf varieties this disease have become the limiting disease of rice production and affects both quality and quantity of rice. Plant height and heading date are the characters which are associated with resistance to sheath blight. Data used for the determination of resistance against sheath blight includes plant height and heading data. Out of six QTLs associated with the disease only one QTL marker is not dependent on height (Li et al., 1995). In rice genome sheath blight related QTLs are present in close association with the QTLs of heading data. Researchers have been producing commercially resistant rice varieties using breeding programmers but these are not very effective. The possible reason for this is because of multiple rice genes control (QTLs) control sheath blast disease resistance. Resistant species can be used for the identification of QTLs related to the sheath blight resistance. New identified QTLs can be joined with already known QTLs to generate rice cultivars having resistance against SBH (Table 7.2). Approximately 20 QTLs have been identified which are related to disease and its resistance. In 1995 (Li et al., 1995) identified the first QTL using RFLP for this disease. Similarly (Pan et al., 1999) identified three QTLs on chromosomes no. 2, 3, and 7. In a study that was focused on the identification of QTLs position six QTLs were found which were related to sheath blight disease by using 128 lines of F2 clonal population by a cross between Jasmine 85 (resistant) and Lemont (Susceptible) (Wen et al., 2015). Kunihiro et al. (2002) obtained four QTLs on chromosome no. 2, 37, and 11 by a cross between Quing and Stato et al. (2004) use 60 lines obtained by a cross between Hinohikari/WSS2/Hinohikari.15 QTLs were derived using a inbreed line population of Tequing and Lemont (Pinson et al., 2010).

7.5.3 TRANSGENIC APPROACH

Some plant pathogenesis-related (PR) genes, such as PR-3 chitinase (Datta et al., 2000) and PR-5 (thaumatin-like protein) (Datta et al., 1999), provide resistance to SBH. Instead of a single PR gene, a combination of two PR genes produces efficient outcomes and a higher level of sheath blight resistance. According to Shah et al. (2009), the Endochitinase gene (cht42) produced in transgenic rice reduces sheath blight disease by up to 62%. So far, no germplasm that is completely resistant to SBH has been discovered. To reduce production losses due to SBH, cultivars resistant to sheath blight must be developed. So far, only moderately resistant genotypes have been reported. These genotypes have variable disease reactivity from season to season, which limits their application in breeding programs. To deal with challenging and complex features like sheath blight, Dey et al. (2019) devised a two-step breeding technique. To begin, genotypes with modest resistance from the identified pool should be intermitted to allow for the accumulation of numerous minor effect QTLs, which would be visible in populations as increased phenotypic effect. Second, an elite vulnerable cultivar should be crossed with a breeding line that is more resistant to sheath blight than either of its parents.

7.6 ROLE OF DEFENSE RESPONSE GENES AGAINST SHEATH BLIGHT RESISTANCE

7.6.1 QUANTITATIVE TRAIT LOCI (QTL) BASED RESISTANCE

For QTL study before 2011, see the review by Srinivasachary et al. (2011) and Jia et al. (2009). For a study, identified new QTLs are tabulated, while the re-established QTLs are excluded.

Superscript I, indica rice genotype, superscript J, *japonica* rice genotype; BIL, backcross inbred; RIL, recombination inbred line, DH, double haploid; CSSL, chromosomal segment substitution lines.

TABLE 7.1 A Summary of Quantitative Trait Loci Detected for Seath Bight Rsistance.

Name of QTLs	Chromosome	Flanking markers/ nearest marker	Mapping population (resistance X susceptible)	Molecular markers (Number)	LOD value/ phenotypic variance	Associated character	Remark	References
qshb1.1	1	RM151-RM12253	210 F2 (ARC 10531' X BPT-5204')	SSR (70)	10.7	Percentage relative lesion height	32 candidate genes identified in the region qShB902	Yadav et al. (2015)
Qshb7.1	7	RM81-RM615			8.8			
Qshb7.2	7	RM10-RM2169			6.7			
Qshb8.1	8	RM21792-RM310			4.2			
qSBL7 (E2)	7	D760-RM248	190 F2 (Yangdao 4' X Lemont')	SSR (52) and InDel (128)	3.12	DR-disease rating	Sheath blight resis-tance is correlated with plant height	Wen et al. (2015)
qSBL7 (E2)	7	D760-RM248			5.07	LH-lesion height		
qSBD-12-2 (E1)	12	RM1246-D1252			3.74	PL-percentage of lesion height		
qHZaLH3	3	RM143-RM514	116 DH (TN1'XCJ06')	SSR (214)		DR, LH	No correlation was found between LH and PH	Zeng et al. (2015)
qHZaLH6	6	WX-RM587						
qHZaLH8	8	RM1376-RM4085						
qHZaDR9	9	RM444-AGPSMA						
qHZbDR5	5	RM3321-RM3616						
qSB-9TO	9	CY-85 and Y86	235 CSSIs (BC6F3) (Teging (TQ) X Lemont)	InDel and CAPS (22)		Sheath blight resistance	Fine mapped (146 Kb covering region), 12 genes were annotated	Zuo et al. (2014a)

TABLE 7.1 *(Continued)*

Name of QTLs	Chromosome	Flanking markers/ nearest marker	Mapping population (resistance X susceptible)	Molecular markers (Number)	LOD value/ phenotypic variance	Associated character	Remark	References
qDR-4	4	RM1155-RM5757	155 RIL F8:11 (RSB02XHH1B)	SSR (163)	2.71	DR-disease rating LL-lesion length	Epistasis and QTL X environment (QE) interaction were studied	Liu et al. (2014)
qRLL-4	4	RM1155-RM5757			5.84	LH-lesion height RLL-relative lesion length		
qRLH-4	4	RM1155-RM5757			4.77	RLH-relative LH		
qSB-11^{LS}	11	Z22-27C and Z23-33C	112 CSSIs (Teging' x Lemont' (LE)	STS and CAPS (26)		ShB resistance	Fine mapped (79 Kb covering region), 11 genes were annotate	Zuo et al. (2013)
qRTL3	3	RM570	BIL (Jarian' x Koshihikari')	SSR (151)	3.5	RTL-Rate of tillers with lesions		Taguchi-Shiobara et al. (2013)
qRTL5	5	RM5784			4.3			
qRTL6	6	RM1161			7.7			
qRTL9	9	RM6251			3.1			
qRTL3	3	RM16200			5.9			
qRTL6	6	RM2615			2.9			
qRTL12	12	RM7025			3.2			
qRTL5	5	RM3286			3.1			
qRTL6	6	RM6395			5.8			

TABLE 7.1 *(Continued)*

Name of QTLs	Chromosome	Flanking markers/ nearest marker	Mapping population (resistance X susceptible)	Molecular markers (Number)	LOD value/ phenotypic variance	Associated character	Remark	References
qRTL9	9	RM3533			3.8			
qShB2-1	2	RM279-RM2	216RIL (Jasmine 85' x Lemont')	SSR (199)	3.7	ShB resistance	The major QTL qShB9-2 was reconfirmed based on the field data	Liu et al. (2013)
ARqShB7-AR	7	RM5711-RM2			4.0			
qShB7 LA	7	RM5711-RM2			6.0			
qShB11-1	11	RM7203-RM536			3.2			
TXqShB11-2-TX	11	RM536-RM229			3.3			
qShB6 (wild1-field 2009)	6	RM3431-RM3183	252 Wild-1 and 253 wild-2 BC2F2 (Oryzanivara x Bengal (O. sativa))	SSR (131)	7.8	ShB resistance	Colocalization of qShB6 with qDH1 and qShB1 with qPH1 revealed the influence of heading date and plant height on resistance	Eizenga et al. (2013)
qShB6 (wild2-field 2009)	6	RM253-RM3431			21.2			
qShB6 (wild2-field 2008)	6	RM253-RM3431			11.1			

TABLE 7.1 *(Continued)*

Name of QTLs	Chromosome	Flanking markers/ nearest marker	Mapping population (resistance X susceptible)	Molecular markers (Number)	LOD value/ phenotypic variance	Associated character	Remark	References
qShB6 (wild 2-field 2008)	1	RM431-RM1361			4.7			
qShB6-mc (wild 1-micro-chamber)	6	RM3183-RM541			3.3			
qsbr_2.1	2	RM8254-RM8252	197 DH (MCR10 x Cocodrie)	SSR (111)	3.4–29.7	SBF-sheath blight		Nelson et al. (2012)
qsbr_2.2	2	RM3857-RM5404			2.9–37.8	Disease severity in field		
qsbr_12.1	12	RM3747-RM27608			49.1	SBM-disease severity in mist chamber		
qSBR1	1	RM11229	217 core collection of USDA	SSR (154)	9.5%	Sheath blight resistance		Jia et al. (2012)
qSBR11	11	RM7203		Indel (1)	1.9%			
qSBR1-1	1	RM5389-RM3825	121 RIL (RSB03 x HH1B	SSR (123)	3.2	DR-disease rating		Fu et al. (2011)
						LL-lesion lenght		
qSBR2-1	2	RM5340-RM521			3.1	LH-lesion height		
qSBR2-2	2	RM110-OSR14			5.2	RLL-relative LL		
qSBR2-3	2	RM7245-RM5303			3.3	RLH-relative LH		
qSBR4	4	RM3288-RM7187			3.8			

TABLE 7.1 *(Continued)*

Name of QTLs	Chromosome	Flanking markers/ nearest marker	Mapping population (resistance X susceptible)	Molecular markers (Number)	LOD value/ phenotypic variance	Associated character	Remark	References
qSBR5-2	5	RM7446-RM3620			4.8			
qSBR7	7	RM1132-RM473			3.3			
qSBR8	8	RM8264-RM1109			4.2			
qSBR9	9	RM23869-RM3769			5.0			
qShB1 (2007/2008)	1	RM431-RM12017	251 DH (Baiyegiu x Maybelle	SSR (227)	5.1–8.03	Sheath blight resistance		Xu et al. (2011)
qShB2	2	RM74-RM145			3.96			
qShB3	3	RM35-RM186			3.42			
qShB5	5	RM18872-RM421			4.35			

TABLE 7.2 Transgenic Rice Against Sheath Blight Pathogen *Rhizoctoniasolani*: A Historical Overview

Rice cultivar	Transformation method	Promotor used	Gene	Origin	Comments	Reference
ChinsurahBoro II	Protoplast	CaMV35S	Chi11	Rice	A class 1 chitinase a PR3 protein	Lin et al. (1995)
ChinsurahBoro II IR72 and IR51500	Protoplast and biolistic	CaMV35S	Tip-D34	Rice	Rice thaumatin like protein-a PR5 protein	Datta et al. (1999b)
Basmati 122, Tulsi and Vaidehi	*Agrobacterium* mediated	CaMV35S	Chi11	Rice	A class 1 chitinase	Datta et al. (2000)
IR72, IR64, IR68899B, MH63 and ChinsurahBoro II	Biolistic	CaMV35S	RC7	Rice	Class 1 chitinase	Datta et al. (2001)
M202	Biolistic	Maize ubiquitin	pinA pinB	Wheat	Puroindoline-antimicrobial peptides	Krishnamurthy et al. (2001)
Swarna	Biolistic to anther derived calli and embryo	CaMV35S	Chi1	Rice	Chitinase	Baisakh et al. (2001)
IR72	Pyramiding of transgenes using marker-assisted selection	CaMV35S for chi1	Chi11, xa21, cry gene	Rice	Rice chitinase, receptor kinase like protein and BT toxin protein	Data et al. (2002)
Kenfong	Biolistic	Rice rbcSpromotor for MOD1 and act1 promotor for RCH10	MOD1 and RCH10	Maize and rice	Maize ribosome inactivating protein and rice basic chitinase	Kim et al. 2003
PusaBasmati 1	*Agrobacterium mediated*	maize ubiquitin	Chi11	Rice	Chitinase	Kumar et al. (2003)

TABLE 7.2 *(Continued)*

Rice cultivar	Transformation method	Promotor used	Gene	Origin	Comments	Reference
Pusa Basmati 1	*Agrobacterium mediated*	maize ubiquitin	Chi11	Rice	Chitinase	Sridevi et al. (2003)
Ishikari-shiroge	*Agrobacterium mediated*	Act1	Ech42, nag70 and gluc78	Trichodermaatroviride	Endochitinase and exo-1,3,-b-glucanase	Liu et al. (2004)
ADT38, ASD16, IR50 and Taipei 309	Biolistic	Maize ubiquitin	tlp and chi11	Rice	Rice thaumatin-like protein and chitinase	Kalpana et al. (2006)
Pusa Basmati 1	*Agrobacterium mediated*	PAL promotor and ubiquitin	Ace-AMP1	*Allium cepa*	A nonspecific lipid transfer protein with antimicrobial property	Patkar and Chatto 2006
Pusa Basmati 1, Co43, white Ponni and ADT38	*Agrobacterium mediated*	Maize ubiquitin	RC7	Rice	Chitinase	Nandakumar et al. (2007)
ASD16, ADT38, IR72, IR64 and White Ponni	Biolistic	Maize ubiquitin for tlp and chi11, Native promoter for xa21	tlp, chi11 and xa21	Rice	Rice thaumatin like protein, chitinase, and serine-threonine kinase	Maruthasalam et al. (2007)
Pusa Basmati 1	*Agrobacterium mediated*	Maize ubiquitin for chi11 and CaMV35S for glucanase	Chi11 and b-1,3 glucanase	Rice and tobacco	Rice chitinase and tobacco b-1,3-glucanase	Sridevi et al. (2008)
Pusa Basmati 1	*Agrobacterium mediated*	Maize ubiquitin	Chi11	Rice	Chitinase	Sripriya et al. (2008)
Pusa Basmati 2	*Agrobacterium mediated*	Maize ubiquitin	Dm-AMP1	Dahlia merckii	Defensing protein	Jha et al. (2009)
JinHui 35	*Agrobacterium mediated*	Maize ubiquitin 1	McCHIT1	Momordicacharantia	A class 1 secretory endochitinase	Li et al. (2009)

TABLE 7.2 *(Continued)*

Rice cultivar	Transformation method	Promotor used	Gene	Origin	Comments	Reference
Pusa Basmati 1	*Agrobacterium mediated*	Maize ubiquitin 1	Dm-AMP1 and Ra-AFP2	*Dahlia merckii*and *Raphanussativus*	Both are plant defensin antimicrobial protein	Jha and Chatto (2009)
Pusa Basmati 1	*Agrobacterium mediated*	CaMV35S	cht42	*Trichodermavirens*	Endochitinase	Shah et al. (2009)
Pusa Basmati 1	*Agrobacterium mediated*	Maize ubiquitin	RS-AFP2	*Raphanussativus*	Plant defenin protein	Jha and Chattoo, (2010)
Pusa Basmati 1	*Agrobacterium mediated*	Maize ubiquitin for chi11 and CaMV35S for ap24	Chi11 and ap24	Rice and tobacco	Chitinase and osmotin	Rao et al. (2011)
Xiushui 11	*Agrobacterium mediated*	Maize ubiquitin	Os WRKY30	Rice	Transcription factor gene	Peng et al. (2012)
Kitaake	*Agrobacterium mediated*	PB21	OsACS2	Rice	1-Amino cyclopropane-1-carboxylic acid synthase	Helliwell et al. (2013)
PusaSungandhi-2	Biolistic	Rice Pd54o544	Osoxo4	Rice	Rice oxalate oxidase	Molla et al. (2013)
Chaitnya and Samba Mahsuri	*Agrobacterium mediated*	CaMV35S	BjNPR1	*Brassica junica*	Nonexpressor of pathogenesis-related gene 1	Sadumpati et al. (2013)
Taipei 309	*Agrobacterium mediated*	CaMV35S	RCH10 and AGLU1	Rice and alfalfa	Basic chitinase and b-1,3-glucanase	Mao et al. (2014)
Zhonghua 11	*Agrobacterium mediated*	CaMV35S	OsPGIP1	Rice	Polygalacturonase inhibiting proteins	Wang et al. (2015)
Pusa Sungandhi-2	*Biolistic*	Rice Po54o544	AtNPR1	Arabidopsis thaliana	Nonexpressor of pathogenesis related gene 1	Molla et al. (2016)
Xiushui 11	*Agrobacterium mediated*	Maize ubiquitin	OsWRKY80	Rice	Transcription factor	Peng et al. (2016)

TABLE 7.2 *(Continued)*

Rice cultivar	Transformation method	Promotor used	Gene	Origin	Comments	Reference
BR-29	*Agrobacterium mediated*	Rice Po54o544 and maize PEPC	OsOXO4 and OsCHI11	Rice	Rice oxalate oxidase 4 and rice chitinase 11	Karmakar et al. (2016)
Xudao 3	*Agrobacterium mediated*	Maize Ubiquitin	OsOSm1	Rice	Rice osmotin	Xue et al. (2016)
Jaldi-13	*Agrobacterium mediated*	Rice Po54o544 and maize PEPC	AtNPR1 and OsCHI11	Arabidopsis thaliana and rice	Nonexpressor of pathogenesis related gene 1 and rice chintinase 11	Karmakar et al. (2017)
Taipei 309 (TP 309)	Biolistic	CaMV35S	Chitinase	Rice	Rice chitinase	Richa et al., (2017)
Nipponbare	*Agrobacterium mediated*	CaMV35S	Bacisubin	B. subtilisdtrain BS-916	Oxalate decarboxylase	Qi et al. (2017)
Zhonghua 11	*Agrobacterium mediated*	Maize Ubiquitin 1 promotor	OsASR2	Rice	Absicisic acid, stress and ripening 2 protein	Li et al. (2017)
Nipponbare	*Agrobacterium mediated*	Maize Ubiquitin 1 promotor	OsBSR2	Rice	Cytochrome P450 protein (CYP78A family)	Maeda et al. (2019)

7.7 CONCLUSION

Lot of study has been conducted on Sheath blight disease of rice so that the effective management could be developed against Sheath blight like development of fungicides, transgenic plants varieties, resistant rice varieties using molecular breeding approach. But till date there is no effective management has been developed for complete removal of *R. solani,* so that rice yield could be increased. Apart from the various above-mentioned approaches against Sheath blight, only molecular breeding approaches, like Marker assisted gene pyramiding approach is most appreciated by which several resistant varieties against Sheath blight have been developed. Development of resistant variety is an effective management against sheath blight. We must encourage the breeding work so that we can develop the more and more resistant varieties.

ACKNOWLEDGMENT

I am grateful for Himalayan University, Itanagar, A.P., INDIA to provide the financial assistance so that the work of manuscript could be completed successfully.

KEYWORDS

- **sheath blight**
- **quantitative trait loci (QTL)**
- **pathogenesis-related (PR)**
- **R. solani**

REFERENCES

Agarwal, M.; Shrivastava, N.; Padh, H. Advances in Molecular Marker Techniques and Their Applications in Plant Sciences. *Plant Cell Rep.* **2008,** *27*, 617–631.

Anees, M.; Edel-Hermann, V.; Steinberg, C. Buildup of Patches caused by Rhizoctonia Solani. *Soil Biol. Biochem.* **2010,** *42*, 1661–1672.

Baisakh, N.; Datta, K.; Oliva, N.; Ona, I.; Rao, G.; Mew, T.; Datta, S. K. Rapid Development of Homozygous Transgenic Rice using Anther Culture Harboring Rice *Chitinase* Gene for Enhanced Sheath Blight Resistance. *Plant Biotechnol.* **2001,** *18*, 101–108.

Bishwas, N.; Sharma, M.; Hasan, A., et al. Improvement of Rice Crop by Marker-assisted Back-cross Method. *Magnesium* **2016,** *8*, 2.

Brooks, S. A. Sensitivity to a Phytotoxin from Rhizoctonia Solani Correlates with Sheath Blight Susceptibility in Rice. *Phytopathology* **2007,** *97*, 1207–1212.

Browse, J. Jasmonate Passes Muster: A Receptor and Targets for the Defense Hormone. *Annu. Rev. Plant. Biol.* **2009,** *60*, 183–205.

Chen, Z. X.; Zhang, Y. F.; Feng, F.; Feng, M. H.; Jiang, W.; Ma, Y. Y.; Pan, C. H.; Hua, H. L.; Li, G. S.; Pan, X. B.; Zuo, S. M. Improvement of Japonica Rice Resistance to Sheath Blight by Pyramiding qSB-9TQ and qSB-7TQ. *Field Crops Res.* **2014,** *161*, 118–127.

Chen, Z. X.; Zou, J. H.; Xu, J. Y.; Tong, Y. H.; Tang, S. Z.; Wang, Z. B.; Jiang, R. M.; Ling, B.; Tang, J.; Pan, X. B. A Preliminary Study on Resources of Resistance to Rice Sheath Blight. *Chin. J. Rice Sci.* **2000,** *14*, 15–18.

Collard, B.; Jahufer, M.; Brouwer, J.; Pang, E. An Introduction to Markers, Quantitative Trait Loci (QTL) Mapping and Marker-assisted Selection for Crop Improvement: the Basic Concepts. *Euphytica* **2005,** *142*, 169–196.

Collard, B. C.; Mackill, D. J. Marker-assisted Selection: An Approach for Precision Plant Breeding in the Twentyfirst Century. *Philos. Trans. R. Soc. B: Biol. Sci.* **2008,** *363*, 557–572.

Datta, K.; Baisakh, N.; Maung Thet, K.; Tu, J.; Datta, S. K. Pyramiding Transgenes for Multiple Resistance in Rice Against Bacterial Blight, Yellow Stem Borer and Sheath Blight. *Theor. Appl. Genet.* **2002,** *106*, 1–8.

Datta, K.; Koukolikova, N. Z.; Baisakh, N.; Oliva, N.; Datta, S. K. Agrobacterium-mediated Engineering for Sheath Blight Resistance of Indica Rice Cultivars from Different Ecosystems. *Theor. Appl. Genet.* **2000,** *100*, 832–839.

Datta, K.; Koukolikova-Nicola, Z.; Baisakh, N.; Oliva, N.; Datta, S. K. *Agrobacterium*-mediated Engineering for Sheath Blight Resistance of Indica Rice Cultivars from Different Ecosystems. *Theor. Appl. Genet.* **2000,** *100*, 832–839.

Datta, K.; Tu, J.; Oliva, N.; Ona, I.; Velazhahan, R.; Mew, T. W.; Muthukrishnan, S. Enhanced Resistance to Sheath Blight by Constitutive Expression of Infection-related Rice Chitinase in Transgenic Elite Indica Rice Cultivars. *Plant Sci.* **2001,** *160*, 405–414.

Datta, K.; Velazhahan, R.; Oliva, N.; Ona, I.; Mew, T.; Khush, G.; Muthukrishnan, S.; Datta, S. K. Over-expression of the Cloned Rice Thaumatin-like Protein (PR-5) Gene in Transgenic Rice Plants Enhances Environmental Friendly Resistance to *Rhizoctonia solani* causing Sheath Blight Disease. *Theor. Appl. Genet.* **1999,** *98*,1138–1145.

Datta, K.; Velazhahan, R.; Oliva, N.; Ona, I.; Mew, T.; Khush, G.; Muthukrishnan, S. Over-Expression of the Cloned Rice *Thaumatin-like protein* (PR-5) Gene in Transgenic Rice Plants Enhances Environmental Friendly Resistance to *Rhizoctonia Solani* Causing Sheath Blight Disease. *Theor. Appl. Genet.* **1999b,** *98*, 1138–1145.

Day, S.; Badri, J.; Ram, K.; Chhabra, A. K.; Janghel, D. K. Current Status of Rice Breeding for Sheath Blight Resistance. *Int. J. Curr. Microbiol. App. Sci.* **2019,** *8* (2), 163–175.

Dey, S.; Badri, J.; Prakasam, V.; Bhadana, V. P.; Eswari, K. B.; Laha, G. S.; Priyanka, C.; Aku, R.; Ram, T. 2016. Identification and Agro-morphological Characterization of Rice Genotypes Resistant to Sheath Blight. *Australas. Plant Pathol.* **2019,** *45*,145–153.

Eizenga, G. C.; Lee, F. N.; Rutger, J. N. Screening Oryzae Species Plant for Rice Sheath Blight Resistance. *Plant Dis.* **2002,** *86*,808–812.

Eizenga, G.; Prasad, B.; Jackson, A.; Jia, M. Identification of Rice Sheath Blight and Blast Quantitative Trait Loci in Two Different *O. Sativa/O. Nivara* Advanced Backcross Populations. *Mol. Breed.* **2013,** *31*, 889–907.

Eizenga, G. C.; Lee, F. N.; Rutger, J. N. Screening Oryza Species Plants for Rice Sheath Blight Resistance. *Plant Dis.* **2002,** *86*, 808–812.

Fu, D.; Chen, L.; Yu, G.; Liu, Y.; Lou, Q.; Mei, H.; Xiong, L., et al. QTL Mapping of Sheath Blight Resistance in a Deep-water Rice Cultivar. *Euphytica* **2011,** *180*, 209–218.

Gangopadhyay, S., Chakrabarti, N. K. Sheath Blight of Rice. *Rev. Plant Pathol.* **1982,** *61*, 451–460.

Glazebrook, J. Genes Controlling Expression of Defense Responses in Arabidopsis-2001 status. *Curr. Opin. Plant Biol.* **2001,** *4*, 301–308.

Glazebrook, J. Contrasting Mechanisms of Defense against Biotrophic and Necrotrophic Pathogens. *Annu. Rev. Phytopathol.* **2005,** *43*, 205–227.

Grichar, W. J.; Jaks, A. J.; Besler, B. A. Response of Peanuts (Arachis hypogaea) to Weather-based Fungicide Advisory Sprays. *Crop Prot.* **2004,** *24*, 349–354.

Growth, D. E.; Rush, M. C.; Hollier, C. A. Prediction of Rice Sheath Blight Severity and Yield Loss Based on Early Season Infection. *La. Agric.* **1992,** *35*, 20–23.

Gullino, M. L.; Leroux, P.; Smith, C. M. Uses and Challenges of Novel Compounds for Plant Disease Control. *Crop Prot.* **2000,** *19*, 1–11.

Han, Y. P.; Xing, Y.; Gu, S.; Chen, Z.; Pan, X., et al. Effect of morphological traits on sheath blight resistance in rice. *Acta Botanica Sinica.* **2003,** *45*, 825–831.

Helliwell, E. E.; Wang, Q.; Yang, Y. Transgenic Rice with Inducible Ethylene Production Exhibits Broad-spectrum Disease Resistance to the Fungal Pathogens *Magnaporthe Oryzae* and *Rhizoctonia Solani*. *Plant Biotechnol. J.* **2013,** *11*, 33–42.

Hitaka, H. Studies on the Lodging of Rice Plants. *Jpn. Agric. Res. Quart.* **1969,** *4*, 1–6.

Huang, X.; Wei, X.; Sang, T.; Zhao, Q.; Feng, Q.; Zhao, Y.; Li, C.; Zhu, C.; Lu, T.; Zhang, Z. Genome-wide Association Studies of 14 Agronomic Traits in Rice Landraces. *Nat. Genet.* **2010,** *42*, 961–967.

Jain, S. M.; Brar, D. S.; Ahloowalia, B. *Molecular Techniques in Crop Improvement*; Springer: The Netherlands, 2010.

Jena, K. K.; Mackill, D. J. Molecular Markers and Their use in Marker-assisted Selection in Rice. *Crop Sci.* **2008,** *48*, 1266–1276.

Jha, S.; Chattoo, B. B. Transgene Stacking and Coordinated Expression of Plant Defensins Confer Fungal Resistance in Rice. *Rice* **2009,** *2*, 143–154.

Jia, L.; Yan, W.; Zhu, C.; Agrama, A. H.; Jackson, A.; Yeater, K.; Li, X.; Huang, B.; Hu, B.; McClung, A.; Wu, D. Allelic Analysis of Sheath Blight Resistance with Association Mapping in Rice. *PLoS One* **2012,** *7* (3), e32703.

Jones, R. K.; Belmar, S. B. Characterization and Pathogenicity of *Rhizoctonia* spp. Isolated from Rice, Soybean, and other Crops Grown in Rotation with Rice In Texas. *Plant Dis.* **1989,** *73*, 1004–1010.

Kalpana, K.; Maruthasalam, S.; Rajesh, T.; Poovannan, K.; Kumar, K. K.; Kokiladevi, E.; Raja, J. A., et al. Engineering Sheath Blight Resistance in Elite Indica Rice Cultivars Using Genes Encoding Defense Proteins. *Plant Sci.* **2006,** *170*, 203–215.

Karmakar, S.; Molla, K. A.; Chanda, P. K.; Sarkar, S. N.; Datta, S. K.; Datta, K. Green Tissue-specific Co-expression of *Chitinase* and *Oxalate Oxidase 4* Genes in Rice for Enhanced Resistance Against Sheath Blight. *Planta* **2016,** *243*, 115–130.

Karmakar, S.; Molla, K. A.; Das, K.; Sarkar, S. N.; Datta, S. K.;Datta, K. Dual Gene Expression Cassette is Superior Than Single Gene Cassette for Enhancing Sheath Blight Tolerance in Transgenic Rice. *Sci. Rep.* **2017,** *7,* 7900.

Kim, J. K.; Jang, I. C.; Wu, R.; Zuo, W. N.; Boston, R. S.; Lee, Y. H.; Ahn, I. P. Co-expression of a Modified Maize Ribosome-inactivating Protein and A Rice Basic Chitinase Gene in Transgenic Rice Plants Confers Enhanced Resistance to Sheath Blight. *Transgenic Res.* **2003,** *12,* 475–484.

Kloepper, J. W.; Schroth, M. N. *Plant Growth Promoting Rhizobacteria on Radish,* Proceedings of the Fourth Conference Plant Pathogenic Bacteria, INRA, Angers, France, 1978.

Knight, S. C.; Anthony, V. M.; Brady, A. M.; Greenland, A. J.; Heaney, S. P. Rationale and Perspectives on the Development of Fungicides. *Annu. Rev. Phytopathol.* **1997,** *35,* 349–372.

Köhl, J.; Postma, J.; Nicot, P.; Ruocco, M.; Blum, B. Stepwise Screening of Microorganisms for Commercial Use in Biological Control of Plant-pathogenic Fungi and Bacteria. *Biol. Control* **2011,** *57,* 1–12.

Krishnamurthy, K.; Balconi, C.; Sherwood, J. E.; Giroux, M. J. Wheat Puroindolines Enhance Fungal Disease Resistance in Transgenic Rice. *Mol. Plant-Microbe Interact.* **2001,** *14,* 1255–1260.

Kumar, K. K.; Poovannan, K.; Nandakumar, R.; Thamilarasi, K.; Geetha, C.; Jayashree, N.; Kokiladevi, E., et al. A High Throughput Functional Expression Assay System for a Defence Gene Conferring Transgenic Resistance on Rice Against the Sheath Blight Pathogen, *Rhizoctonia Solani*. *Plant Sci.* **2003**, *165*, 969–976.

Kunihiro, Y.; Qian, Q.; Sato, H.; Teng, S.; Zeng, D. L. QTL Analysis of Sheath Blight Resistance in Rice (Oryza sativa L.). *Chin. J. Genet.* **2002,** *29,* 50–55.

Kunkel, B. N.; Brooks, D. M. Cross Talk Between Signaling Pathways in Pathogen Defense. *Curr. Opin. Plant Biol.* **2002,** *5,* 325–331.

Lee, F. N., Rush, M. C. Rice Sheath Blight: A Major Rice Disease. *Plant Dis.* **1983,** *67,* 829–832.

Leegood, R. C., Evans, J. R., Furbank, R. T. Food Security Requires Genetic Advances to Increase Farm Yields. *Nature* **2010,** *464,* 831–831.

Li, N.; Wei, S.; Chen, J.; Yang, F.; Kong, L.; Chen, C.; Chu, Z. Os ASR 2 Regulates the Expression of a Defence-related Gene, Os2H16, by Targeting the GT-1 Cis-element. *Plant Biotechnol.* **2018,** *J. 16,* 771–783.

Li, P.; Pei, Y.; Sang, X.; Ling, Y.; Yang, Z.; He, G. Transgenic Indica Rice Expressing a Bitter Melon (*Momordica charantia*) Class I Chitinase Gene (*McCHIT1*) Confers Enhanced Resistance to *Magnaporthe Grisea* and *Rhizoctonia Solani*. *Eur. J. Plant Pathol.* **2009,** *125,* 533–543.

Lin, W.; Anuratha, C.; Datta, K.; Potrykus, I.; Muthukrishnan, S.; Datta, S. K. Genetic Engineering of Rice for Resistance to Sheath Blight. *Nat. Biotechnol.* **1995,** *13,* 686–691.

Liu, G.; Jia, Y.; McClung, A.; Oard, J.; Lee, F.; Correll, J. Confirming QTLs and Finding Additional Loci Responsible for Resistance to Rice Sheath Blight Disease. *Plant Dis.* **2013,** *97,* 113–117.

Liu, M.; Sun, Z. X.; Zhu, J.; Xu, T.; Harman, G.; Lorito, M. Enhancing Rice Resistance to Fungal Pathogens by Transformation with Cell Wall Degrading Enzyme Genes from *Trichoderma Atroviride*. *J. Zhejiang Univ. Sci.* **2004,** *5,* 133–136.

Liu, Y.; Chen, L.; Fu, D.; Lou, Q.; Mei, H.; Xiong, L.; Li, M. Dissection of Additive, Epistatic Effect and QTL9 Environment Interaction of Quantitative Trait Loci for Sheath Blight Resistance in Rice. *Hereditas* **2014,** *151*, 28–37.

Maeda, S.; Dubouzet, J. G.; Kondou, Y.; Jikumaru, Y.; Seo, S.; Oda, K.; Mori, M. The Rice CYP78A Gene BSR2 Confers Resistance to *Rhizoctonia Solani* and Affects Seed Size and Growth in Arabidopsis and Rice. *Sci. Rep.* **2019,** *9*, 587.

Mao, B.; Liu, X.; Hu, D.; Li, D. Co-expression of RCH10 and AGLU1 Confers Rice Resistance to Fungal Sheath Blight *Rhizoctonia Solani* and Blast *Magnorpathe Oryzae* and Reveals Impact on Seed Germination. *World J. Microbiol. Biotechnol.* **2014,** *30*, 1229–1238.

Maruthasalam, S.; Kalpana, K.; Kumar, K.; Loganathan, M.; Poovannan, K.; Raja, J.; Kokiladevi, E. Pyramiding Transgenic Resistance in Elite Indica Rice Cultivars against the Sheath Blight and Bacterial Blight. *Plant Cell Rep.* **2007,** *26*, 791–804.

McCouch, S. R.; Teytelman, L.; Xu, Y.; Lobos, K. B.; Clare, K.; Walton, M.; Fu, B.; Maghirang, R.; Li, Z.; Xing, Y. Development and Mapping of 2240 new SSR Markers for Rice (Oryza sativa L.). *DNA Res.* **2002,** *9*, 199–207.

Meng, Q. Z.; Liu, Z. H.; Wang, H. Y.; Zhang, S. S.; Wei, S. H. Research Progress in Rice Sheath Blight (in Chinese with English Abstract). *J. Shenyang Agric. Univ.* **2001,** *32*, 376–381.

Miah, G.; Rafii, M. Y.; Ismail, M. R. A Review of Microsatellite Markers and their Applications in Rice Breeding Programs to Improve Blast Disease Resistance. *IJMS* **2013,** *14*, 22499–22528.

Miyaki, I. Studies under Die Resistance in Japan. *J. Cell Agric. Imp Univ.* **1910,** 2, 237–276.

Molla, K. A.; Karmakar, S.; Chanda, P. K.; Ghosh, S.; Sarkar, S. N.; Datta, S. K.; Datta, K. Rice Oxalate Oxidase Gene Driven by Green Tissue-specific Promoter Increases Tolerance to Sheath Blight Pathogen (*Rhizoctonia solani*) in Transgenic Rice. *Mol. Plant Pathol.* **2013,** *14*, 910–922.

Molla, K. A.; Karmakar, S.; Chanda, P. K.; Sarkar, S. N.; Datta, S. K.; Datta, K. Tissue-specific Expression of *Arabidopsis NPR1* Gene in Rice for Sheath Blight Resistance Without Compromising Phenotypic Cost. *Plant Sci.* **2016,** *250*, 105–114.

Monna, L.; R. Ohta, H. Masuda, A. Koike, Y.; Minobe, Y. Genome-wide Searching of Single-nucleotide Polymorphisms Among Eight Distantly and Closely Related Rice Cultivars (*Oryza sativa L.*) and a Wild Accession (*Oryza rufipogon* Griff.). *DNA Res.* **2006,** *13* (3), 43–51.

Li, Z.; Pinson, S. R. M.; Marchetti, M. A.; Stansel, J. W.; Park, W. D. Characterization of Quantitative Trait Loci (QTLs) in Cultivated Rice Contributing to Field Resistance to Sheath Blight (Rhizoctonia solani). *Theor. Appl. Genet.* **1995,** *91*, 382–388.

Nagarajkumar, M.; Bhaskaran, R.; Velazhahan R. Involvement of Secondary Metabolites and Extracellular Lytic Enzymes Produced by Pseudomonas Fluorescens in Inhibition of Rhizoctonia Solani, the Rice Sheath Blight Pathogen. *Microbiol Res.* **2004,** *159*, 73–81.

Nagarajkumar, M.; Jayaraj, J.; Muthukrishnan, S.; Bhaskaran, R.; Velazhahan, R. Detoxification of Oxalic Acid by Pseudomonas Fluorescens Strain PfMDU2: Implications for the Biological Control of Rice Sheath Blight Caused by Rhizoctonia Solani. *Microbiol. Res.* **2005,** *160*, 291–298.

Nandakumar, R.; Babu, S.; Kalpana, K.; Raguchander, T.; Balasubramanian, P.; Samiyappan, R. *Agrobacterium*-Mediated Transformation of Indica Rice with Chitinase Gene for Enhanced Sheath Blight Resistance. *Biol. Plant.* **2007,** *51*, 142–148.

Nelson, J. C.; Oard, J. H.; Groth, D.; Utomo, H. S.; Jia, Y.; Liu, G.; Moldenhauer, K. A. K. Sheath-blight Resistance QTLS in Japonica Rice Germplasm. *Euphytica* **2012,** *184*, 23–34.

Ogoshi, A. *Introduction of the Genus Rhizoctonia: Taxonomy, Molecular Biology, Ecology, Pathology and Disease Control*; Kluwer Academic Publishers: Dordrecht, NL, 1996.

Oka, K.; Amano, Y.; Katou, S.; Seo, S.; Kawazu, K.; Mochizuki, A.; Kuchitsu, K., et al. Tobacco MAP Kinase Phosphatase (NtMKP1) Negatively Regulates Wound Response and Induced Resistance Against Necrotrophic Pathogens and Lepidopteran Herbivores. *Mol. Plant Microbe Interact.* **2013,** *26*, 668–675.

Pan, X. B.; Zou, J.; Chen, Z.; Lu, J.; Yu, H.; Li, H.; Wang, Z.; Pan, X.; Rush, M. C.; Zhu, L. Tagging Major Quantitative Trait Loci for Sheath Blight Resistance in a Rice Variety, Jasmine 85. *Chin. Sci. Bull.* **1999,** *44*, 1783–1789.

Patkar, R. N.; Chattoo, B. B. Transgenic Indica Rice Expressing ns-LTP-like Protein Shows Enhanced Resistance to Both Fungal and Bacterial Pathogens. *Mol. Breed.* **2006,** *17*, 159–171.

Peng, X.; Hu, Y.; Tang, X.; Zhou, P.; Deng, X.; Wang, H.; Guo, Z. Constitutive Expression of Rice *WRKY30* Gene Increases the Endogenous Jasmonic Acid Accumulation, PR Gene Expression and Resistance to Fungal Pathogens in Rice. *Planta* **2012,** *236*, 1485–1498.

Peng, X.; Wang, H.; Jang, J.-C.; Xiao, T.; He, H.; Jiang, D.; Tang, X. OsWRKY80-OsWRKY4 Module as a Positive Regulatory Circuit in Rice Resistance Against *Rhizoctonia Solani*. *Rice* **2016,** *9*, 63.

Pinson, S. R. M.; Shahjahan, A. K. M.; Rush, M. C.; Groth, D. E. Bacterial Panicle Blight Resistance QTLs in Rice and Their Association with Other Disease Resistance Loci and Heading Date. *Crop Sci.* **2010,** *50*, 1287–1297.

Qi, Z.; Yu, J.; Shen, L.; Yu, Z.; Yu, M.; Du, Y.; Zhang, R., et al. Enhanced Resistance to Rice Blast and Sheath Blight in Rice (*Oryza sativa* L.) by Expressing the Oxalate Decarboxylase Protein Bacisubin from Bacillus Subtilis. *Plant Sci.* **2017,** *265*, 51–60.

Ramonell, K. M.; Somerville, S. The Genomics Parade of Defense Responses: To Infinity and Beyond. *Curr Opin Plant Biol.* **2002,** *5*, 291–294.

Rao, M. V. R.; Parameswari, C.; Sripriya, R.; Veluthambi, K. Transgene Stacking and Marker Elimination in Transgenic Rice by Sequential *Agrobacterium*-mediated Co-transformation with the Same Selectable Marker Gene. *Plant Cell Rep.* **2011,** *30*, 1241–1252.

Richa, K.; Tiwari, I. M.; Devanna, B. N.; Botella, J. R.; Sharma, V.; Sharma, T. R. Novel Chitinase Gene LOC_Os11g47510 from Indica Rice Tetep Provides Enhanced Resistance Against Sheath Blight Pathogen *Rhizoctonia solani* in Rice. *Front. Plant Sci.* **2017,** *8*, 596.

Sadumpati, V.; Kalambur, M.; Vudem, D. R.; Kirti, P. B.; Khareedu, V. R. Transgenic Indica Rice Lines, Expressing *Brassica Juncea* Nonexpressor of Pathogenesis-related Genes 1 (BjNPR1), Exhibit Enhanced Resistance to Major Pathogens. *J. Biotechnol.* **2013,** *166*, 114–121.

Sato, H.; Ideta, O.; Ando, I.; Kunihiro, Y.; Hirabayashi, H.; Iwano, M.; Miyasaka, A.; Nemoto, H.; Imbe, T. Mapping QTLs for Sheath Blight Resistance in the Rice Line WSS2. *Breed. Sci.* **2004,** *54*, 265–271.

Sattari, A.; Fakheri, B.; Noroozi, M.; Gudarzi, K. M. Review: Breeding for Resistance to Sheath Blight in Rice. *Intl. J. Farm & Alli. Sci.* **2014,** *3* (9), 970–979.

Shah, S. M.; Shabir, G.; Aslam, K.; Sabar, M.; Arif, M. Identification of SSR Markers to Find Adulteration in Elite Basmati Rice Varieties. *Environ. Plant Sys.* **2015b,** *1*, 4–15.

Shah, J. M.; Raghupathy, V.; Veluthambi, K. Enhanced Sheath Blight Resistance in Transgenic Rice Expressing an Endochitinase Gene from Trichoderma Virens. *Biotechnol. Lett.* **2009,** *31*, 239.

Sharma, A.; McClung, A. M.; Pinson, S. R. M.; Kepiro, J. L.; Shank, A. R.; Tabien, R. E.; Fjellstrom, R. Genetic Mapping of Sheath Blight Resistance QTLs within Tropical Japonica Rice Cultivars. *Crop Science* **2009,** *49*, 256–264.

Singh, A.; Singh, V. K.; Singh, S. P.; Pandian, R. T. P.; Ellur, R. K.; Singh, D.; Bhowmick, P. K.; Gopala Krishnan, S.; Nagarajan, M.; Vinod, K. K.; Singh, U. D.; Prabhu, K. V.; Sharma, T. R.; Mohapatra, T.; Singh, A. K. Molecular Breeding for the Development of Multiple Disease Resistance in Basmati Rice. *AoB Plants* **2012,** pls029; DOI: 10.1093/aobpla/pls029.

Sridevi, G.; Parameswari, C.; Sabapathi, N.; Raghupathy, V.; Veluthambi, K. Combined Expression of *Chitinase* and b-*1, 3-Glucanase* Genes in Indica Rice (*Oryza sativa* L.) Enhances Resistance Against *Rhizoctonia Solani*. *Plant Sci.* **2008,** *175*, 283–290.

Sridevi, G.; Sabapathi, N.; Meena, P.; Nandakumar, R.; Samiyappan, R.; Muthukrishnan, S.; Veluthambi, K. Transgenic Indica Rice Variety Pusa Basmati 1 Constitutively Expressing a Rice *Chitinase* Gene Exhibits Enhanced Resistance to *Rhizoctonia solani*. *J. Plant Biochem. Biotechnol.* **2003,** *12*, 93–101.

Sripriya, R.; Raghupathy, V.; Veluthambi, K. Generation of Selectable Marker-free Sheath Blight Resistant Transgenic Rice Plants by Efficient Co-transformation of a Cointegrate Vector T-DNA and a Binary Vector T-DNA in One *Agrobacterium Tumefaciens* Strain. *Plant Cell Rep.* **2008,** *27*, 1635–1644.

Sumner, D. R. *Sclerotia Formation by Rhizoctonia Species and their Survival: Taxonomy, Molecular Biology, Ecology, Pathology and Disease Control*. Kluwer Academic Publishers: Dordrecht, NL. 1996.

Taguchi-Shiobara, F.; Ozaki, H.; Sato, H.; Maeda, H.; Kojima, Y.; Ebitani, T.; Yano, M. Mapping and Validation of QTLs for Rice Sheath Blight Resistance. *Breed. Sci.* **2013,** *63*, 301.

Tang, Q.; Peng, S.; Buresh, R. J.; Zou, Y.; Castilla, N. P., et al. Rice Varietal Difference in Sheath Blight Development and its Association with Yield Loss at Different Levels of N Fertilization. *Field Crops Res.* **2007,** *102*, 219–227.

Van Loon, L. C.; Bakker, P. A.; Pieterse, C. M. Systemic Resistance Induced by Rhizosphere Bacteria. *Annu. Rev. Phytopathol.* **1998,** *36*, 453–483.

Vikal, Y.; Bhatia, D. Genetics and Genomics of Bacterial Blight Resistance in Rice. In *Advances in International Rice Research*; Li, J. Q., Ed.; InTech: London (UK), 2017; pp 175–213.

Wang, R.; Lu, L.; Pan, X.; Hu, Z.; Ling, F.; Yan, Y.; Liu, Y., et al. Functional Analysis of OsPGIP1 in Rice Sheath Blight Resistance. *Plant Mol. Biol.* **2015,** *87*, 181–191.

Watson, I. W.; Singh, D. The Future of Rust Resistant Wheat in Australia. *J. Aust. Inst. Agric. Sci.* **1952**, *28*, 190–197.

Webster, R. K.; Gunnell, P. S. *Compendium of Rice Diseases*. American Phytopathological Society: St. Paul, M. N., USA, 1992.

Wen, Z. H.; Zeng, Y. X.; Ji, Z. J.; Yang, C. D. Mapping Quantitative Trait Loci for Sheath Blight Disease Resistance in Yangdao 4 Rice. *Genet. Mol. Res.* **2015,** *14*, 1636–1649.

Wen, Z. H.; Zeng, Y. X.; Ji, Z. J.; Yang, C. D. Mapping Quantitative Trait Loci for Sheath Blight Disease Resistance in Yangdao 4 Rice. *Genet. Mol. Res.* **2015,** *14*, 1636–1649.

Wu, W.; Huang, J.; Cui, K.; Nie, L.; Wang, Q. Sheath Blight Reduces Stem Breaking Resistance and Increases Lodging Susceptibility of Rice Plants. *Field Crops Res.* **2012,** *128*, 101–108.

Xie, X. W.; Xu, M. R.; Zang, J. P.; Sun, Y.; Zhu, L. H. Genetic Background and Environmental Effects on QTLs for Sheath Blight Resistance Revealed by Reciprocal Introgression Lines in Rice. *Acta Agronomica Sinica.* **2008,** *34*, 1885–1893.

Xu, Q.; Yuan, X.; Yu, H.; Wang, Y.; Tang, S.; Wei, X. Mapping Quantitative Trait Loci for Sheath Blight Resistance in Rice using Double Haploid Population. *Plant Breed* **2011,** *130*, 404–406.

Xue, X.; Cao, Z. X.; Zhang, X. T.; Wang, Y.; Zhang, Y. F.; Chen, Z. X.; Zuo, S. M. Overexpression of OsOSM1 Enhances Resistance to Rice Sheath Blight. *Plant Dis.* **2016,** *100*, 1634–1642.

Yadav, S.; Anuradha, G.; Kumar, R. R.; Vemireddy, L. R.; Sudhakar, R.; Donempudi, K.; Venkata, D. Identification of QTLs and Possible Candidate Genes Conferring Sheath Blight Resistance in Rice (*Oryza sativa* L.). *Springer Plus.* **2015,** *4*, 175.

Zeng, Y. X.; Xia, L. Z.; Wen, Z. H.; Ji, Z. J.; Zeng, D. L.; Qian, Q.; Chang-deng, Y. Mapping Resistant QTLs for Rice Sheath Blight Disease with a Doubled Haploid Population. *J. Integr. Agric.* **2015,** *14*, 801–810.

Zhang, C. Q.; Liu, Y. H.; Ma, X. Y.; Feng, Z.; Ma, Z. H. Characterization of Sensitivity of Rhizoctonia solani, causing Rice Sheath Blight to Mepronil and Boscalid. *Crop Protection* **2009,** *28*, 381–386.

Zhou, J.; You, A.; Ma, Z.; Zhu, L.; He, G. Association Analysis of Important Agronomic Traits in Japonica Rice Germplasm. *Afr. J. Biotechnol.* **2012,** *11*, 2957–2970.

Zuo, S.; Yin, Y.; Pan, C.; Chen, Z.; Zhang, Y.; Gu, S.; Zhu, L. Fine Mapping of qSB-11 LE, the QTL that Confers Partial Resistance to Rice Sheath Blight. *Theor. Appl. Genet.* **2013,** *126*, 1257–1272.

Zuo, S.; Zhang, Y.; Yin, Y.; Li, G.; Zhang, G.; Wang, H.; Chen, Z. Fine-mapping of qSB-9 TQ, A Gene Conferring Major Quantitative Resistance to Rice Sheath Blight. *Mol. Breed.* **2014a,** *34*, 2191–2203.

CHAPTER 8

The Current Scenario, Progress, and Prospects of Bioinformatics for Rice Sheath Blight Disease Resistance

BALWINDER KAUR[1], KARANSHER SINGH SANDHU[2], and JAGMOHAN SINGH[3,4]

[1]*Department of Entomology and Nematology, University of Florida, Gainesville, FL, USA*

[2]*Department of Crop and Soil Sciences, Washington State University, Pullman, WA, USA*

[3]*Indian Agricultural Research Institute, New Delhi, India*

[4]*Institute of Molecular Plant Science, University of Edinburgh, Edinburgh EH9 3BF, UK*

ABSTRACT

Sheath blight of rice is a devastating disease of cultivated rice caused by soil-borne fungus *Rhizoctonia solani* worldwide. Previously, a range of cultural practices have been recommended to manage this disease. Furthermore, screening of rice germplasm to identify genetic resistance in cultivars for better disease control has been studied but has not been very successful. No major resistance gene against sheath blight of rice was found, except some minor quantitative trait loci, suggesting it to be a quantitative disease. In recent years, advancements in genomic technologies have generated a large amount of data which can generate new insights on hosts and pathogens.

Fungal Diseases of Rice and Their Management. Deepti Srivastava, Md. Shamim, Malik Mobeen Ahmad, and R. S. Upadhyay (Eds.)

Bioinformatics tools provide enormous information using genomic and phenotypic information and can help in understanding rice sheath blight resistance for designing better control strategies. As there is an exponential growth of genomic, transcriptomic, and proteomic data, there is a parallel growth in bioinformatics using computer science, information technology, visualization, analysis, modeling, and prediction. Bioinformatics can play a significant role in designing better disease diagnostic tools, analysis of the interactions among pathogenesis-related genes in host and pathogen, understanding disease genetics, and pathogenicity to in-depth knowledge of molecular mechanisms of resistance in rice sheath blight disease. All these aspects of bioinformatics open up the profound scope of this field and provide a platform for the development of new tools for disease diagnosis and prediction. This chapter provides information on the present status, progress, and future prospects of bioinformatics tools for managing rice sheath blight disease.

8.1 INTRODUCTION

In recent years, genomic studies utilizing plant genetic models have produced an enormous amount of biological data. Bioinformatics is an interdisciplinary area of science and technology for interpreting and handling biological data with the help of computer science and information technology. It can help to understand plant–pathogen interactions, quality traits of plants, and yield traits for developing new cultivars with improved traits. Bioinformatics allows the scientific community to decipher, manage, analyze, visualize, model, and predict gigantic biological data using various computational tools and databases. Various bioinformatics tools are utilized for sequence, structural, and functional analysis of genome and molecules (Tramontano, 2009). Understanding the enormous genetic and molecular information on species, biological processes in plants and pathogens can assist in developing disease-resistant plants (Koltai and Volpin, 2003).

Rice is an essential staple food crop globally (Khush, 2005). Numerous plant pathogens including fungi, bacteria, viruses, and nematodes attack rice and reduce the crop production (Saha et al., 2015). Sheath blight (ShB) of rice, caused by a soil-borne fungus *Rhizoctonia solani* Kühn, is a serious disease of rice worldwide (Zheng et al., 2013). Rice sheath blight occurs in East Asia and the southern USA (Chen et al., 2012) causing heavy yield losses from 5–50% depending on the stage of crop, severity of disease, and environmental conditions at the onset of disease (Singh et al., 2015). *R.*

solani is a hemi-biotrophic pathogen with broad host range (Kouzai et al., 2018) and survives between crops in the form of "sclerotia"—hard brown structures measuring approximately one-eighth of an inch in diameter. Thisungus also survives as mycelia in the rice plant or other plant residues during its necrotrophic phase (Wamishe et al., 2013). The management of disease has been done using application of chemical fungicides in combination with cultural control strategies.

Creating durable resistant varieties using plant breeding programs is an economical and environment-friendly approach for managing plant diseases (Miah et al., 2013). To our knowledge, due to the polygenic nature of ShB, neither the rice cultivars showing complete resistance nor major resistant genes to ShB have not been documented yet. However, few cultivars with different levels of resistance have been documented (Xue et al., 2016). This chapter describes the information available on ShB resistance in rice with some key concepts and introduces potential bioinformatics tools in relevance to sheath blight disease resistance.

8.2 CONVENTIONAL APPROACHES FOR MANAGING SHB RESISTANCE

8.2.1 QTLS FOR SHEATH BLIGHT RESISTANCE

Sheath blight of rice is usually shown to be associated with polygenic trait with few exceptions which suggests it is controlled by a single major gene. Quantitative trait loci (QTL) mapping involves precise mapping of a particular locus which is responsible for controlling resistance to ShB in a particular genotype. As QTL mapping highly depends upon the genotype and the environment where it is expressed, these factors should be considered and reported very attentively. The positions of QTLs vary depending upon the environment and the genotype being tested. The expression of QTLs depends upon the interaction of the said two, which can help in answering the question of why the expression of a particular QTL is prominent or even present in one environment in comparison to the other. Until now, around 50 QTLs have been mapped for resistance to rice ShB on all its 12 chromosomes (Srinivasachary, 2011). Furthermore, several studies have even categorized the genes coding for kinase, glutathione transferase, glucanase, and chitinase proteins and are known to be responsible for ShB disease resistance.

Li et al. (1995) were the first to report the QTLs associated with rice ShB using restriction fragment length polymorphism markers. Several QTLs for rice ShB have been mapped using recombinant inbred lines, double haploids, and association mapping population on chromosomes 2, 3, 6, 7, 10, and 11 under the field conditions (Table 8.1). All these QTLs are mapped on different landraces or local accessions of *O. sativa* L. ssp. *japonica* and *O. sativa* L. ssp. *indica*, although none of these genotypes are completely resistant to all prevalent races of rice ShB but provide a broader variation for selection. In general, *indica* rice has more resistance loci mapped compared to *japonica* subspecies (Sharma et al., 2009). Pathogenesis-related proteins or resistant candidate genes can be explored once these QTLs are validated under different mapping studies for providing the quantitative disease resistance against the rice ShB.

Different gene mapping studies have observed a strong association between QTLs controlling ShB resistance and other plant morphological traits such as height of plant, days to heading, and culm height. This association can be attributed to either the colocalization of the genes controlling both traits or the pleiotropic effects of those QTLs. In many cases, the peak association of logarithm of odds (LOD) scores of these morphological traits colocalized with the peak of LOD scores for QTL for rice ShB (Pinson et al., 2005). It has been suggested that these morphological traits do not provide the real cause for resistance, as they to confounding effect. For example, plants heading late might not be resistant genetically, but they showed resistant phenotypes because all the plants in the field were inoculated on the same day, that is how late heading plants could escape the disease pressure. Furthermore, rice plants having longer duration of growth are resistant because they grow slowly and provide less favorable environment for the disease development.

Few rice varieties like jasmine 85, Tetep, and LSBR-5 have been reported to be more resistant to ShB under the field conditions compared to other varieties, although this high resistance cannot be associated with any particular factor (Khush, 1977). Some studies have shown that the increased resistance can be attributed to different ecological and morphological characteristics, in spite of just the production of the pathogenesis-related proteins for regulating the host plant defense mechanism (Channamallikarjuna et al., 2010). Rapid progress in the identification of QTLs due to reduction in genotyping cost has led a way for the positional cloning of the said gene. This can ultimately aid in the understanding of the biochemical and molecular mechanisms involved in the host plant resistance mechanism.

TABLE 8.1 QTLs Mapped for Rice Sheath Blight Resistance and the Associated Traits

QTL	Chromosome no.	Percentage of variation explained	Traits associated with resistance	References
qSB-2	2	5.5	Plant height	Li et al. (1995)
qSB-3	3	20	Plant height and heading date	Li et al. (1995)
qSB-12	12	3.8	NA	Li et al. (1995)
qSB-3	3	27	NA	Zou et al. (2000)
qSB-11	11	26	NA	Zou et al. (2000)
qSB-3	3	19.5	Culm length	Sato et al. (2004)
qSB-9	9	15	NA	Tan et al. (2005)
qSB-1	1	7.9	Heading date	Pinson et al. (2005)
qSB-3	3	18	Plant height and heading date	Pinson et al. (2005)
qSB-12	12	9	Heading date	Pinson et al. (2005)
qSB-9	9	7.5	NA	Sharma et al. (2009)
qSB-5	5	5	Not determined	Liu et al. (2009)
qSB-9	9	25	Not determined	Liu et al. (2009)
qSB-7	7	20	NA	Channamallikarjuna et al. (2010)
qSB-11	11	22	NA	Channamallikarjuna et al. (2010)

8.2.2 GENE PYRAMIDING AND TRANSGENIC APPROACHES FOR SHB RESISTANCE

Gene pyramiding methods provide an alternative breeding method/procedure for resistant varieties by introgressing multiple QTLs into the variety. Since ShB of rice is polygenic, thus genetic transformation with single gene does not provide the durable resistance. Hence, gene pyramiding with different QTLs helps in providing a broader level of resistance. The success of gene pyramiding depends on the combination of genotypes selected for gene introgression. For example, Kim et al. (2003) pyramid the rice chitinase gene and maize ribosome-inactivating gene for enhancing ShB resistance in rice cultivars. Similarly, Kalpana et al. (2006) cotransformed thaumatin-like protein and rice chitinase into *O. sativa* L. ssp. *indica* rice for providing higher level of resistance to rice ShB.

Gene pyramiding has previously been utilized in rice breeding programs for developing disease resistance against bacterial blight (*Xanthomonas oryzae* pv.) and bacterial blast (*Magnaporthe oryzae*) (Jia et al., 2012). Furthermore,

Datta et al. (2002) developed transgenic lines for pyramiding genes resistant to rice ShB, bacterial blight, and insect resistance. Pathogenesis-related genes also called PR proteins (PR-3 proteins), thaumatin-like proteins, and chitinases are the most used effective sources of resistance against ShB of rice. All these genes have been isolated from different species of plant and introgressed into rice cultivars for developing resistance cultivars (Rajesh et al., 2020). Additionally, some putative-resistant alleles associated with ShB resistance have been documented (Jia et al., 2012). Hence, pyramiding the putative-resistant alleles in a plant cultivar using marker-assisted breeding can contribute to greater ShB resistance.

8.3 NONCONVENTIONAL APPROACHES FOR MANAGING SHB RESISTANCE

8.3.1 STUDYING PATHOGEN STRAINS CONFERRING HOST–PATHOGEN INTERACTION

The occurrence of rice ShB involves various complex interactions between the landing pathogen strain and the target genotype. These host–plant interaction studies are important for providing the valuable and timely disease management guidelines by the pathologists. Previously, these recommendations or studies were based on histology, microbiology, plant breeding, and plant physiology approaches, but nowadays with development of multi-omics and computational tools, such studies were conducted based on advanced genomics, molecular biology, bioinformatics, biochemistry, and -omics-based approaches (Ojo and Omabe, 2011). Each pathogen has a specific response which is identified by the host for initiating the forest line of defense. These responses from pathogens are known as pathogen-associated molecular patterns (PAMPs) which are detected by the surface molecules of host plants called Pattern Recognition Receptors which initiate the first line of defense by PAMP-triggered immunity (PTI). Plant pathogens which are able to fight with this PTI can successfully cause the infection and are called effectors (Koltai and Volpin, 2003). Plants also keep surveillance of effector proteins by monitoring the production of active defense proteins (R genes) for controlling the downstream production of pathogen severity. This continuous production of active defense proteins is called effector-triggered immunity (ETI) (Rao et al., 2008). With recent development of genomics and bioinformatics tools, the molecular basis of understanding of PTI and ETI is effectively understood.

There are different races identified for the *R. solani*, their genome complexity and interactions with host plants will aid in the selection of resistance cultivars. Zheng et al. (2013) were the first to sequence the whole genome of *R. solani* using the next-generation Illumina Genome Analyzer. The pathogen encodes the diverse set of proteins, enzymes for primary and secondary metabolism, and various transporters that cause the severe damage to the plant. This study identified 25 effector proteins that cause rice ShB and validated 3 effector proteins which initiate the crop defenses by interacting with the host plant. Various effector proteins and virulence factors in *R. solani* can be predicted and identified using bioinformatics tools as provided in Table 8.2. Various beta-glucanase and chitinase genes have been identified which can degrade the cell wall of *R. solani*, hence preventing the colonization of pathogen on the host plant (Mao et al., 2014). Cell wall in rice serves as a first line of defense against rice ShB pathogen. During the initiation of infection, *R. solani* produces hydrolytic enzymes such as glucanases, xylanases, and polygalacturonases for degrading the plant cell wall. Resistance rice varieties secrete enzymes such as beta-glucanase and chitinase proteins for preventing the degradation of plant cell wall of rice (Shanmugam, 2005).

TABLE 8.2 Bioinformatics Tools Available for Prediction of Effectors Proteins, Secretory Proteins, and Virulence Factors in Fungal Genomes

Database/Tools	Application	Reference/Website
SignalP3.0, Phobius	Signal peptide cleavage-site prediction	http://www.cbs.dtu.dk/services/SignalP-3.0/ http://phobius.binf.ku.dk/
ProtComp	Comparison of known homologous proteins	http://linux1.softberry.com
Big-PI	GPI-anchored proteins identification	http://mendel.imp.ac.at/gpi/fungi_server.html
PHYLIP	Protein distances calculation	http://evolution.genetics.washington.edu/phylip/
Pfam	Proteins domain prediction	http://pfam.sanger.ac.uk/
FunSecKB (the fungal secretome knowledge base)	Secretory proteins identification	http://proteomics.ysu.edu/secretomes/fungi.php
DFVF (database of fungal virulence factors)	Information on host fungal virulence factors	http://sysbio.unl.edu/DFVF/ Lu et al. (2012)

8.3.2 HOST PLANT ALTERATIONS FOR SHB RESISTANCE

Plant pathogenic fungi produce numerous extracellular enzymes including polygalacturonases (PGs) to degrade the plant cell wall in order to enter the plant tissue. However, polygalacturonase-inhibiting proteins (PGIPs) present in the cell wall of plants can specifically recognize and inhibit the activity of extracellular PGs, allowing the accumulation of elicitor-active oligogalacturonides conferring plant defense (Federici et al., 2006). In rice, PGIP1 (OsPGIP1) inhibits the PGase activity in pathogen *R. solani* and thus *OsPGIP1* gene has high utility in rice-breeding programs for regulating ShB resistance (Wang et al., 2015). The expression profile of *OsPGIP* family has been identified using microarray data (Lu et al., 2012). Various bioinformatics tools have been utilized to analyze different aspects of *OsPGIP1* protein as described in Table 8.3.

TABLE 8.3 Potential Bioinformatics Tools for Managing Rice Sheath Blight Resistance

Bioinformatic tools	Application	References/Website
ExPASy	Amino acid sequence deduction, signal peptides, and glycosylation sites prediction, conserved domain analysis	http://us.expasy.org/tools/dna.html
Vector NTI Suite 8.0	For determining molecular weight, isoelectric point (pI), and amino acid composition	https://vector-nti-suite.software.informer.com/8.0/
TMHMM Server v. 2.0	Transmembrane domain analysis	http://www.cbs.dtu.dk/services/TMHMM/
SignalP-5.0 Server	Signal peptide cleavage site	http://www.cbs.dtu.dk/services/SignalP/
The NetPhos 3.1 Server	Determining phosphorylation sites	http://www.cbs.dtu.dk/services/NetPhos/
BLAST	To search or compare subject proteins	https://blast.ncbi.nlm.nih.gov/Blast.cgi

8.4 NEW APPROACHES TOWARD MANAGEMENT OF SHB RESISTANCE

8.4.1 STUDYING PATHOGENESIS FOR SHB RESISTANCE

Upon acknowledgement of presence of pathogen(s), the host plant attacks the pathogen via either suppressing induced defenses or modulating the defense-signaling pathways. For example, introduction of *Brassica juncea* "non-expressor of pathogenesis-related genes 1" (*BjNPR1*) in rice led to improved ShB resistance (Sadumpati et al., 2013). Furthermore, -omics studies have been done to understand the response of genes in host plants under pathogen attack. Recently, proteometabolic alterations of *AtNPRI* complemented transgenic rice lines in response to *R. solani* infestation have been reported (Karmakar et al., 2019). A total of 38 differentially expressed proteins and 40 differentially accumulated metabolites were detected in the rice lines before and after the *R. solani* infection (Karmakar et al., 2019). These findings showed that plants adjust the energy and carbohydrate metabolism upon *R. solani* infection (Karmakar et al., 2019).

8.4.2 NOVEL TOOLS FOR STUDYING SHB RESISTANCE IN RICE

Novel tools including high-throughput marker technologies, single-nucleotide polymorphism, and diversity array technology markers can be utilized for exploring ShB resistance (Wittenberg et al., 2005). By the arrival of next-generation sequencing, fungus genomic information is available readily on commonly used databases including Genomes OnLine Database (Reddy et al., 2014), NCBI Genome (https://www.ncbi.nlm.nih.gov/genome), and MycoCosm (Grigoriev et al., 2012, 2013). Furthermore, different sequencing technologies are available such as Pacific Biosciences, single-molecule real-time sequencing, and BioNano Genomics (BioNano) (Wee et al., 2018).

High-resolution mapping such as association mapping can provide new sights on ShB resistance along with being less time intensive (Gupta et al., 2005). Furthermore, identifying donor resources for ShB resistance from non-*indica* type-rice might provide a wide genetic base for ShB resistance. Knowledge of GxE interactions (genetic and environmental risk factors) along with QTL mapping and QTL expression would help in documenting the severity of the disease.

8.5 CONCLUSIONS

This study provides a comprehensive overview of the rice sheath blight disease. A vast number of small-effect QTLs have been documented to be known for controlling the resistance against the *R. solani*, suggesting the polygenic nature of the disease. These QTLs and several resistant proteins such as chitinases, thaumatin-like proteins, and PR proteins have been utilized for pyramiding for developing resistance cultivars. Application of bioinformatics is at its infancy stages for managing the rice sheath blight, but this study provides a detailed overview about the various bioinformatics tools which potentially could be used for understanding and managing this disease.

ACKNOWLEDGMENTS

We thank Dimple Sharma and Gurlaz Kaur for their valuable suggestions that greatly added to the manuscript.

KEYWORDS

- **bioinformatics**
- **genomics**
- **host pathogen interaction**
- **QTL mapping**
- **rice sheath blight**

REFERENCES

Channamallikarjuna, V.; Sonah, H.; Prasad, M.; Rao, G. J. N.; Chand, S.; Upreti, H. C.; Singh, N. K.; Sharma, T. R. Identification of Major Quantitative Trait Loci qSBR11-1 for Sheath Blight Resistance in Rice. *Mol Breed.* **2010,** *25*, 155–166.

Chen, Y.; Zhang, A. F.; Wang, W. X.; Zhang, Y.; Gao, T. C. Baseline Sensitivity and Efficacy of Thifluzamide in Rhizoctonia solani. *Ann. Appl. Biol.* **2012,** *161*, 247–254.

Datta, K.; Baisakh, N.; Thet, K. M.; Tu, J.; Datta, S. K. Pyramiding Transgenes for Multiple Resistance in Rice Against Bacterial Blight, Yellow Stem Borer and Sheath Blight. *Theor. Appl. Genet.* **2002,** *106*, 1–8.

Federici, L.; Di Matteo. A.; Fernandez-Recio, J.; Tsernoglou, D.; Cervone, F. Polygalacturonase Inhibiting Proteins: Players in Plant Innate Immunity? *Trends Plant Sci.* **2006,** *11*, 65–70.

Grigoriev, IV.; Nikitin, R.; Haridas, S.; Kuo, A.; Ohm, R.; Riley, R.; Salamov, A.; Zhao, X.; Korzeniewski, F.; Smirnova, T. MycoCosm Portal: Gearing up for 1000 Fungal Genomes. *Nucleic Acids Res.* **2013.** DOI**:** 10.1093/nar/gkt1183.

Grigoriev, IV.; Nordberg, H.; Shabalov, I.; Aerts, A.; Cantor, M.; Kuo, A.; Minovitsky, S.; Nikitin, R.; Ohm, R.A.; Otillar, R. The Genome Portal of the Department of Energy Joint Genome Institute. *Nucleic Acids Res.* **2012,** *40*, D26–D32.

Gupta, P. K.; Rustgi, S.; Kulwal, P. L. Linkage Disequilibrium and Association Studies in Higher Plants: Present Status and Future Prospects. *Plant Mol. Biol.* **2005,** *57*, 461–485.

Jia, L.; Yan, W.; Zhu, C.; Agrama, H. A.; Jackson, A.; Yeater, K.; Li, X.; Huang, B.; McClung, A.; Wu, D. Allelic Analysis of Sheath Blight Resistance With Association Mapping in Rice. *Plos One*. **2012,** *7*, e32703.

Kalpana, K.; Maruthasalam, S.; Rajesh, T.; Poovannan, K.; Kumar, K. K.; Kokiladevi, E.; Raja, J. A. J.; Sudhakar, D.; Velazhahan, R.; Samiyappan, R.; Balasubramanian, P. Engineering Sheath Blight Resistance in Elite Indica Rice Cultivars Using Genes Encoding Defense Proteins. *Plant Sci*. **2006,** *170*, 203–215.

Karmakar, S.; Datta, K.; Molla, K. A.; Gayen, D.; Das, K.; Sarkar, S. N.; Datta, S. K. Proteo-metabolomic Investigation of Transgenic Rice Unravels Metabolic Alterations and Accumulation of Novel Proteins Potentially Involved in Defence Against Rhizoctonia Solani. *Sci. Rep*. **2019,** *9*, 1–16.

Khush, G. S. Disease and Insect Resistance in Rice. *Adv. Agron*. **1977,** *29*, 265–341.

Khush, G. S. What it Will Take to Feed 5.0 Billion Rice Consumers in 2030. *Plant Mol. Biol.* **2005,** *59*, 1–6.

Kim, J. K.; Jang, I. C.; Wu, R.; Zuo, W. N.; Boston, R. S.; Lee, Y. H.; Ahn, I. P.; Nahm, B. H. Co-expression of a Modified Maize Ribosome-Inactivating Protein and a Rice Basic Chitinase Gene in Transgenic Rice Plants Confers Enhanced Resistance to Sheath Blight. *Transgenic Res.* **2003,** *12*, 475–484.

Koltai, H.; Volpin, H. Agricultural Genomics: An Approach to Plant Protection. *Eur. J. Plant Pathol*. **2003,** *109*, 101–108.

Kouzai, Y.; Kimura, M.; Watanabe, M.; Kusunoki, K.; Osaka, D.; Suzuki, T.; Matsui, H.; Yamamoto, M.; Ichinose, Y.; Toyada, K.; Matsuura, T. Salicylic Acid-Dependent Immunity Contributes to Resistance Against Rhizoctonia Solani, a Necrotrophic Fungal Agent of Sheath Blight, in Rice and Brachypodium Distachyon. *New Phytol*. **2018,** *217*, 771–783.

Li, Z.; Pinson, S. R.; Marchetti, M. A.; Stansel, J. W.; Park, W. D. Characterization of Quantitative Trait Loci (QTLs) in Cultivated Rice Contributing to Field Resistance to Sheath Blight (Rhizoctonia solani). *Theor. Appl. Genet*. **1995,** *91*, 382–388.

Liu, G.; Jia, Y.; Correa-Victoria, F. J.; Prado, G. A.; Yeater, K. M.; McClung, A.; Correll, J. C. Mapping Quantitative Trait Loci Responsible for Resistance to Sheath Blight in Rice. *Phytopathology* **2009,** *99*, 1078–1084.

Lu, L.; Zhou, F.; Zhou, Y.; Fan, X.; Ye, S.; Wang, L.; Chen, H.; Lin, Y. Expression Profile Analysis of the Polygalacturonase-Inhibiting Protein Genes in Rice and Their Responses to Phytohormones and Fungal Infection. *Plant Cell Rep*. **2012,** *31*, 1173–1187.

Mao, B.; Liu, X.; Hu, D.; Li, D. Co-expression of RCH10 and AGLU1 Confers Rice Resistance to Fungal Sheath Blight Rhizoctonia Solani and Blast Magnorpathe Oryzae and Reveals Impact on Seed Germination. *World J. Microbiol. Biotechnol*. **2014,** *30*, 1229–1238.

Miah, G.; Rafii, M. Y.; Ismail, M. R.; Puteh, A. B.; Rahim, H. A.; Asfaliza, R.; Latif, M. A. Blast Resistance in Rice: A Review of Conventional Breeding to Molecular Approaches. *Mol. Biol. Rep*. **2013,** *40*, 2369–2388.

Ojo, O. O.; Omabe, M. Incorporating Bioinformatics Into Biological Science Education in Nigeria: Prospects and Challenges. *Infect. Genet. Evol*. **2011,** *11*, 784–787.

Pinson, S. R. M.; Capdevielle, F. M.; Oard, J. H. Confirming Qtls and Finding Additional Loci Conditioning Sheath Blight Resistance in Rice Using Recombinant Inbred Lines. *Crop Sci.* **2005,** *45*, 503–510.

Rajesh, T.; Maruthasalam, S.; Kalpana, K.; Poovannan, K.; Kumar, K. K.; Kokiladevi, E.; Sudhakar, D.; Velazhahan, R.; Balasubramanian, P. Pyramiding Insect and Disease Resistance in an Elite Indica Rice Cultivar asd16. *Biol. Plant.* **2020**, *64*, 77–86.

Rao, V. S.; Das, S. K.; Rao, V. J.; Srinubabu, G. Recent Developments in Life Sciences Research: Role of Bioinformatics. *Afr. J. Biotechnol.* **2008**, *7*.

Reddy, T. B.; Thomas, A. D.; Stamatis, D.; Bertsch, J.; Isbandi, M.; Jansson, J.; Mallajosyula, J.; Pagani, I.; Lobos, E. A.; Kyrpides, N. C. The Genomes OnLine Database (GOLD) v. 5: a Metadata Management System Based on a Four Level (meta) Genome Project Classification. *Nucleic Acids Res.* **2015,** *43*, D1099–D1106.

Sadumpati, V.; Kalambur, M.; Vudem, D. R.; Kirti, P. B.; Khareedu, V. R. Transgenic Indica Rice Lines, Expressing Brassica Juncea Nonexpressor of Pathogenesis-Related Genes 1 (BjNPR1), Exhibit Enhanced Resistance to Major Pathogens. *J. Biotechnol.* **2013,** *166*, 114–21.

Saha, S.; Garg, R.; Biswas, A.; Rai, A. B. Bacterial Diseases of Rice: An Overview. *J. Pure Appl. Microbiol.* **2015**, *9*, 725–736.

Sato, H.; Ideta, O.; Ando, I.; Kunihiro, Y.; Hirabayashi, H.; Iwano, M.; Miyasaka, A.; Nemoto, H.; Imbe, T. Mapping Qtls for Sheath Blight Resistance in the Rice Line WSS2. *Breed Sci.* **2004**, *54*, 265–271.

Shanmugam, V. Role of Extracytoplasmic Leucine Rich Repeat Proteins in Plant Defence Mechanisms. *Microbiol. Res.* **2005**, *160*, 83–94.

Sharma, A.; McClung, A. M.; Pinson, S. R. M.; Kepiro, J. L.; Shank, A. R.; Tabien, R. E.; Fjellstrom, R. Genetic Mapping of Sheath Blight Resistance QTLs Within Tropical *Japonica* Rice Cultivars. *Crop Sci.* **2009,** *49*, 256–264.

Singh, R.; Sunder, S.; Kumar, P. Sheath Blight of Rice: Current Status and Perspectives. *Indian Phytopathol.* **2016**, *69*, 340–351.

Srinivasachary, L. Resistance to Rice Sheath Blight (Rhizoctonia solani Kühn) [(Teleomorph: Thanatephorus cucumeris (A. B. Frank) Donk.] Disease: Current Status and Perspectives. *Euphytica.* **2011**, *178*, 1–22.

Tan, C. X.; Ji, X. M.; Yang, Y.; Pan, X. Y.; Zuo, S. M.; Zhang, Y. F.; Zou, J. H.; Chen, Z. X.; Zhu, L. H.; Pan, X. B. Identification and Marker-Assisted Selection of Two Major Quantitative Genes Controlling Rice Sheath Blight Resistance in Backcross Generations. *Acta Genet. Sin.* **2005,** *32*, 399–405.

Tramontano, A. Bioinformatics. eLS. 2009.

Wamishe, Y.; Cartwright, R.; Lee, F. Management of Rice Diseases. *Arkansas Rice Production Handbook;* Little Rock, Arkansas, 2013; vol 72204, pp 126–133.

Wang, R.; Lu, L.; Pan, X.; Hu, Z.; Ling, F.; Yan, Y.; Liu, Y.; Lin, Y. Functional Analysis of OsPGIP1 in Rice Sheath Blight Resistance. *Plant Mol. Biol.* **2015,** *87*,181–191.

Wee, Y.; Bhyan, S. B.; Liu, Y.; Lu, J.; Li, X.; Zhao, M. The Bioinformatics Tools for the Genome Assembly and Analysis Based on Third-Generation Sequencing. *Brief. Funct. Genom.* **2019,** *18*, 1–2.

Wittenberg, A. H.; Van Der Lee, T.; Cayla, C.; Kilian, A.; Visser, R. G.; Schouten, H. J. Validation of the High-Throughput Marker Technology DArT Using the Model Plant Arabidopsis Thaliana. *Mol. Genet. Genom.* **2005**, *274*, 30–39.

Xue, X.; Cao, Z. X.; Zhang, X. T.; Wang, Y.; Zhang, Y. F.; Chen, Z. X.; Pan, X. B.; Zuo, S. M. Overexpression of OsOSM1 Enhances Resistance to Rice Sheath Blight. *Plant Dis.* **2016,** *100*, 1634–1642.

Zheng, A.; Lin, R.; Zhang, D.; Qin, P.; Xu, L.; Ai, P.; Ding, L.; Wang, Y.; Chen, Y.; Liu, Y.; Sun, Z. The Evolution and Pathogenic Mechanisms of the Rice Sheath Blight Pathogen. *Nat. Commun.* **2013,** *4*, 1–10.

Zou, J. H.; Pan, X. B.; Chen, Z. X.; Xu, J. Y.; Lu, J. F.; Zhai, W. X.; Zhu, L. H. Mapping Quantitative Trait Loci Controlling Sheath Blight Resistance in Two Rice Cultivars (Oryza Sativa L.). *Theor. Appl. Genet*. **2000,** *101*, 569–573.

CHAPTER 9

An Overview of Ecology, Epidemiology, and Identification Measures for Rice Sheath Rot Disease

MOHD. SAID[1], SABA SIDDIQUI[1], SALMAN AHMAD[1], NADEEM KHAN[1], DEEPTI SRIVASTAVA[1], MOHD. HARIS SIDDIQUI[1,2], and MALIK M. AHMAD[1]

[1]*Department of Agriculture, Integral Institute of Agricultural Science and Technology (IIAST), Integral University, Lucknow, India*

[2]*Department of Bioengineering, Faculty of Engineering, Integral University, Lucknow, India*

ABSTRACT

Rice is one of the most important staple food crop that has predominantly been grown in almost all the continents of the planet. About a century ago, fungal sheath rot disease was characterized primarily by rotting of sheaths. The causal pathogen was known as *Acrocylindrium oryzae,* which was later characterized as *Sarocladium oryzae.* It was thought to be a seed-borne disease and had been a source of a major production limitation in some rice-growing countries. It produces a wide array of phytotoxins that generally induces necrosis in plants, chaffiness, grain discoloration, and sterility leading to yield drop, to a maximum extent of about 85%. The incubation period of this disease was 3–4 days, which may become more intense if the ambient temperature is between 20–30°C with a relative humidity of around 65–85%. *S. oryzae* was classified using cultural and morphological

Fungal Diseases of Rice and Their Management. Deepti Srivastava, Md. Shamim, Malik Mobeen Ahmad, and R. S. Upadhyay (Eds.)

characteristics including which appeared as compact, confined growth and hyaline, smooth, single-celled conidia. Colony characteristics of *Sarocladium* sp. were cottony, produces mycelia color varies from white to light yellow, sometimes turn into pale orange with age and pale orange color. Generally, the overall size of *Sarocladium* conidia ranged from 2.40–7.20 × 1.20–2.40 μm. Sheath rot-infected rice seeds showed a progressive and considerable drop in total sugar, reducing sugar, and nonreducing sugar content, while the phenol content increases. This review mainly focuses on the most commonly reported rice sheath rot pathogens and strives by improving our understanding of the ecology and identification of rice sheath rot disease. Morphological data are not enough as a basis for determining a species because it can lead to improper identification. Implementation of molecular fungal identification needs to be done to identify fungal species through phylogenetic analysis to get correct results.

9.1 INTRODUCTION

Rice is a major crop in the globe, providing as a staple food for half of the world's population and also being used in industry and animal feed. It is grown in different agroecological zones in tropical and subtropical regions, particularly in Asia, which accounts for 90% of global production (IRRI, 2015). Besides pests, rice crop is generally attacked by various types of attacked by fungal, bacterial, and viral pathogens leading to the yield loss of upto 40% per annum (Srinivasachary et al., 2002). Fungal blast, brown spot bacterial blight, and tungro viral diseases are believed to be major diseases of rice crop. Sheath rot disease of rice has not been studied well but is highly ubiquitous in all rice-growing nations of Asia and United States, causing a considerable injury to the rice crops (Ou, 1985). Sheath rot, a seed-borne disease, caused by *Sarocladium oryzae* (Sawada) W. Gams and D. Hawks, was first isolated in 1922 in Taiwan. Until 1973, there were no significant reports of rice sheath rot caused by *S. oryzae* from Bangladesh (Shahjahan et al., 1994). The random nature of various aspects acting in different environmental conditions explains why losses attributed to *S. oryzae* can vary from 20 to 85% (Sakthivel, 2001). Rice sheath rot was once thought to be a small and geographically limited disease. It is only recently that it has acquired a thrust and spread widely. Rice farming systems have changed dramatically since the green revolution, especially with the use of high-yielding varieties, sowing methods, greater chemical fertilizer use, efficient water usage systems, and so on. Crop intensification strategies such as increased plant density, high nitrogenous fertilizer rates, and the adoption of semidwarf and photoperiod-insensitive cultivars have promoted rice crop susceptibility to pathological problems, including the sheath rot complex. It is also thought that new photoperiod-insensitive cultivars have lost their ability to prevent flowering during high temperature and humidity environmental conditions, which have been advantageous to disease attacks (Mew et al., 2004). Furthermore, increasing international trade of planting materials may also have added in the advancement of disease incidence. This elucidates why practically no detailed studies revealing the link in between the existence of pathogen and magnitude of disease inoculum and yield loss (Mew and Gonzales, 2002).

Because *S. oryzae* is one of the major rice sheath rot phytopathogen, more attention will be laid on the symptoms of rice sheath rot produced by *S. oryzae* in this review. Besides, it will also investigates at the evidence on causal agents and ecology, host range, morphological, physiological, biochemical, and molecular effects, virulence factors, synergism and relationships among various causal factors, causal agents, ecology, epidemiology, spatial distributions and interactions with farming methods, and identification methods.

9.2 ECOLOGY

S. oryzae was at first described as *Acrocylindrium oryzae* (Bigirimana et al., 2015; Ayyadurai et al., 2005). The genus *Sarocladium* was established in 1975 (Gams and Hawksworth, 1975) and currently it covers 16 species including phytopathogens, saprophytes, endophytes, mycoparasites, and opportunistic human pathogens (Giraldo et al., 2015). The genus comes under the order of Hypocreales which belongs to the Ascomycota Phylum. One of the *Sarocladium* species, *S. attenuatum* was previously characterized as a separate species is nowadays considered as a synonym of *S. oryzae* (Bridge et al., 1989). *S. oryzae* grows at a very slow rate (about 2.5 mm/day on potato dextrose agar (PDA) at 28°C) and produces a sparsely branched mycelium while the colony characteristics of *Sarocladium* sp. were cottony, produces mycelia color varies from white to light yellow, sometimes turn into pale orange with age and pale orange color of reverse views (Pramunadipta et al., 2020) while the mycelial colonies on host plant were pale pinkish or whitish and sporulation is profusely rich at the center of the colony. The pathogen produces verticilliate condiophore with one or two phialides having hyaline, fusiform, often curved, and single-celled conidia of size 3.5–9.0 × 1–2.5

μm (Ou, 1985). Chowdhury et al. (2015) showed that conidia size of 29 isolates of *S. oryzae* ranged with a size of 2.40–7.20 μm and 1.20–2.40 μm. Fungal infestation in rice grains causes significant alterations in the amounts of sugars and starch. During the storage of infected rice grains, seed protein levels dropped dramatically (Bilgrami et al., 1979). With varied degrees of seed infection induced by *S. oryzae*, there was a progressive and considerable decline in total sugar, reducing sugar as well as nonreducing sugar content of rice seeds whereas the phenolic content increases noticeably (Gopalakrishnan et al., 2010). It is also believed that the elevated level of helvolic acid induces a greater chance of sheath rot disease causing incidence (Ayyadurai et al., 2005).

All sheath rot pathogens cause a similar greyish-brown necrotic lesion on the uppermost leaf sheath surrounding the young panicles of the plant. The lesions caused by the fungi appear as oblong or irregular spots in shape with a size ranging from 0.5 to 1.5 cm long, with grey at centers and brown at margins or greyish brown throughout; which expands and often coalesce and cover most of the leaf sheath, ensuing in unemerged or moderately emerged young panicles having unfilled or partially filled rice grains with a copious whitish powdery growth may be found inside compromised sheaths and juvenile panicles are decaying. So, we can say that one of the key features of sheath rot disease is rotting and discoloration of the sheath and grains, viability, and nutritional value of seeds are reduced due to chaffiness and sterility of the resultant grains (Sakthivel, 2001; Gopalakrishnan et al. 2010). Diseased plants produce brownish, infertile, or empty seeds or no panicle at all, depending on the extent of the infection (Weeraratne et al., 2019). Insect infestation, the availability of entry points, heavy nitrogen treatment, high relative humidity, dense crop growth and leaf canopy, and the incidence of temperatures between 20–28°C at heading to maturity of the rice crop are all predisposing factors for sheath-rot infection. When applied to the rice sheath, the fungus releases two phytotoxins, helvolic acid and cerulenin, which have been shown to mimic part or all of the sheath-rot symptoms (Gnanamanickam and Mew, 1991; Sakthivel and Gnanamanickam, 1987).

9.3 EPIDEMIOLOGY

So far *S. oryzae* has been accounted in countries of Mexico, USA, Argentina, Brazil, Venezuela, Bangladesh, China, India, Indonesia, Japan, Malaysia, Nepal, Pakistan, Philippines, Sri Lanka, Thailand, Vietnam, Burundi,

Cameroon, Côte d'Ivoire, Gambia, Kenya, Madagascar, Niger, Nigeria, Senegal, Tanzania, Brunei Darussalam, Saudi Arabia, Tajikistan, Uzbekistan, and Australia (CABI, 2007). It can be found in all the rice-growing countries of the world, being extremely common in wet seasons (Mew and Gonzales, 2002), mostly in lowland locations (Pearce et al., 2001), and hot and humid weather supports the disease (Sakthivel, 2001). *S. oryzae* infections were reported below 1250 m in Nepal, according to Sharma et al. (1997). Generally, relative humidity of 65–85% and temperatures of 20–30°C favors the development of sheath rot disease (Sakthivel, 2001). The disease development can be at its peak when the temperature ranges in between 17–20°C with a relative humidity of 40–50% at flowering stage. Mohan and Subramanian (1977) observed that relative humidity in association with the number of cold days (nicto-temperature below 20°C) favored the disease development. Singh and Mathur (1993) observed that seed transmission of the pathogen was more in plants grown in greenhouse than in climate room. They reported 85.5 and 45.1% of seed infection with a highest infection level of 94.4 and 81.8% in empty and filled seeds, respectively. However, Singh and Raju (1981) observed that in the regions of north India, the mycelial threads or spores can survive for 4 months in seeds, 7 months in leaf sheaths at room temperature, and for 10 months in leaf sheaths in crop field. These can significantly highlight the importance of infected stubble and straw, which can further act as primary inoculum source. This made clear that when environmental conditions are favorable, the pathogen can not only survive in infected seeds, plant straws, or stubbles, but also in soil, water, or weeds.

The fungus can attack the host plant at different time intervals of growth stages. It can enter from cut-wounds and/or stomata, and can be most destructive after booting stage. The access of *S. oryzae* in rice plant is generally assisted by insects and mites and/or by damaging or deteriorating of the plant cell walls by additional existing pathogens. Secondary infections may be wind-borne from injured tissues. However, not much information is available about the transmission of infection via seed-borne method. Although, rice crop is considered to be the key host of *S. oryzae* but the pathogen has also been found in bamboo plants (Pearce et al., 2001), grasses, and sedges. Several alternate and collateral hosts have also been reported for *S. oryzae*, which are responsible for spreading the inoculum to host plants such as *Oryza rufipogon* Grieff., *E. colona* (L.) Link (Dekka and Phookan, 1992), *Eleusine indica* (L.) Gaertn., *Monochoria vaginalis* (Burm. f.) Persl., *Cyperus teneriffae* Poir, *C. iria* L., *C. difformis* L., *Echinochloa crusgalli* (L.) Beauv. (Balakrishnan and Nair, 1981), *Bambusa balcooa*, *B. vulgaris* (Boa and Brady, 1987), *M. vaginalis* (Burm. f.) Persl., *Hymenachne assamica* USA (Hook f.) Hitchc., *Leersia hexndra* Swartz., *Panicum walense* Mez, *Eleusine indica* (L.) Gaertn (Rehman et al., 1982; Singh and Dodan, 1995).

In a study, Estrada et al. (1984) found the yield loss to be 52.8% in Philippines, 9.6–26% in India (Chakravarty and Biswas, 1978), and in Thailand, it was around 35% (Surin et al., 1988). In a different study, Shahjahan et al., (1994) showed the total yield loss in Bangladesh was found to be 18.9–47.4% in different seasons. Depending on the pathosystem conditions, the yield loss can be from 20% to as high as 85% (Sakthivel, 2001; Bigirimana et al., 2015). As a consequence, it can be said that this sheath rot disease is a very vicious to rice crops.

9.4 IDENTIFICATION METHODS

9.4.1 MORPHOLOGICAL METHODS

Fungal identification based on phenotypic characters is the most traditional and widespread method used for characterizing fungi and was further developed with advancement in microscopes during 19th century. Conventionally, fungi are grown at temperatures at around 20–30°C with alternating light and dark or completely dark condition in incubators. Growth of culturable fungi are generally expressed in terms of colony-forming units obtained from plate counts which has been developed from the exponential multiplication of a single fungal cell or by collection of fungal cells (Macher, 2001).

Ou et al. (2020) observed 102 strains of *Sarocladium* associated with rice crop. The researchers had found that *Sarocladium attenuatum* and *S. oryzae* are two distinct species and also they found two additional species *S. sparsum* sp. nov. and *S. spirale* sp. nov., can be identified. *S. sparsum* is closely related to *S. oryzae* and *S. attenuatum* and can cause the disease symptom of sheath rot during pathogenicity tests whereas *S. spirale* is more closely related to *S. hominis* and *S. spinificis* and is nonpathogenic to rice plants. Afifah et al. (2020) had observed that isolated fungi from symptomatic sheaths and grains (KP, KP2, KP3, PW03, and PW3) varied in colors and mycelial growth. The front sides of all colonies were white, while the reverse sides of KP3 and PW3 colonies have yellowish color with faster mycelium growth while the rest of the fungal morphotype colonies have orange color with slower mycelium growth. Peeters et al. (2020) described *S. oryzae* isolates from various rice producing areas can be assembled into three distinct phylogenetic lineages. Different isolates of sheath rot pathogen were cultured on PDA plates for 15 days at 28°C. Group 1 including RFRG2 and IBNG0013 isolate was powdery with brightest colored colonies, radially folded, pale to dark orange with white borders. Group 2 isolates had powdery to wet aerial mycelium with radially folded/striated, white or pale salmon colonies while Group 3 colonies were morphologically powdery, dark, or pale orange colonies on PDA. In a study performed in Indonesia investigating sheath rot pathogen, Pramunadipta et al. (2020) isolated six isolates of *S. oryzae,* namely SO2, SO3, SO5, SO8, and SO13. Morphological colony characteristics of these *Sarocladium* isolates were cottony with mycelial color varying from white to light yellow, at times turn into pale orange with age and pale orange color of reverse side. Microscopic view of vegetative hyphae and was hyaline, smooth, septate, and thin-walled while conidiphores were hyaline, smooth-walled, and can be branched or simple. Conidia are generally hyaline, cylindrical, and aseptate, 1.6–4.8 μm × 0.6–1.7 μm in size and positioned in slimy heads. Sporodochia, a type of conidiophore made up of aggregated conidiophores, was discovered for the first time. *S. attenuatum* and *S. oryzae* generate sporodochia phialides, which are noticeably shorter but are subulate in *S. attenuatum* while in *S. oryzae*, they are flask-shaped (Ou et al., 2020). In an attempt to morphologically identify three different *S. oryzae* isolates, Sobanbabu et al. (2018) showed that *S. oryzae* isolate ASD1 showed a good mycelial growth than isolate KKM1 and KKM2. The colony emerged as compact with constricted growth while the reverse side of the plate showed orange coloured discolouration. Hittalmani et al. (2016) identified *S. oryzae* Saro-13 isolate by observing mycelial morphology and its colony characteristics wherein the fungus produced sparingly branched mycelium with orange pigmentation on PDA plates while the conidium was hyaline, cylindrical, and single-celled in nature while in another study, Giraldo et al. (2015) while explaining the phylogeny of *Sarocladium* described its phenotypic characters as white to orange-white colonies when grown on PDA medium at 25°C with simple and branched conidiophores, cylindrical phialides of up to 60 μm long and cylindrical conidia. Bigirimana et al. (2015) while describing the sheath rot disease showed that *S. oryzae* is a slow-grower fungus. The two different isolates RFNG33 and RFRG2, grown on PDA medium at 28°C for 14 days, produced a sparsely branched white to slightly pinkish mycelium while the reverse view is generally light orange to dark orange in colour. While first reporting of sheath rot disease in Western Australia, Lanoiselet et al. (2012) showed usual phenotypic conidiophore and conidia characteristics of *S. oryzae*.

9.4.2 BIOCHEMICAL METHODS

The quality of rice seeds degrades when they experience biochemical changes during pathogenesis. Sugars are vital in disease progression as pathogens require carbon from sugars to grow, and they also serve as precursors for the production of various polyphenolics, phytoalexins, and protein compounds. With the augmentation of sheath rot infection in host plant, reducing, nonreducing as well as total sugar content starts decreasing (Gopalakrishnan et al., 2009). This suggests that *S. oryzae* utilizes different sugars present in rice seeds well for its growth (Reddy et al., 2000). The enhancement in phenolic content is generally due to infection of *S. oryzae* in infected seeds while increment in phenolase activity in the plant tissues especially in and around diseased locations is a response which shows the advancement of disease (Gopalakrishnan et al., 2010). The decrease in starch concentration, according to Mirocha and Zaki (1966), could be due to rapid breakdown of starch degrading enzymes synthesized by the pathogens, or consumption of starch molecules, or both. Vidhyasekeran et al. (1984) had shown that the decline of protein content of host plant at preliminary stage may be due to their hydrolysis to simpler forms by fungal proteolytic enzymes. Joe and Manibhushanrao (1995) found a positive correlation with isolates of *S. oryzae* and *S. attenuatum* by observing the in vitro production of CMCase, PTE, invertase, and protease enzymes. Both researchers had also observed that *S. oryzae* was most aggressive species in comparison to *S. attenuatum* while causing the sheath rot disease.

S. oryzae also is identified to generate antimicrobial secondary metabolites like cerulenin and helvolic acid (Ghosh et al., 2002; Hittalmani et al., 2016). Peeters et al. (2020) compared 10 *S. oryzae* isolates of three different phylogenic groups for *in vitro* toxin production and in the host plant. This might be due to the four phases of fungal growth: lag, log, stationary, and death, where the toxins increased to a maximum during the lag phase, and subsequently decreased with fungus aging (Hafiz et al., 2009). Bodalkar and Awadhiya (2014) proved the pathogenicity of different *S. oryzae* isolates by inoculating them on rice plants of various stages to instigate grain discoloration, which had started due to production of helvolic acid and cerulenin from the pathogen (Ghosh et al., 2002).

9.4.3 MOLECULAR METHODS

Identification of correct fungal species based on morphological and/ or biochemical character is somewhat tiresome and old-fashioned way.

Nowadays, recent advances in science have driven taxonomic conclusions toward DNA-based approaches and thus, species concepts must therefore rely on the use of molecular-based methods which are quickest and most reliable. The analysis of internal transcribed spacer (ITS) sequences of rDNA is one of the most frequent methods for identifying plant pathogenic fungus. Prasannakumar et al. (2021) devised a LAMP-based molecular identification method for *S. oryzae* isolates in just 30 min with a sensitivity of 100 fg. Pramunadipta et al. (2020) isolated a total of 24 fungal species based on sheath rot symptoms from six different provinces of Indonesia and performed ITS rDNA gene sequencing for molecular confirmation of fungal species. It was observed that only six of the isolates were *S. oryzae* species causing sheath rot disease. Ou et al. (2020) based on *ACT, ITS, LSU,* and *TUB2* genes observed 102 strains of *Sarocladium* which were grouped into Clades I (*S. attenuatum*), II (*S. oryzae*), III (*S. sparsum*), and IV (*S. spirale*). In order to uncover the genetic variability of sheath rot disease causing fungus, Sharma et al. (2018) using RAPD technique classified *S. oryzae* isolates into two major clusters. Sobanbabu et al. (2018) identified the ASD1 as the most virulent strain and maximum disease-causing capacity than isolate KKM1 and KKM2 using ITS primers. Saleh et al. (2016) during her study determined molecular variability of sheath rot pathogen *S. oryzae* by SCAR and SRAP primers, which were able to differentiate the fungal isolates collected from diverse locations. The isolated were categorized into high and low virulent clusters. Ayyadurai et al. (2005) observed biological and molecular variability of *S. oryzae* via RAPD which had been isolated from various locations of North East and South India. These scientists had also found that isolates producing higher concentration of helvolic acid generally induces a higher percent incidence of sheath rot.

9.5 CONCLUSION

Rice sheath rot is gaining momentum as a new fatal rice disease, yet scientific knowledge of the disease is still limited. Since the initial reports of rice sheath rot symptoms in Taiwan in 1922, which were linked *to S. oryzae*, there have been numerous reports of identical or identical illness symptoms in various regions of the world. The hypothesis that "everything is everywhere, but, the environment selects" (de Wit and Bouvier, 2006) relates to rice sheath rot; there are numerous organisms that can possibly cause rice sheath rot, and they can be found almost anywhere but the environment chooses the ones that can adapt to the local environmental conditions. As a result of this circumstance, symptoms in the rice sheath rot disease complex overlap, especially at the rice reproductive stage (Johanson et al., 1998; Hu et al., 2008). The morphological, biochemical, and molecular identification methods are only performed so that they may be further used in rice improvement programmes, economic, and sustainable disease management strategies. Indeed, with the above scientific information available in the public domain, new breed of researchers will carry out at faster rate and coherent experiments to analyze the rice-*Sarocladium* interactions, which may further help to develop better disease control measures.

KEYWORDS

- **ecology**
- **epidemiology**
- **identification techniques**
- **rice**
- **sheath rot disease**

REFERENCES

Afifah, K.; Wiyono, S.; Yuliani, T. S. Rice Sheath Rot Disease Etiology and Characterization of the Pathogen. Southeast Asia Plant Protection Conference; *IOP Conf. Ser. Earth Environ. Sci.* **2020,** *468,* 012041.

Ayyadurai, N; Kirubakaran, S. I.; Srisha, S.; Sakthivel, N. Biological and Molecular Variability of *Sarocladium Oryzae*, the Sheath Rot Pathogen of Rice (*Oryza Sativa* L.). *Curr. Microbiol.* **2005,** *50,* 319–323.

Balakrishnan, B.; Nair, M. C. Weed Hosts of *Acrocylindrium Oryzae* Sawada, a Sheath Rot Pathogen of Rice. *Int. Rice Res. Newslett.* **1981,** *6,* 13.

Bigirimana, V. de P.; Hua, G. K.; Nyamangyoku, O. I.; Höfte, M. Rice Sheath Rot: An Emerging Ubiquitous Destructive Disease Complex. *Front. Plant Sci.* **2015,** *6,* 1066.

Bilgrami, K. S.; Jamaluddin, R. K. S.; Sinha, K.; Prasad, T. Changes in Seed Contents of Paddy Due to Fungal Flora. *J. Phytopathol.* **1979,** *96,* 9–14.

Boa, E. R.; Brady, B. L. *Sarocladium Oryzae* Associated With Blight of Bambusa Species in Bangladesh. *Trans. Brit. Mycol. Soc.* **1987,** *89,* 161–166.

Bodalkar, C.; Awadhiya, G. K. Assessment of Percent Grain Discoloration in Important Rice Varieties. *Int. J. Curr. Res. Biosci. Plant Biol.* **2014,** *1,* 61–64.

Bridge, P. D.; Hawskworth, D. L.; Kavishe, D. F.; Farnell, P. A. A Revision of the Species Concept in *Sarocladium*, the Causal Agent of Sheath-Rot in Rice and Bamboo Blight, Based on Biochemical and Morphometric Analyses. *Plant Pathol.* **1989,** *38,* 239–245.

CABI. Crop Protection Compendium; CAB International: Wallingford, 2007.

Chakravarty, D. K.; Biswas, S. Estimation of Yield Loss in Rice Affected by Sheath Rot. *Pl. Dis. Reptr.* **1978,** *62,* 226–227.

Chowdhury, M. T. I.; Mian, M. S.; Mia, M. A. T.; Rafii, M. Y.; Latif, M. A. Agro-ecological Variations of Sheath Rot Disease of Rice Caused by *Sarocladium Oryzae* and DNA Fingerprinting of the Pathogen's Population Structure. *Genet. Mol. Res.* **2015,** *14,* 18140–18152.

de Wit, R.; Bouvier, T. Everything is Everywhere, but, the Environment Selects; What did Baas Becking and Beijerinck really say? *Environ. Microbiol.* **2006,** *8,* 755–758.

Dekka, A. K.; Phookan, A. K. Some Common Weed Hosts of *Sarocladium Oryzae* in Assam, India. *Int. Rice Res. Newsl.* **1992,** *17,* 25.

Estrada, B. A.; Torres, C. C.; Bonman, J. M. Effect of Sheath Rot on Some Yield Component. *Int. Rice Res. Newsl.* **1984,** *9,* 14.

Gams, W.; Hawksworth, D. L. Identity of *Acrocylindrium Oryzae* Sawada and a Similar Fungus Causing Sheath-Rot of Rice. *Kavaka* **1975,** *3,* 57–61.

Ghosh, M. K.; Amudha, R.; Jayachandran, S.; Sakthivel, N. Detection and Quantification of Phytotoxic Metabolites of *Sarocladium Oryzae* in Sheath Rot-Infected Grains of rice. *Lett. Appl. Microbiol.* **2002,** *34,* 398–401.

Giraldo, A.; Gené, J.; Sutton, D. A.; Madrid, H.; de Hoog, G. S.; Cano, J.; Decock, C.; Crous, P. W.; Guarro, J. Phylogeny of *Sarocladium* (Hypocreales). *Persoonia* **2015,** *34,* 10–24.

Gnanamanickam, S. S.; Mew, T. W. Interactions Between *Sarocladium Oryzae* and Stem Attacking Fungal Pathogens of Rice. *Plant Soil* **1991,** *138,* 213–219.

Gopalakrishnan, C.; Kamalakannan, A.; Valluvaparidasan, V. Effect of Seed-Borne *Sarocladium Oryzae*, the Incitant of Rice Sheath Rot on Rice Seed Quality. *J. Plant Prot. Res.* **2010,** *50,* 98–102.

Hafiz, M.; Khan, J. A.; Naz, S.; Khan, S. N.; Akram, M. Grain Discoloration Disease Complex: A New Threat for Rice Crop and its Management. *Pak. J. Phytopathol.* **2009,** *21,* 31–36.

Hittalmani, S.; Mahesh, H. B.; Mahadevaiah, C.; Prasannakumar, M. K. De Novo Genome Assembly and Annotation of Rice Sheath Rot Fungus *Sarocladium Oryzae* Reveals Genes Involved in Helvolic Acid and Cerulenin Biosynthesis Pathways. *BMC Genom.* **2016,** *17,* 1–13.

Hu, C. J.; Li, Y. R.; Wei, Y. W.; Huang, S. L. A PCR-Based Method to Detect *Sclerotium Hydrophilum* in Infected Rice Leaf Sheaths. *Australas. Plant Pathol.* **2008,** *37,* 40–40.

IRRI. World Rice Statistics in 2015; [Online] 2015. http://ricestat.irri.org.

Johanson, A.; Turner, H. C.; McKay, G. J.; Brown, A. E. A PCR-Based Method to Distinguish Fungi of the Rice Sheath-Blight Complex, *Rhizoctonia Solani, R. oryzae* and *R. oryzae-sativae*. *FEMS Microbiol. Lett.* **1998**, *162,* 289–294.

Lanoiselet, V.; You, M. P.; Li, Y. P.; Wang, C. P.; Shivas, R. G.; Barbetti, M. J. First Report of *Sarocladium Oryzae* Causing Sheath Rot on Rice (*Oryza Sativa*) in Western Australia. *Plant Dis.* **2012,** *96,* 1382.

Mew, T. W.; Gonzales, P. *A Handbook of Rice Seedborne Fungi*; Science Publishers: Enfield, NH, 2002.

Mew, T. W.; Leung, H.; Savary, S.; Cruz, C. M. V.; Leach, J. E. Looking Ahead in Rice Disease Research and Management. *CRC Crit. Rev. Plant Sci.* **2004,** *23,* 103–127.

Mirocha, C. J.; Zaki A. I. Fluctuation in Amount of Starch in Host Plants Invaded by Rust and Mildew Fungi. *Phytopatholo*gy **1966,** *56,* 1220–1224.

Mohan, R.; Subramanian, C. L. Studies on the Epidemiology of Sheath Rot Disease of Rice. *Food Fmg. Agric.* **1977**, *9,* 71–72.

Ou, S. H. Rice Diseases; CAB International: Wallingford, UK, 1985.

Ou, J.; Lin, G.; Chen, C. *Sarocladium* Species Associated With Rice in Taiwan. *Mycol. Prog.* **2020,** *19,* 67–80.

Pearce, D. A.; Bridge, P. D.; Hawksworth, D. L. Species Concept in *Sarocladium*, The Causal Agent in Sheath Rot in Rice and Bamboo Blight. In *Major Fungal Diseases of Rice: Recent Advances*; Sreenivasaprasad, S., Johnson, R., Eds.; Springer Netherlands, 2001; pp 285–292.

Peeters, K. J.; Haeck, A.; Harinck, L.; Afolabi, O. O.; Demeestere, K.; Audenaert, K.; Höfte, M. Morphological, Pathogenic and Toxigenic Variability in the Rice Sheath Rot Pathogen *Sarocladium oryzae*. *Toxins* **2020,** *12,* 109.

Pramunadipta, S.; Widiastuti, A.; Wibowo, A.; Suga, H.; Priyatmojo, A. *Sarocladium Oryzae* Associated With Sheath Rot Disease of Rice in Indonesia. *Biodiversitas* **2020,** *21,* 1243–1249.

Prasannakumar, M. K.; Parivallal, P. B.; Pramesh, D.; Mahesh, H. B.; Raj, E. LAMP-Based Foldable Microdevice Platform for the Rapid Detection of *Magnaporthe Oryzae* and *Sarocladium Oryzae* in Rice Seed. *Sci. Rep.* **2021,** *11,* 178.

Reddy, M. M.; Reddy, C. S.; Gopal, S. B. Effect of Sheath Rot Disease on Quality Characters of Rice Grain. *J. Myc. Plant Pathol.* **2000,** *30,* 68–72.

Rehman, M. M.; Shahjahan, A. K. M.; Miah, S. A. *Echinochloa Colona*, an Alternate Host of *Sarocladium Oryzae* Causing Shaeath Rot of Rice. *Intern. Rice Res. Newsl.* **1982,** *7,* 16–17.

Sakthivel, N. Sheath Rot Disease of Rice: Current Status and Control Strategies. In *Major Fungal Diseases of Rice: Recent Advances*; Sreenivasaprasad, S., Johnson, R., Eds.; Springer Netherlands, 2001; pp 271–283.

Sakthivel, N.; Gnanamanickam, S. S. Evaluation of *Pseudomonas Fluorescens* for Suppression of Sheath Rot Disease and for Enhancement of Grain Yields in Rice (*Oryza sativa* L.). *Appl. Environ. Microbiol.* **1987,** *53,* 2056–2059.

Saleh, M.; Khatab, I.; El-Shafey, R. Biological and Molecular Variability of Rice Sheath Rot Pathogen *Sarocladium Oryzae* Using SCAR and SRAP Markers. *Egypt. J. Phytopathol.* **2016,** *44,* 157–173.

Shahjahan, A. K. M.; Akanda, S. I.; Mondal, A. H.; Nahar, N. S. Relationship of Sheath Rot (*Sarocladium Oryzae*) Severity to Yield of Rice. *Bangladesh J. Bot.* **1994,** *23,* 211–215.

Sharma, L.; Sharma, K. K.; Sinha, A. P. Molecular Characterization of *Sarocladium Oryzae* Through RAPD Technique: Causal Agent of Sheath Rot Disease of Rice. *Ann. of Pl. Protec. Sci.* **2018,** *26,* 91–97.

Sharma, S.; Sthapit, B.; Pradhanang, P.; Joshi, K. Bacterial Sheath Brown Rot of Rice Caused by Pseudomonas Fuscovaginae in Nepal. In *Rice Cultivation in Highland Areas*; Poisson, C., Rakotoarisoa, J., Eds.; Proceedings of the CIRAD Conference: Antananarivo, Madagascar, 1997; pp 107–112.

Singh, R.; Dodan, D. S. Sheath Rot of Rice. *Int. J. Trop. Plant Dis.* **1995,** *13,* 139–152.

Singh, K.; Mathur, S. B. Further Evidence of Transmission of *Sarocladium Oryzae* Through Rice Seeds and its Quarantine Significance. *Indian Phytopath.* **1993,** *45,* 454–456.

Singh, R. A.; Raju, C. A. Studies on Sheath Rot of Rice. *Int. Rice Res. Newsl.* **1981,** *6,* 11–12.

Sobanbabu, G.; Sabarinathan, K. G.; Parthiban, V. K.; Ramamoorthy, V. Isolation, Screening and Identification of Virulent Isolates of *Bipolaris Oryzae* Causing Rice Brown Spot and *Sarocladium Oryzae* Causing Sheath Rot Disease. *Int. J. Curr. Microbiol. App. Sci.* **2018,** *7,* 930–939.

Sri S., Hittalmani, S.; Kumar, K. G.; Shashidhar, H. E.; Vaishali, M. G. Identification of Quantitative Trait Loci Associated With Sheath Rot Resistance (*Sarocladium oryzae*) and Panicle Exsertion in Rice (*Oryza sativa* L.). *Curr. Sci.* **2002,** *82,* 133–135.

Surin, A.; Arunyanart, P.; Dhitikiattipong, R.; Rojanahusdin, W.; Disthaporn, S.; Soontrajarn, S. Rice Yield Loss to Sheath Rot (ShR). *Int. Rice Res. Newslt.* **1988,** *13,* 34–35.

Vidhyasekaran, P.; Ranganathan, K.; Rajamanickam. Quality of Rice Grains from Sheath Rot Affected Plant. *Int. Rice Res. Newslett.* **1984,** *12,* 17.

Weeraratne, N.; Stodart, B. J.; Venturi, V.; Hofte, M.; Hua, G. K. H.; Ongena, M.; Savocchia, S.; Steel, C. C.; Ash, G. J. Syringopeptin Contributes to the Virulence of *Pseudomonas fuscovaginae*, Based on sypA Biosynthesis Mutant Analysis. *Phytopathology* **2019,** *110,* 780–789.

CHAPTER 10

Research Status and Prospectus of Sheath Rot Disease Resistance in Rice

ANURAG MISHRA[1], VANDANA SHARMA[1], RAJAT CHAUDHARY[1], and PRASHANT YADAV[2]

[1]*Division of Genetics, I.A.R.I., New Delhi, India*

[2]*ICAR-Directorate of Rapeseed Mustard Research, Bharatpur, Rajasthan, India*

ABSTRACT

Rice (*Oryza sativa*) is a staple food of the world, and its production is reduced due to abiotic and biotic stresses. Sheath rot is one of the biotic stresses that is caused by *Sarocladium oryzae* (Sawada); it reduces yield and affects the grain quality of rice as well as the protein and starch content of the grain. It has been reported from all rice-growing areas of the world, including India. The mycelia of *S. oryzae* are branched and septate, colonies are pale pinkish or whitish, and sporulation is abundant in the center of the host, appearing as greyish brown lesions on upper leaf sheaths enclosing panicles ranging from 0.5 to 15 cm long and 0.2 to 0.5 cm wide. Disease sensitivity mainly depends on the environmental condition and is highly devastating when the temperature lies between 17 and 20 °C and the relative humidity is between 40 and 50% at the flowering stage of the plant. This disease can be controlled by field management (more spacing and less densely planted fields) and chemical management (spraying fungicides and pesticides). QTL "*qShR-12*" has been detected for sheath rot resistance.

Fungal Diseases of Rice and Their Management. Deepti Srivastava, Md. Shamim, Malik Mobeen Ahmad, and R. S. Upadhyay (Eds.)

10.1 INTRODUCTION

Rice is one of the major staple foods of the world; half of the world's population consumes it as food on a daily basis. Rice production during 2018–2019 was 110 million tones in India (Commodity Profile for Rice–March, 2019). Rice production is drastically decreased due to many biotic (insects, pests, diseases) and abiotic (drought, submergence, salinity, heat stress) factors. The rice crop is highly damaged due to 70 different diseases that are caused by bacteria, fungi, viruses, and nematodes (Zhang et al., 2009). Such diseases are brown leaf spot, leaf blast, sheath blight, sheath rot, bacterial leaf blight, neck blast, bakanae, foot rot, stem rot, bunt, and smut are the most serious diseases of irrigated rice. *Sarocladium oryzae* (Sawada) Gams and Hawksworth caused sheath rot disease, which is a highly devastating disease among the above-mentioned rice diseases. It is present in most of the rice-growing countries worldwide, particularly in rain-fed rice ecosystems, and is more prevalent during wet than dry seasons. It was first identified in Sawada (1922), and later on it was reported from all rice-growing areas of the world, including China, India, Bangladesh, Japan, Korea, Cameroon, the Philippines, Peru, South East Asia, Taiwan, Venezuela, Thailand, Vietnam, Brazil, the USA, Kenya, and Nigeria. The yield loss is varying from 20 to 85% due to the unpredictable pathosystem of *S. oryzae*, and it is also affecting the grain quality of rice; the seeds are chaffy, partially fielded, and have reduced protein and starch content (Sakthivel et al., 2001). *S. oryzae* has a 32.78 Mb genome that contains 10,526 protein-coding genes (Table 10.1). According to functional annotation of protein-coding genes, the *S. oryzae* genome has evolved with many expanded gene families of the major superfamily: proteinases, zinc finger proteins, sugar transporters, dehydrogenases/reductases, cytochrome P450, WD domain G-beta repeats, and FAD-binding proteins (Hittalmani et al., 2016).

10.2 RECENT ADVANCES IN RESEARCH OF SHEATH ROT

Loop-mediated isothermal amplifcation (LAMP) is a primer-based assay, which has been widely used to detect viruses, bacteria, and fungal pathogens. On the basis of the principles of LAMP: a microdevice platform has been designed to detect the *S. oryzae* in seeds of riceand it could detect the presence of *S. oryzae* at a concentration down to 100 fg within 20 min at 60°C (Prasanna Kumar et al., 2021).

10.3 CAUSAL ORGANISM AND SYMPTOMS

Acrocylindriumoryzae (Sawada, 1922) was the causative organism for sheath rot, and it was later called Sarocladiumoryzae (Sawada) Gams and Hawksworth (Gams and Hawksworth, 1975). Sarocladium was classified into two species, *S. attenuatum* (Agnihothrudu, 1973) and S. oryzae, based on the shapes and sizes of the conidia (Gams and Hawksworth, 1975). The verticillate conidiophores and conidia of *S. attenuatum* are longer and narrower than those of S. oryzae. The fungus' taxonomy is very similar, and it is not always clear which reports relate to which Sarocladium species (Ou, 1985). As a result, *S. attenuatum* may be thought of as a synonym for *S. oryzae* (Bridge et al., 1989). *S. oryzae* mycelia are branched septate, with pale pinkish or whitish colonies and abundant sporulation in the middle of the host. *S. oryzae* has verticilliate condiophores with three to four phialides, which contain fusiform, often curved, hyaline, single cell conidia measuring 3.5 to 9 × 1 to 2.5 Jlm (Ou, 1985). Rice sheath rot was once thought to be a minor disease, but after the green revolution in the 1960s, it became prevalent due to changes in rice crop farming practices. Usage of high-yielding crops, a lot of fertilizers, and water conservation, seedling method, increase plant density in transplanting plants have less space, a high amount of nitrogen being used, and semidwarf and photoperiod insensitive plants being preferred, which make favorable conditions for susceptibility for many diseases, sheath rot is one of them. First-time sheath rot was reported in Taiwan (1922), and the upper leaf sheaths enclosing panicles produce greyish brown lesions. The color of the lesion ranges from grey brown to purple brown, depending on the cultivar (Shahjahan et al., 1977). The lesions in the entail stage vary in size from 0.5 to 15 cm long and 0.2 to 0.5 cm wide, with a chocolate brown and grey colour in the middle and a green sheath around the lesions (Amin et al., 1974). The lesions change form from oblong to linear and have irregular margins. They enlarge, coalesce, and cover the entire sheath with an abundant powdery mass of conidia found in the center of the lesion seen within the leaf sheath, and young panicle sheaths rot. The infection results browning of grains, sterile chaffy grains, unexerted panicles, and a reduction in yield, nutritional value, and quality of grain (Chien and Huang, 1979; Ou, 1985; Sakthivel, 2001; Gopalakrishnan et al., 2010).

10.4 EPIDEMIOLOGY AND DISEASE CYCLE

S. oryzae is present in all rice-growing countries around the world, and it is particularly prevalent during rainy seasons (Mew and Gonzales, 2002). This fungus is mainly found in lowland areas (Pearce et al., 2001), and the disease thrives in hot and humid conditions (Sakthivel, 2001). Infections caused by *S. oryzae* were discovered below 1250 m in Nepal (Sharma et al., 1997). The temperature for infection is 20–30°C, with a relative humidity level of 65–85% favoring the growth of sheath rot (Sakthivel, 2001). When environmental conditions are favorable, the pathogen lives in infected plants, plant residues, soil, water, and weeds. The fungus infects rice plants at various stages of development, with the booting stage being the most damaging. It enters through the stomata or wounds, and is aided by insect and mite damage, as well as other pathogens weakening the plant (Fig. 10.1) (Pearce et al., 2001). Secondary infections could be transmitted by the wind, and little is known about disease transmission through seeds. Cerulenin and helvolic are secondary metabolites produced by the pathogen. The whole genome sequencing of *S. oryzae* revealed that the genes involved in cerulenin and helvolic acid biosynthesis are responsible for the greyish brown necrotic lesion on the flag leaf sheath.

The disease is transmitted primarily by airborne conidia, as well as seed-borne conidia. Singh and Mathur (1993) reported 85.5 and 45.1% seed infection with a highest infection level of 94.4 and 81.8% in empty and filled seeds. The transmission of infection takes place through stomata or wounds. Infected plant debris is the most common source of inoculum. Airborne conidia produced on the leaf sheath are used for secondary spread. The most common source of inoculum is contaminated plant debris. Secondary spread is accomplished by airborne conidia formed on the leaf sheath.

10.5 FIELD MANAGEMENT OF RICE SHEATH ROT DISEASE

Disease sensitivity mainly depends on the environmental condition of rice cultivation how disease damage the crop by causal pathogen, which is intensely influenced by the temperature and relative humidity. Disease is highly devastating when the temperature is lies between 17°C to 20°C and the relative humidity is also minimal at 40–50% at flowering stage of the plant (Singh and Raju, 1981). In cold days, temperatures below 20°C are highly favourable for the disease's development (Mohan and Subramanian, 1977). The disease chances increased when the planted seedlings have very little spacing and are dense. Fifty days old seedlings are more susceptible then 25 days old seedlings as observed by Pal and Das (1976). Mukerjee (1990) reported that the pathogen infected floral and vegetative parts of the plant but did not affect the ovary directly. Infection was severe in poorly exerted panicles due to the presence of half-opened leaf sheaths that favoured infection (Vinod et al., 1990). After the flowering stage, the embryo and

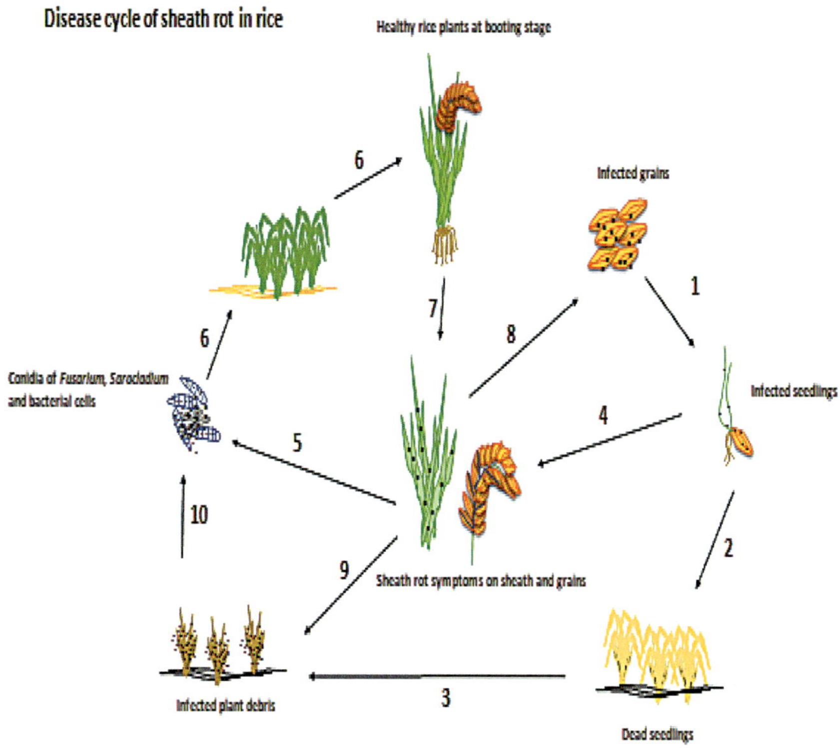

FIGURE 10.1 Cycle of sheath rot disease. Pathogens that cause sheath rot are transmitted through seed, resulting in infected seedlings (1). Infected seedlings can either die (2) or survive (3), resulting in infected plant debris. *P. fuscovaginae* can colonize the entire plant as an endophyte, or it can survive epiphytically and infect inflorescences during the booting stage. The fungal pathogens' seedling transmission is less well known (4). Conidia or bacterial cells produced by infected plants cause secondary infections (5). Wind or rain disperse conidia, or bacterial cells, to healthy plants. Infected plants at the booting stage are particularly vulnerable. Insects and mites, in the case of *S. oryzae*, can also spread conidia and promote infection by making wounds (6). The sheath enclosing the young panicles rots, causing the grains on the affected tillers to become chaffy and discolored. Mycotoxins can be present in grains that have been contaminated with Fusarium sp (7). Infected grains can spread pathogens to new fields (8). Infected plant debris can remain in the field after harvesting (9) and act as inoculum for the next growth cycle (10).

endosperm did not get infected. Only the husk was found to be infected in a late stage infection.

At the boot-leaf stage and 15 days later, spray chemicals like carbendazim (500 g), edifenphos (1 L), or mancozeb (2 kg) per hectare. Gypsum (500 kg/ha) was applied to the soil in two phases. Many other chemicals have been used to control sheath rot, including dithane M-45, benlate, carbendazim, tridemorph, bavistin, brestanol, dithane, hinosan, diflotaton, and gypsum (Rajan and Nair, 1978; Mathur, 1981; Raju and Singh, 1981; Narashimhan et al., 1994). Fungicides are not much affective for the sheath rot disease. Plant extracts have been shown to be harmful to many rice fungi, including *S. oryzae* (Miah et al., 1990). Sheath rot has been treated with neem, punna, and rubber cakes. A high nitrogen dose is assisting cultivars in developing sheath rot (Misra and Mathur, 1983; Akanda et al., 1984). A heavy dose of nitrogen in cultivars should be avoided. Sheath rot was more common in densely planted fields than in fields with wider spacing, according to several reports (Agnihothrudu, 1973; Amin et al., 1974; Amin, 1976; Agarwal etal., 1989). The spacing of 15 × 15 cm was found to significantly reduce disease spread and increase grain and straw yields (Kanagasabai, 1980).

10.6 CHEMICAL METHODS FOR SHEATH ROT DISEASE MANAGEMENT

Using fungicides spray at the time of Panicle emergence is the effective for controlling the disease. The combination of systemic pesticides like Tridemorph (a fungicide) and phosphamid (an insecticide) protected the plants from sheath rot. At the booting stage of the plant, foliar spraying with edifenphos, mancozeb, or carbendazim was found to be effective against the sheath rot disease. Spray with a concentration of carbendazim (250 g), chlorothalonil (1 kg), or edifenphos (1 lit/ha). Foliar sprays of benomyl and copper oxychloride were also found to be effective. Seed treatment with mancozeb and benomyl effectively eliminates seedborne inoculum.

10.7 SHEATH ROT DISEASE RESISTANCE

Disease resistance is a complex phenomenon that influenced by many factors like environment, host, and pathogen (Agriose, 2005). Application of potassium (K) fertilizer enhance the tolerance to sheath rot disease and reduce grain yield loss in rice (Zhang et al., 2021). OsWRKY13, WRKY12, and PR2 TF gene in rice plant is known to play a regulatory role of signaling in physiological pathways related to either development or disease resistance. They have higher gene expression against sheath blight and sheath rot-resistant variety (Lilly et al., 2019). QTL "qShR-12" detected for sheath rot lesion length and simplified severity scores located on chromosome-12 flanked by SSR marker RM7315 and RM-28118 with intermarker distance of 15.3 cM, explaining 5.51 and 6.27% of phenotypic variation, respectively (Mahadevaiah et al., 2017). Other QTLs are also reported for sheath rot resistance in case of rice (Table 10.2) (Hittalmaini et al., 2002)

TABLE 10.1 Whole Genome Assembly Features and Details of Coding Genes, Exon, and Intron of *S. oryzae*.

A	**Assembly parameters**	**Value**
	Total genome size (bp)	32,778,109
	Number of contigs	5856
	Maximum contig length (bp)	89796
	Minimum contig length (bp)	209
	Average contig length (bp)	5597
	Total number of non-ATGC characters	0
	Depth of genome coverage (x)	82
	N50 value (bp)	18070
	GC (%)	53.49
B	**Gene details**	**Value**
	Number of coding genes	10,526
	Average gene length (bp)	1689.11
	Gene density (number of genes per Mb)	294
	Average distance between genes (Kb)	1.08
	GC content (%) in coding region	56.61
	Mean protein length (amino acids)	506.25
C	**Exon details**	
	Number of exon	29,293
	Total exon length (Mb)	15.97
	Average exon length (bp)	545.74
	Number of exons per gene	2.78
D	**Intron details**	
	Number of intron	18,769
	Total intron length (Mb)	1.76
	Average intron length (bp)	93.85
E	**Protein annotation**	
	Total number of proteins/genes predicted	10,526
	Total number of annotated proteins/genes	9658
	Total number of un-annotated proteins/genes	868

TABLE 10.2 QTLs Identified for Sheath Rot Resistance.

QTL	Chromosome	Flanking markers	Peak lod	Additive effect	Variation (%)
1	1	RG612-RG140	3.83	8.67	10.50
2	1	RZ276-CDO9029	2.72	7.30	8.00
3	1	RG1O9-RG236	6.13	11.82	8.80
4	2	RG1O2-RG73	3.19	7.60	8.60
5	4	RG498-RG864	2.29	8.23	7.40
6	5	RG360-RG573	2.46	9.32	8.40
7	6	RG192-RG172	3.52	9.63	12.40
8	7	RG272-RG511	2.25	9.62	8.10
9	8	RG2O-RG333	2.75	7.15	8.80
10	1	RG690-RZ8O1	2.17	−0.81	9.40
11	7	Est-9PGMS0.7	2.39	−0.83	10.00

QTL 1 to 9 are in CO39 × Moroberekan RIL popuration at Ponnampet, India QTL10 and 11 are in IR64 × Azuccena DH popuration at Hebbal, Bangalore, India

10.8 CONCLUSION

Since the initial reports of rice sheath rot symptoms, which were linked to *S. oryzae*, there have been numerous reports of identical or related illness symptoms in various regions of the world. Now, Rice sheath rot disease is gaining traction as an emerging destructive disease of rice. Various strategies have been used for the management of rice sheath rot disease but still there is limited scientific understanding on this. So, more research focused on genomics and proteomics is needed to better understand the disease process and identify candidate genes for sheath rot resistance in rice. Sheath rot disease-resistant cultivars were developed using these potential genes.

KEYWORDS

- **rice**
- **biotic stress**
- ***Sarocladium oryzae***
- **sheath rot disease**

REFERENCE

Agnihothrudu, V. *Acrocylindrium oryzae* Sawada-Sheath Rot on Paddy. *Kavaka* **1973,** *1*, 68–71.

Agrawal, P. C.; Mortensen, C. N.; Mathur, S. B. *Seedborne Diseases and Seed Health Testing of Rice*; CAB International Mycological Institute: Kew, Surrey, U.K, 1989.

Agrios, N. G. *Plant Pathology,* 5th ed.; Elsevier: Amsterdam, 2005,: p 635.

Akanda, S. I.; Shajahan, A. K. M.; Miah, S. A. Nitrogen Fertilization and Sheath Rot (ShR) Development in Rice. *Int. Rice Res. Newslett.* **1984,** *9*, 12–13.

Amin, K. S. Sources of Resistance to Acrocylindrium Oryzae Sheath Rot of Rice. *Plant Dis. Reptr.* **1976,** *60*, 72–73.

Amin, K. S.; Sharma, B. D.; Das, C. R. Occurrence of Sheath Rot of Rice in India Caused by *Acrocylindrium. Plant Dis. Reptr.* **1974,** *58*, 358–360.

Bridge, P. D.; Hawksworth, D. L.; Kavishe, D. F.; Farnell, P. A. A Revision of the Species Concept in Sarocladium, the Causal Agent of Sheath-Rot in Rice and Bamboo Blight, Based on Biochemical and Morphometric Analyses. *Plant Pathol.* **1989,** *38*(2), 239–245.

Chien, C. C.; Huang, C. H. The Relation Between Sheath Rot and the Sterility of Rice Plant. *J. Agric. Res. China* **1979,** *29*, 7–16.

Gams, W.; Hawksworth, D. L. The Identity of Acrocylindrium Oryzae Sawada and a Similar Fungus Causing Sheath-Rot of Rice. *Kavaka* **1975,** *3*, 57–61.

Gopalakrishnan, S.; Kannan, I. G. K.; Alekhya, G.; Humayun, P.; Meesala, S. V.; Kanala, D. Efficacy of Jatropha, Annona and Parthenium Biowash on Sclerotium Rolfsii, Fusarium Oxysporum f. sp. Ciceri and Macrophomina Phaseolina, Pathogens of Chickpea and Sorghum. *Afr. J. Biotechnol.* **2010,** *9* (47), 8048–8057.

Hittalmani, S.; Mahesh, H. B.; Mahadevaiah, C.; Prasannakumar, M. K. De Novo Genome Assembly and Annotation of Rice Sheath Rot Fungus Sarocladium Oryzae Reveals Genes Involved in Helvolic Acid and Cerulenin Biosynthesis pathways. *BMC Genom.* **2016,** *17* (1), 1–13.

Hittalmani, S.; Shashidhar, H. E.; Bagali, P. G.; Huang, N.; Sidhu, J. S.; Singh, V. P.; Khush, G. S. Molecular Mapping of Quantitative Trait Loci for Plant Growth, Yield and Yield Related Traits Across Three Diverse Locations in a Doubled Haploid Rice Population. *Euphytica* **2002,** *125* (2), 207–214.

Kanagasabai, R. Studies on an Integrated Approach for the Control of Sheath Rot Disease of Rice Caused by Acrocylindrium Oryzae Sawada. M.Sc. (Ag) Thesis, Annamalai University, India, 1980.

Lilly, J. J.; Subramanian, B. Gene Network Mediated by WRKY13 to Regulate Resistance Against Sheath Infecting Fungi in Rice (*Oryza sativa* L.). *Plant Sci.* **2019,** *280*, 269–282.

Mahadevaiah, C.; Prasanna Kumar, M. K.; Hittalmani, S. Dissecting Parameters Associated With Sheath Rot (Sarocladium Oryzae [(Sawada) W. Gams & D. Hawksw]) Disease in Rice (Oryza Sativa L.) *Curr. Sci.* **2017,** 151–155.

Mathur, S. C. Observations on Diseases of Dry Land Rice in Brazil. *Int. Rice Res. Newslett.* **1981,** *6*, 11–12.

Mew, T. W.; Gonzales, P. *A Handbook of Rice Seedborne Fungi*. International Rice Resource Institute, 2002.

Miah, M. A. T.; Ahmed, H. U.; Sharma, N. R.; Ali, A.; Miah, S. A. Antifungal Activity of Some Plant Extracts. *Bangladesh J. Bot.* **1980,** *19*, 5–10.

Misra, A. K.; Mathur, S. C. Sheath Rot Incidence in Cultivars Grown at Different Nitrogen Levels. *Int. Rice Res. Newslett.* **1983,** *8*, 5.

Mohan, R.; Subramanian, C. L. Influence of Nitrogen Nutrition and Sheath Rot Disease on Sugar Contents of Two Paddy Varieties. *Curr. Sci.* **1977,** *46* (21), 753–755.

Mukerjee, P. Histological Studies of Sheath Rot of Rice. In Extended Summary, Proceeding of International Symposium Rice Research; *New Frontiers*, 1990; p 241.

Narasimhan, V.; Ramadoss, N.; Sridhar, V. V.; Kareem, A. A. Using Gypsum to Manage Sheath Rot in Rice, *Int. Rice Res. Newslett.* **1994,** *19*, 27–28.

Ou, S. H. Rice Diseases, CAB International Mycological, Institute Kew, Survey, UK, 1985.

Pal A.; Das, C.R. Preliminary Observations on the Incidence of Some Rice Diseases in Relation to Planting Dates and Growth Stages. *Int. Rice Res. Newsl.* **1976,** *1* (2), 14.

Pearce, R. S. Plant Freezing and Damage. *Ann. Bot.* **2001,** *87* (4), 417–424.

Prasannakumar, M. K.; Parivallal, P. B.; Pramesh, D.; Mahesh, H. B.; Raj, E. LAMP-Based Foldable Microdevice Platform for the Rapid Detection of Magnaporthe Oryzae and Sarocladium Oryzae in Rice Seed. *Sci. Rep.* **2021,** *11* (1), 1–10.

Rajan, K. M.; Nair, P. V. Fungicidal Control of Sheath Rot and Glume Discolouration in Rice. *Indian J. Mycol. Plant Pathol.* 1978**,** *8*, 85–87.

Raju, C. A.; Singh, R. A. Development of Inoculation Technique and Varietal Reaction to Sheath Rot of Rice. *Philipp. Phytopathol.* **1981,** *17*, 67–73.

Sakthivel, N.; Mortensen, C.; Mathur, S. Detection of *Xanthomonas oryzae* pv. *oryzae* in Artificially Inoculated and Naturally Infected Rice Seeds and Plants by Molecular Techniques. *Appl. Microbiol. Biotechnol.* **2001,** *56*, 435–441.

Sawada, K. *Descriptive Catalogue of Formosan Fungi II*; Report of the Government Research Institute, Department of Agriculture, Formosa, 1922; vol 2, pp 1–135.

Shahjahan, A. K. M.; Harahap, Z.; Rush, M. C. Sheath Rot of Rice Caused by Acrocylindrium Oryzae in Louisiana. *Plant Dis. Reptr.* **1977,** *61* (4), 307–310.

Sharma, A.; Sharma, U. S. Liposomes in Drug Delivery: Progress and Limitations. *Int. J. Pharm.* **1997,** *154* (2), 123–140.

Singh, K.; Mathur, S.B. Further Evidence of Transmission of Sarocladium Oryzae Through Rice Seeds and its Quarantine Significance. *Indian Phytopathol.* **1993,** *45,* 454–456.

Singh, R. A.; Raju, C. A. Studies on Sheath Rot of Rice Int. *Rice Res. News* **1981,** *6* (2), 11.

Vinod, K. K.; Vivekanandan, P.; Subramanian, P. Effect of Cytoplasmic Male Sterility on Panicle Exertion and Sheath Rot Incidence in F2 Rice Hybrids. Int. Rice Res. Newslett. **1990,** *15* (2), 5.

Zhang, J.; Li, Y.; Guo, L.; Cao, R.; Zhao, P.; Jiang, W.; Ma, Q.; Yi, H.; Li, Z.; Jiang, J.; Wu, J.; Wang, Y.; Si, S. DH166, a Beta-Carboline Derivative, Inhibits the Kinase Activity of PLK1. *Cancer Biol. Ther.* **2009,** *8* (24), 2374–2383.

CHAPTER 11

Rice Brown Spot Disease (*Helminthosporium oryzae*): Ecology, Epidemiology, and Identification Measures

AMIT KUMAR MAURYA[1], VINNY JOHN[1], HEMLATA PANT[2], and D. K. SRIVASTAVA[3]

[1]*Department of Plant Pathology, Sam Higginbottom University of Agriculture, Technology and Sciences, Prayagraj, Uttar Pradesh, India*

[2]*Department of Zoology, C. M. P. PG. College, Prayagraj, Uttar Pradesh, India*

[3]*Council of Science and Technology, Lucknow, Uttar Pradesh, India*

ABSTRACT

Rice (*Oryzae sativa* L.) comes under essential commodity overtaken other crops all over the world for decades. It is grown in about 11% of the total cultivated land area globally and serves as the major food for nearly half of the world's population and two-thirds of Indians. Rice crop is damaged by many diseases caused by fungi leads to considerable yield loss. Brown spot disease of rice is considered an economically important fungal disease amid other diseases and is generally caused by *Helminthosporium oryzae* is one of the worldwide occurrences. It could result in significant quantitative and qualitative losses in paddy and milled rice grain yields. The disease is most prevalent when low-input management is used and when direct sowing is used in upland areas. The seed-borne pathogen produces inter- and intra-cellular mycelium, circular to oval or ellipsoidal brown spots, most

Fungal Diseases of Rice and Their Management. Deepti Srivastava, Md. Shamim, Malik Mobeen Ahmad, and R. S. Upadhyay (Eds.)

conspicuous (around 2.8 mm × 0.5 mm) on leaves with a yellow halo and a black or dark brown mat on the affected tissues. There is a requirement to work into the epidemiological reason which initiates and enlargement of disease. The sources and contribution of primary inoculums in the disease's occurrence are poorly known. *Helminthosporium oryzae* occurs primarily in two phases. These phases are asexual also known as anamorph dark brown or olive green sporophores emerges on the lateral branches from hyphae, 8–15 μm or more in diameter and the sexual phase also known as perfect stage with conidial length 35–170 × 11–17 μm. usually bent conidia, wide at the middle and tapered toward the apex, with thin peripheral wall fully mature conidia are brown.

11.1 INTRODUCTION

Rice (*Oryzae sativa* L.) is the very vital crop that dominates other crops. Rice is grown about 11% of the cultivated land around the world as the major food crop half of the world's population depend on rice including two-third Indian (Rout and Tiwari, 2012). There is a continuing increase in worldwide demand for rice, which is predicted to reach 852 million tonnes by 2035 from its current production status of 676 million tonnes, which is why there is a need to raise rice productivity by 10 to 12.5 tonnes per hectare to satisfy this deficiency of 176 million tonnes (Khush, 2011). More than 90% of the world's rice is grown and consumed in Asia, where 60% of calories are consumed by 3 billion Asians, (Khush, 1997). India is the largest rice-producing country in the world. Rice is farmed on 43.86 million hectares in India, with a yield of 104.80 million tonnes and productivity is about 2390 kg/ha, accounting for 21.81% of world rice production during 2015–2016 as per the data of Government of India. However, the world highest production is China with 205.21 million tonnes with a productivity of 6710 kg/ha, accounting for 27.70% of world rice production. In Bangladesh, about 92% of farmers grow rice covering 75% of the total cropped area (Rekabder, 2004). About 450,000 hectares of arable land in the European Union are devoted to rice and almost 25% of it is in Spain. One of the most important areas devoted to this crop is the Marismasdel Guadaquivir in southern Spain (38,000 ha) and accounts for 40% of the total rice production of Spain (Aguilar, 2001). The share of other continents in which rice was grown is Africa (7.78%), South America (6.4%), North America (1.4%), and Viraktamath (2007). In India, based on rice production during 2015–2016, the top rice-producing states are West Bengal with 15.75 million tonnes over 5.46 million hectares of cultivable area followed by Uttar Pradesh with 12.5 million hectares over 5.86 million areas. Punjab holds third position with 11.82 million tonnes over 2.97 million hectares area followed by Tamil Nadu and Haryana (India Today Web Desk New Delhi, September 18, 2018. https://www.indiatoday.in).

Brown spot of rice caused by *Helminthosporium oryzae* (Breda de Haan) was known in Japan during 1900's with a Japanese name as nai-yake, which means seedling blight/sesame leaf spot. Brown spot disease has been found in most of the rice-growing countries including Japan, Bangladesh, China, Sri Lanka, Iran, Burma, Australia, Malaysia, Africa, South America, Russia, North America, Philipines, Saudi Arabia, and Thailand (Ou, 1985; Khalili et al., 2012). Brown spot is also one of the most important rice diseases in Florida, adversely affecting the yield and milling quality of the grain, Datnoff et al. (1994) and one of the most common rice diseases in the Louisiana state of United States (Groth and Hollier, 2013). In India, brown spot disease of rice was the first report by Sundraraman in 1919 from Madras and is reported to present in all the rice producing states of India. In the states of Bihar, Chhatisgarh, Madhya Pradesh, Orissa, Assam, Jharkhand, and West Bengal, it was found to be more severe in dry and direct-seeded rice (Gangopadhyay, 1983; Ou, 1985; Sunder et al., 2014). The pathogen fungus *Helminthosporium oryzae* was initially described by Breda de Haan in 1900, which was rechristened as *Drechlera oryzae* by Subramanian and Jain in 1966. In 1959, Shoemaker renamed it as *Bipolaris oryzae* since most conidia germinate from two end cells. The fungus was classified as belonging to the genus *Cochliobolus* by Drechsler (1934) but after discovering the teleomorphic stage of the fungus during its growth, Ito and Kuribayashi (1927) renamed the *Helminthosporium oryzae* as *Cochliobolus miyabeanus*. The fungus creates inter and intra-cellular mycelium, which grows greyish brown to the dark brown mycelial mat on the infected tissues. The culture of *Helminthosporium oryzae* fungi appears grey to olive or black in color in petri plate. The sporophores are thick, erect, geniculate, dark olivaceous at bottom and lighter near tips, emerge through stomata in clusters of 3–5. Conidia are, often somewhat curved and broadest in the center, five to ten septate and are arranged in acropetal succession. Conidia that are fully matured are fuliginous or brownish, and they germinate with two polar germ tubes, one from each of the thin-walled areas, whereas less mature, subhyaline spores may produce germ tubes from transitional segments. Conidiophores and conidia have been seen to range in size from 68–688 × 7.6–20 μm and 15–132 × 10–26 μm in Japan; 70–175 × 5.6–7 and 45–106 × 14–17 μm in India, 99–345 × 7.11 μm and 24–122 × 7–23 μm in China, and 150–600 × 4-8 μm and 35–170

× 11–17 μm in the USA (Ou, 1985). The optimal temperature for mycelia growth and germination of *Helminthosporium oryzae* is 27–38°C, while conidial growth and germination needs a favorable temperature of 25–30°C (Ou, 1985). However, conidia can grow at 5°C and 35–38°C, optimum being 25°C (Ou, 1985; Kumari et al., 1997) at a pH range of 2.6–10.9 (Ou, 1985), optimum at 6.8–7.0 (Kumar and Mishra, 1993). Both light and dark periods were mandatory for the sporulation of *Bipolaris oryzae*, and stimulated near UV light, while blue light inhibited sporulation (Ou, 1985).

Brown spot fungus has been found to thrive for up to 2 years in soil and infected plant parts such as stubbles, straw, and grains, which can serve as the principal source of inoculums. (Ou, 1985). In 1996, Wesley et al. shown that the disease can survive in seeds also. The primary infection is frequently instigated by the infected seed (Sharma and Maheshwari, 1982; Damicone et al., 2001) as necrotic lesions on coleoptiles and inital leaf sheaths develop. Secondary infection by airborne spores formed on primary lesions causes the subsequent lesion on leaves (Ou, 1985). *Bipolaris oryzae* spores were generally found in the air throughout the year (Ghose et al., 1960; Kulkarni et al., 1982), however, during the cooler months with high humidity, they were found in much more abundance (Kulkarni et al., 1982). According to Dasgupta and Chatopadhyay (1977), greater temperatures and relative humanity have a substantial impact on disease progression, and prolonged exposure to a saturated environment before and after inoculation reduced disease growth. Datnoff and Lentini, (1994), also reported that disease development was favoured by relative humidity between 86 and 100% and temperature between 20 and 25.5°C.

11.2 SYMPTOMATOLOGY

The disease symptom of the brown spot was studied accordingly as per Ramakrishnan and Subramanian (1977), on infected plants from the seedling stage up to crop maturity both in the pot culture and in-vivo experiment covered by photography. *Helminthosporium oryzae* pathogen produces minute spots on the coleoptiles, leaf blade, leaf sheath, and glumes, and has been found most noticeable on leaf blades and glumes. Brown spots emerged on the surface of the leaves with grey or ashy center, oval with a yellow halo, initial spots are small, round and dark brown to purple-brown spots leads to dryness of leaves. The burned appearance of the impacted nursery may also be seen from afar due to the mortality of the seedlings. The spots on susceptible cultivars are substantially larger, measuring up to 1 cm or more in length.

FIGURE 11.1 Healthy crop.

FIGURE 11.2 Infected crop of rice.

11.3 ECOLOGY

Helminthosporium oryzae mycelium was grey to dark grey conidial colonies that developed and spread quickly. Aerial mycelium was cushioned, furry, and grey olive with brownish. Conidiophores were single or in small groups, straight to flexuous, sometimes geniculate, pale to mid-brown or olivaceous brown, whitish near the apex, septate 430–580 × 4–7 µm (average 500 × 5 µm). Conidia were usually curved, obclavate, sometimes almost cylindrical, pale to a mid-golden brown, smooth, 5–12 distosepted, 46.5–125 × 10–26 µm and hillum were minute dark or light, often protruding, and slightly papillate.

11.4 MORPHOLOGIC FEATURES

Mycelium: On culture plate, the mycelium appears initially hyaline and later turn into dark brown or olivaceous, septate, profusely branching with constriction at the point of branching, branching angle were found irregular.

Conidiophores: The sporophores are dark olivaceous, thick, erect, and geniculate at the base and lighter toward the tips.

Conidia: The conidia are globules with bulged at the center and tapering at both ends, occasionally cylindrical was also found. Length of conidia is measured 108.3–122.2 µm and breadth is 49.5–54.4 µm. Conidia are having 4–5 pseudoseptation with pale to dark brown.

11.5 DISEASE CYCLE

The optimum temperature for growth and germination of *Helminthosporium oryzae* is 27–38°C and 25–30°C, respectively (Ou, 1985), conidia are formed between 5 and 38°C, optimum being 25°C (Ou, 1985; Kumari et al., 1997). Conidia germinate at a pH range of 2.6–10.9 (Ou, 1985), optimum at 6.8–7.0 (Kumar and Mishra, 1993). Both light and dark periods were required for sporulation of *Bipolaris oryzae*, and stimulated near UV light, while blue light inhibited sporulation (Ou, 1985).

The pathogen is pulverized in soil for 2–3 years and infects plant parts leaves, stalks, silage, and cereal; these plant parts acts as the primary source of inoculum (Ou, 1985). Common weed crops grown with the rice crop have also been regarded as a source of inoculums (Biswas et al., 2008). Kulkarni et al. (1981) and Hiremath et al. (1980). A technique is known as the baiting

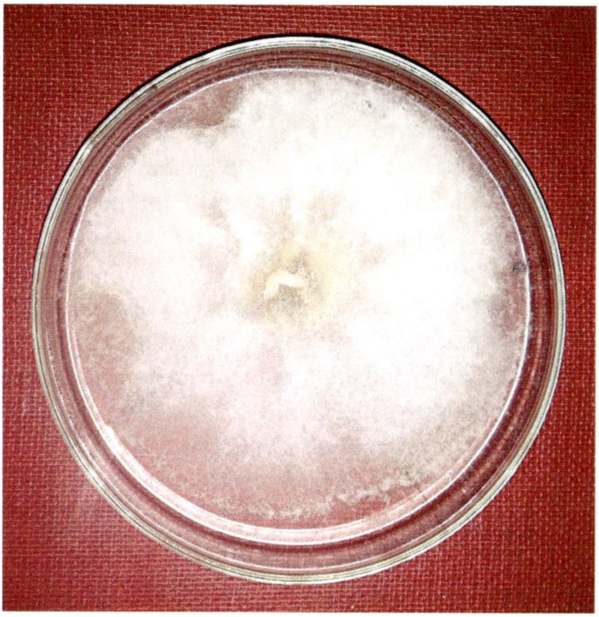

FIGURE 11.3 Pure culture of *Helminthosporium oryzae.*

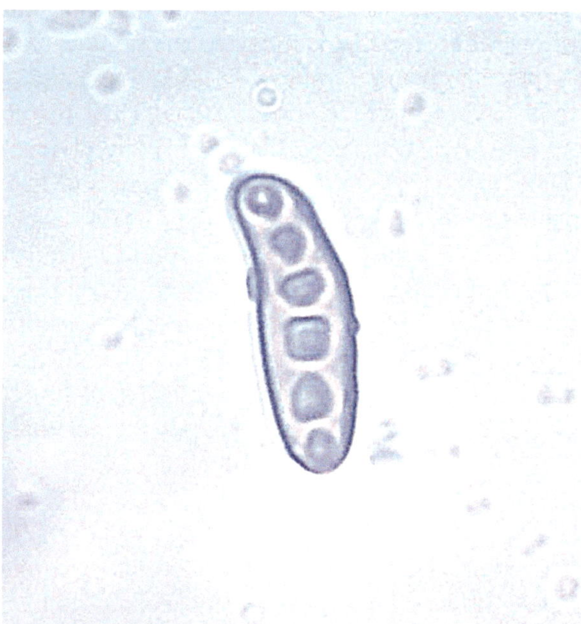

FIGURE 11.4 Conidia of *Helminthosporium oryzae.*

technique usually used for the quantitative assay of the pathogen *B. oryzae* from the soil. It has been reported that the pathogen is enveloped in seed as well (Wesely et al., 1996). It is observed that in most cases infected seeds initiate initial infection (Bernaux, 1981; Sharma and Maheshwari, 1982; Damicone et al., 2001). Necrotic lesions on the sheath of first leaves are found followed by a lesion on leaves, which mainly emerges from a secondary infection that is via, air-borne spores spread from initial or primary lesions (Ou, 1985).

The host epidermis is punctured during infection by infection pegs that emerge from the appressoria (Ou, 1985). Rice leaf exudates contain aspartic acid, glutamic acid, alanine, and methionine, which promote the pathogen to colonize host leaves (Purkayastha and Mukhopadhyay, 1974). The fungus has been found in the seed coat and endosperm (Vidhyasekaran, 1980). When compared, Vu-Van and Sangchote (2006) discovered the highest incidence of *B. oryzae* in rachilla while comparing with sterile lemma, palea, embryo, and endosperm of rice kernels with typical brown spot. In the sand, blotter paper, and tube agar methods, the extend of *B. oryzae* from infected kernels to seedlings ranged from 57 to 77%, with symptoms appearing on coleoptiles and roots after 7–14 days and on the first leaf after 3 to 4 weeks, finally leading to wilting and death of severely infected seedlings.

Various factors affecting disease development have been observed, including plant age, soil and fertilizer, seeding methods, sowing time, temperature, light, and moisture. During the dough and mature stages, the plants are more sensitive (Ou, 1985; Jha, 2001). The disease is known to occur in nutrient lacking soils (Ou, 1985), the disease incidence is correlated with low available soil Si, Mn, or K (Wang et al., 1980), (Kaur et al., 1982) reported that maximum disease intensity was completely correlated with N, P, and Mn content and negatively with Ca, Mg, K, Fe, and K/N and Fe/Mn ratio of the plant. It has been observed that NO3 induces greater susceptibility than NH4 (Dasgupta and Chattopadhyay, 1977). A significant lessening in both leaf spot and stalk rot of brown spot increase in N levels from 0–180 kg/h, irrespective of cultivars has been calculated (Sunder et al., 2005).

11.6 METHODS FOR IDENTIFICATION/DETECTION

Disease detection methods have significant importance in management of disease. Sethy et al. discussed image processing techniques for the diagnosis of rice plant disease. For the detection of bacterial light, blast and brown spot disease of rice, a green pixel masking with Naïve Bayes' classifier is used

(Islam et al., 2018). Similarly, Tejas et al. (2021) discussed various methods for the detection of rice plant disease such as SVM, ANN, etc. AlexNet is another efficient method of disease detection of rice (Matin et al., 2020).

11.7 CONCLUSION

Rice is the universally grown cereal crop that command over other crops. The main emphasis of this study is to focus on the behaviour of the BS pathosystem of rice brown disease with management and the relationship between pathogen, host and their environment with various factors involved. The main concern is the measurement of the inactive period and the environmental and variability factors. The estimated incubation period gives scope for quantitative study of the monocycle in BS, with other aspects, among all a complete operational clarification for the resistant components is occurring. A further area of concern is the complex water-soil condition or physiology of plant nutrients susceptibility of plant interaction with a pathogen, stages of biological incorporation inscribe the host crop, pathogen, physiology, and environmental interactions of both host crop and the pathogen. The constituents present are so numerous, interactions are so dismayed and usually unsystematic, therefore a strong modeling system can detect progress. The experiment done by Carvalho et al. (2010) revealed that an entire plant preposition is important and obtained results observed both in the breeding field and bio-control measures.

11.8 FUTURE PERSPECTIVES

Application of biological control of BS is practiced by many farmers in India and Asia for the diseases of rice caused by several fungi, which includes seed treatment with bio-control agents. Researchers are working on decoding the interaction between bio-control agents and host plant resistance and physiology host plant, an association of bio-control agents for nutrient uptake, growth improvement and also aware farmers to adopt new practices toward recent research of plant–pathogen interactions, protest to environmental stress. This chapter mainly focused on brown spot disease that causes economical loss to rice crops and it reflects the number of factors responsible for the epidemics of disease and their outcomes. A restricted consideration of the life cycle of pathogen its impact on the environment contributes to the limited information toward this research but more research are in progress, the result of global change and the switch in socioeconomic contexts regulates the management of crop makes BS a marker for ecological changes. Rice brown spot disease is still present in some rice-growing areas, but not in others. There is a need to look into the epidemiological elements that cause the disease to start and spread. The origins of primary inoculum and their involvement in the disease's occurrence remain unknown. Physical variables that influence the host physiology and are linked to a higher frequency of brown spots must be addressed, especially in light of recent improvements in the use of various molecular tools.

KEYWORDS

- **brown spot**
- **ecology**
- **epidemiology**
- ***Helminthosporium oryzae***
- **identification measures**
- **rice**

REFERENCE

Aguilar, M.; *Cultivo Del Arroz en el Sur de Espana*; Caja de Ahorros El Monte: Sevila.

Ba, V. V.; Sangchote, S. Seed Borne and Transmission of *Bipolaris Oryzae*, the Causal Pathogen of Brown Spot of Rice. *Kasetsart J. Natl. Sci.* **2006,** *40,* 353–360.

Biswas, S. K.; Ved, R.; Srivastava, S. S. L.; Singh, R. Influence of Seed Treatment With Biocides and Foliar Spray With Fungicides for Management of Brown Leaf Spot and Sheath Blight of Paddy. *Indian Phytopath.* **2001,** *61,* 55–59.

Biswas, S. K.; Ved, R.; Srivastava, S. S. L.; Singh, R. Influence of Seed Treatment With Biocides and Foliar Spraywith Fungicides for Management of Brown Leaf Spot and Sheath Blight of Paddy. *Indian Phytopath.* **2008,** *61,* 55–59.

Bernaux, P. Evaluation of the Susceptibility of Rice Glumes to *Pyricularia Oryzae* Cav. and *Drechslera Oryzae* (Breda de Haan) Subr. & Jain Consequences for Disease Transmission. *Agronomie* **1981,** *1,* 261–264.

Carvalho, M. P.; Rodrigues, F. A.; Silveira, P. R.; Andrade, C. C. L.; Baroni, J. C. P.; Paye, H. S. Rice Resistance to Brown Spot Mediated by Nitrogen and Potassium. *J. Phytopathol.* **2010,** *158,* 160–166.

Chattopadhyay, S. B.; Chakrabarti, N. K.; Ghose, A. K. Estimation of Loss in Yield of Rice Due to Infection of Brown Spot Incited by *Helminthosporium Oryzae. FAO Int. Rice Comm. Newsl.* **1975,** *24,* 67–70.

Damicone, J.; Moore, B.; Fox, J.; Sciumbato, G. Rice Diseases in Mississippi: A Guide to Identification; Mississippi State University Extension Service, 2001.

Dasgupta, M. K.; Chattopadhyay, S.B. Effect of Different Doses of N and P on the Susceptibility of Rice to Brown Spot Caused by *Helminthosporium Oryzae. Z. Pftanzenkrankh. Pftlanzensch* **1977,** *84,* 276–285.

Datnoff, L. E.; Lentini, R. S. Brown Spot in Florida Rice; 1994. pp 128. http://edis.ifas.ufl.edu

Drechsler, C. Phytopathological and Taxonomic Aspects of Ophiobolus, *Pyrinophora, Helminthosporium* and a New Genus, *Cochliobolus. Phytopathology* **1934,** *24,* 953–985.

Gangopadhyay, S. Current Concepts on Fungal Diseases of Rice. Today and Tomorrow's Printers & Publishers: New Delhi, 1983; pp 349.

Ghose, R. L. M.; Ghatge, M. B.; Subramanian, V. Rice in India, revised ed.; ICAR: New Delhi, 1960; p 474.

Groth, D.; Clayton, H. Brown Spot of Rice. Louisiana Plant Pathology (Disease Identification and Management Series); LSU Ag Centre Research and Extension; Pub.3108. [Online] (2013). www.Isuagcenter.com

Hiremath, P. C.; Hegde, R. K.; Kulkarni, B. G. A New Technique for Quantitative Assay of *Drechslera Oryzae* From Soil. *Indian Phytopath.* **1980,** *33,* 7–9.

Ito, S.; Kuribayashi, K. Production of the Ascigerous Stage in Culture of *Helminthosporium oryzae. Ann. Phytopathol. Soc. Japan.* **1927,** *2,* 1–8.

Islam, T.; et al. In *A Faster Technique On Rice Disease Detection Using Image Processing of Affected Area In Agro-Field,* Proceedings of the 2nd International Conference on Inventive Communication and Computational Technologies (ICICCT), Coimbatore, , 2018; pp 62–66.

Jha, A. C. Development and Management Brown Spot of Rice Caused by *Drechslera Oryzae* (Breda de Haan) Subramanian and Jain. Ph.D Thesis, RAU Bihar, Pusa, Samastipur, 2001.

Kaur, P.; Kaur, S.; Padmanabhan, S.Y. Relationship of Phosphorus Concentration in the Host and Brown Spot Disease of Rice. *Indian Phytopath.* **1982,** *35,* 393–398.

Khalili, E.; Sadravi, M.; Naeimi, S.; Khosravi, V. Biological Control of Rice Brown Spot With Native Isolates of Three Trichoderma Species. *Braz. J. Microbiol.* **2012,** *43,* 297–305.

Khus, N. Assessment of Organic and Inorganic Fertilization and Plant Density on Production Potential of Rice-Rice- Cropping System. *Madras Agric. Jour.* **2011,** *85* (12), 604–607

Khush, G. S. Origin, Dispersal, Cultivation and Variation of Rice. *Plant Mol. Biol.* **1997,** *35,* 25–34.

Kulkarni, S.; Ramakrishnan, K.; Hegde, R. K. A Baiting Technique for Detecting *Drechslera Oryzae* From Soil. *Indian Phytopath.* **1981a,** *34,* 228–229.

Kulkarni, S.; Ramakrishnan, K.; Hegde, R. K. Epidemiology and Control of Brown Spot of Rice in Karnataka. IX. Atmospheric Variation in Spore Load of *Drechslera Oryzae. Indian Phytopath.* **1982,** *35,* 80–82.

Kumar, R. N.; Mishra, R. R. Effect of Various Factors on Spore Germination and Growth of Brown Spot Pathogen *Drechslera oryzae. Indian Phytopath.* **1993,** *46,* 405–406.

Kumari, V.; Jariwala, S.; Rai, B. Effect of Physic-Chemical Factors on Growth of Drechslera Oryzae. *J. Mycopathol. Res.* **1997,** *35,* 131–136.

Matin, M.; Khatun, A.; Moazzam, M.; Uddin, M. An Efficient Disease Detection Technique of Rice Leaf Using AlexNet. *J. Comput. Commun.* **2020,** *8,* 49–57. DOI: 10.4236/jcc.2020.812005.

Motlagh, M. R. S.; Kaviani, B. Characterization of New Bipolaris spp.: The Causal Agent of Rice Brown Spot Disease in the North of Iran. *Int. J. Agric. Bio.* **2008,** *10* (6), 638–642.

Ou, S. H. Rice Diseases, 2nd ed.; CMI: Kew, England, 1985; p 370.

Purkayastha, R. P.; Mukhopadhyay, R. Factorsaffecting Colonization of Rice Leaves by *Helminthosporium Oryzae*. *Trans. British Mycol. Soc.* **1974,** *62,* 402–406.

Ramakrishnan, G.; Subramanian, C. L. Studies on Helminthosporiose of Rice: Involvement of Different Species in Disease Development. *Madras Agric. J.* **1977,** *64*, 60–61.

Rekabder, M. F. H. Dhan Chasher Nana Katha, Krishikatha, 2004; vol *64* (2), pp 39–40.

Rout, S.; Tewari, S. N. Amalab-e, a Formulated Botanical Product Potential Against Rice Blast Instant *Pyricularia Grisea. Bioscan*, **2012,** *7,* 547–552.

Sethy, P. K.; Barpanda, N. K.; Rath, A. K.; Behera, S.K. Image Processing Techniques for Diagnosing Rice Plant Disease: A Survey.

Procedia Comput. Sci. **2020,** *167,* 516–530.

Sharma, S. K.; Maheshwari, S. K. Efficacy of Fungicides for Control of Brown Spot. *Int. Rice Res. Newsl.* **1982,** *7,* 17–18.

Shoemaker, R.A. Nomenclature of Drechslera and Bipolaris, Grass Parasites Segregated From *Helminthosporium*. *Canadian J. Bot.* **1959,** *37,* 879–887.

Subramanian, C. V.; Jain, B. L. A Revision of Some Graminicolous Helminthosporia. *Curr. Sci.* **1966,** *35,* 352–355.

Sunder; Singh, R.; Agarwal, R. Brown Spot of Rice: An Over View. *Indian Phytopath.* **2014,** *67* (3), 201–215.

Sunder, S.; Singh, R.; Dodan, D. S.; Mehla, D. S. Effect of Different Nitrogen Levels on Brown Spot (*Drechlera Oryzae*) of Rice and its Management Through Host Resistance and Fungicides. *Pl. Dis. Res.* **2005,** *20,* 111–114.

Tawde, T.; Deshmukh, K.; Verekar, L.; Reddy, A.; Aswale, S.; Shetgaonkar, P. Rice Plant Disease Detection and Classification Techniques: A Survey. *Int. J. Eng. Res. Technol.* **2021,** *10* (07).

Wesely, E. G.; Ebenezar, E. G.; Rajarathinam, K.; Sekar, T.; Jayabalan, M. Survival of Two Fungal Pathogens Associated With the Grain Discoloration of Rice. *J. Ecobiol.* **1996,** *8,* 29–32.

Vidhyasekaran, P. The Use of Dichloromethane to Incorporate Fungicides Into Rice Seeds for Control of *Drechslera Oryzae*. *Seed Sci. Technol.* **1980,** *8,* 357–362.

Viraktamath, B.C. In Rice Research in India; Current and Future Prospects. Proceedings of 2nd Workshop Cum Training Programme on DUS Tests in Rice; Directorate of rice Research: Hyderabad, 2007; pp 1–3.

Wang, T. S. L.; Hsu, L. S.; Lin, C. H. Improvement of Soil Fertility for Reducing Brown Spot Incidence in Paddy Fields of Eastern Taiwan. *J. Agric. Res. China.* **1980,** *29,* 35–36.

CHAPTER 12

Comparative Analysis of Conventional and Molecular Methods for Rice Brown Spot Disease Resistance

SHIVI RATHORE[1] and RASHMI MAURYA[2]

[1]*Department of Material Science and Engineering, National Taiwan University of Science and Technology, Da'an District, Taipei, Taiwan*

[2]*Department of Botany, University of Lucknow, Lucknow, Uttar Pradesh, India*

ABSTRACT

Investigations into various fungal diseases including brown spots in rice have been reported. In ancient times, people were using many conventional methods to avoid brown spot disease from rice. Further, breeders change the genetic makeup of the rice plants by simple hybridization methods to develop resistance against different diseases. In this chapter, we are focusing on the comparative analysis of the conventional and molecular methods for brown spot disease resistance of rice. The development of new molecular markers and other methods for disease management is currently used for disease resistance in rice. Molecular methods are extensively used for the early and rapid detection of pathogens. Some desired resistance genes and the incorporation of resistance gene(s) have been successfully performed by applying molecular techniques in rice for brown spot disease. Genetic markers flanking with QTLs and resistance donors will be the most important factors for marker-assisted selection to improve brown spot resistance breeding.

Fungal Diseases of Rice and Their Management. Deepti Srivastava, Md. Shamim, Malik Mobeen Ahmad, and R. S. Upadhyay (Eds.)

12.1 INTRODUCTION

Bipolaris oryzae is known as the main fungus to cause rice brown spot disease in plants. This disease was the primary route of the Bengal famine in 1943. This fungus was used by the USA as a biological weapon in World War II against Japan. Rice is the main staple food for 1.6 million people in the world. Presently, this disease spread widely over the rice-growing regions, resulting in significant yield loss next to Blast disease (Sharma et al., 2008). When timely control measures are absent, yield loss can range from 18.75 to 22.50% (Kamal and Mia 2009). The brown spot is a widespread occurrence in India (Reddy et al., 2010). Additionally, there is evidence that the brown spot is getting worse in locations that are prone to drought (Savary et al., 2005). According to Sunder et al. (2014), the disease is more severe in direct-seeded rice tracts in the Indian states of Bihar, Jharkhand, Madhya Pradesh, Orissa, Chhattisgarh, Assam, and West Bengal. By using appropriate control strategies, such as resistant cultivars, chemical, biological, cultural, and other methods, either separately or in combination, the yield loss can be minimized. Additionally, it is important to do molecular studies to identify and characterize the QTLs underlying brown spot resistance. Only a small number of biological control methods have been developed so far, and because of their limited field applicability, it is necessary to conduct more research before advocating their use for commercial purposes.

12.2 FACTORS AFFECTING RICE BROWN SPOT DISEASE

12.2.1 BIOTIC AND ABIOTIC STRESS

Several environmental factors and agro–ecological conditions are also responsible for the rice brown spot disease. The severity of this disease is faster in the initial period and reached a maximum level at a 5.18% periodical increment in percentage disease index (PIPDI) and a 12.7% cumulative increment in percentage disease index (CIPDI) during the third and fourth weeks of September. In this season, the maximum temperature was reported around 32.6–34.21°C and the minimum average temperature is 23.61–26.18°C and humidity varied between 80.71 and 85.81% (Biswas et al., 2018). Different abiotic stress is also induced for rice brown spot disease (Stephen et al., 2019).

12.3 MANAGEMENT OF BROWN SPOT DISEASE

We can avoid brown spot disease to take some precautions (Bohra et al., 2018):

(A) Maintaining soil nutrients such as P, N, and S.
(B) Avoiding water stress and drought conditions.
(C) Use resistant tolerant cultivars; they can be used against the most economical means of control.
(D) Using fungicide for seed treatment as fungus can survive in seeds for up to 4 years. Chitosan, carbendazim, or mancozeb have been found to reduce seedling infections.
(E) Avoiding rice debris and weeds in the field after harvesting the crop as these act as sources of inoculums.
(F) By seed treatment at a high temperature (53–54°C) temperature for 10–12 min may also be effective before sowing.

12.4 CULTURAL METHOD FOR MANAGEMENT OF RICE BROWN SPOT DISEASE

In ancient times, cultural methods were used to control rice brown spot diseases such as spacing between two plants and timing. Some primary nutrients are very essential such as nitrogen (N) and silica (Si) which affect disease incidence and development (Monisha et al., 2019). Silica storage in rice tissue shows disease resistance and the best result against insects. It increases the leaves' erectness which increases photosynthesis and decreases heavy metal toxicity, water storage, and cuticular transpiration. Silica does not support crop yield but showed potential against disease in many rice species (Siregar et al., 2016). Currently, fungicides are the most used method to repress the disease. Some farmers in developing countries cannot use pesticides due to the cost involved, and besides, these pesticides create a lot of environmental hazards. In the past time, people used well-known antibiotics for rice blast, for example, antimycin-A, cephalothecium, antiplasmin, blastmycin, blasticidin-A. A spray of cow urine is also used as an effective fungicidal agent. It is cost-effective, can reduce agricultural produce, and maintain agricultural sustainability. It is previously recorded that cow urine was effective as pest control. Cow urine contains physiologically active substances, for example, growth regulators. These growth regulators increase the fertility and yield of grain. Grain contains a high protein yield recording maximum at the highest fertility level.

12.5 MANAGEMENT THROUGH PHENOLIC AGENTS

Phenolic agents were used against the rice brown spot disease to control the disease and these phenolic compounds showed a significant effect on increasing the photosynthetic pigment in rice leaves, total carbohydrate content, protein contents and dry weight increased after using these phenolic agents. Benzoic acid, salicylic acid, and hydroquinone (3, 6, 6, 9, 12, and 15 mM) were tested against the linear growth of *B. oryzae*. Benzoic acid and salicylic acid are the most-effective inhibitors for the growth of *B. oryzae*. Salicylic acid and benzoic acid showed complete inhibition at 9 mm. Under the field condition disease severity (DS), disease incidence (DI), and grain yield, the grain quality was estimated 3 times at 7 days intervals. After 7 days of treatment, there was no significant difference among all treatments. After 14 days application, a significant difference in DS was reported with benzoic acid 20 mM. After 21 days, rice brown spot was treated with benzoic acid at 20 mM and showed the best inhibition of the lowest value of DS (2.5) concerning DI. Benzoic acid and SA were found to be the most effective inhibitors for the growth of *B. oryzae* under in-vitro conditions (Shabana et al., 2008).

In vitro and *in vivo* evaluation of 12 fungicides was done against brown spot disease; these 12 fungicides were Carbendazim 50% wettable powder (WP), Propineb 70%, Hexaconazole 50% emulsifiable concentrate (EC), Tricyclozole 75% WP, propiconazole 25% EC, Kresoxim-methyl 44.3% suspension concentrate (SC), Isoprothiolane 49% EC, Difeno-conazole 25% EC, Zineb 68% + Hexaconazole 4% WP, Tebuconazole 50% + Tebuconazole 25% + water-dispersible granules (WG), and Carbendazim 25% + Mancozeb 50% WP, which were used at different concentrations such as 50, 100, 250, 500, 1000, 1500 ppm by poison food technique. In vitro and in vivo evaluations of 12 fungicides were carried out, among the 12 fungicides, Hexaconazole 5% EC and Tebuconazole 25% + Trifloxystrobin 50% WP, and Zineb 68% + Hexaconazole 25% WG have the competency even at 50 ppm concentration with 100% inhibition over control (Monisha et al., 2019). Alginate encapsulation with *Trichoderma harizanum* against brown spots showed inhibition of disease. *Trichoderma* antibiotics inhibit disease in more indirect ways. Three treatments of alginate encapsulation were performed against rice seedlings. This encapsulation showed a prominent control over the rice brown spot disease (Anuar et al., 2019).

12.6 RESISTANCE SOURCES AGAINST BROWN SPOT IN RICE

To examine and screen out advanced hybrid lines that are resistant/tolerant to brown spot and blast disease, Hossain et al. (2004) evaluated 29 different genotypes/advanced breeding lines. Four cultivars/checks of rice, BR 11, BR 22, BR 25, and BRRI dhan 38, were chosen for the study. In the T. aman season, one accession displayed a resistant (R) reaction, while in the boro season, 29 cultivars displayed a moderately resistant (MR) reaction to brown spots. In the T. aman season, 13 accessions were discovered to be moderately resistant (MR), three breeding lines/genotypes were shown to be highly resistant (HR), and five were discovered to be resistant (R) to blast. To produce rice, these lines and cultivars may be used as breeding stock. Channakeshava and Pankaja (2018) explored rice varieties and found that 31 genotypes had a moderately resistant response to the infection. Additionally, 11 genotypes were found to be the disease-resistant sources, including Rasi, JGL-1798, BR-2655, Raksha, KMP-201, BI-33, Sagbatta, Honnekattu, Klame, Kavekantak, and Togarshi. These genotypes could be used by breeders to increase the yield of rice crops because they remained disease-free throughout the cropping period.

12.7 MOLECULAR APPROACHES FOR RICE BROWN SPOT DISEASE MANAGEMENT

Molecular marker/DNA-based marker is a sequence of DNA. These markers are located in the genome. Molecular markers are phenotypically neutral. These markers are useful to scan the sequence of the entire genome of the plants. Molecular markers are used in various molecular techniques, for example, sequence-characterized amplified regions (SCARs), Restriction fragment length polymorphism (RFLP), cleaved amplified polymorphic sequences (CAPs), and other polymerase chain reaction (PCR). Molecular markers are random. These markers are polymorphic and quickly distributed in the whole genome and easy to detect. In the first generation of the RFLP technique, a labeled DNA probe is used to select the gene of interest from a digested DNA sample by hybridization. The DNA fragments are separated by gel electrophoresis. Genetic variation is visualized by hybridization bands. The same genotypes have the same banding pattern and different genotypes have different banding patterns. This process involves DNA extraction, digestion of the isolated DNA with a restriction enzyme, and hybridization of a digested DNA fragment with a labeled DNA probe. The hybridization band is separated by the agarose polyacrylamide gel electrophoresis. The PCR-based marker does not require a DNA hybridization probe. The PCR-based markers are used in random amplified polymorphic DNA (RAPD), amplified fragment length polymorphism (AFLP), microsatellite or SSRs, single-nucleotide polymorphism (SNP), and expressed sequenced tag (EST). Genetic engineering and bioinformatics play an important role to deal with rice brown spot disease. Before going into detail about molecular approaches to control rice blast resistance, one should have to know about the mechanism behind the disease. Pathogen-associated molecular pattern (PAMP) is recognized as a receptor. This receptor leads to an immune response called PAMP-triggered immunity (PTI). First, bacterial, viral, and fungal PTI include host defense which is known as effectively triggered susceptibility (ETS). Second, high polymorphic protein is related to effector-triggered immunity (ETI). In plants, RNAi recognizes the viral double-stranded RNA (dsRNA). This dsRNA is linked to PAMP (bacteria and fungus). This dsRNA may have a similar blocking function as PTI. After plant–viral infection, this plant viral suppressant of RNAi is linked to the host RNAi (like ETS). Plant R protein can recognize the virulent protein that can mediate a strong defence as ETI (Liu et al., 2016; Chukwu et al., 2019).

12.7.1 MARKER-ASSISTED SELECTION AND GENE PYRAMIDING FOR RESISTANCE IN RICE

A molecular marker is used in efficient conventional breeding. It is a tool for present germplasm that will be used in breeding activities. Marker-assisted selection and gene pyramiding are used for the development of disease resistance and quality improvements in several crops. Gene pyramiding is used to study the genetic diversity of DNA (polymorphism) and help to identify the special breed of genes present in the chromosome sequences. Different types of markers are present to study the genome sequences (Chukwu et al., 2019). The main target of MABB is to transfer the gene by minimizing the donor segment and recovering the parent character (genes). Also, MABB is highly suitable for conventional backcross because they are highly accurate and efficient as well as time-saving. This method is used to transfer other resistance genes in other popular varieties of plants. The MABB technique has been reported previously for the development of new rice varieties (Srivastava et al., 2017). Agricultural scientists from the last few decades are using marker-assisted breeding (MAB) as an alternative, faster, and more accurate method to get beyond the constraints of conventional breeding strategies (Balachiranjeevi et al., 2018; Jairin et al., 2017). The DNA marker is based on different polymorphisms. Sequencing promoted gene mapping and marker-assisted selection breeding (MAS). Specific length amplified sequence (SLAF-seq) and reduced representation library (RRL) construction have an efficient method of large-scale SNP discovery and genotype. The entire genome sequencing of individuals tools (pool-seq) provides a cost-effective alternative method for gene identification. The pool technique is useful for the rice blast QTL mapping (Hasan et al., 2015; Chen et al., 2016).

The development of genomic methods and sequences, the availability of genome sequences, internet databases, and numerous bioinformatic tools all support the MAB approach. Since it primarily focuses on enhancing the overall performance, yield stability, traits including tolerances to numerous stresses, and adoption by farmers, the MAB strategy has become an indispensable tool throughout time (Balachiranjeevi et al., 2018; Nelson et al., 2018; Shakoor et al., 2017). Matsumoto et al. (2021) altered the genetic background of the high-yielding but susceptible "Mienoyume" to incorporate a chromosomal segment containing a quantitative trait locus (QTL) for BS resistance, qBSfR11, derived from the BS-resistant local resource "Tadukan". In a 1.3-Mbp area, a single recessive gene regulates resistance. This gene was given the name "bsr1" (brown spot resistance 1).

This is the initial report on the marker-assisted selection-bred BS resistant rice variety. In fields where BS was mild, the near-isogenic line bsr1-NIL outperformed Mienoyume in yield, grain width, and other agronomic traits; in fields where BS was more severe, it outperformed Mienoyume in yield by 28.8% and demonstrated resistance to multiple isolates of the BS fungus. It shown consistent resistance to BS and exhibited superior agronomic qualities when exposed to BS. The bsr1-NIL with BS resistance was created by us, and we registered it as "Mienoyume BSL" with the Ministry of Agriculture, Forestry, and Fisheries in Japan. By conducting field resistance testing using recombinant inbred lines obtained from crossings between the resistant "Tadukan" and the susceptible "Hinohikari," we discovered a significant BS resistance QTL, qBSfR11, on chromosome (Chr.) 11 (Sato et al. 2015). In the "Koshihikari" background, the Tadukan allele at the QTL also provided BS resistance (Sato et al. 2015). Near qBSfR11, two other QTLs were found: qBSR11-kc, derived from ""CH45" (Matsumoto et al. 2017), and BSq11.2v, derived from IR62266 (Katara et al. 2010). There have not been any reports of the creation of BS-resistant cultivars employing BS resistance QTLs, and there are not many QTLs or genes for BS resistance either. The donor parent (R307-48-9) was a NIL created by Sato et al. (2015) in which the main resistance QTL (qBSfR11) on Chr. 11 was transferred into the genetic background of "Koshihikari" (susceptible), deriving from the indica "Tadukan" (resistant) variety. Through consecutive backcrossing, the qBSfR11 gene was introduced into the Mienoyume background. Promising individuals or lines were chosen by MAS with simple sequence repeat (SSR) markers (McCouch et al. 2002) based on the target region on Chr. 11 during backcrossing or selfing from 2014 to 2015. In 2016, the size of the replaced segments was confirmed using six SSR markers at the qBSfR11 locus.

12.7.2 QTLS FOR BROWN SPOT RESISTANCE

Researchers reported various qualitative and quantitative genes for the brown spot resistance in rice (Table 12.1). The majority of QTL are related to qualitative genes. Resistance in the host plant is a crucial strategy for controlling the rice brown spot disease and has been crucial in maintaining rice yield. According to Sato et al. (2008, 2015), the quantitative aspect of BS resistance makes it susceptible to genotype by environment interactions. Screening cultivars for BS resistance has been done in a number of studies (Pantha et al., 2017). The cultivars "Tadukan" and "Tetep" have shown quantitative resistance to BS (Sato et al. 2008, 2015; Mizobuchi et al. 2016). To diversify the current gene pool for brown spot resistance, new sources of resistance could be integrated into rice genotypes (Mizobuchi et al. 2016). From the 110 F5 lines created from crossings between Tadukan (a resistant indica rice variety) and Hinohikari (a susceptible japonica rice variety), developed from various F2 plants by single seed descent, three QTLs, namely, qBS2, qBS9, and qBS11 were found. On chromosomes 2, 9, and 11, respectively, these QTLs were found. While the peak of qBS2 could only be discovered by CIM analysis, the qBS9 and qBS11 could be detected by both IM and CIM analysis (Sato et al., 2008). In 2017, Matsumoto and co-workers reported four QTLs on chromosomes 2 (qBSR2-kc), 7 (qBSR7-kc), 9, and 11 (qBSR11-kc) that contributed to BS resistance. The hybrid between the indica Koshihikari (susceptible) and the japonica CH45 (resistant) resulted in a rise in the "Koshihikari" allele at qBSR9-kc as well as the "CH45" allele at qBSR2-kc, qBSR7-kc, and qBSR11-kc. A whopping 23.0–25.9% of the overall phenotypic variance was explained by the main QTL qBSR11-kc. The detection of two QTLs (qBSR9-kc and qBSR11-kc) occurred in both years, whereas the detection of the other two QTLs occurred only in 2016. The four QTLs flanked by genetic markers will be effective instruments for

TABLE 12.1 Molecular Method Used to Control Rice Brown Spot Disease.

Method/modern inhibitors	Area of action	Result
RAPD/inter simple sequence repeat (ISSR) (Kumar et al., 2016)	In this work, different isolates were taken from the leaf and seeds of paddy to study the variability. Also, cluster analysis was carried out of 20 RAPD and 20 ISSR primers, and 5 RAPD and ISSR primers gave very prominent patterns. Isolates were characterized for morphological traits like colony character, colony diameter, and sporulation	Moderate and fast-growing isolates chosen from different genetic variations of 20 isolates of *B. oryzae* were analyzed by RAPD and ISSR markers. Twenty RAPD and ISSR primers were screened, 11 RAPD and 10 ISSR markers were amplified but only five RAPD and five ISSR markers gave a very good reproducibility banding pattern RAPD marker showed polymorphism between 54.0 and 100% ISSR marker showed polymorphism between 49.2 and 100%
QTL mapping (Mondal et al., 2017).	QTL mapping for brown spot tolerance using RIL population of rice. In this investigation RIL genotypic and phenotypic data of 122 RIL populations derived from Danteshwari and Daga Deshi were used to identify QTL for brown spot disease tolerance. Total of 3 QTL "qBS1.1," qBS5.1, qBS5.2 were mapped for brown spot disease tolerance on chromosomes 1, 5, and 5, respectively	QTL were identified on chromosome No. 5 with high phenotypic variance. This work is not reported earlier. QTL qBS1.1 was identified on chromosome No. 1 for brown spot tolerance. This tolerance was reported around 9.15%
QTL mapping (Sato et al., 2015)	In this work, quantitative resistance in local verities and QTL analysis of RILs from Tadukan (resistance) × Hinohikari (susceptible). The parent and later generation of RILs were transplanted into paddy fields where brown spots severely occurred. Some QTL (qBS2, qBS9, qBS11) showed resistance in seedlings in a greenhouse. Three new QTL (qBSfR1, BSFR4, and qBSfR11 were detected on chromosomes 1, 4, and 11, respectively)	Tadukan alleles qBSfR1 and BSfR11 and hinohikari alleles qBSfR4 increased resistance. Non-isogenic lines (NILs) and qBSfR1 and BSfR11 "Tadukan" resistance allele with one or both QTLs in a susceptible background (Koshihikari) showed efficacy under the field condition. qBSfR11 showed the confirmed effectiveness. This BSfR11 will be used as an effective marker-assisted selection to improve brown spot disease

TABLE 12.1 *(Continued)*

Method/modern inhibitors	Area of action	Result
QTL mapping (Matsumoto et al., 2017).	Local cultivars CH45 (resistance) and Japanese Koshihikari (susceptible). In this study, back cross inbred lines (BILs) of brown spot resistance are analyzed with SSR marker and these markers are polymorphic marker	QTL contributing to BS were identified on chromosome 2 (qBSK2-kc), 7 (qBSK7-kc), 9 (qBSK9-kc), and 11 (qBSK11-kc). The CH45 alleles at qBSR2-kc, qBR7-kc, and qBSR11-kc and Koshihikari alleles at qBSR9-kc increased resistance in plants infected with brown spots. Major QTL qBSR11-kc showed a total phenotypic variation 23.0–25.9%
QTL mapping (brown spot, bacterial seedling rot, and bacterial grain rot) (Mizobuchi et al., 2016)	QTL resistance for BS, BSR, and BGR has been reported to improve disease resistance and promote molecular genetics	In this work, molecular genetics and evaluation methods for disease resistance have facilitated the detection of QTL along with resistance. A seedling evaluation method for BS resistance was developed that found several QTL was reconfirmed as qBSfR11, a QTL for field resistance to BS. Also, RBG2 confirms resistance against BGR was modified by a cut-panicle inoculation method. No agronomic trait is linked with RBG2, this agronomic trait might be detected regarding to the environmental condition. Chromosome segment substitution lines (CSSLs) are very useful regarding the genetic analysis for the substituted segment to detect QTL; CSSLs are used for the detection of RBG1 (resistance to BGR) and QTL of BS resistance. Isolation of RBG1 and other genes will shed light on the genetic mechanism of resistance. qBS11 have reconfirmed as qBSfR11 field resistance to BS. RBG1 for resistance showed multiple resistances to several strains indicating that RBG1 is a QTL without race specificity of RBG2 (for resistance to BGR is not known). BGR and BSR are caused by same the pathogen (*Burkholderia glumae*). RBG1 and RBG2 characterization is on the way. Several QTLSv have been discovered, such as qBS9 and Bsq9.1, qBS11 and qBSfR11 for BS, RBG1 for BSR and RBG2 for BGR; all of these can explain over 30% of phenotypic variation in QTL analysis. So, it is necessary to identify some new QTL for gene pyramiding
Gene pyramiding (Mizobuchi et al., 2016).	Blast disease resistance in rice (BS, BGR, and BSR) was reported in a different type of rice species such as *Oryza sativa*	Factors linked with BSR and BGR are different so here is the need to discover new additional QTL along with RBG1and RBG2. The reason behind this new discovery is that BS, BGR, and BSR are common diseases but these diseases took a major role in the resistance of rice plants due to global warming. The optimum temperature is around 30°C. Pyramid QTL is necessary to develop a bewildering germplasm against resistance to all of these important diseases

TABLE 12.1 *(Continued)*

Method/modern inhibitors	Area of action	Result
MALDI-TOF MS analysis	In this work, the researchers used *Bacillus amyloliquefaciens* strain BS5 against brown spot disease of rice and host–pathogen bioagent interactions through a proteomic approach. Forty *Bacillus* spp. have the antagonistic potential against *B. oryzae* under in vitro. Based on the antagonistic approach, 10 strains were selected and characterized at the molecular level. Antibiotic biosynthetic genes producing bacilomycin, bacilysin, iturin, surfactin, subtilin, mersacidin, ericin, mycosubtilin, and fengycin from the effective strain detected by PCR analysis. Third, focus of the researcher was about protein identification among the nine different expressed proteins	Protein isolation was performed and this protein was used for the proteomic study. Before MALDI-TOF MS and 2D-PAGE has been performed to study the changes in the protein expression of tissue under different stresses. After 2D-PAGE, MALDI-TOF MS study focused on the nine different expressed proteins. They divided these proteins into three sub-groups: Up-regulated protein but no change in the interaction (UN), down-regulated protein but no change in the interaction (DN), and protein with no change in pathogen but up-regulated in interaction (NU). A total of nine proteins (S1–S9) were sent for MALDI-TOF analysis. The MALDI-TOF MS data searched in MASCOT software demonstrated the homology for the following proteins: 2-*cys*-peroxiredoxin, ribulose 1,5 bisphosphate carboxylas, trehalose-phosphate phosphatase, ATP synthase, 50 s ribosomal protein serine/threonine protein kinase. The functions were evaluated by comparing the protein sequences to the public protein database NCBInr. The function of protein is related to plant metabolism, defense response, and disease resistance
Genetic fingerprinting (Kamal et al., 2009)	In this technique, researchers used 152 isolates. These isolates were characterized on the basis of DNA fingerprinting, tandem repeat analysis, and evaluated phylogenetic identity, diversity, and pathogenicity	DNA was isolated from 306 isolates including 2 isolates of IRRI. The DNA concentration was estimated 0.8% agarose gel electrophoresis. The BR24 rice variety showed the pathogenicity test 20.5–84.5% germination failure, 2.50–12.50, 5.84–11.67, and 7.50–14.17% post emergency seedling death at 15, 30, and 45 days. After transplanting 15.75–22.50% yield loss reported
ITS sequence (Sobanbabu et al., 2018)	Three different isolates of *B. oryzae* and *S. oryzae* were used in this studyname as: *B. oryzae* three isolates are ASD1, KKM1, and CBE1. *S. oryzae* isolates are: KKM1, KKM2, and ASD	KKM1 based *B. oryzae* and *S. oryzae*-based ASD1 grew steadily and brought about the maximum disease incidence virulence isolates of *B. oryzae* and *S. oryzae* could be used as potential isolates on screening resistance rice cultivars. ITR sequence is used molecular method for the identification at species level. ITS 1 and ITS primer using Ia thermocycler and product visualized as a single band in agarose gel. ITS sequence of *B. oryzae-based* KKM1 matched with the National Center for Biotechnology Information (NCBI) data, and virulent *B. oryzae*-based KKM1 used in this study was confirmed as *B. oryzae*

marker-assisted selection to increase BS resistance. A cross between the American rice cultivars "Dawn" (resistant) and "Koshihikari" (susceptible backcross) revealed four QTLs for BS resistance on chromosomes 3, 6, and 7 (qBSR3.1-kd and qBSR3.2-kd), respectively. Each QTL accounted for 6.2 to 20.3% of the phenotypic variation. The BS resistance was influenced by the "Dawn" allele for all four QTLs. One QTL, qBSR6-kd, which was connected to qDTH6-kd, where the "Dawn" allele reduced DTH, accounted for 15.1–20.3% of the total phenotypic variation.

12.7.3 ASSOCIATION MAPPING AND ALLELE MINING

Association mapping is used to scanning the entire genome of SNPs. Association mapping talks about the linkage disequilibrium. It is directly related to the gene/QTL governing trait. Allele mining has been used a gene of interest. An allele is already characterized by a large collection of germplasm. Allele mining is used for crop improve- ment. Allele mining is used for the scanning of diverse gene varieties. A sequence is difficult between polymorphisms and putative function of allele mining for R (resistance) genes (Srivastava et al., 2017). Currently, a significant part of plant breeding involves mining advantageous alleles (Li et al., 2013). The standard method of QTL discovery by linkage mapping, however, typically employs populations with little parental variation (Leung et al., 2015). Opportunities to get around this weakness exist with GWAS. The efficiency of QTL mapping is significantly increased and the relatively high false-positive rate of GWAS is greatly reduced when GWAS is conducted in conjunction with linkage mapping (Lu et al., 2010). This method has been used to analyze complicated biotic stress features in rice with success (Zheng et al., 2012). For rice breeding, simultaneous investigation of natural variations would be very beneficial. In the current rice improvement initiative, allele mining is also being used to identify a limited number of genomes and accessions from a huge collection based on either sequencing or phenotypic data. There are already many genes in rice that are resistant to serious maladies like blast and bacterial blight. However, in order to control disease epidemics and newly developing diseases, a continuing search for a variety of resistance mechanisms is required due to the evolution of pathogen populations and the dynamic nature of host–pathogen interaction (Leng et al., 2015). Because incidence, severity, and soil, nutrient, and drought conditions are all interconnected, disease expression is complicated. A QTL on chromosome 12 was discovered in a prior study utilizing a biparental hybrid between the landraces Denorado (resistant) and IR36 (sensitive)

(Banu et al. 2008). This QTL was identified through screening with a single virulent strain of *B. oryzae* and confirmed using the association mapping method in a multi-parent population (indica MAGIC) and a 2,000 diversity panel. On chromosome 12, a QTL with strong SNP markers was found to span a 1.3 Mb area. Local haplotypes made up of 7 SNPs in this area were linked to resistance. To confirm this QTL in a range of backgrounds, the 3K panel was investigated. To confirm this QTL in a range of backgrounds, the 3K panel was investigated. In the 3K genomes, accessions containing the resistant haplotype were found using the SNP haplotype. Of the 3000 accessions, 2777 (or 93%) had the identified haplotype. It seems surprising that the R haplotype is so prevalent considering the absence of active selection for brown spot resistance in most breeding programs. A possible explanation is that landraces or traditional varieties were selected for disease resistance or other traits tightly linked to this region by farmers long ago before modern rice breeding; thus most landraces carry the R-haplotype. This idea seems to be supported by examining the region using the QTL genome viewer (Yonemaru et al., HYPERLINK "https://thericejournal.springeropen.com/articles/10.1186/s12284-015-0069-y#ref-CR43"2010) which shows the presence of QTLs for 1000 grain weight, blast disease, and soil stress tolerance. The R haplotype's high prevalence is surprising given that most breeding programs do not actively select for brown spot resistance. Examination of the area with the QTL genome viewer was conducted by Yonemaru et al. (2010), and reveals the existence of QTLs for 1000 grain weight, blast disease, and soil stress tolerance, which seems to confirm this theory. The preservation of Pup1, a gene for improved phosphorus uptake, in upland and drought-tolerant cultivars is another illustration of this (Chin et al. 2011). In the absence of focused selection, the Pup1 gene is found often. Furthermore, it is possible that more than one locus can provide brown spot protection. A large population of sequenced genomes is required to identify some alleles since they may be naturally rare and appear at low frequency (Leng et al., 2015).

12.8 RECENT ADVANCES IN MOLECULAR TECHNIQUES AND TOOLS

Morphological markers are the consecutive markers that are used to identify the desirable and non-desirable gene map. Uses of such markers are roust-about intensive, time, and space consuming.

12.8.1 miRNA

The micro-RNA (miRNA) is usually found in eukaryotes but some pieces of evidence have been reported. Also, miRNA played an important role in immunity against the host–pathogen plant. Small RNAs are non-coding, short RNAs that control chromatin methylation, mRNA decay, and translational repression to direct gene silence (Seo et al., 2013). Based on their biogenesis and place of origin, they are divided into microRNAs (miRNAs) and small interfering RNAs (siRNAs). Biological processes such as development and reactions to biotic and abiotic stimuli are regulated by both miRNAs and siRNAs (Khraiwesh et al., 2012). From computational prediction to experimental identification and functional characterization, the research of miRNAs in rice (*Oryza sativa*) has advanced (Jones-Rhoades and Bartel, 2004; Liu et al., 2005). In 2004 (Wang et al., 2004), the first batch of rice miRNAs to be experimentally discovered was reported. Repression of ethylene signalling has been demonstrated to promote resistance against the necrotrophic rice brown spot fungus, while the buildup of ethylene appears to be necessary for resistance against *M. oryzae* (Iwai et al., 2006). Discovering the molecular pathways by which MIR166k-166h activity is incorporated in the intricate regulatory systems that fine-tune adaptive responses to fungus infection will be a significant future for the control of brown spot disease in rice (De Vleesschauwer et al. 2010).

12.8.2 CRISPR/CAS9

The use of genome editing technologies, such as CRISPR-Cas9, provides a precise and efficient way to introduce specific genetic changes into the rice genome. This technique can be employed to modify or disrupt genes that play a role in disease susceptibility, thereby enhancing resistance to brown spot disease. Genome editing enables precise modifications without introducing foreign genes, which may be advantageous in terms of public acceptance and regulatory considerations. CRISPR is a prokaryotic DNA that contains a combination of proto-spacer and genetic code. In this technique, a single sequence nuclease breaks the cleavage of genomic DNA. The introduction of foreign DNA is now feasible for the majority of important crops including rice (Jung and Seo, 2017; Mishra et al., 2018; Kumar et al., 2022). In the current scenario, DNA can be specifically mutated using CRISPR/Cas9 genome editing technology (Baltes and Voytas, 2015). It has been used more frequently in a variety of organisms, including crops that were modified for trait enhancements, due to its simplicity, viability, and adaptability (Ahmadpour et al., 2018; Liu et al., 2021; Verma et al., 2023). This CRISPR DNA is the remainder of the genetic code that is obtained from previous bacteria (Hsu et al., 2014; Hille and Charpentier, 2016). Cas9 is an endonuclease attached to CRISPR; Cas9 contains two active domains, HNH and RuvC, targeting double-helix DNA. Furthermore, CRISPR works like a DNA scissor and Cas9 is like a protein that opens the unlock DNA (Rani et al., 2016). Additionally, numerous features might be improved at the same time in the genome targeting capability for the disease resistance in rice (Oliva et al., 2019; Tang et al., 2017; Wang et al., 2016; Xu et al., 2019; Zhang et al., 2018; Zhou et al., 2016).

TABLE 12.2 Molecular Diagnosis of Phytopathogenic Problems in Fungi.

Method	Mode of action	Limitations
RNA sequencing	Sensitivity and specificity improve	Needs expensive pieces of equipment and bioinformatics knowledge in data analysis
Northern blotting	Detection of RNA size	Small sample of the gene is required
Sage	Genome knowledge is not required	A specialty of tag sequences
RNA interference (RNAi)	Thousands of genes can be interrogated	The non-specificity of reagents variability and incompleteness of knockdowns
NASBA	Better than RT-PCR	Specialty is depending upon the thermolabile enzyme. Reaction temperature cannot overrun above 42°C
LAMP	Rapid, sensitive, and highly specific	Only one pathogen can recognize, contamination chances are present, primer designing is complex
Microarray	It does not require large DNA sequencing, easily use	Large amount of mRNA is required
FISH	Useful for non-dividing cells	Probe preparation is very difficult because it is necessary to tailor the probe to identify the particular sequence of DNA
In situ hybridization	Short supply tissue is majorly used	Difficulty to identify how low the RNA and DNA copies
Real-time PCR	Post-amplifications are not necessary for analysis	Thermal cycling and fluorescence detection increase cost and complexity
Reverse transcriptase- real-time polymerase chain reaction (RT-PCR)	Provides quantitative data and is more sensitive than conventional PCR	Formation of each assay is expensive and time-consuming, requiring the expensive equipment and reagents

TABLE 12.2 *(Continued)*

Method	Mode of action	Limitations
Multiplex PCR	Different primer pairs are used in the same reaction	Primers and probes interfere which reduce sensitivity
Nested PCR	Two sets of primers increase the yield and specificity and amplification of the targeted DNA	Two cycles of amplification that can cause contamination
Conventional PCR	Give rapid and precise results when using the primer of specific species	Labor and cost are very high

12.9 CONCLUSION

The fungus *B. oryzae* causes the most disturbing disease which reduces the yield in rice plants. Conventional control methods are playing an important role to stop disease in rice plants but these conventional methods cannot provide full resistance against disease-causing pathogens. Biocontrol methods and disease forecasting are effective but their use on a large area is labor-intensive as well as costly. Researchers are using molecular methods to induce resistance against disease. Managing brown spot disease in rice through molecular approaches involves utilizing genetic and molecular techniques to develop resistant rice varieties or deploy molecular markers for disease monitoring and management. It is important to note that molecular approaches should be complemented with other disease management practices such as cultural practices (crop rotation, field sanitation), chemical control (fungicides), and agronomic practices (appropriate planting density, nutrient management), to achieve effective and sustainable control of brown spot disease in rice. Additionally, regulatory considerations and public acceptance of genetically modified organisms should be taken into account when employing genetic modification techniques.

12.10 FUTURE PERSPECTIVE

Rice crops play an important role in good food security for the growing population of the world. Resistance and tolerance of abiotic and biotic is a significant target of plant breeders. Both conventional and molecular methods hold promising future prospects for managing brown spot disease in rice. Conventional breeding methods have been successful in developing rice varieties with resistance to brown spot disease and continuous breeding efforts can lead to the development of high yielding rice cultivars with durable resistance which can provide long term management of rice. It also allows the combination of multiple resistance genes through crossbreeding leading to enhanced resistance against brown spot disease. This approach can improve the durability and effectiveness of resistance in rice varieties. On the other hand, molecular methods enables the rapid identification of resistance genes and which can be used in various breeding programs. Additionally genetic modification and genome editing technologies like CRISPR-Cas9 offer the potential to introduce specific genetic changes into rice plants. These techniques can be used to enhance resistance by modifying or disrupting genes associated with disease susceptibility. Overall, future lies in integrating conventional breeding methods with molecular techniques. Combining the strengths of both approaches can expedite the development brown spot disease resistance rice varieties with improved yield potential, nutritional qualities and adaptability to changing environmental conditions.

KEYWORDS

- **advanced hybrid lines**
- **disease incidence**
- **disease resistance**
- **molecular marker**
- **resistance gene (s)**
- **recombinant inbred lines (RIL)**
- **rice brown spot disease**
- **QTL (quantitative trait loci)**

REFERENCES

Anuar, I. S. M.; Ku sulong, K. A.' Gani, H. A.; Safi, N. N.; Samsudin, K. A., Wahab, M. Z. Alginate Encapsulation of Trichoderma harzianum as Biocontrol Agent Against Brown Spot Disease on Rice (Oryza sativa) in Vitro Assays. *Int J. Ad. Sci. Ang. Inform. Technol.* **2019,** *9* (2), 461–466.

Aslam, S.; Tahir, A.; Aslam, M. F.; Aslam, M. W.; Shedayi, A. A.; Sadia, S. Recent Advances in Molecular Technique for the Identification of Phytopathogenic Fungi. *J. Plant Interact.* **2017,** *12* (1), 493–504.

Balachiranjeevi, C.; Naik, B. S.; Kumar, A. V.; Harika, G.; Swamy, M. H.; Masood, H. S.; Kumar, D. T.; Miriyala, A.; Kale, R. Yugender, A. Marker-assisted Pyramiding of Two Major Broad-Spectrum BacTerial Blight Resistance Genes, Xa21 and Xa33 into an Elite Maintainer Line of Rice, DRR17B. bioRxiv: **2018,** 368712.

Banu, S. P.; Meah, B.; Ali, A.; Brar, D. S.; Leung, H.; Cruz, C. V. Inheritance and Molecular Mapping for Brown Spot Disease Resistance in Rice. *J. Plant Pathol* **2008,** *90* (S2), 295.

Biswas, M. K.; Quiry, S.; Ghosh, T. Role of Abiotic Factors on the Development Rice Brown Spot Caused by B. Oryzae. *Bionature* **2018,** *38* (2), 86–95.

Bohra, J. S.; Sadhukhan, R.; Pal, V. Management of Brown Spot Disease in Rice (Helminthosporium oryzae) by Spraying of Cow Urine. *IJCS* **2018,** *6* (1), 1721–1723.

Channakeshava, C.; Pankaja, N. S. Performance of Paddy Varieties Against Brown Leaf Spot Disease Under Flooded Conditions in Mandya District, Karnataka, India. *Int J. Curr Microbiol App Sci* **2018,** *7* (12), 33–38.

Chen, S.; Wang, W.-J.; Su, J.; Wang, C.-Y.; Feng, Ai.-q.; Yang, J. Y.; Zeng, L. X.; Zhu, X.-Y. Rapid Identification of Rice Blast Resistance Gene by Specific Length Amplified Fragement Sequencing. *Taylor Francis* **2016,** 462–468.

Chin, J. H.; Gamuyao, R.; Dalid, C.; Bustamam, M.; Prasetiyono, J.; Moeljopawiro, S.; Wissuwa, M.; Heuer, S. Developing Rice with High Yield Under Phosphorus Deficiency: Pup1 Sequence to Application. *Plant Physiol.* **2011,** *156,* 1202–1216.

Chukwu, S, C.; Rafii, Md. Y.; Ramlee, S. I.; Ismail, S. I.; Oladosu, Y.; Okporie, E.; Onyishi, G.; Okporic, E.; Onyishi, G.; Utobo, E.; Ekwu, L.; Swaray, S.; Jalloh, M. Marker-Assisted Selection and Gene Pyramiding for Resistance to Bacterial Leaf Blight Disease of Rice (Oryza sativa L.). *Taylor Francis* **2019,** 440–455.

Cohen, S. P.; Leach, J. T. Abiotic and Biotic Stress Induce a Core Transcriptome Response in Rice. *Nature* **2019,** *9,* 6273.

De Vleesschauwer, D.; Yang, Y.; Cruz, C. V.; Höfte, M. Abscisic Acid-Induced Resistance against the Brown Spot Pathogen Cochliobolus miyabeanus in Rice Involves MAP Kinase-Mediated Repression of Ethylene Signaling. *Plant Physiol.* **2010,** *152,* 2036–2052.

Hahn, A.; Kilian, J.; Mohrholz, A.; Ladwig, F.; Peschke, F.; Dautel, R.; Harter, K.; Berendzen, K. W.; Wanke, D. Plant Core Environmental Stress Response Genes are Systemically Coordinated During Abiotic Stresses. *Int. J. Mol. Sci.* **2013,** *14,* 7617–7641.

Harish, S.; Sarvanakumar, D.; Radjacommare, R.; Ebenzar, E. G.; Seetharaman, K. Use of Plant Extracts and Biocontrol Agent for the Management of Brown Spot Disease in Rice. *Bio. Control* **2008,** *53,* 555–567.

Hasan, M. M.; Rafii, M. Y.; Ismail, M. R.; Mahmood, M.; Rahim, H. A.; Alam, M. D.-A.; Ashkani, S.; Malek, Md.-A.; Latif, M. A. Marker Assisted Back Crossing: A Useful Method for Rice Improvement. *Taylor Francis* **2015,** 1314–3530.

Hille, F.; Charpentier, E. CRISPR-Cas: Biology, Mechanisms and Relevance. *Philos. Trans. R. Soc. Lond. B Biol. Sci.* **2016,** *371,* 20150496.

Hossain, M.; Khalequzzaman, K. M.; Mollah, M. R. A.; Hussain M. A.; Rahim, M. A. Reaction of Breeding Lines/Cultivars of Rice Against Brown Spot and Blast under Field Condition. Asian J. Plant Sci. 2004, 3, 614–617.

Hsu, P. D.; Lander, E. S.; Zhang, F. Development and Applications of CRISPR-Cas9 for Genome Engineering. Cell **2014,** 157 (6), 1262–1278.

Iwai, T.; Miyasaka, A.; Seo, S.; Ohashi, Y. Contribution of Ethylene Biosynthesis for Resistance to Blast Fungus Infection in Young Rice Plants. *Plant Physiol.* **2006,** *142,* 1202–1215.

Jairin, J.; Kotchasatit, U.; Saleeto, S.; Jearakongman, S.; Srivilai, K.; Chamarerk, V.; Kothcharerk, J.; Pattawatang, P.; Korinsak, S.; Wongsaprom, C. Application of Marker-Assisted Breeding to Improve Biotic Stress Resistance for Rrainfed Lowland Rice in Northeastern Thailand. *SABRAO J. Breed Genet.* **2017,** *49,* 168–178.

Jatoi, G. H.; Ueerio, A. U.; Abdulle, Y. A.; Qiu, D. Effect of Selected Fungicides and Bio-Pesticides on the Mycelial Colony Growth of the Helminthosporium oryzae Brown Spot of Rice. *Acta Ecol. Sin.* **2018,** 1872–2032.

Jones-Rhoades, M. W.; Bartel, D. P. Computational Identification of Plant microRNAs and Their Targets, Including a Stress-induced miRNA. *Mol. Cell* **2004,** *14,* 787–799.

Kamal, M. M.; Mia, M. A. T. Diversity and Pathogens of the Rice Brown Spot Pathogens Bipolaris oryzae (Breda De Haan) Shoem. In Bangladesh Assessed by Genetic Fingerprint Analysis. *Bang. J. Bot.* **2009,** *38* (2), 119–125.

Kamal, M. M.; Mia, M. A. T. Diversity and Pathogenicity of the Rice Brown Spot Pathogen, Bipolaris oryzae (Breda de Haan) Shoem. *Bang J. Bot.* **2009,** *38* (2), 119–125.

Katara, J. L.; Sonah, H.; Deshmukh, R. K.; Chaurasia, R.; Kotasthane, A. S. Molecular analysis of QTLs associated with resistance to brown spot in rice (Oryza sativa L.). *Indian J. Genet 2010,* **70,** 17–21.

Khraiwesh, B.; Zhu, J. K.; Zhu, J. Role of miRNAs and siRNAs in Biotic and Abiotic Stress Responses of Plants. Biochim. *Biophys. Acta* **2012,** 1819, 137–148.

Kumar, A.; Solanki, I. S.; Akhtar, J.; Gupta, V. Morpho-Molecular Diversity of B. oryzae Causing Brown Spot of Paddy. *Indian J. Agric. Sci.* **2016,** *86* (5), 615–620.

Kumar, D.; Yadav, A.; Ahmad, R,; Dwivedi, U. N.; Yadav, K. CRISPR-Based Genome Editing for Nutrient Enrichment in Crops: A Promising Approach Toward Global Food Security. *Front. Genet.* **2022,** *13,* 932859.

Leung, H.; Raghavan, C.; Zhou, B.; Oliva, R.; Choi, I. R.; Lacorte, V.; Jubay, M. L.; Cruz, C. V.; Gregorio, G.; Singh, R. K.; Ulat, V. J.; Borja, F. N.; Mauleon, R.; Alexandrov, N. N.; McNally, K. L.; Sackville Hamilton R. Allele mining and enhanced genetic recombination for rice breeding. *Rice* **2015,** *8,* 34.

Li, Z. K.; Zheng, T. Q. Utilization of Exotic Germplasm In: Zhang QF, Wing R, editors. Genetics and Genomics of Rice: Springer Science + Business Media, Inc.; **2013,** pp. 349–361.

Liu, B.; Li, P.; Li, X.; Liu, C.; Cao, S.; Chu, C.; Cao, X. Loss of Function of OsDCL1 Affects microRNA Accumulation and Causes Developmental Defects in Rice. *Plant Physiol.* **2005,** *139,* 296–305.

Liu, Q.; Yang, F.; Zhang, J.; Liu, H.; Rahman, S.; Islam, S.; Ma, W.; She, M. Application of CRISPR/Cas9 in Crop Quality Improvement. *Int J. Mol Sci.* **2021,** *22* (8), 4206.

Liu, W.; Wang, G.-L. Plant Innate Immunity in Rice: A Defense Against Pathogen Infection. *Natl. Sci. Rev.* **2016,** *3,* 295–308.

Lu, Y.; Ye, X.; Guo, R.; Huang, J.; Wang, W.; Tang, J.; Tan, L.; Zhu, J. K.; Chu, C.; Qian, Y. Genome wide Targeted Mutagenesis in Rice Using the CRISPR/Cas9 System. *Mol Plant* **2017,** *10* (9), 1242–1245.

Lu, Y.; Zhang, S.; Shah, T.; Xie, C.; Hao, Z.; Li, X.; Farkhari, M.; Ribaut, J. M.; Cao, M.; Rong, T.; Xu, Y. Joint Linkage-Linkage Disequilibrium Mapping is a Powerful Approach to

Detecting Quantitative Trait Loci Underlying Drought Tolerance in Maize. *Proc Natl Acad Sci. USA* **2010,** *107,* 19585–19590.
Matsumoto, K.; Ota, Y.; Seta, S.; Nakayama, Y.; Ohno, T.; Mizobuchi, R.; Sato, H. Identification of QTLs for Rice Brown Spot Resistance in Backcross Inbred Lines Derived From a Cross Between Koshihikari and CH45. *Breed Sci.* **2017,** *67* (5), 540–543.
Matsumoto, K.; Ota, Y.; Yamakawa, T.; Ohno, T.; Seta, S.; Honda, Y.; Mizobuchi. R.; Sato, H. Breeding and Characterization of the World's First Practical Rice Variety with Resistance to Brown Spot (Bipolaris oryzae) Bred Using Marker-Assisted Selection. *Breed Sci.* **2021,** *71* (4), 474–483.
McCouch, S. R.; Teytelman, L.; Xu, Y.; Lobos, K. B.; Clare, K.; Walton, M.; Fu, B.; Maghirang, R.; Li, Z.; Xing, Y. Zhang, Q.; Kono, I.; Yano, M.; Fjellstrom, R.; DeClerck, G.; Schneider, D.; Cartinhour, S.; Ware, D.; Stein, L. Development and Mapping of 2240 New SSR Markers for Rice (Oryza sativa L.). *DNA Res.* **2002,** 9, 199–207.
Meng, X.; Yu, H.; Zhang, Y.; Zhuang, F.; Song, X.; Gao, S.; Gao, C.; Li, J. Construction of a Genome-Wide Mutant Library in Rice Using CRISPR/Cas9. *Mol Plant* **2017,** *10* (9), 1238–1241.
Mishra, R.; Joshi, R. K.; Zhao, K. Genome Editing in Rice: Recent Advances, Challenges, and Future Implications. *Front Plant Sci.* **2018,** *9,* 1361.
Mizobuchi, R.; Fukocoka, S.; Tsushima, S.; Yano, M.; Sato, H. QTLS for Resistance to Major Rice Disease Exacerbated by Global Warming: Brown Spot, Bacterial Seedling Rot and Bacterial Grain Rot. *Rice (NY)* **2016,** *9,* 23.
Mondal, L.; Verma, S. K.; Kotasthane, A. S.; Verulkar, S. B. QTL Mapping for Brown Spot Tolerance Using RIL Population of Rice. *EJBB* **2017,** 2321–9122.
Monisha, S.; Praveen, N. M.; Ramanathan, A. Isolation, Characterization and Management of Brown Spot Disease of Rice. *JPP* **2019,** *8* (3), 4539–4545.
Monisha, S.; Praveen, N. M.; Ramanathan, A. Isolation, Characterization and Management of Brown Spot Disease of Rice. *J. Pharmacogn Phytochem* **2019,** *8* (3), 4539–4545.
Narsai, R.; Wang, C.; Chen, J.; Wu, J.; Shou, H.; Whelan, J. Antagonistic, Overlapping and Distinct Responses to Biotic Stress in Rice (Oryza sativa) and Interactions With Abiotic Stress. *BMC Genom.* **2013,** *14,* 93.
Nelson, R.; Wiesner-Hanks, T.; Wisser, R.; Balint-Kurti, P. Navigating Complexity to Breed Disease-Resistant Crops. *Nat Rev Genet* **2018,** *19,* 21.
Ota, Y.; Matsumoto, K.; Nakayama, Y.; Yamakawa, T.; Ohno, T.; Mizobuchi, R.; Sato, H. QTL Analysis for Brown Spot Resistance in American Rice Cultivar 'Dawn'. *Breed Sci.* **2021,** *71* (4), 491–495.
Prabhukarthekeyan, S. R.; Yadav, M. K.; Anandan, A.; Arvinda, S.; Keerthana, U.; Raghu, S.; Baite, M. S.; Parmeswaran, C.; Panneerselvam, P.; Rath, P. C. Bio-Protection of Brown Spot Disease of Rice and Insight into the Molecular Basis of Interaction Between Oryza sativa, B. oryzae and B. Amyloliquefaciens. *Biol Control* **2019,** *137,* 104018.
Rani, R.; Yadav, P.; Barbadikar, K. M.; Baliyan, N.; Malhotra, E. V.; Singh, B. K.; Kumar, A.; Singh, D. CRISPR/Cas9: A Promising Way to Exploit Genetic Variation in Plants. *Biotechnol Lett* **2016,** *11* (1), 1–8.
Reddy, C. S.; Laha, G. S.; Prasad, M. S.; Krishnaveni, D.; Castilla, N. P.; Nelson, A.; Savary S. Characterizing Multiple Linkages Between Individual Diseases, Crop Health Syndromes, Germplasm Deployment and Rice Production Situations in India. *Field Crops Res* **2010,** *120,* 241–253.

Sadhukhan, R.; Bohra, J. S. Management of Brown Spot Disease in Rice (Helmintho oryzae) by Spraying of Cow Urine. *IJCS* **2018,** *6* (1), 1721–1723.
Saghai-Maroof, M. A.; Soliman, K. M.; Jorgensen, R. A.; Allard, R. W. Ribosomal DNA Spacer-Length Polym

WEB RESOURCES

"https://bsppjournals.onlinelibrary.wiley.com/authored-by/ContribAuthorRaw/Ahmadpour/A." Ahmadpour, A.;
"https://bsppjournals.onlinelibrary.wiley.com/authored-by/ContribAuthorRaw/Castell%E2%80%90Miller/C." Castell-Miller, C.;
"https://bsppjournals.onlinelibrary.wiley.com/authored-by/ContribAuthorRaw/Javan%E2%80%90Nikkhah/M." Javan-Nikkhah, M.;
"https://bsppjournals.onlinelibrary.wiley.com/authored-by/ContribAuthorRaw/Naghavi/M.+R." Naghavi, M. R.;
"https://bsppjournals.onlinelibrary.wiley.com/authored-by/ContribAuthorRaw/Dehkaei/F.+P." Dehkaei, F. P.;
"https://bsppjournals.onlinelibrary.wiley.com/authored-by/ContribAuthorRaw/Leng/Y." Leng, Y.; "https://bsppjournals.onlinelibrary.wiley.com/authored-by/ContribAuthorRaw/Puri/K.+D." Puri, K. D.;
"https://bsppjournals.onlinelibrary.wiley.com/authored-by/ContribAuthorRaw/Zhong/S." Zhong S. Population Structure, Genetic Diversity, and Sexual State of the Rice Brown Spot Pathogen Bipolaris oryzae From Three Asian Countries. Plant Pathol 2018
"https://bsppjournals.onlinelibrary.wiley.com/toc/13653059/2018/67/1"67 (1), 181–192.

CHAPTER 13

Ecological Status and Biology of Rice Kernel Smut Pathogen

MAHESH KUMAR[1], MD. SHAMIM[1], ABHISHEK KUMAR[1], SANJEEV KUMAR[2], and SANTOSH KUMAR[3]

[1]*Department of Molecular Biology and Genetic Engineering, Kalam Agricultural College (Bihar Agricultural University, Sabour), Kishanganj, Bihar, India*

[2]*Department of Plant Pathology, Bihar Agricultural University, Sabour, Bhagalpur, Bihar, India*

[3]*Department of Plant Pathology, Mandan Bharti Agricultural College, Agwanpur, Saharsa (Bihar Agricultural University, Sabour) Bhagalpur, Bihar, India*

ABSTRACT

Rice is considered one of the major staple cereal crops for the world population in terms of food as well as nutritional calories. Biotic stress is playing a major role in rice production and productivity. Around 37% of crops worldwide are lost due to pests and diseases, of which the damage caused by disease alone accounts for 24%. Some of the rice diseases that are responsible for the economic losses in Asia are bacterial leaf streak, blast (leaf, neck, nodal, and collar), false smut, brown spot, eyespot, crown sheath rot, downy mildew, leaf scald, narrow brown leaf spot, pecky rice (kernel spotting), root rot, and kernel smut. In this chapter, we have briefly elaborated and discussed various aspects, such as the historical aspect of the Genus

Fungal Diseases of Rice and Their Management. Deepti Srivastava, Md. Shamim, Malik Mobeen Ahmad, and R. S. Upadhyay (Eds.)

Tilletia and systematic. In addition to that, we have elaborated biology of the *Tilletia horrida* and its management.

13.1 INTRODUCTION

Rice (*Oryza sativa* L.) is grown and consumed by more than 2500 million people in about 112 countries. Asia produces and consumes around 90% of the world's rice. Rice kernel smut disease is caused by a smut fungus that belongs to the phylum Basidiomycota's class Ustilaginomycetes. The term comes from the Latin word "ustulatus," which means "burned," referring to the blackened look of diseased plants. Bunt fungi are cereal-infecting *Tilletia* species that generate teliospores in their hosts' ovaries and are considered to be a derivative of the phrase burnt Kernel smut, also known as black smut or bunt of rice. It was first reported from Japan by Takahashi (Takahashi, 1896) followed by Anderson (Anderson, 1899) from South Carolina, USA. Later on, its occurrence was reported in rice-growing countries such as China, Burma, Formosa, British Guyana, India, Cuba, Fiji, Australia, Greece, Indo-china, Indonesia, Japan, Java, Malaysia, Mexico, Nepal, Pakistan, Philip-pines, Siam, Sierra Leone, Taiwan, Thailand, Togo, Trinidad, The United States, Venezuela, Vietnam (Miura et al., 1982; Padwick, 1950; Padwick et al., 1944; Singh, 1990). Surinam (IS), Brazil (Ribeiro et al., 1973), and Malaya (Thompson and Johnston, 1953). In India, the disease is known to occur in Haryana, Punjab, Uttar Pradesh, Gujarat, Madhya Pradesh (Butler, 1913; Butler, 1918), Andhra Pradesh (Reddy and Reddy, 1992), Orissa, Tamil Nadu, West Bengal, Assam (Rathaiah, 1988).

13.2 HISTORICAL BACKGROUND OF THE GENUS *TILLETIA*

Tilletia and *Neovossia* K orn are home to the majority of bunt fungi. *Tilletia caries*, the type species, is a systemic wheat pathogen with light yellow to reddish–brown reticulate teliospores mixed with hyaline thick-walled sterile cells (Duran and Fischer, 1961). Nielsen (1963) reported that trimethylamine compound production during field infection leads to the release of a rotten smell. The genus *Neovossia* was named after Vossia molinae, a species having pitted, elongated teliospores. The distinction between *Neovossia* and *Tilletia* has been disputed. Tulli and Johnson (1952) studied the *T. horrid* and *Neovossia barclayana* and amended the *Neovossia* description to include species that produce hyaline teliospores, fragmentary appendages that may or may not persist after maturity, large numbers of nonconjugated primary basidiospores, and two types of sporidia. They also stated incorrectly that *Tilletia* species only generate one sort of sporidium. Other groups of scientists have utilized this concept to justify the placement of *Tilletia horrida* and *Tilletia indica* in *Neovossia* (Singh and Pavgi, 1972; Whitney, 1989). The production of two kinds of sporidia is characteristic of *Tilletia* species, according to Goates and Hoffmann (1979). Differences in the quantity and fusion of primary basidiospores have also been used to support *T. horrida* and *T. indica*'s placement in *Neovossia* (Joshi et al., 1983; Singh and Pavgi, 1972; Singh et al., 1979; Whitney, 1989). Vanky (2002) defined *Neovossia* as possessing lengthy hyaline appendages on the teliospores, as well as a trimethylamine odor, and causing localized infection. Castlebury performed molecular studies on large subunit nuclear rDNA sequences, and the results show that there is no evolutionary justification for keeping *Neovossia* apart from *Tilletia* (Castlebury et al., 2005). Although they support four primary lineages within the genus *Tilletia*, none of them solely conforms to taxa with the morphological features employed by Vanky (Vanky, 2002) to designate *Neovossia.* Furthermore, locally infecting taxa with significant numbers of nonconjugating basidiospores do not form a single phylogenetic unit, suggesting that these traits are not useful at the generic level.

13.3 RICE KERNEL SMUT – *T. horrida*

Rice kernel smut is caused by the *T. horrida* plant disease and is also known as caryopsis smut, black smut, grain smut, or bunt of rice (*Oryza saliva* L.). *T. horrida*, like *T. indica*, creates a partial bunt that reduces yield and productivity. Infected grain has extremely low quality. The synonyms of the *T. horrida* are *Tilletia ajrekarii, Tilletia pulcherrima, N. barclayana, Neovossia horrida, and Neovossia pulcherrima.* It was first reported in Japan by Takahashi (Takahashi, 1896) followed by Anderson (Anderson, 1899.) from South Carolina, USA. Later on, its occurrence was reported in rice-growing countries around the world such as China, Burma, Formosa, British Guyana, India, Cuba, Fiji, Australia, Greece, Indochina, Indonesia, Japan, Java, Malaysia, Mexico, Nepal, Pakistan, Philippines, Siam, Sierra Leone, Taiwan, Thailand, Togo, Trinidad, the USA, Venezuela, Vietnam (Miura et al., 1982; Padwick, 1950; Padwick et al., 1944; Singh, 1990). Surinam (IS), Brazil (Ribeiro et al., 1973), and Malaya (Thompson and Johnston, 1953).

Rice kernel smut may now be found in Asia, Australasia, and Oceania, as well as in Europe (Greece), Central America (Belize, Cuba, Nicaragua, Panama, and Trinidad), South America (Brazil, Guyana, Surinam, and Venezuela), North America (Mexico, United States), and Africa (Sierra Leone) (CMI, 1991; Webster, 1992). Rice kernel smut is a chronic but small disease that affects rice growers all over the world (Biswas, 2003; Ou, 1985). Long-grain rice varieties, which are the most common in the United States, are the most vulnerable to *T. horrida*, while short-grain rice is the least vulnerable, and medium-grain rice is in the middle (Biswas, 2003). The infection is ubiquitous, resulting in significant qualitative and quantitative losses. Excessive nitrogen fertilizer has a big impact on the severity in grower fields, especially on newer cultivars like Cypress and LaGrue (Cartwright et al., 1996).

As only a few grains in each ear are attacked, and frequently only partially, the damage produced by *Tilletia barclayana* was not deemed significant in early investigations on the condition, according to Padwick (1950). In Myanmar, Su (1933) recorded a loss of 2–5% of the grain in a year. According to Reyes (1933), the damaged plants' growth is stunted, which may exacerbate the loss. The presence of spores caused considerable discoloration of rice flour, according to Fulton (1908). These spores were found in a sample of rice that included 3–4% of the afflicted grains (Fulton, 1908).

13.4 SYMPTOMS

Black powered mass of the spore is responsible for the development of the symptom. After the infection of the spore; only the diagnostic symptoms are visible. Only a few grains in an ear and a few ears in a stool are infected. The degree of transformation of the kernel into bunt sorus possibly depends on the stage of infection as the development of a fertilized ovary and the fungus occurs simultaneously. Generally, whole grain is not transformed into a sorus and the sori remain hidden by the glumes. However, in severe infections, whole grain is destroyed and a beak/spur-like outgrowth is produced on the rupturing glumes. The black powdery mass may be seen adhering to awns and adjacent leaves due to the release of spores from rupturing sori. Awns of partially infected spikelets remain greenish-grey for a longer period than on healthy grains (Singh and Pavgi, 1973). In mild infections, seeds may germinate but the seedlings are stunted and have a lesser number of tillers and poor yield.

Rice with more than 3% bunted kernels is graded "smutty" (USDA, 2005) and punished in the mill, according to the United States Standards for Rice, Washington, DC. Because a maximum of 0.5% infection is allowed in certified seed in India, even a low frequency of infection can result in significant economic losses in hybrid rice seed production (Biswas, 2003). Although the pathogen is widespread the disease has not been considered of much economic importance (Biswas, 2003). In recent years, its incidence has increased with the popularization of high-yielding varieties, hybrid rice cultivation, intensive cropping patterns, and the use of more nitrogenous fertilizers.

13.5 DISEASE CYCLE

The disease was first thought to be systemic, as in the case of other diseases caused by species of *Tilletia.* The seedling infection has been reported by several workers (Agarwal et al., 1989; Anderson, 1902; Tullis and Jhonson, 1952). Primary infection originates from teleutospores surviving in the soil or on the seeds (Xuan-Keng, 1995). The teleutospores get mixed into the soil at the time of harvesting, threshing and winnowing and survive for more than 2 years. They germinate by a germ tube bearing abundant sporidia which detach forcibly, become air-borne and cause local infection in opening flowers (Chowdhury, 1946; Chowdhury, 1951; Singh, 1975; Templeton, 1961; Templeton et al., 1960; Templeton et al., 1961; Whitney et al., 1975). Xuan-Keng (Xuan-Keng, 1995) reported that teliospores survive for more than 3 years under indoor conditions and for 15 months under outdoor conditions in Changsha, China, Singh, and Pavgi (Singh and Pavgi, 1973b) traced the development of sorus of the fungus and reported that violently discharged air-borne sporidia lodge on the feathery stigma of a floret and the infection hyphae penetrate down through the style and nucellus reaching the chalazal end of the ovary and the funicle. The mycelial hyphae remain between the aleurone layer and the seed coat. Initially, the sorus develops as a small cavity at the base of the ovary. Later due to the ramification of mycelium and digestion of endosperm, the sorus develops from three sides of the endosperm which shrinks in towards the embryo. Finally, digestion of the entire endosperm occurs leaving the remnant portion of the aleurone layer near the embryo. In 1918, Butler explained that spore mass replaces the starch to form a layer in the endosperm. However, Singh and Pavgi observed that sorus never forms a layer in the endosperm but digests the latter by absorbing nutrients through the aleurone layer (Singh and Pavgi, 1973a).

Initial infection by *T. indica* begins when sporidia deposited on spikes germinate to create hyphae that enter stomata in the case of the Karnal bunt disease cycle begins when teliospores at or near the soil surface germinate, creating basidiospores that develop Hyphae next extend intercellularly to the floret's base, where they penetrate the periderm of embryonic kernels through the funiculus (Goates, 1988). The rachis's stomata may also be critical for early infection, according to scanning electron microscope research (Dhaliwal, 1989). However, data on hyphae mobility inside spikes suggests that the rachis is not a common location for initial infection (Rattan and Aujla, 1991). Basidiospores or hyphae of *T. indica* routinely merge to create a dikaryotic mycelium before infection of plants, according to Nagarajan (Nagarajan et al., 1987), and basidiospore fusion during culture has also been documented. During infection research in *T. indica*, evident hyphal anastomosis was only seen a few times before the plant was penetrated, and it was not regarded as a regular method of dikaryon production (Goates, 1988).

Although Singh and Pavgi suggested that sporidia lodge on the feathery stigma and enter through the style to the chalazal end of the ovary, the manner of penetration and time of dikaryon production in *T. horrida* remain unknown (Singh and Pavgi, 1973a). Hyphae, allantoid sporidia, and sporidia of succeeding generations of *T. horrida* teliospores float to the surface of irrigation water and germinate after the rice is sown in rice fields (Whitney et al., 1975). In *T. horrida*, sporidia and hyphae establish themselves in the crop canopy (Bains and Dhaliwal, 1990) or create a thick mat on the surface of the rice field water (Templeton et al., 1960; Webster, 2005). Some form of "safety mechanism," such as a feature that causes teliospores to germinate while the plant is heading, has been proposed to ensure the existence of adequate basidiospores at the time of heading (Whitney and Frederiksen, 1975).

The illness manifests itself in rice kernel smut by the production of teliospores in the pericarp's inner layers, which split and expand as endosperm tissue decreases (Singh and Pavgi, 1973a). Although the embryo is destroyed in severe infections, the embryo is not colonized and contaminated seed can often germinate (Singh and Pavgi, 1973b). Teliospores are generated by sporogenous cells forming a thin hymenial layer on the surfaces of cavities created by the separation of the inner and outer layers of the pericarp, as well as the separation of the inner pericarp from the seed coat (Roberson and

Luttrell, 1987). *T. indica* teliospores. The dikaryotic cytoplasm is demarcated by a septum at the terminal end of sporogenous hyphae, and karyogamy occurs during the expansion of teliospore starting cells (Roberson and Luttrell, 1987). The teliosporogenesis process in Temperature affects *T. indica*. Even after effective infection induction, maximum daily temperatures of 35 to 40 degrees Celsius following the grain-filling stage dramatically limit teliosporogenesis (Kumar et al., 2003).

13.6 BIOLOGY OF RICE KERNEL SMUT

13.6.1 MORPHOLOGICAL CHARACTERS

Morphology of the teleutospores is globose or sub-globose, pulverulent, light brown or vandyke brown to black, measuring between 15 and 32 μm diameter or 22.5–28.7 μm in size with or without an appendage/apiculus. Along with these spores, sterile cells are also found in the powdery mass which is more or less globose, hyaline to yellowish tinted, 2–4 μm thick, and 10–30 μm diameter (Ou, 1985; Singh, 1990). The teliospores have three distinct wall layers, the perisporium/sheath, the exosporium, and the endosporium. The perisporium is thin, delicate, and hyaline. The exosporium/episporium is deep live, thickly covered with regularly arranged conspicuous projections which are hyaline or slightly colored and curved/pointed at the apex, irregularly polygonal at the base, measuring 2–5 μm in height and 1.5–2.0 μm apart at their free ends (Singh, 1990).

Nawaz and Hess (1987) were observed rod-let-like projections on the surface of teleutospores which are individually distinct and not grouped into fascicles, measuring 14.3–22.5 nm width and 80–110 nm length. They also discovered that hibernating teleutospores have strong spore walls that are divided into four separate layers. The partition layer is made up of wavy lamellar surfaces. As in another smut fungus, the plasma membrane had invaginations and many particles. The promycelium is mostly simple, nonseptate to 1–3 septate, 7 μm diameter, and 35–525 μm length. The initial sporidia are 35–60 × 1.3–2 μm long, cylindrical to filiform or needle-shaped with a flat base, and a pointy tip. The secondary sporidia are hyaline and curved, measuring around 7.5–14 × 1.2–20 μm (Singh and Pavgi, 1972). Matsumoto et al. (1985) discovered significant diversity in teliospores of several *N. horrida* isolates in California. These physical distinctions and ranges in variation were utilized to provide a particular criterion for the identification of this species that quarantine workers may employ.

13.6.2 DORMANCY OF TELIOSPORE

T. indica has been shown to have three forms of teliospore dormancy. The first kind of teliospore hibernation is postharvest dormancy. Teliospores germinated relatively poorly under this kind of dormancy when obtained from newly harvested grain compared to bunted grains held for many months to a year (Bansal et al., 1983; Mitra, 1935; Rattan and Aujla, 1990; Smilanick et al., 1985; Thinggaard and Leth, 2003). *T. horrida* has this sort of teliospore dormancy as well (Chahal et al., 1993). Germination of teliospores ranges between 15% and 30% in the second kind of dormancy and seldom surpasses 50% even under perfect conditions when teliospores are held for more than a year. This illustrates the second kind of dormancy, which is long-lasting and most likely helps to teliospore survival in the field. The chilly temperature induces the third form of dormancy.

When Zhang et al. (1984) experimented on *T. indica*, they found that dry teliospores stored at −18°C progressively lost their capacity to germinate during 12 weeks of treatment (Zhang et al., 1984). In these tests, the teliospores were plated immediately after treatment, and nongerminability was assumed to imply fatality. In later tests, however, the suppression of germination was removed if teliospores were plated after a 20-day "thawing" interval at 22°C rather than immediately following the cold treatment (Chahal and Mathur, 1992). Cold-induced dormancy has also been demonstrated in trials conducted at 0°C for 4 to 8 days. *T. horrida* exhibits similar dormancy (Chahal and Mathur, 1992). *T. controversa* Kuhn, the dwarf bunt pathogen, also undergoes cold-induced dormancy in natural field situations in temperate latitudes every year (Hoffmann and Goates, 1981).

13.6.3 TELIOSPORE GERMINATION

Most locally infecting bunt fungi observed contain echinulate, verrucose, or tuberculate teliospores that germinate at 20–25°C. The impact of physical variables on teliospore germination in *T. indica* and *T. horrida* has been widely explored (Chahal et al., 1992; Dupler et al., 1987; Lin, 1936; Smilanick et al., 1985; Zhang et al., 1984). Germinating teliospores are incredibly robust and tenacious during the process, even when subjected to extreme fluctuations in pH, temperature, and soil moisture, including freezing and desiccation (Smilanick et al., 1985; Whitney and Frederiksen, 1975). The basidium (also known as a "promycelium") bursts through the ruptured wall of the teliospore and either quickly generates basidiospores or elongates to over 500 m length, depending on the species and circumstances. Retraction septa are formed by elongating basidia, restricting the cytoplasm to the apical area. There occurs the formation of a terminal whorl of 10–150 primary basidiospores, also known as primary sporidia. The amount of basidiospores produced per basidium varies greatly between taxa, and not all locally infecting species generate a substantial number of basidiospores (Whitney and Frederiksen, 1975).

Within a species, the number of basidiospores can also vary greatly (Teng, 1931). Basidiospores are hyaline, aseptate, filiform to fusiform in shape, and range in size from 1.0–2.5 m in width and 15–80 min length (Peterson and Bonde, 1998). Basidiospores may stretch somewhat after separation and become curved or undulate before germinating to generate hyphae or a sterigma-like structure on which an asymmetrically formed haploid, uninucleate, allantoid sporidium is generated (Singh and Pavgi, 1972). *T. indica*, *T. horrida*, and other locally infecting *Tilletia* species investigated produced allantoids sporidia with the same morphology and forceful discharge as *T. caries* (Goates and Hoffmann, 1986). Allantoid sporidia can germinate directly through germ tubes or repeatedly to create more sporidia (Ingold, 1996.). Filiform sporidia are generated in culture from short, lateral sporogenous cells on the hyphae and are passively distributed (Ingold, 1996.). *T. indica and T. horrida* are the principal infective agents of allantoid sporidia, but they have not been examined in other locally infecting bunts. The role of the filiform, passively disseminated sporidia in the infection process is unknown.

There are several factors responsible for the germination of teliospores of *T. horrida.* The fresh teleutospores of kernel smut do not germinate readily and require a period of dormancy (Lin, 1936; Xuan-Keng, 1995). Singh (1975) reported higher germination in 12–15 months old spores as compared to fresh teliospores stored at room temperature. Takahashi (1896), Teng (1931), and Wei (1934) reported that light and oxygen are essential for chlamydospore germination. Short-wavelength light has been found strikingly superior for spore germination than long-wavelength light. In blue light, 10–14% germination has been recorded as against no germination in total darkness. Exposure to UV light for 10–30 min at an unspecified intensity increased the germination from a trace to 10% and also induced germination in fresh teliospores. Chlamydospore germination was also promoted by exposing them to a fluorescent lamp of 50–60 candle power for 4–6 h of direct sunlight· for 2 h (Lin, 1936; Lin, 1955; Ribeiro et al., 1973).

Kumar and Singh (1987b) observed that teliospore germination is significantly increased by soaking them in water for 3 days at 30°C and pH 5. So, the maximum spore germination was recorded in 73 weeks old teleutospores at 29°C under a 12 h photo-period with artificial daylight (cool fluorescent light). There was a marked increase in germination rate when teliospores in bunted grains were pretreated for 30 min at 60°C. The 17 weeks old teliospores and those buried in dry or moist sand at 2 mm depth did not germinate (Chahal et al., 1993). A prolonged thawing period, hydrochloric acid, sodium hydroxide, potassium hydroxide, potassium dichromate, potassium permanganate, mannitol, glycine, cysteine, lysine, and alanine resulted in enhanced spore germination (Kumar and Singh, 1987a).

13.6.4 CYTOLOGY

Singh and Pavgi (1972) reported that a mature spore contains a single diploid nucleus that enlarges and undergoes successive divisions in the spore itself. The exosporium cracks open giving rise to a short, thick, hyaline, and cylindrical, 1–3 septate-tipped promycelium. The four nuclei in the spore migrate into the promycelium. Further mitotic divisions occur during the process of elongation of promycelium resulting in eight nuclei. With the full development of promycelium, subsequent mitotic divisions increase the number of nuclei to 40 or more which move towards the tip of the promycelium where small projections have already appeared.

The projections ultimately elongate to form sporidia. One nucleus with a small amount of cytoplasm moves into each sporidium. There are 32–76 sporidia in a whorl at the tip of promycelium which do not fuse in pairs or form an H-shaped structure. The single nucleus undergoes a mitotic division making the primary sporidium binucleate which passes into secondary sporidia. These nuclei undergo a conjugate division to form more nuclei for subsequent secondary sporidia. On germination, these sporidia produce infection hyphae which are dikaryotic (Cralley, 1934; Singh, 1975).

13.7 RICE KERNEL SMUT MANAGEMENT

Kernel smut of rice is difficult to manage due to the mode of the perpetuation of the pathogen and the nonavailability of resistant cultivars. However, some strategies to combat the disease have been suggested by several researchers as given below.

13.7.1 HOST RESISTANCE

Several resistant sources have been identified in different countries which may be utilized in resistance breeding programs. Susceptibility to *T. barclayana* varies by rice variety. Short-grain cultivars are the most resistant, while long-grain cultivars are the most sensitive, and medium-grain cultivars are in the middle of the resistance spectrum. The duration of the anthesis and the size of the floret opening seems to be connected to susceptibility in part. Long-grain florets have the largest floret opening.

In the Philippines, Reyes (1933) discovered that numerous rice cultivars, including Palagad, Sipot, and Apostol, were vulnerable to *T. barclayana*, whereas Elon-elon appeared to be immune. Early-maturing cultivars were more vulnerable than late-maturing cultivars in India, according to Kameswar Row (1962) and Chauhan and Verma (1964). In the United States, the rice cultivars Nova and others were sensitive, but Arkose, Zenith, and Saturn were resistant. In the southern United States, LaGrue, Priscilla, Cypress, Cocodrie, Newbonnet, and Jefferson were recently discovered to be very sensitive, whereas Wells, Lemont, Lacassine, and Madison were resistant (Cartwright et al., 1999).

When 16 rice cultivars were tested for *T. barclayana* resistance using chlamydospores and the partial vacuum technique at two sites in Pakistan, two were resistant (less than 10% infection) at both locations and two exhibited intermediate resistance (more than 10% infection) (Akhtar et al., 1987; Srinivasan, 1982). During an unusually strong outbreak of *T. barclayana* in 1976, the response of 10 early-maturing cultivars and nine later-maturing cultivars to spontaneous infection by the disease were studied. Cultivars with a shorter growth season were more vulnerable in general than cultivars with a longer growing season (Muthusamy and Amed, 1977).

During 1991, Fusarium sheath rot caused by *Fusarium moniliforme* [*Gibberella fujikuroi*] and *T. barclayana* was found on CMS lines and their respective maintainers. A total of 100 tillers were randomly picked from five evenly spaced locations to measure disease incidence. In comparison to their maintainers, restorers, and inbred rice cultivars, the illnesses were often more severe on the CMS lines and hybrids. *T. barclayana* had a 9.5–15% incidence and *G. fujikuroi* had a 33–42% incidence with CMS lines, compared to 0.3–1.2 and 15–19%, respectively, for the other lines. The CMS genotypes' longer panicle sheaths give more favorable circumstances for disease development.

T. barclayana had a 9.5–15% incidence and *G. fujikuroi* had a 33–42% incidence with CMS lines, compared to 0.3–1.2 and 15–19%, respectively, for the other lines. The CMS genotypes' larger panicle sheaths give more favorable circumstances for disease development, and this degree of vulnerability may be a limit on hybrid seed production in the future (Sharma et al., 1993). Differences in the number of grains per plant and the percentage of panicles damaged by *T. barclayana* were significant for varieties, treatments, and their interaction when 17 rice varieties were cultivated under circumstances of high and low nitrogen, phosphorus, and potassium (NPK) application. The cultivars 9435, IR506, RP5B, IR8, and Vijaya were the least affected, suggesting that they might be useful in resistance breeding. According to Kumar et al. (1978), significant volumes of fertilizer should be utilized while testing for resistance to *T. barclayana*.

13.7.2 GENETICS OF RESISTANCE

Templeton (1968), Templeton (1971), Templeton (1962), Templeton (1965), Temlpeton and Johnston (1970) reported that common parent to bunt resistant varieties viz; Arkose, Zenith, Saturn, Cl 9647, Cl 9657, Cl 9662, and Cl 9694 was a blue rose, indicating that the variety is carrying one or more gene for resistance to kernel smut pathogen. For the breeding of resistant cultivars and effective screening purposes, the heterothallic nature of the fungus should be studied

13.7.3 CHEMICAL, CULTURAL CONTROL, BIOLOGICAL CONTROL

Seed selection with 10% salt water or muddy water resulted in a 93.2 and 80.4% reduction in kernel smut incidence (Hu and Xue, 1987). Walker (1899) recommended the rejection of Light grains floating on the surface of water and soaking of heavier grains settling at the bottom in 0.2% potassium sulfide solution for 2 h. Quarantine measures should be performed strictly to avoid the introduction of the disease into new areas (Xuan-Keng, 1995; Thrimurty et al., 1982).

The disease can be managed by hot water treatment and seed dressing chemicals like copper carbonate, formalin, organomercurials, triadimefon, and benomyl (Reyes, 1950; Xuan-Keng, 1995). Contrary to this, Templeton et al. (1960) and Whitney (1977) observed that neither seed treatment nor the foliar application of benomyl, maneb, chlorothalonil, and Triphenyl tin hydroxide at the heading stage could control the disease. Singh (1975) recommended 3 years of crop rotation to minimize the disease as teleutospores do not survive in the soil for more than 27 months. However, Ospishchev (1980) advocated that besides crop rotation, effective prophylactic measures should be taken with benlate or thiophanate methyl during anthesis after careful weather forecasting diagnosis of the disease in the previous crop. In China, spraying the crop with Triadimefon, Topsin M, and carbendazim at initiation and full heading and flowering stages is recommended for effective control of kernel smut (Xuan-Keng, 1995). Application of carbendazim at early and maximum flowering period and Bayleton (Tridemorph) @ 105 g a.i.lha at early heading proved most effective and gave 70–80% disease control along with a significant increase in yield (Hu and Xue, 1987; Whitney and Frederiksen, 1971). *Bacillus pumilus* is a bioagent, used for the efficient control of the kernel smut of rice.

13.8 CONCLUSION

The disease of rice is a most important threat to the production of rice in many regions of rice-growing countries. Among the above fungal disease of the rice; kernel smut is coming under minor disease. *T. horrida* is a causal agent of the rice kernel smut. The genus *Tilletia* is a group of smut fungi that infects grasses either systemically or locally. There is scanty literature available on the web on the recent advancement of rice kernel smut.

KEYWORDS

- **fungal disease**
- **kernel smut**
- **yield loss**
- **rice**
- **biotic stress**

REFERENCES

Agarwal, P. C.; Mortensen, C. N.; Mathur, S. B. Seed-Borne Diseases and Seed Health Testing of Rice. In *Phytopathological – Papers*, 1989; vol 30, p 106.

Akhtar, M. A.; Sarwar, M. Incidence of Rice Kernel Smut in Pakistan. *Int. Rice Res. Newsl.* **1987,** *12* (4), 15.

Anderson, A. P. A New *Tilletia* Parasitic on *Oryza sativa* L. *Bot. Gaz.* **1899,** *27,* 467–472.

Anderson, A. P. *Tilletia horrida* Tak. on Rice in South Carolina. *Bull. Torrey Bot. Club* **1902,** *29,* 35–36.

Bansal, R.; Singh, D. V.; Joshi, L. M. Germination of Teliospore of Karnal Bunt of Wheat. *Seed Res.* **1983,** *11,* 258–261.

Butler, E. 1. Diseases of Rice. *Bull. Pusa Agric. Res. Inst.* **1913,** *34.*

Butler, E. J. *Fungi and Diseases in Plants.* Thacker, Spink & Co.: Calcutta, 1918; pp 226–228.

Cartwright, R. D.; Lee, F. N.; Parsons, C.E.; Ross, W.J.; Vann, S.R.; Overton, R. Monitoring of Rice Disease and On-Farm Evaluation of Rice Varieties in Arkansas. B.R. Research Series 468; Norman, R.J., Johnston, T. H., Eds.; Arkansas Agriculture Experimental Station: Fayetteville, 1999; pp 148–156.

Castlebury, L. A.; Carris, L. M.; V´anky, K. Phylogenetic Analysis of *Tilletia* and Allied Genera in Order Tilletiales (Ustilaginomycetes; Exobasidiomycetidae) Based on Large Subunit Nuclear rDNA Sequences. *Mycologia* **2005,** *97*, 888–900.

Chahal, S. S.; Mathur, S. B. Germination of Deep Frozen *T: Letia indica* and *Tilletia barclayana* Teliospores. *FAO PI. Prot. Bull.* **1992,** *40* (1–2), 31–35.

Chahal, S. S.; Aulakh, K. S.; Mathur, S. B. Germination of Teliospores of *T. barclayana,* the Causal Agent of Kernel Smut of Rice, in relation to Some Physical Factors. *J. Phytopathol.* **1993,** *137* (4), 301–308.

Chauhan, L. S.; Verma, S. C. Bunt resistant Paddy Varieties in Uttar Pradesh. *Sci.Cult.* **1964,** *30,* 20.

Chowdhury, S. Mode of Transmission of the Bunt of Rice. *Curr. Sci.* **1946,** *15,* 111.

Chowdhury, S. Studies in the Bunt Rice *(Oryza sativa* L.) *Indian Phytopathol.* **1951,** *4,* 25–27.

Cralley, E. M. Rice Disease Investigations. *Ark. Agric. Expt. Sta. (Ann. Rep.* 46) *Bull.* **1934,** *312,* 52.

Duran, R.; Fischer, G. W. *The Genus Tilletia.* Washington State University, 1961; 138, pp 64.

Goates, B. J.; Hoffmann, J. A. Somatic Nuclear Division in *Tilletia* Species Pathogenic on Wheat. *Phytopathology* **1979,** *69,* 592–598.

Goates, B. J.; Hoffmann, J. A. Formation and Discharge of Secondary Sporidia of the Bunt Fungus, *Tilletia foetida. Mycologia* **1986,** *78*, 371–379.

Hoffmann, J. A.; Goates, B. J. Spring Dormancy of *Tilletia controversa* Teliospores. *Phytopathology* **1981,** *71,* 881.

Hu, M.; Xue, T. P. Study on the Occurrence of the Kernel Smut of Rice Caused by *Neovossia horrida* and its Control. *Acta Phytopath. Sin.* **1987,** *14* (2), 87–92.

Joshi, L. M.; Singh, D. V.; Srivastava, K. D.; Wilcoxson, R. D. Karnal Bunt: A Minor Disease That is Now a Threat to Wheat. *Bot. Rev.* **1983,** *49*, 309–330.

Kameswar Row, V. S. R. Incidence of Bunt Disease of Rice. *Sci. Cult.* **1962,** *28,* 534–535.

Kumar, A.; Singh, R. A. Effect of Some Physical and Chemical Factors on Teliospore Germination of *Neovossia horrida. Indian Phytopathol.* **1987a,** *40,* 337–341.

Kumar, A.; Singh, R. A. Serological Differentiation Between *Neovossia indica* and *Neovossia horrida. Indian Phytopathol.* **1987b,** *40,* 114–115.

Kumar, I.; Kang, M. S.; Saini, S. S. Fertilizer Levels and Incidence of Bunt Disease in Rice in India. *Int. Rice Res. Newsl.* **1978,** *3* (4), 5.

Kumar, M.; Luthra, O. P.; Chawla, V.; Yadav, N. R.; Kumar, R.; Khar, A. Genetic Analysis of Karnal Bunt (*Neovossia indica*) Resistance in Wheat. *J. Biosci.* **2003,** *28,* 199–203.

Lin, C. K. Factors Affecting the Germination of Chlamydospores of *Tilletia horrida Tak. Bulletin, College of Agriculture and Forestry, University of Nanking*,1936; vol 45, p 11.

Lin, C. K. Studies on the Nature of Light Reaction in Chlamydospore Germination of Rice Kernel Smut. *Acta Phytopath. Sin.* **1955,** *1*, 183–190.

Matsumoto, T. T.; Whitney, N. G.; Bonde, M. R.; Royer, M. H. Morphological Variation of Teliospores of *N. horrida. Phytopathology* **1985,** *75,* 1365 (Abstr.).

Miura, Y.; Tsuji, H.; Ishigaki, M. On the Outbreak of the Kernel Smut *(T. horrida* Takahashi) of Rice Plant in Miyagi Prefecture. *Ann. Rep. Soc. Pl. Prot. N. Japan* **1988,** *39*, 94–97.

Mitra, M. Stinking Smut (Bunt) of Wheat With Special Reference to *Tilletia indica* Mitra. *Indian J. Agric. Sci.* **1935,** *1,* 51–74.

Muthuswamy, M.; Jaleel Ahmed, N. Varietal Susceptibility of Rice to Kernel Smut. *Madras Agric. J.* **1977,** *64,* 692.

Nagarajan, S.; Aujla, S. S.; Nanda, G. S.; Sharma, I.; Goel, L. B.; et al. Karnal Bunt (*Tilletia indica*) of Wheat—A Review. *Rev. Plant Pathol.* **1997,** *76,* 1207–1214.

Nawaz, M. S.; Hess, W. M. Ultrastructure of *N. horrida* Teliospores. *Mycologia* **1987,** *79* (2), 173–179.

Nielsen, J. Trimethylammonium Compounds in *Tilletia* Spp. *Can. J. Bot.* **1963,** *41*, 335–339.

Ospishchev, A. L. Treatment Only After Forecasting. *Zashchita Rastenii* No. **1980,** *4,* 20–21.

Ou, S. H. *Rice Diseases*; CMI: Kew, 1985; pp 311–315.

Padwick, G. W. *Manual of Rice Diseases;* CM1: Kew, 1950; p 198.

Padwick, G. W.; Khan, A. Notes on Indian Fungi. 11. *Mycological Papers*; Imp. Mycol. Inst.: Kew, 1944; vol 10, pp 2–4.

Rattan, G. S.; Aujla, S. S. Survival of Karnal Bunt (*Neovossia indica*) Teliospores in Different Types of Soil at Different Depths. *Indian J. Agric. Sci.* **1990,** *60,* 616–618.

Rattan, G. S.; Aujla, S. S. Distribution of Infection in Karnal Bunt Infected Wheat Spike. *Ann. Biol.* **1991,** *7,* 225–228.

Rathaiah, Y. Diseases of Deep Water Rice in Jorhat District, Assam, India. *Int. Rice Res. News* **1988,** *13* (4), 32.

Reddy, C. S.; Reddy, S. M. Smuts of Rice. Indian 1. *Mycol. Pl. Pathol.* **1992,** *22,* 125–138.

Reyes, G. M. The Black Smut or Bunt of Rice *(Oryza sativa* L.) in the Philippines. *Philipp. 1. Agric.* **1933,** *4,* 241.

Reyes, G. M. Seed Borne Diseases of Rice. *Plant Ind. Digest (Bur. Pl. Ind. Philipp.)* **1950,** *13* (6), 7–9.

Ribeiro, M. I. S. D.; Costa, W. F. D. A.; Drummond, O. A. Occurrence of Kernel Smut or Bunt of Rice *(Oryza saliva* L.) in the State of Rio de Janeiro. In *Federal Rural de Rio de Janeiro*; Arquivos da Universidade, 1973**;** vol *3* (2), pp 1–6.

Roberson, R. W.; Luttrell, E. S. Ultrastructure of Teliospore Ontogeny in *Tilletia indica. Mycologia* **1987,** *79,* 753–763.

Sharma, R. C.; Sidhu, G. S.; Bharaj, T. S.; Sharma, H. L. Pathological Constraints on Hybrid Rice Production Technology. *Int. Rice Res. Newsl.* **1993,** *18* (3), 15.

Singh, R. A. Sources of Inoculum and Epidemiology of Bunt of Rice. *Jl Riso* **1975,** *25,* 77–80.

Singh, R. A.; Pavgi, M. S. Cytology of Teliospore Germination and Development in *N. horrida. Jl Riso* **1972,** *21,* 259–268.

Singh, R. A.; Pavgi, M. S. Artificial Culture and Chlamydospore Development of *N. horrida. Nova Hedwigia* **1973a,** *24* (2/4), 487–492.

Singh, R. A.; Pavgi, M. S. Development of Sorus in Kernel Bunt of Rice. *Jl Riso* **1973b,** *22,* 243–250.

Singh, R. A.; Whitehead, M. D.; Pavgi, M. S. Taxonomy of *Neovossia horrid* (Ustilaginales). *Sydowia* **1979,** *32*, 305–308.

Singh, R. S. Bunt of Rice. In *Plant Diseases,* VI ed.; Oxford and IBH Publishing Co. Pvt. Ltd.: New Delhi; pp 313–316.

Smilanick, J. L.; Hoffmann, J. A.; Royer, M. H. Effect of Temperature, pH, Light, and Sesiccation on Teliospore Germination of *Tilletia indica. Phytopathology* **1985,** *75,* 1428–1431.

Srinivasan, S. Varietal Resistance to Kernel Smut. *Int. Rice Res. News* **1982,** *7* (1), 6.

Su, M. T. *Report of the Mycologist, Burma, Mandalay for the Year Ended 31st March, Xuan-Keng, Li.* 1933; vol 19953, p 102.

Takahashi, Y. On *Usti/ago Virens* Cooke and a New Species of *Tilletia* Parasitic on Rice Plant. *Bot. Mag. Tokyo* **1896,** *10,* 16–20.

Templeton, G. E. Local Infection of Rice Florets by the Rice Kernel Smut Organism, *T. horrida. Phytopathology* **1961,** *51,* 130–131.

Templeton, G. E. Research on Rice Diseases in Arkansas. *The Rice* 1. **1962,** *65* (7), 28.

Templeton, G. E.; 19Ribeiro; et al. Kernel Smut of Rice as Affected by Nitrogen. *Arkansas Farm Res.* **1973a,** *12* (5), 12. *(The Rice* 1 **1964,** *67* (3), 39).

Templeton, G. E. Rice Disease Research. *The Rice* 1 1965, *68* (7), 44.

Templeton, G. E. Rice Disease Studies - 1968. *The Rice* 1 **1968,** *72* (7), 41.

Templeton, G. E. Rice Disease Research in Arkansas - 1970. *The Rice* 1 **1971,** *74* (6), 37.

Templeton, G. E.; Johnston, T. H. A Source of Resistance to the Rice Kernel Smut Organism, *T. bare/ayana.* 1971. *The Rice* 1 **1970,** *73* (7), 74.

Templeton, G. E.; Johnston, T. H.; Henry, S. E. Kernel Smut of Rice. *Arkansas Far Res.* **1960,** *9* (6), 10.

Templeton, G. E., Johnston, T. H.; Henry, S. E. Kernel Smut of Rice. *Rice. J.* **1961,** *64* (2), 24.

Teng, S. C. Observations on the Germination of the Chlamydospores of *T. horrid* Tak. *Contr. Biol. Lab. Sci. Soc. China, Bot. Ser.* **1931,** *6,* 111–114.

Thinggaard, K.; Leth, V. Use of the Fluorochrome Vital Dye Acridine Orange to Determine Viability and Germination of *Tilletia indica* Teliospores in Soil. *Seed Sci. Technol.* **2003,** *31,* 329–340.

Thompson, A.; Johnston, A. A Host List of Plant Diseases in Malaya. *Mycological Papers* No., 1953, 52, p 38.

Thrimurty, V. S.; Singh, A. K. Incidence of Kernal Smut in Wetland Rice Varieties. *Int. Rice Res. Newsl.* **1982,** *7* (6), 5.

Tullis, E. C.; Johnson, A. G. Synonymy of *T. horrida* and *N. Bare/Ayana. Mycologia* **1952,** *44,* 773–788.

V´anky, K. *Illustrated Genera of Smut Fungi.* APS Press: Minneapolis, MN, 2002; p 238.

Walker, E. A Preliminary Report Upon Treatment for Rice Smut. *S. Carolina Agric. Apt. Sta. Bull.* **1899,** *41,* 15–31.

Webster, R. K. Being at the Right Place, at the Right Time, for the Right Reasons—Plant Pathology. *Annu. Rev. Phytopathol.* **2005,** *43*, 1–24.

Webster, R. K.; Gunnell, P. S. Eds.; *Compendium of Rice Diseases.* APS Press: St. Paul, MN, 1992; p 62.

Wei, C. T. Rice Diseases. *Nanking College of Agriculture and Forestry Bulletin (New Series)*, 1934; vol 16, p 40, 28 figs.

Whitney, N. G. Effect of Fungicide Application on Kernel Smut of Rice. *Pl. Dis. Reptr.* **1977,** *61,* 379–381.

Whitney, N. G. Taxonomy of the Fungus Causing Kernel Smut of Rice. *Mycologia* **1989,** *81* (3), 468–471.

Whitney, N. G.; Frederiksen, R. A. Kernel Smut of Rice. In 4th *Proc. Ann. Tex. Con! Insect Plant Dis. Weed Brush Control*, 1971; pp 131–1J2.

Whitney, N. G. and Frederiksen, R. A. 1975. Kernel smut of rice. *Texas Agric. Expt. Sta.* MP-1231.

X.-K., Li. Managements on Kernel Smut **in** Hybrid Rice Seed Production. In *4th International Hybrid Rice Training Course,* held at Hunan Hybrid Rice Research Center, Changsha, Hunan, China, Sept 26–Aug 12, 1995; pp 94–96.

Zhang, Z.; Lange, L.; Mathur, S. B. Teliospore Survival and Plant Quarantine Significance of *Tilletia indica* (Causal Agent of Karnal Bunt) Particularly in Relation to China. *EPPO Bull.* **1984,** *14,* 119–128.

CHAPTER 14

Current Scenario of Progress and Prospects for the Management of Rice Kernel Smut Disease Resistance

MOHD. SAID, MOHD. RAMEEZ, NADEEM KHAN, SALMAN AHMAD, and MALIK M. AHMAD

Department of Agriculture, Integral Institute of Agricultural Science and Technology (IIAST), Integral University, Lucknow, India

ABSTRACT

Rice kernel smut usually belongs to a minor category of plant disease but it can become epidemic in local areas especially in the rice-growing regions. It, also known as black smut or bunt of rice, was first illustrated by Takahashi from Japan and Anderson from the USA. This disease is caused by the fungus *Tilletia horrida* syn. *Neovossia horrida.* This disease exclusively affects grains. Although, only a few rice grains in the entire panicle are compromised by this disease. And, in most occasions, only a fraction of the grain is converted to black powder. Damage is caused in the rice grain endosperm being is generally substituted by masses of black fungal spores. This causes poor milling, grey-colored milled grain, and significant discoloration during the parboiling process. The pathogen is widespread and causes considerable qualitative and quantitative losses. Its damage can occur even at low levels of the disease in parboiling processes in which teliospores of the pathogen strain the grains, and thus making them not fit for human consumption. Yield losses caused by kernel smut ranges from <1 to 15% and can significantly reduce the grain quality. Quality issues with this disease focus on

Fungal Diseases of Rice and Their Management. Deepti Srivastava, Md. Shamim, Malik Mobeen Ahmad, and R. S. Upadhyay (Eds.)

the appearance of the rice grains, and especially those used for parboiling. Physical, cultural, and chemical control measures are some strategies that can be used to manage the smut disease along with host resistance.

14.1 INTRODUCTION

Rice (*Oryza sativa* L.) is the staple food of more than half of the world's population, accounting for more than half of the calories consumed (Sharif et al., 2014). Kernel smut of rice is a fungal disease caused by *Neovossia horrida* (Takah.) Padwick and A. Khan (syn. *Tilletia barclayana* (Bref.) Sacc. and P. Syd.) Tilletiales (Castlebury et al., 2005), which, like *T. indica* produces partial bunt, which has an impact on yield and quality. Kernel smut and false smut are fungal diseases of rice that cause partial or absolute replacement of kernels with spore masses, reducing grain yield and quality (Webster et al., 1992). Rice kernel smut that is caused by *Tilletia horrida* was first described in Japan in 1896. It was believed to be an endemic disease that had spread to epidemic proportions, a chronic problem, and was persistent in rice-growing areas in the southern United States, particularly with parboiled rice. In addition, it is also considered a minor disease, although, in some rice-producing locations, such as the Southern United States and India, it has resulted in significant yield losses (Brooks et al., 2009; Carris et al., 2006). Kernel smut is considered as minor with sporadic nature but became important economical rice fungal diseases with severe quality and yield losses in Texas (Uppala et al., 2017). Rice kernel smut, recognized as caryopsis smut and black smut (Biswas, 2003), was widespread and presents almost in all rice-grown ecosystems worldwide. It has found in upland or irrigated rice-growing countries (Elshafey, 2013; Biswas, 2003). The pathogen spreads through air and soil. Teliospores acting as primary inoculum infect the rice crop through soil (Whitney, 1992) whereas the secondary sporidia were forced into the air and infected floral tissues through open florets. It was estimated that the direct loss of grain yield through this disease can reach around 15% besides a significant loss of rice grain quality (Gravois and Bernhardt, 2000).

Pathogen survival is important in epidemiology and disease progression, and teliospores, as a key source of infection with thick-walled spores, offered the ability to survive at least wintering from one season to the next. As a result, the most important aspect of management is to begin rice growing with healthy seeds (Chahal, 2001). It causes a partial bunt in rice, causing black, sooty masses of powdery spores to replace all or part of the grain, reducing productivity and quality. The pathogen is most commonly found in nature as dark-brown tuberculate teliospores that overwinter in the soil and are found outside or inside the seeds of the host plant. The spores can live in the soil for more than a year and 3 years within the body (Wang et al., 2015). These diseases have still been largely uninvestigated because of the historical focus on main diseases like a blast and bacterial blight by scientists. Furthermore, the tag of "minor disease" for smuts (Leung et al., 2003), environmental inconsistency, and the time required to study these persistent adult plant diseases have resulted in a general lack of progress in disease control.

Fungi causing smut diseases belong to class Ustilaginomycetes, phylum Basidiomycota. The class name had been derived from the Latin word "ustulatus" which means "burned," about the blackened appearance of infected plants (Carris, 2001) while the cereal-infecting *Tilletia* species that generate teliospores in their hosts' ovaries are generally known as bunt fungi, and are assumed to be a derivative of the word "burned" (Duran et al., 1961). *Tilletia horrida* is a highly specific fungus on its hosts and organs that may infect numerous kinds of grasses in addition to rice. Its infection is confined to the florets and ovaries (Singh and Pavgi, 1972; Whitney, 1992). In this situation, plants start to protect themselves against invader diseases by using structural barriers or antimicrobial compounds to prevent the colonization of tissues, according to research. Hypersensitive reactions, reactive oxygen species (ROS) generation, pathogenesis-related proteins, and ion fluxes across the plasma membrane are all examples of induced defense responses (Hassan et al., 2014). However, there is a need for an integrated approach to managing these critical diseases that involve disease resistance and cultural management.

Fungicides are effective on rice smuts (Tsuda et al., 2006) but their use is not practical for rice production. Neither disease produces visual symptoms until the grain begins to fill, at which time a fungicide application is too late to prevent yield loss. At growth stages when control with fungicides is achievable, there is no way to determine the degree of disease severity and, thus, to determine whether the application could be aided by reviewing under natural infection while providing a source of inoculum which could improve the finding of field resistant germplasms. Kernel smut could move from a minor to a major disease based on the results in a change in climatic conditions, thereby causing serious difficulties. As a result, understanding the biology of the pathogen was crucial in the host–pathogen interaction and integrated disease management. Therefore, the integration of disease-resistant germplasm and cultural management techniques are needed to control these diseases at a production scale (Brooks et al., 2009).

14.2 HOSTS, SPREAD, AND DISTRIBUTION OF PATHOGENS

Kernel smut infects paddy crop and other grasses of genera *Brachiaria*, *Digitaria*, *Eriochloa*, *Panicum*, and *Pennisetum*. It is said that because the florets of long-grain rice cultivars open broader and for a longer time duration during flowering, they may become more susceptible to kernel smut. Besides this, airborne distribution of the pathogenic spores can infect the rice crop and re-infect the same rice panicles within the same crop and other neighboring rice crops. Long-distance dissemination of fungal spores occurs through infected/contaminated grains, equipment, and/or machinery. These spores may also be spread as they can float on the water surface. The disease generally occurs in most rice-producing countries such as China, India, Indonesia, Japan, Philippines, Burma, Korea, Senegal, Mexico, and the USA.

14.3 MORPHOLOGICAL CHARACTERISTICS OF *T. horrida*

Promycelium is often, long branched, non-septate, and 34–523 × 6–7.3 μm in size while the basidiospores are aseptate, hyaline in appearance with filiform to spine-like shape and 52–75 × 1.1–2 μm in size. Generally, the 20–100 basidiospores are found on the terminal whorls of primary sporidia (Elshafey, 2018) but they may vary in different isolates (Castalbury et al., 2005; Carris et al., 2006). Teliospores of kernel smut pathogen are generally of size 14–36 μm (<25 μm are generally considered as mature teliospores), light to dark chestnut brown, echinulate to sometimes cerebriform ridges, curved (Turgay et al., 2020) but can also be semi-opaque. They can be of globose to subglobose in shape. The sterile teliospores are occasionally hyaline in color and with a size of 18–31 μm. To protect the spore, teliospores have four distinct layers; the outer layer, episporium, is relatively transparent and gelatinous while the perisporium, the middle layer is dense which gives a reticulate pattern and the last inner layer wall is known as endosporium (Nawaz and Hess, 1987; Elshafey, 2018). The episporium layer is covered with a dense and thick-walled sheath which turns into spines to protect the teliospores and maintain the spore dormancy state. Filiform secondary sporidia grow after 3–4 days of teliospores germination while allantoid sporidia grow after 7 days of teliospores germination. The colony appears as cottony with white powdery color and the mycelia are septate, 37.5–62.5 × 2.4–2.6 μm, while allantoid secondary sporidia are dark purple pigmented, septate, and have falcate shape, 10–21.2 × 2.9–3.8 μm in size (Carris et al., 2006).

14.4 SYMPTOMS

Diagnostic symptoms of the disease become evident only at the time of grain formation due to the presence of black powdery mass dispersed on the glumes. Not all the ears of stool in rice crops carry the infection and only a few grains in the ears are found diseased. The degree of transformation of the kernel into bunt sorus possibly depends on the stage of infection as the development of a fertilized ovary and the fungus occurs simultaneously. We are presenting some of the aspects of one of the diseases entitled the smut, smut is the Greek word and they represent the dark black colors in the reproductive bodies; here, generally the reproductive organ is replaced by the dark full patches of the dark spores of the carbon color and when the spores are crushed they appears like the ash (Agarwal, 2017). Kernel smut of rice is most evident at crop maturity. The rice grains are replaced by olive-black masses of spores that rupture out through the hulls. This black powder gives a foul smell due to the biosynthesis of trimethylamine chemicals. The disease is noticeable at crop maturity. Most of the infected plants display only a few smutted grains. Black pustules of teliospores push through the glumes, especially if they swell from moisture provided by overnight dew.

14.5 DISEASE CYCLE OF *T. horrida*

Rice fungal kernel smut is not systemic. The fungus thrives as spores in the soil as well as in infected seed grains. *T. horrida* teliospores had also been found to float on the surface of irrigated waters and germinate after the rice is planted in rice fields (Whitney and Frederiksen, 1975). *T. horrida* sporidia and hyphae establish themselves on the rice crop canopy (Bains and Dhaliwal, 1990) or form a dense mat on the surface of the rice paddy water (Templeton et al., 1960; Webster, 2005). Once infected, the germ tube from the teliospores emerges and basidiospores and promycelium grow. Spores were pushed forcibly upward to infect the open florets of the immature growing rice grains. The endosperm of the grain is replaced by a mass of black spores of *T. horrida*. Wind-borne spores infect flowers individually, and just a few florets in a panicle may be infected.

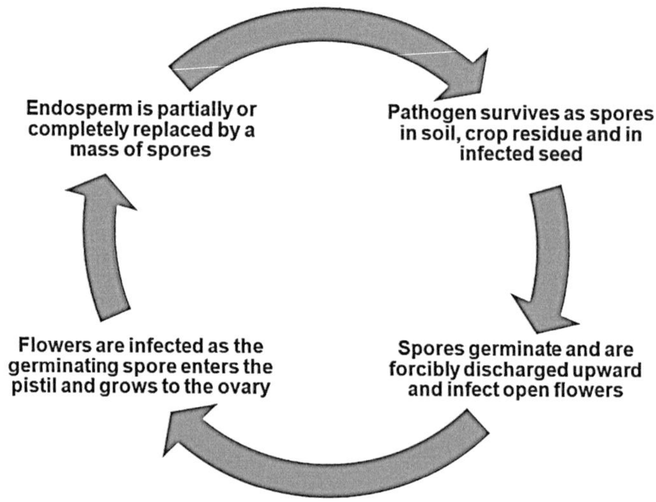

FIGURE 14.1 Disease cycle of kernel smut.

14.6 ECONOMIC IMPORTANCE

Although the pathogen is widespread and the disease has not been considered of much economic importance. In recent years, its incidence has increased with the popularization of high yielding varieties, hybrid rice cultivation, intensive cropping patterns, and the use of more nitrogenous fertilizers. Under congenial conditions for disease development, considerable quantitative and qualitative losses have been reported by various workers as given below.

14.7 FAVORABLE CONDITIONS FOR THE KERNEL SMUT DEVELOPMENT

Certain conditions can favor the development of kernel smut disease in rice crops. They are:

1. to spread the pathogen needs temperature ranging from 25 to 30°C.
2. presence of heavy rainfall followed by hot sunshine immediate after sowing are conducive to seedling infection and high humidity (>85%).

3. applications of higher dose of nitrogenous fertilizers at heading stage followed by lodging of the crop at blossom period predisposes the host more to this disease.
4. disease incidence is more in the fields sown with early duration varieties year after year than mid and late maturing due to heavy build-up of inoculum and synchronization of sporidial discharge with their anthesis.
5. with more seedling infection in damp/waterlogged soils, the disease is predominantly more.
6. irrigated and semi-dry soils under waterlogging impair the spore viability, germination, and discharge of sporidia to cause floral infection. The disease is favored by light sandy loam soils which facilitate the teliospore germination and easy emergence of the promycelium.

14.8 DISEASE ASSESSMENT AND ITS MANAGEMENT

14.8.1 DISEASE ASSESSMENT

Crop disease evaluation entails the measurement and quantification of plant disease and is thus critical in the research and analysis of plant disease outbreaks. Several novel ways to disease evaluation have recently joined traditional methodologies which had been possible due to the significant improvements in computer technology. In addition, recent assays like molecular and immunological or a combination of these methods have been widely used for the detection, identification, and quantification of different phytopathogenic microorganisms. The use of remote sensing and its image analysis has also given new concepts to assess the severity of plant diseases. Usually, to estimate severity and percentage of infection, plant pathologists use 50 panicles of rice plants of the mature stage to examine them and record the infected plants and calculate the infection percentage according to the following equations (Slaton et al., 2004):

$$Disease\ severity = \frac{No.\ of\ infected\ grains}{Total\ no.\ of\ grains} \times 100$$

$$Percentage\ of\ infection = \frac{No.\ of\ infected\ grains}{Total\ no.\ of\ rice\ panicles} \times 100$$

14.9 DISEASE MANAGEMENT

Kernel smut of rice is hard to control due to the manner of maintenance of the pathogen and the lack of resistant rice cultivars. However, few disease controlling methods have been proposed by several researchers as given below:

14.9.1 CHEMICAL AND CULTURAL CONTROL

Walker (1899) recommended the rejection of light grains floating on the surface of water and soaking of heavier grains settling at the bottom in 0.2% potassium sulfide solution for 2 h. Seed selection with 10% salt water or muddy water resulted in 93.2% and 80.4% reduction in kernel smut incidence. Quarantine measures should be performed strictly to avoid the introduction of the disease into new areas. The disease can be managed by hot water treatment and seed dressing chemicals like copper carbonate, formalin, organomercurials, triadimefon, and benomyl. Contrary to this, Templeton et al. (1960) and Whitney (1992) observed that neither seed treatment nor the foliar application of benomyl, maneb, chlorothalonil, and Triphenyl tin hydroxide at the heading stage could control the disease. Singh et al. (1998) recommended 3 years of crop rotation to minimize the disease as teleutospores do not survive in the soil for more than 27 months. However, Ospishchev (1980) advocated that besides crop rotation, effective prophylactic measures should be taken with benlate or thiophanate methyl during anthesis after careful weather forecasting diagnosis of the disease in the previous crop. In China, spraying the crop with Triadimefon, Topsin M, and carbendazim at initiation and full heading and flowering stages is recommended for effective control of kernel smut. Application of carbendendazim at early and maximum flowering period and Bayleton (Tridemorph) @ 105 g a.i./ha at early heading proved most effective and gave 70–80% disease control along with a significant increase in yield.

Field screening of rice germplasm for reaction to kernel smut of rice continued at the UAPB Field Station near Lonoke, AR and the Pine Tree Station near Colt, AR. A total of 170 cultivars and breeding lines were evaluated but hot, dry weather prevented uniform kernel smut infection. Commercial cultivars with greater than 5% kernel smut were Priscilla, LaGrue, Cypress, and Cocodrie while Wells, Drew, Lemont, and Madison had only traces of the disease. These results were consistent with previous observations. Preventative applications of 4 fl oz/A Tilt fungicide (propiconazole) during the boot stage of susceptible rice varieties provided the same reduction in kernel smut (88%) at higher rates. Statewide disease monitoring determined that kernel smut was less severe in 1999 than in any of the previous six growing seasons (Cartwright et al., 1998).

14.9.2 CROP ROTATION STRATEGIES

Multi-cropping-systems rotation studies had been performed by various scientists to evaluate their effect on rice kernel smut disease. Brooks et al. (2009) showed that disease severity was approx. 60% reduced under normal application of fertilizers than the enhanced fertility while no-tillage reduced the disease severity up to 34% with the continuous rice rotation. This suggests that an increase in nitrogen fertilizer dosage would favor the disease development.

14.10 CONCLUSION

Tilletia is a monophyletic group of species having reticulate, smooth, or tuberculate ornamentation on teliospores, local, or systemic infection and conjugating or non-conjugating basidiospores (Castlebury et al., 2005; Vankey, 2002). *T. horrida* causes rice kernel smut, which results in a partial bunt that reduces production and quality. This fungus was purified and identified as *T. horrida* after 46 fungal isolates were obtained from diseased rice plants showing evident indications of black smut. This fungus was also formerly reported to be the causal agent of rice kernel smut disease. Kernel smut is widespread in approximately all the rice-growing regions of the globe. The pathogen converts the grains into black powdery mass resulting in considerable loss both in yield and quality of the produce.

KEYWORDS

- **kernel smut disease**
- **management**
- **rice (*Oryza sativa* L.)**
- ***Tilletia barclayana***
- ***Tilletia horrida* syn. *Neovossia horrida* (Tak.)**

REFERENCES

Agarwal, T. Smut of Crops: A Review. *J. Pharmaco. Phytochem.* **2017,** *5,* 54–57.

Bains, S. S.; Dhaliwal, H. S. Production of *Neovossia Indica* Sporidia on the Host and Nonhost Plants. *Plant Soil* **1990,** *126,* 85–89.

Biswas, A. Kernel Smut Disease of Rice: Current Status and Future Challenges. *Environ. Ecol.* **2003,** *21,* 336–351.

Brooks, S. A.; Anders, M. M.; Yeater, K. M. Effect of Cultural Management Practices on the Severity of False Smut and Kernel Smut of Rice. *Plant Dis.* **2009,** *93,* 1202–1208.

Carris, L. M. Smut Fungi, 2001**;** pp 919–921.

Carris, L. M.; Castlebury, L. A.; Goates, B. J. Nonsystemic Bunt Fungi *Tilletia Indica* and *T. horrida*: A Review of History, Systematics and Biology. *Ann. Rev. Phytopathol.* **2006,** *44,* 113–133.

Cartwright, R. D.; Lee, F. N.; Parsons, C. E.; Ross, W. J.; Vann, S. R.; Overton, R. Monitoring of Rice Diseases and On-Farm Evaluation of Varieties in Arkansas. In *B.R. Wells Rice Research Studies 1998*; Norman, R., Johnston, T., Eds.; Arkansas Agricultural Experiment Station Research Series 468, University of Arkansas: Fayetteville, 1998; pp 148–156.

Castlebury, L. A.; Carris L. M.; V′anky, K. Phylogenetic Analysis of *Tilletia* and Allied Genera in Order Tilletiales (Ustilaginomycetes; Exobasidiomycetidae) Based on Large Subunit Nuclear rDNA Sequences. *Mycologia* **2005,** *97,* 888–900.

Chahal, S. S. Epidemiology and Management of Two Cereal Bunts. *Ind. Phytopathol.* **2001,** *54,* 145–157.

Duran, R.; Fischer, G. W. The Genus *Tilletia.* Washington State University: Pullman, Washington, 1961; pp 138.

Elshafey, R. A. S. Evaluation of Varietal Resistance and Physiological Characters of False and Kernel Smut Diseases of Rice in Egypt. *J. Plant Prot. Path., Mansoura Univ.* **2013,** *4,* 125–142.

Elshafey, R. A. S. Biology of Rice Kernel Smut Disease Causal Organism *Tilletia barclayana* and its Molecular Identification. *J. Phytopath Pest Manag.* **2018,** *5,* 108–128.

Gravois, K. A.; Bernhardt, J. L. Heritability and Genotype x Environment Interactions for Discolored Rice Kernels. *Crop Sci.* **2000,** *40,* 314–318.

Hassan, N.; Elsharkawy, M. M.; Shivanna, M. B.; Meera, M. S.; Hyakumachi, M. Elevated Expression of Hydrolases, Oxidase, and Lyase in Susceptible and Resistant Cucumber Cultivars Systemically Induced With Plant Growth-Promoting Fungi Against Anthracnose. *Acta Agric. Scand. B Soil Plant Sci.* **2014,** *64,* 155–164.

Leung, H.; Zhu, Y.; Revilla-Molina, I.; Fan, J. X.; Chen, H.; Pangga, I.; Vera Cruz, C.; Mew, T. W. Using Genetic Diversity to Achieve Sustainable Rice Disease Management. *Plant Dis.* **2003,** *87,* 1156–1169.

Nawaz, M. S.; Hess, W. M. Ultrastructure of *Neovossia Horrida* Teliospores. *Mycologia* **1987,** *79,* 173–179.

Ospishchev, A. L. Treatment Only After Forecasting. *Zashchita Rastenii* **1980,** *4,* 20–21.

Sharif, M. K.; Butt, M. S.; Anjum, F. M.; Khan, S. H. Rice Bran: A Novel Functional Ingredient. *Crit. Rev. Food Sci. Nutr.* **2014,** *54,* 807–816.

Singh, R.; Dodan, D. S.; Sheoran, O. P. Kernel Smut of Rice: Present Status. *Int. J. Trop. Plant Dis.* **1998,** *16,* 149–167.

Singh, R.; Pavgi, M. Cytology of Teliospore Germination and Development of *Neovossia Horrida. Riso* **1972,** *21,* 259–268.

Slaton, N. A.; Gbur, E. E. J.; Cartwright, R. D.; DeLong, R. E.; Norman, R. J.; Brye, K. R. Grain Yield and Kernel Smut of Rice as Affected by Preflood and Midseason Nitrogen Fertilization in Arkansas. *Agron. J.* **2004,** *96,* 91–99.

Templeton, G. E.; Johnson, T. H.; Henry, S. E. Kernel Smut of Rice. *Arkan. Farm Res.* **1960,** *9,* 10.

Tsuda, M.; Sasahara, M.; Ohara, T.; Kato, S. Optimal Application Timing of Simeconazole Granules for Control of Rice Kernel Smut and False Smut. *J. Gen. Plant. Pathol.* **2006,** *72,* 301–304.

Turgay, E. B.; ÇelikOğuz, A.; Ölmez, F. Karnal Bunt (*Tilletia Indica*) in wheat. In *Climate Change and Food Security With Emphasis on Wheat*; Ozturk, M., Gul, A., Eds.; Academic Press. 2020; pp 229–241.

Uppala, S.; Liu, G.; Liu, B.; Wu, M.; Zhou, X. G. Texas Rice Kernel Smut Research, [Online] 2017. *https://beaumont.tamu.edu/eLibrary/Newsletter/2016_Highlights_in_Research.pdf.*

Vanky, K. Illustrated Genera of Smut Fungi. APS Press: Minneapolis, MN, 2002; pp 238.

Walker, E. A Preliminary Report Upon Treatment for Rice Smut. *S. Carolina Agric. Apt. Sta. Bull.* **1899,** *41,* 15–31.

Wang, N.; Ai, P.; Tang, Y.; Zhang, J.; Dai, X.; Li, P.; Zheng, A. Draft Genome Sequence of the Rice Kernel Smut *Tilletia Horrida* Strain QB-1. *Genome Announc.* **2015,** *3,* e00621–15.

Webster, R. K. Being at the Right Place, at the Right Time, for the Right Reasons—Plant Pathology. *Annu. Rev. Phytopathol.* **2005,** *43,* 1–24.

Webster, R. K.; Gunnell, P. S. Compendium of Rice Diseases. American Phytopathological Society: St. Paul, MN, USA, 1992.

Whitney, N. G. Kernel Smut. In *Compendium of Rice Diseases*; Webster, R. K., Gunnel, P. S., Eds.; American Phytopathological Society: St. Paul, MN, USA, 1992; pp 28–29.

Whitney, N. G.; Frederiksen, R. A. Kernel Smut of Rice. *Tex. Agric. Exp. Stn. Misc. Publ.* **1975,** 1231.

CHAPTER 15

Identification and Pathogenic Diversity of Rice False Smut Pathogen and Their Resistance Resources for Future Breeding

MD. SHAMIM[1], DEEPTI SRIVASTAVA[2], MAHESH KUMAR[1], DEEPAK KUMAR[3], ANURAG MISHRA[4], PRAKASH SINGH[5], PRAMILA PANDEY[6], SANJEEV KUMAR[7], MEENAKSHI BISHT[2], and V. B. JHA[8]

[1]*Department of Molecular Biology and Genetic Engineering, Dr. Kalam Agricultural College (Bihar Agricultural University, Sabour, Bhagalpur), Kishamnganj, Bihar, India*

[2]*Department of Agriculture, Integral Institute of Agricultural Science and Technology, Integral University, Lucknow, Uttar Pradesh, India*

[3]*Divison of Genetics, Indian Agriculture Research Institute, New Delhi, India*

[4]*Department of Manufacturing and Development, Nextnode Bioscience Pvt. Ltd., Kadi, Gujarat, India*

[5]*Department of Plant Breeding and Genetics, V.K.S. Agricultural College (Bihar Agricultural University, Sabour, Bhagalpur), Dumraon, Bihar, India*

[6]*Department of Botany, Nappa Sri Narain Vocational P.G. College, Station Road Charbagh, Lucknow, Uttar Pradesh, India*

[7]*Department of Plant Pathology, Bihar Agricultural University, Sabour, Bhagalpur, Bihar, India*

Fungal Diseases of Rice and Their Management. Deepti Srivastava, Md. Shamim, Malik Mobeen Ahmad, and R. S. Upadhyay (Eds.)

[8]*Department of Plant Breeding and Genetics, Dr. Kalam Agricultural College (Bihar Agricultural University, Sabour, Bhagalpur), Kishamnganj, Bihar, India*

ABSTRACT

Ustilaginoidea virens is a rice fungus that becomes as most destructive in the world. Rice false smut (RFS) pathogen causes substantial yield drop and quality of grain. The disease-causing fungus developed mycotoxins poses a threat to food safety and human as well as animal health. Recent advances in the study of *U. virens'* life cycle, their infection process mechanisms, pathogenic gene, genome, and bio-molecules produced were investigated. The present chapter outlined false smut disease resistance in rice along with various potential directions and main queries that require to be answered to provide insights into the mechanism involved at molecular levels. In addition to these various other aspects of rice false smut pathogen and management practices are also discussed.

15.1 INTRODUCTION

Ustilaginoidea virens (Cook) Takahashi (teleomorph: *Villosiclava virens*), a pathogenic ascomycete fungus, causes rice grain disease known as rice false smut (RFS). Rice false smut disease used to be a sporadic occurrence in rice-growing areas. Presently, RFS disease is one of the most destructive diseases of rice which is reported in various countries. It is estimated that between 2015 and 2017, it occurred in 2.4 million hectares in China (Jiehua et al., 2019). Anhui, Hunan, Hubei, Jiangsu, Jiangxi, and Zhejiang provinces in China have the highest prevalence (Lu et al., 2018). Furthermore, RFS has freshly been recognized in rice-growing areas around the world with rising frequency (Guo et al., 2012; Ladhalakshmi et al., 2012; Jecmen and Tebeest, 2015). False smut balls (FSB) infect rice grains with are in yellowish orange to green in color, which are before becoming a greenish-black color (Nessa et al., 2015; Kumagai et al., 2016).

Rice false smut was first identified in Tirunelveli district of Tamil Nadu State of India in 1878 (Cooke, 1878). Further RFS disease was also reported in northeastern Arkansas counties in 1997, caused by *Ustilaginoidea* virens (Cke.) Takahashi (1896). Ascomycete fungus *Villosiclava virens* Tanaka and Tanaka (anamorph: *Ustilaginoidea virens* Takahashi) is the causal pathogen of RFS disease (Tanaka et al., 2008). Symptoms of this disease appear immediately after flowering; individual spikelets infection has been reported by fungus infestation and reinstates the seed with a huge orange to green ball recognized as a FSB (Hu et al., 2014). In recent years, this disease causes major concerns in rice growing areas of the World (Nessa et al., 2015).

U. virens infects flowers of rice and colonizes inner organs of flowers with mycelia (Tanaka et al., 2008; Tang et al., 2013), which are ultimately changed into FSB sheltered through powdery chlamydospores (Fig. 15.1a–15.1f). Color of FSB varies from yellowish brown to greenish black (Fan et al., 2016) and they produce sclerotia when they large temperature differences between day and night (Fan et al., 2016). Rice grain filling is also affected by FSB as they increase the sterility of the adjacent spikelets which results in the loss of yield (Dhua et al. 2015). Additionally, mycotoxins are also produced by FSB which are hazardous to both human and animal health (Koiso et al., 1994; Nakamura et al., 1994; Zhou et al., 2012).

Various environmental and soil conditions such as high relative humidity (>90%), high rainfall, low sunlight hours, temperatures (25–30°C), high fertile soil and more nitrogenous fertilizer input and crop sowing conditions such as late sowing or maturity period enhances the rate of infection with *U. virens* (Li et al., 1986; Singh and Khan, 1989; Sugha et al., 1992; Chen et al., 1994; Ahonsi et al., 2000; Yashoda et al., 2000; Nessa et al., 2015c). The other factors, that is, high application of fertilizers and added irrigation and utilization of high yielding rice varieties also increases the infection rate of false smut disease (Singh et al., 1987; Biswas, 2001), (Deng, 1989; Zhou et al., 2004; Lu et al., 2009). The above factors have been confirmed as major aspects for extensive dispersal of this fungal disease.

Presently, RFS disease management can be managed through cultivation practices, chemical control methods, and through host resistance. Though fungicide management at the crucial stage when infection of *U. virens* was efficient in controlling the disease, the best time to spray fungicides before symptoms occur is difficult to predict. Furthermore, the disease has developed resistance to a number of fungicides (Bagga and Kaur, 2006). Fungicide misuse could also pollute the environment and disrupt ecosystems. Cultivation strategies like furrow irrigation may give a rapid and gainful solution for false smut management, but they may also produce other issues that outweigh of false smut reduction and management (Brooks et al., 2010). In disease management, host resistance has been identified as cost-effective,

FIGURE 15.1 Disease and typical symptoms caused by *Ustilaginoidea virens* in rice (a) Infected rice spikelet with the *U. virens* and false smut ball development, (b) false smut balls produced by *U. virens* in rice panicles, (c) a typical diseased panicles of rice showing yellow colour false smut balls, (d) rice panicle with greenish-black balls of false smut in a rice grains, (e) mature false smut ball with a sclerotium covered with greenish-black chlamydospores, (f) heavy infected paddy field by *U. virens*. (Photo taken by Md. Shamim and Prakash Singh).

ecologically friendly, and effective options. However, for the better control of this important disease, the common feature and process of infection and diversity among the pathogen is essential.

15.2 FEATURES OF THE *U. virens* AND THE FS-RICE PATHOSYSTEM

False smut spore balls or sclerotia (sclerotia = pseudomorph) are found growing in association with *Oryza sativa*, on *Zea mays* L. and a few other tropical Graminae hosts that have no significant economic importance (Abbas, 2002). Spore balls are made up of a proliferation of branching, radial, and compacted hyphae that give rise to warty olivaceous chlamydospores are spherical to elliptical (3–5 × 4–6 m) at the terminus (Ou, 1985). Initially, chlamydospores are smooth, round to elliptical, but on maturity, they are warty, spiky, and yellow to orange colored. As the chlamydospores mature, they become olivaceous, globose to erratically spherical, with conspicuous spines on the surface (Kim and Park, 2007). Chlamydospores develop into fine germ tubes.

The chlamydospores develop into fine germ tubes, which results in one to three tiny ovoid secondary conidia (Hashioka et al., 1951). *Ustilaginoidea virens* can be cultured on potato dextrose agar (PDA) or potato dextrose sucrose agar (PDSA), but the fungus is slow-growing and has two different colony types. The two colony types are either green, hardened sclerotia-like after incubation for 2 weeks or white, spreading mycelia after incubation for 3 weeks.

15.3 FALSE SMUT SYMPTOMS IN INFECTED RICE PLANT

Rice infected with *U. virens* has no symptoms until the panicles begin to flower and on rice grain filling stage rice grains are replaced by spore balls that emerge through the glumes. The spore balls, comprised of expanding mycelium and layers of chlamydospores, are at first covered by the pale white expanded testa and aleurone membranes of the seed. With further expansion and development of the chlamydospores, the membranes rupture to expose the chlamydospores which appear orange then become yellow-orange to olive-green then to dark green or nearly black over time. During panicle expansion, the false smut pathogen, *U. virens*, infects rice plants and affects the young ovary of the spikelet. The infection progresses to the point when it forms thick, yellow to soft green balls (smut balls). Pathogen-caused soft green symptoms are noticeable after the milky phase has passed. The smut balls are little at first, and they remain between the glumes. The symptom-causing balls gradually grow in size and encircle the floral elements. Individual grains mature into yellow smut balls that range in color from yellowish orange to green, olive green, and greenish-black. Smut balls burst open, releasing powdery dark green spores (Biswas, 2001; Atia, 2004). If the infection develops before fertilization, the majority of the glumes remain sterile, with no visible signs of infection. When infection occurs after conception, huge, velvety green smut balls form. The conidial, pseudosclerotial, and sclerotial stages of the fungus are represented by the fructifications that have replaced the grains. Mycelial tissue and spore masses, remains of anthers and pieces of palea and lemna make up the pseudosclerotia (green smut balls). In usually, only a few grains are engaged in a panicle, yet in cases of high disease incidence, the number can reach 100 (Ladhalakshmi et al., 2012).

15.3.1 DISEASE CYCLE

Basically, two routes of disease cycle have been proposed. Primary infections occur when chlamydospores and/or sclerotia form airborne conidia and/or ascospores that develop in delayed summer or before time and land on rice inflorescences throughout the flowering stage, according to the original proposed model (Fig. 15.1a). Another theory proposes that the fungi stay alive in the winter season by producing spore balls and/or sclerotia which further infect rice plants (Fig. 15.1f). Chlamydospores unconfined as starting spore balls on seed and/or in or on soil germinate, and the hypha colonizes seedlings immediately after germination. Prior to the formation of the panicle, the fungus colonizes the apical meristematic tissue and then blossoms in the boot. Hashioka et al. (1951) found that 99.6% of the spore balls had undamaged anthers, implying that the majority of infections occur before flowering. Raychaudhuri (1946) identified fragmentation of the ovary, while the style, stigmas, and the anther lobes containing pollen grains are entirely surrounded by the growing fungus which gives rise to the spore ball structure. The sources of inoculum have been hypothesized to be airborne conidia or ascospores produced from germinated chlamydospores or sclerotia that make contact and infect flowers after germination or alternatively, infection and transmission may occur from dormant mycelium associated with the embryo at planting (Kulkarni and Moniz, 1975; Raychaudhuri, 1946; Singh and Dubey, 1984; Sreeramulu and Vittal, 1996).

15.3.2 INFECTION PROCESSES OF *U. virens*

False smut pathogen observed in tropical Asia, Australia, Central and South America, Egypt, Italy, India, and has been in the United States of America since the late 1890's (Cooke, 1878; Lee and Gunnell, 1992; Atia, 2004). Rice is the most important host for *U. virens* (Cooke and Takah) (Lee and Gunnell, 1992). False smut has turn into an ever more important disease of rice since its discovery in Louisiana in 1908, (Fulton, 1908) and in 17 fields in four counties of northeast Arkansas (Cartwright, 1999). It is now widespread in northeastern and central Arkansas Counties (Cartwright et al., 2002; Cartwright and Lee, 2006).

In order to research and treat RFS disease, it is crucial to understand the *U. virens* disease development. Coleoptiles are the first site where *U. virens* initiates the infection and spread throughout the seed germination phase, as well as the roots of seedlings (Zheng et al., 2009; Andargie et al., 2015; Prakobsub and Ashizawa, 2017). However, some research suggested that there is possibility that infections of coleoptile and roots may not spread to the spikelet and, therefore, unable to cause the characteristic symptom of false smut (Tang et al., 2013; Yong et al., 2018b). Some researchers reported that first site infection might be originated from rice pistils and later spread in other parts of plants (Chao et al., 2014; Andargie et al., 2016).

Several reports have revealed that before the rice flower opening, *U. virens* first infects rice filaments (Tang et al., 2013; Hu et al., 2014; Zhang et al., 2014; Fan et al., 2015; Song et al., 2016). *U. virens* infects rice spikelets at the time of booting stage, ensuing in disease development in grains according to normal inspection. These studies have been also confirmed by the artificial inoculation assessment, and serial histology examinations (Shen, 2004). According to Fan et al. (2014), *U. virens* germinates the conidia and forms a huge amount of secondary conidia and hyphae. Under favorable conditions, secondary conidia, and hyphae spreads on the surface of rice spikelet and after 4 dpi (days post inoculation), small gaps between the lemma and palea helps the hyphae to reach the inner spaces of the spikelets which further infects stamen filaments, stigma, or lodicules (Ashizawa et al., 2012; Li et al., 2013; Tang et al., 2013; Hu et al., 2014; Fan et al., 2015; Song et al., 2016).

Transmission electron microscope observation has shown persistent hyphae growth of *U. virens* beside the filaments cell gaps which devoid of penetrating the cell wall of host (Tang et al., 2013). Fungal infection prevents pollination and imitates fertilization of rice ovaries in order to take over the supply of nutrient to rice (Fan et al., 2015; Song et al., 2016). This infection process shows attribute feature of *U. virens* (Talbot, 2003). Yu et al. (2013) reported that at 10 dpi, *U. virens* hyphae covered the stamens, stigma, style and then start to grow out of the spikelet. At 15 dpi, a ball like colonies were formed (Yu et al., 2013). However, rice ovaries remain alive in the rice smut ball which indicates that *U. virens* is a biotrophic parasite and host cell was no killed at the time of infection (Tang et al., 2013).

15.3.3 MYCOTOXINS PRESENT IN *U. virens*

Mycotoxins such as ustiloxins and ustilaginoidins are produced by *U. virens* fungus. The toxins produced by the *U. virens* metabolites are toxic to human and animal ultimately causes disease (Wang et al., 2017). *U. virens* produces large amount of secondary metabolites, including various types of myco-toxins such as ustiloxins and ustilaginoidins (Zhou et al., 2012). Infected rice spikelets and grains with FSBs secrete ustiloxins a type of cyclopeptide mycotoxins. Ingestion of these causes a variety of symptoms, that is, diarrhea, insufficient growth, atrophy of ovary, abortion, liver damage, heart, and dysfunction of kidneys (Zhou, 2012). To date, 7 ustiloxin and 27 ustilaginoidin derivatives have been reported (Sun et al., 2020).

The ustiloxins viz. A, B, C, D, E, F, and G have a 13-membered cyclic core structure with a phenol ether linkage in *U .virens* (Koiso et al., 1994; Koiso et al., 1998; Wang et al., 2017; Lin et al., 2018) and among these A and B are the major abundant ustiloxins detected in RFS balls which are mostly dispersed in the middle layer of young FSB (Wang et al., 2017). *U. virens* ustiloxins inhibits formation of microtubule assembly and cytoskeleton in eukaryotes (Koiso et al., 1994; Koiso et al., 1998; Wang et al., 2017; Lin et al., 2018). Ustilaginoidins are a form of bis-naphtho-pyrone with cytotoxic properties against cancer cells and inhibitory effects on rice seed radicle elongation (Lu et al., 2015; Meng et al., 2015; Wang et al., 2016; Sun et al., 2017).

In *U. virens*, 26 ustilaginodins have been identified, namely A, B, C, D, E, E1, F, G, H, I, J, K, L, M, N, O, P, Q, R, S, T, U, V, and W, isochaetochromin B2, and 2,3-dihydroustilaginoidin (Lu et al., 2015; Meng et al., 2015; Sun et al., 2016; Sun et al., 2017). There are several analytical methods used for the identification and quantification of these mycotoxins such as high performance liquid chromatography (HPLC), liquid chromatograph–mass spectrometry (LC–MS), enzyme linked immunosorbent assay (ELISA), and lateral flow immunoassays (LFIA) (Shan et al., 2012; Bian et al., 2015; Fu et al., 2015; Fu et al., 2015a; Fu et al., 2015b; Fu et al., 2017; Cao et al., 2016). While several mycotoxins have been recognized in *U. virens* but some features still need to be investigated further such as activities and toxicities of each mycotoxin in humans and animals, function of mycotoxins throughout the infection by *U. virens*, variations in mycotoxins in various *U. virens* isolates, mechanism on molecular levels for mycotoxin production in *U. virens* and fresh kind of mycotoxins apart from ustiloxins and ustilaginoidin.

15.4 METHODS FOR CONTROLLING FALSE SMUT IN RICE

There are several methods reported for the control of false smut in rice such as cultural control, control by botanicals, biocontrol agent, and chemical control methods.

15.4.1 CULTURAL METHODS

Rice transplanted early time reported higher disease incidence in comparison to the late-planted (Chhottaray, 1991; Dodan and Ram Singh, 1995) while Sanne (1980) studied that early sowing helps to avoid false smut. In vulnerable cultivars, conservation tillage, continuous rice cropping, and moderate nitrogen fertilizer doses reduced false smut disease (Brooks et al., 2009). To escape heavy loss, rice sowing timing and heading period could be planned in such a way that that could not meet raining at the flowering time. Collection and destruction of infected straw is also good cultural control method and is recommended to decrease the disease incidence. Use of sclerotic free seed of rice for sowing and clear out of bunds of the field may assist the farmers to decrease the early incidence of the disease infection. In respect of cultivation practices, furrow-irrigated rice cultivation system has less disease severity compared to flooded fields. When the plants are damp, avoid field activities. Smut balls are less abundant in early planted crops than in late-planted crops. Diseased plants should be removed and destroyed to prevent sclerotia from falling into the field. To avoid alternate hosts, field bunds and irrigation ditches should be kept clean. Excess nitrogen fertilizer application should be avoided. During the Rabi season, it is critical to keep track on disease incidence on a regular basis (Brooks et al., 2010).

15.4.2 BIOCONTROL OF RICE FALSE SMUT DISEASE

Liu et al. (2017) reported that *Bacillus subtilis* is highly efficient against RFS disease (Liu et al., 2007). Under in vitro conditions nine isolates of *Trichoderma viride*, *Trichoderma virens*, *Trichoderma harzianum,* and *Trichoderma reesei* were effective and have antagonistic activity against rice false smut pathogen. However, maximum antagonistic potential shown by *T. viride* isolate Kannahi et al. (2016). For the control of false smut in rice, *Antennariella placitae* a potential fungal endophyte has been reported as this fungus has adverse effect on disease development both in vitro and in vivo condition (Andargie et al., 2017).

15.4.3 CHEMICAL CONTROL METHOD

Chemical or synthetic fungicides are the one of the most important methods of the RFS disease. Several copper fungicides namely, copper oxychloride and copper hydroxide proved highly effective (Dodan and Singh, 1997) in disease control. Liang et al. (2014) reported that the reduction efficiency is 91.92% by spraying 2.5% suspension of *Bacillus subtilis* (Wenquning), in solution of validamycin through 4.5 L/ha at 6 days earlier than heading. Timings of fungicides application are reported for the management and fever disease developments during the initial infection timing of *U. virens* in rice, as it is initiated from the stage of booting to flowering (Ashizawa et al., 2012). Evaluation of Fujione 40 EC (0.1, 0.2, and 0.3%) and Carbenadazim 50% WP (0.1 %) was done at the booting stage. Treatment with copper oxychloride (Blitox) 50 WP (0.3%) and Propiconazole (tilt) 25 EC (0.1%) on rice crop in field condition were reported highly effective for false smut disease (Bagga and Kaur, 2006). In vitro and in vivo tests of the fungicides triuoxystrobin 25% + tebuconazole 50% and propiconazole 25 EC showed that fungal mycelium growth was completely inhibited. RFS was successfully controlled by prochloraz + carbendazim followed by chlorothalonil (Mohiddin et al., 2012). According to Raji et al. (2016), propiconazole 25 EC (0.1%) caused the least amount of disease in compared to other treatment methods, trailed by trifloxistrobin along with tebuconazole 75 WG when sprayed at booting or 50% panicle emergence. Treatment of Propiconazole 25 EC and trifloxystrobin with tebuconazole 75 WG at the booting phase resulted in elevated yield. Fungicides azoxystrobin (18.2%) SC with difenconozole (11.4%) SC and metiram (55%) WG with pyraclostrobin (5%) WG @ 0.1% reported the smallest amount of infection rates as of 1.85% and 2.52%, respectively (Paramjit et al., 2006; Muniraju et al., 2917). Several other fungicides were reported for RFS disease control in field situation, that is, carbendazim and propiconazole, carbendazim, propiconazole, carbendazim and tebuconazole, propiconazole, carbendazim, tebuconazole, and carbendazim with mancozeb, trifloxystrobin with tebuconazole, and propiconazole (Dodan and Singh, 1997; Hegde et al., 2000; Bagga and Kaur, 2006; Chen et al., 2013; Ladhalakshmi et al., 2014; Shivamurthy, 2017). To suppress the growth of FS on panicles, fungicide applications before to the booting or heading stage have been examined (Cartwright et al., 2002; Cartwright and Lee, 2006; Chen et al., 2013; Hopkins et al., 2002; Tsuda et al., 2006). Tsuda et al. (2006) conducted field studies to investigate the effect of fungicide applications to flooded paddy water on FS control. The results reveal that applying simeconazole granules to paddy water 2 to 5 weeks prior to rice heading stage was extremely effective against FS, with the 3-week application being the most effective. Others (Biswas, 2001; Brooks et al., 2009) reported that water management and fertilization affected FS incidence. Several practices, that is, furrow irrigation, and reduced nitrogen treatments are among the cultural practices that have been studied to reduce false smut infestation (Brooks et al., 2009, 2010). Mew et al. (2004) and Ou (1985) pointed out that variations in disease occurrence and brutality of the infection in different rice varieties grown in identical site or regions do occur. A number of cultivars showed reduced level of infection (Biswas, 2001; Brooks et al., 2009; Brooks et al., 2010; Pan et al., 1999; Singh and Gangopadhyay, 1981; TeBeest et al., 2011), but complete resistance against false smut disease is not reported yet (Cartwright et al., 2000; Cartwright et al., 2002; Cartwright and Lee, 2006; Parsons et al., 2003; Wilson et al., 2003; Wilson et al., 2005).

15.4.4 HOST RESISTANCE

Although cultivated rice varieties of *O. sativa* have not reported with elevated intensity of resistance to *U. virens* (Biswas, 2001; Li et al., 2008); however, few rice genotypes showed significant variation in quantitative resistance (Table 15.1) to *U. virens* after disease resistance screening (Biswas, 2001; Li et al., 2008). Twenty-seven paddy genotypes were evaluated by Singh and Singh (2005) to false smut from the set of 98 rice germplasm collected from different agroclimatic regions. Mohiddin et al. (2012) also evaluated four rice cultivars and one rice genotypes namely HRI 119 showed highest resistance with low false smut disease incidence. Yan et al. (2014), who reported some hybrids with significant level of resistance. Kaur et al. (2015) also evaluated 125 rice cultivars by artificial inoculation method against false smut and reported nine rice hybrids showed complete resistance.

15.4.5 QUANTITATIVE TRAIT LOCI (QTLS) REPORTED FOR FALSE SMUT IN RICE CVS

There are very few resistances to rice false smut are reported although substantial efforts on screening of on-field resistance and quantitative trait loci (QTL) analysis are done (Guo et al., 2012; Huang et al., 2015; Zhou et al., 2014). Two key genes regulating RFS using polygene were also mapped by mixed model and the heritability of the said gene was also conducted (Li et al., 2008). Two false smut disease resistance QTLs were reported (Xu et al., 2002) with the help of isogenic introgression line. Li et al. (2014) identified eight QTLs on chromosomes 1, 2, 4, 8, 10, 11, and 12 namely; qFsr1,

TABLE 15.1 Resistance Rice for False Smut.

Sr. no	Rice cvs reported	Reference
Cultivated/land races rice cvs.		
1.	Lemont	Xu et al. (2002)
2.	HRI 119	Mohiddin et al. (2012)
3.	Jallahari	Upadhyay and Singh (2013)
4.	IR28	Han et al. (2005)
5.	Erra Malallu, JR 201, JGL 3828, Kavya, Kudrat, Varalu, and WGL 32183	Vakiti et al. (2017)
6.	R18/2348, CGY2348, GY66, YLY973, and N5Y828	Li et al. (2017)
7.	MR183–2	Han et al. (2020)
Hybrids		
1.	GK-5025, HRI-140, IRH-74, KRH-4, KPH-467, PRSH-9018, 27P64, RH-10428, and VNR-211	Kaur et al. (2015)

qFsr2, qFsr4, qFsr8, qFsr10a, qFsr10b, qFsr11, and qFsr12, respectively. The identified QTLs reported for the control of false smut resistance. Stable inheritance of false smut-resistant loci on chromosomes 10 and 11 over the years have been reported (Li et al., 2011). Random introgression lines (ILs) along with field-based phenotyping and rice QRLs providing resistance to *U. virens* were recognized. Two QTLs that contribute to RFS resistance namely, qFsr10 and qFsr12, on chromosomes 10 and 12, respectively, were also reported (Xu et al., 2002). Four QRLs namely qFSR-6-7, qFSR-10-5, qFSR-10-2, and qFSR-11-2 have been identified with moderately bigger and reliable possessions transversely the two testing sites (Zhou et al., 2014). Sensitive landrace Daguandao (*O. sativa* subsp. japonica) and a resistant cv. IR28 (*Oryza sativa* subsp. indica), were crossed. Further 157 recombinant inbred lines (RILs) populations were studied (Li et al., 2008). Following that, several RILs and parents were examined in the field after effective artificial inoculation. Further research discovered seven resistance QTLs on chromosomes 1, 2, 4, 8, 10, 11, and 12, with phenotypic variance ranging from 9.8 to 22.5% (Li et al., 2011). By using a natural infection approach, the inheritance of resistance in two-line hybrid rice was examined (Tang et al., 2014). Because natural RFS infection varies by field and season, an effective artificial inoculation procedure is strongly required for validating the disease phenotype. Although significant progress has been made in terms of screening for resistant materials and detecting resistance inheritance, the genes that cause resistance remain unknown. In the RIL population, five QTLs with significant levels of resistance were discovered on chromosomes

2 (qFsr2-1), 4 (qFsr4-1), 8 (qFsr8-1 and (qFsr8-2)), and 11 (qFsr11–1). Among these QTLs, qFsr8–1 on chromosome 8 in a tiny region is a large quantitative locus that explains the most phenotypic variance (Han et al., 2020).

15.5 MOLECULAR CHARACTERIZATION AND DETECTION OF U. VIRENS IN INFECTED RICE PLANTS

The development of the RFS pathosystem has been vulnerable by the lack of an efficient diagnostic and disease evaluation protocol or technique. Previous reports remarked that *U. virens* hyphae bother and inhabit at rice coleoptiles during the seedling germination stage and maintain an asymptomatic association with rice sieve tube elements (Ikegami, 1963). Nested PCR is used for the detection of *U. virens* and other plant pathogens where the fungal DNA is present in infected tissues (Willits and Sherwood, 1999; Zhou et al., 2003; Ashizawa and Kataoka, 2005; Ditmore and TeBeest, 2006). When two groups of scientists independently designed nested-PCR primers specific to *U. virens* rDNA, the developed protocols were used to detect *U. virens* rDNA in total DNA extracted from panicles prior to emergence from the boots (Ashizawa and Kataoka, 2005; Zhou et al., 2003) or in DNA from rice spikelet, leaf sheath, and stem tissues after artificially inoculating rice (Fujita et al., 1989; Zhou et al., 2003). The development of agrobacterium and electroporation-mediated transformation of *U. virens* has opened up new options for researching plant–pathogen interactions (Zhang et al., 2006; Tanaka et al., 2011).

A study of intraspecific ribosomal DNA (rDNA) internal transcribed spacer (ITS) sequences to create precise primers and developed a scheme for the identification of *U. virens* with nested PCR (Zhou et al. 2003). At the early reproductive development stage of rice, *U. virens* was found attached or infected on the auricle of flag leaves; simultaneously, *U. virens* also was found in the flag leaf ear of rice during early reproduction stage and duckweed in the field (Zhou et al., 2006). A transgenic *U. virens* expressing the green fluorescent protein gene (GFP) to investigate the early infection in rice panicles prior to heading was also studied (Ashizawa et al., 2012). At the booting stage, Ashizawa et al. (2005) found *U. virens* in both inoculated and non-inoculated rice, demonstrating that *U. virens* spores could naturally penetrate into the spike bud outer rice husk, attach to, and infect juvenile glume, implying that the early and late booting stages of rice were a phase of validity for *U. virens* conidia infection (Chen et al., 2013). The development and implementation of these technologies established the groundwork for a more in-depth investigation of the regularity of *U. virens* infection, as well as the rapid and accurate diagnosis and prediction of RFS (Zhou, 2004). Ditmore and TeBeest (2006) amplified *U. virens* rDNA from seed infested with *U. virens* chlamydospores. The above-said reports indicate that *U. virens* rDNA is present in rice. Thus, nested-PCR primers could be used to detect and quantify *U. virens* rDNA from colonized rice tissues at different stages of rice development and determine the distribution of the fungus within specific tissues of plants including spikelets of panicles prior to emergence from the boot.

15.5.1 STUDY OF *U. virens* GENOME AND GENETICS

U. virens genome was sequenced in 2014 and draft genome sequences are available which include 8426 expected genes and about 25% repetitive sequences (Zhang et al., 2014). Surprisingly, *U. virens* has a close evolutionary relationship with the entomopathogenic *Metarhizium* spp. *U. virens* genome data can provide insights of disease-causing ability such as secretome of up to 628 proteins has been predicted which might expressed during the early stages of infection (Zhang et al., 2014). Additionally, as compared to majority of other sequenced ascomycetes, the *U. virens* genome has fewer genes linked to polysaccharide degradation, G-protein receptors and transporters, and secondary metabolism, which may be due to its biotrophic lifestyle. A predicted protein–protein interaction network was prepared on the basis of the availability of genome sequences and gene expression profiles at the time of *U. virens* infection (Zhang et al., 2017). The previously mentioned protein–protein interaction network database, which provides insights into *U. virens* protein functions (Zhang et al., 2017). Because of geographical conditions rather than rice cultivars, *U. virens* isolates from various rice production areas exhibit considerable genetic diversity (Pan et al., 2007; Li et al., 2013; Lu et al., 2013; Sun et al., 2013; Wang et al., 2014; Wang et al., 2014b). Diverse regional populations of *U. virens* isolates show major genetic variations, whereas isolates from diverse rice cultivars in the same areas show greater genetic similarity (Sun et al., 2013). Furthermore, the availability of the *U. virens* genome sequence, as well as able approaches for targeted gene erasure and improved artificial inoculation methods, have drastically changed the methods used to explore the biology of RFS disease (Hu et al., 2014, Jia et al., 2015; Liang et al., 2018). The connection between *U. virens* pathogenicity and pathogen diversity is still unknown.

15.6 CONCLUSION AND FUTURE PROSPECTS

A number of studies are currently being conducted in order to classify rice resistance genes and for RFS disease. Several crucial questions could be resolved in the near future such as mechanism of infection and resistant gene for false smut in rice which can be utilized in breeding programs. Many other research areas are also currently being studied, including how *U. virens* takeovers the nutrient supply in rice, role of effectors, and pathogenic genes during *U. virens* infection processes. The discovery of vital genetic features of *U. virens* which are conserved in all the strains will help in the development of new fungicides. However, more study is needed for the identification and forecasting of false smut disease.

KEYWORDS

- **ascomytete fungus**
- **balls**
- **false smut**
- **rice**
- **fungicide**

REFERENCES

Abbas, H. K.; Cartwright, R. D.; Sciumbata, G. L.; Keeling, B.; Vesonder, R. F.; Shier, W. T. Mycotoxins in False Smut Balls From Rice. *Phytopathology* **2000,** *Abstract.90* (supplement), S1.

Ahonsi, M. O.; Adeoti, A. A.; Erinle, I. D.; Alegbejo, T. A.; Singh, B. N.; Sy, A. A. Effect of Variety and Sowing Date on False Smut Incidence in Upland Rice in Edo State, Nigeria. IRRI Notes **2000,** *25* (1), 14.

Andargie, M.; Congyi, Z.; Yun, Y.; Li, J. Identification and Evaluation of Potential Bio Control Fungal Endophytes Against *Ustilaginoidea Virens* on Rice Plants. *World J. Micrbiol. Biotechnol.* **2017,** *33* (6), 120–125.

Andargie, M.; Li, L. Y.; Feng, A. Q.; Zhu, X. Y.; Li, J. X. Colonization of Rice Roots by a Green Fluorescent Protein-Tagged Isolate of *Ustilaginoidea Virens*. *Am. J. Plant Sci.* **2015,** *6* (14), 2272–2279.

Andargie, M.; Yang, C.; Li, J. X. Generation of β-Glucuronidase Reporter-Tagged Strain to Monitor *Ustilaginoidea Virens* Infection in Rice. *J. Microbiol. Methods* **2016,** *131*, 148–155.

Ashizawa, T.; Kataoka, Y. Detection of *Ustilaginoidea Virens* in Rice Panicles Before and After Heading in the Field Using Nested-PCR Technique With Species-Specific Primers. *Jpn. J. Phytopathol.* **2005,** *71* (1), 16–19.

Ashizawa, T.; Takahashi, M.; Arai, M.; Arie, T. Rice False Smut Pathogen, *Ustilaginoidea Virens*, Invades Through Small Gap at the Apex of a Rice Spikelet Before Heading (Fungal Diseases). *J. Gen. Plant Pathol.* **2012,** *78*, 255–259.

Ashizawa, T.; Takahashi, M.; Moriwaki, J.; Hirayae, K. Quantification of the Rice False Smut Pathogen U. Virens From Soil in Japan Using Real-Time PCR. *Eur. J. Plant Pathol.* **2010,** *128,* 221–222.

Atia, M. M. M. Rice False Smut (*Ustilaginoidea Virens*) in Egypt. *J. Plant Dis. Protect.* **2004,** *14,* 71–82.

Bagga, P. S.; Kaur, S. Evaluation of Fungicides for Controlling False Smut (*Ustilaginoidea Virens*) of Rice. *J. Indian Phytopath.* **2006,** *59,* 115–117.

Bian, Y. F.; Yu, S. S.; Mou, R. X.; Cao, Z. Y.; Sun, W. H.; Yang, H.; Lin, X. Y.; Chen, M. X. Identification of Ustiloxins in False Smut Balls of Rice Based on High Performance Liquid Chromatography-High Resolution Mass Spectrometry. *Chin. J. Chromat.* **2015,** *33* (10), 1046–1050. (In Chinese With English Abstract)

Biswas, A. False Smut Disease of Rice: A Review. *Environ. Ecol.* **2001,** *19,* 67–83.

Biswas, A. Field Reaction of Hybrid Rice Varieties to False Smut (FS) and Kernel Smut (KS) Disease in West Bengal, India. *Environ. Ecol.* **2001,** *19,* 229–230.

Brooks, S. A.; Anders, M. M.; Yeater, K. M. Effect of Cultural Management Practices on the Severity of False Smut and Kernel Smut of Rice. *Plant Dis.* **2009,** *93,* 1202–1208.

Brooks, S. A.; Anders, M. M.; Yeater, K. M. Effect of Furrow Irrigation on the Severity of False Smut In Susceptible Rice Carieties. *Plant Dis.* **2010,** *94,* 570–574.

Cao, Z. Y.; Sun, L. H.; Mou, R. X.; Lin, X. Y.; Zhou, R.; Ma, Y. N.; Chen, M. X. Analysis of Ustiloxins in Rice Using Polymer Cation Exchange Cleanup Followed by Liquid Chromatography-Tandem Mass Spectrometry. *J. Chromat. A* **2016,** *1476,* 46–52.

Cartwright, R. D.; Lee, F. N. *Arkansas Plant Disease Control Products Guide MP 154*; Cartwright, R., Ed.; Arkansas Cooperative Extension Service: Little Rock, AR, 2006; p 29.

Cartwright, R. D.; Parsons, C. E.; Sutton, E. A.; Lee, F. N. Disease Monitoring and Evaluation of Rice Varieties on Arkansas Farms. In *B.R. Wells Rice Research Studies 2002*; Norman, R. J., Meullenet, J. F., Moldenhauer, K. A. K., Eds.; Arkansas Agricultural Experiment Station Research Series, 2002; vol 504, pp 219–228.

Cartwright, R. D.; Sutton, E. A.; Parsons, C. E.; Dodgen, B. J.; Clark, S. Evaluation of Foliar Fungicides for Control of False Smut of Rice. Fungicide and Nematicide Tests. *Report* **2000,** *56,* CF2.

Chao, J. Q.; Jin, J.; Wang, D.; Han, R.; Zhu, R. S.; Zhu, Y, G.; Li, S. Q. Cytological and Transcriptional Dynamics Analysis of Host Plant Revealed Stage-Specific Biological Processes Related to Compatible Rice-*Ustilaginoidea Virens* Interaction. *PLoS One* **2014,** ***9*** (3), e91391.

Chen, F.; Lin, T.; Gan, L.; Du, Y.; Ruan, H.; Yang, X. SCAR Marker and PCR Detection of *Ustilaginoidea Viren. Acta Phytopathol. Sinica* **2013,** *40* (6), 481–487

Chen, Z. Y.; Yin, S. Z.; Chen, Y. L. Biological Characteristics of *Ustilaginodea Virens* and the Screening *In Vitro* of the Causal Pathogen of False Smut of Rice and Fungicides. *Chinese Rice Res. Newsl. (CRRN)* **1994,** *2* (1), 4–6.

Chen, Y.; Zhang, Y.; Yao, J.; Li, Y. -F.; Yang, X.; Wang, W. -X.; Zhang, A. -F.; Gao, T. -C. Frequency Distribution of Sensitivity of Ustilaginoidea Virens to Four EBI Fungicides,

Prochloraz, Difenoconazole, Propiconazole and Tebuconazole and Their Efficacy in Controlling Rice False Smut in Anhi Province of China. *Phytoparasitica* **2013,** 1–8.

Chen, Y.; Zhang, Y.; Yao, J.; Li, Y. F.; Yang, X.; Wang, W. X.; Zhang, A. F.; Gao, T. C. Frequency Distribution of Sensitivity of *Ustilaginoidea Virens* to Four EBI Fungicides, Prochloraz, Difenconazole, Propiconazole and Tebuconazole and Their Efficacy in Controlling False Smut in Anhui Province of China. *Phytoparasitica* **2013,** *14* (3), 277–284.

Chhottaray, P. K. Doctor of Philosophy Thesis, Orissa University of Agriculture and Technology, Bhubaneswar, Orissa, 1991, 175.

Cooke, M. C. Some Extra–European Fungi. *Grevillea* **1878,** *7,* 13–15.

Deng, G. S. Present Status of Research on False Smut in China. *Plant Protect.* (Chinese) **1989,** *15* (6), 39–40.

Dhua, U.; Dhua, S. R.; Sahu, R. K. Precise Disease Severity Assessment for False Smut Disease of Rice. *J. Phytopathol.* **2015,** *163,* 931–940.

Ditmore, M.; TeBeest, D. O. Detection of Seed-Borne Ustilaginoidea Virens by Nested PCR. In *B.R. Wells Rice Research Studies 2005*; Norman, R. J., Meullenet, J. F., Moldenhauer, K. A. K., Eds.; Arkansas Agricultural Experiment Station Research Series 2006, vol 550, pp 121–125.

Dodan, D. S.; Singh, R. Effect of Planting Time on the Incidence of Blast and False Smut of Rice in Haryana. *Indian Phytopathol.* **1995,** *48,* 185–186.

Dodan, D. S.; Singh, R. Evaluation of Fungi Toxicants Against False Smut of Rice. *J. Mycol. Pl. Pathol*. **1997,** *27* (1), 32–34.

Fan, J.; Guo, X. Y.; Li, L.; Huang, F.; Sun, W. X.; Li, Y.; Huang, Y. Y.; Xu, Y. J.; Shi, J.; Lei, Y.; Zheng, A. P.; Wang, W. M. Infection of *Ustilaginoidea Virens* Intercepts Rice Seed Formation but Activates Grain-Filling-Related Genes. *J. Integr. Plant Biol*. **2015,** *57* (6), 577–590.

Fan, J.; Yang, J.; Wang, Y. Q.; Li, G. B.; Li, Y.; et al. Current Understanding on *Villosiclava Virens*, A Unique Flower-Infecting Fungus Causing Rice False Smut Disease. *Mol. Plant Pathol.* **2016,** *17,* 1321–1330.

Fan, L. L.; Yong, M. L.; Li, D. Y.; Liu, Y. J.; Lai, C. H.; et al. Effect of Temperature on the Development of Sclerotia in *Villosiclava Virens. J. Integr. Agric*. **2016,** *15,* 2550–2555.

Fu, X. X.; Wang, X. H.; Cui, Y. L.; Wang, A. L.; Lai, D. W.; Liu, Y.; Li, Q. X.; Wang, B. M.; Zhou, L. G. A Monoclonal Antibody-Based Enzyme-Linked Immunosorbent Assay for Detection of Ustiloxin A in Rice False Smut Balls and Rice Samples. *Food Chem*. **2015b,** *181,* 140–145.

Fu, R. T.; Wang, J.; Lu, D. H.; Zhang, H.; Gong, X. S.; Chen, X. J.; Ren, H. Z.; Mao, J. H. Resistance Identification and Influence Factor of Rice False Smut. *Chin. Agric. Sci. Bull.* **2015,** *31* (18), 266–272. (In Chinese With English Abstract)

Fu, X. X.; Wang, A. L.; Wang, X. H.; Lin, F. K.; He, L. S.; Lai, D. W.; Liu, Y.; Li, Q. X.; Zhou, L. G.; Wang, B. M. Development of a Monoclonal Antibody-Based icELISA for the Detection of Ustiloxin B in Rice False Smut Balls and Rice Grains. *Toxins* **2015a,** *7* (9), 3481–3496.

Fu, X. X.; Xie, R. S.; Wang, J.; Chen, X. J.; Wang, X. H.; Sun, W. B.; Meng, J. J.; Lai, D. W.; Zhou, L. G.; Wang, B. M. Development of Colloidal Gold-Based Lateral Flow Immunoassay for Rapid Qualitative and Semiquantitative Analysis of Ustiloxins A and B in Rice Samples. *Toxins* **2017,** *9* (3), 79–90.

Fujita, Y.; Sonoda, R.; Yaegashi, H. Inoculation With Conidiospores of False Smut Fungus to Rice Panicles at the Booting Stage. *Ann. Phytopath. Soc. Japan* **1989,** *55*, 629–634.

Fulton, H. R. Diseases Affecting Rice in Louisiana. Bulletin of the Louisiana Agricultural Expt. Station, 1908; pp 1–28.

Guo, X.; Li, Y.; Fan, J.; Li, L.; Huang, F.; Wang, W. Progress in the Study of False Smut Disease in Rice. *J. Agr. Sci. Tech.* **2012,** *2,* 1211–1217.

Han, Y.; Li, D.; Yang, J.; Huang, F.; Sheng, H.; Sun, W. Mapping Quantitative Trait Loci for Disease Resistance to False Smut of Rice. *Phytopathol. Res.* **2020,** *2*, 20.

Han, Y.; Zhang, K.; Yang, J.; Zhang, N.; Fang, A.; Zhang, Y.; Liu, Y.; Chen, Z.; Hsiang, T.; Sun, W. Differential Expression Profiling of the Early Response to Ustilaginoidea Virens Between False Smut Resistant and Susceptible Rice Varieties. *BMC Genom.* **2015,** *16*, 955.

Hashioka, Y.; Yoshino, M.; Yamamoto, T. Physiology of the Rice False Smut, Ustilaginoidea Virens (Cke.) Tak. Research Bulletin, Saitama Agricultural Experiment Station, 1951, vol 2, pp 1–20.

Hegde, Y. R.; Anahosur, K. H., Kulkarni, S. Chemical Control of False Smut of Rice Caused by *Claviceps Oryzae-Sativae* Hashioka. *Karnataka J. Agric. Sci.* **2000,** *13* (3), 623–627.

Hopkins, A.; Bowman, R.; Parsons, C. E.; Clark, S. D.; Cartwright, R. D. Evaluation of Foliar Fungicides for Control of False Smut of Rice; Fungicide and Nematicide Tests, Report 2002; vol 58, p C069.

Hu, M. L.; Luo, L. X.; Wang, S.; Liu, Y. F.; Li, J. Q. Infection Processes of *Ustilaginoidea Virens* During Artificial Inoculation of Rice Panicles. *Eur. J. Plant Pathol.* **2014,** *139* (1), 67–77.

Hu, M.; Luo, L.; Wang, S.; Liu, Y.; Li, J. Infection Processes of Ustilaginoidea Virens During Artificial Inoculation of Rice Panicles. *Eur. J. Plant Pathol.* **2014,** *139,* 67–77.

Huang, F.; Li, Y.; Shi, J.; Fan, J.; Xu, Y. J.; Wang, W. M. Screening and Polymorphism Analysis of Rice Germplasms for Resistance to False Smut Disease in Sichuan Province. *Acta Phytopathol. Sin.* **2015,** *46* (2), 247–257.

Ikegami, H. Studies on the False Smut of Rice. X. Invasion of Chlamydospores and Hyphae of the False Smut Into Rice Plants. Research Bulletin of the Faculty of Agriculture, Gifu University: Naka, Gifu, 1963; vol 18, pp 54–60.

Jecmen, A. C.; Tebeest, D. O. First Report of the Occurrence of a White Smut Infecting Rice in Arkansas. *J. Phytopathol.* **2015,** *163* (2), 138–143.

Jia, Q.; Lv, B.; Guo, M. Y.; Luo, C. X.; Zheng, L.; Hsiang, T.; Huang, J. B. Effect of Rice Growth Stage, Temperature, Relative Humidity and Wetness Duration on Infection of Rice Panicles by *Villosiclava Virens*. *Eur. J. Plant Pathol.* **2015,** *141* (1), 15–25.

Jiehua, Q.; Shuai, M.; Yizhen, D.; Shiwen, H.; Yanjun, K. *Ustilaginoidea Virens*: A Fungus Infects Rice Flower and Threats World Rice Production. *Rice Sci.* **2019,** *26*(4), 199–206.

Kannahi, M.; Dhivya, S.; Senthil Kumar, R. Biological Control on Rice False Smut Disease Using *Trichoderma* Species. *Int. J Pure App. Biosci.* **2016,** *4* (2), 311–316.

Kaur, Y.; Lore, J. S.; Pannu, P. P. S. Evaluation of Rice Genotypes for Resistance Against False Smut. *Plant Dis. Res.* **2015,** *30* (1), 46–49.

Kim, K. W.; Park, E. W. Ultrastructure of Spined Conidia and Hyphae of the Rice False Smut Fungus Ustilaginoidea Virens. *Micron* **2007,** *38,* 626–631.

Koiso, Y.; Li, Y.; Iwasaki, S.; Hanaoka, K.; Kobayashi, T.; et al. Ustiloxins, Antimitotic Cyclic Peptides From False Smut Balls on Rice Panicles Caused by *Ustilaginoidea Virens. J. Antibiot.* **1994,** *47*, 765–773.

Koiso, Y.; Li, Y.; Iwasaki, S.; Hanaoka, K.; Kobayashi, T.; Sonoda, R.; Fujita, Y.; Yaegashi, H.; Sato, Z. Ustiloxins, Atimitotic Cyclic Peptides From False Smut Balls on Rice Panicles Caused by *Ustilaginoidea Virens*. *J Antibiot.* **1994,** *47* (7), 765–773.

Koiso, Y.; Morisaki, N.; Yamashita, Y.; Mitsui, Y.; Shirai, R.; Hashimoto, Y.; Iwasaki, S. Isolation and Structure of an Antimitotic Cyclic Peptide, Ustiloxin F: Chemical Interrelation With a Homologous Peptide, Ustiloxin B. *J. Antibiot.* **1998,** *51* (4), 418–422.

Kulkarni, C. S.; Moniz, L. False Smut of Rice-an Airborne Disease. *Curr. Sci.* **1975,** *44,* 483–484.

Kumagai, T.; Ishii, T.; Terai, G.; Umemura, M.; Machida, M.; Asai, K. Genome Sequence of *Ustilaginoidea Virens* IPU010, A Rice Pathogenic Fungus Causing False Amut. *Genome Announc* **2016,** *4* (3), e00306–16.

Ladhalakshmi, D.; Laha, G. S.; Krishnaveni, D.; Prakasam, V.; Prasad, M. S. *Evaluation of Selected Fungicides Against Rice False Smut Disease.* 3rd Int. Conference on Agric & Horti.; Hyderabad International Convention Centre: India, 2014.

Ladhalakshmi, D.; Laha, G. S.; Singh, R.; Karthikeyan, A.; Mangrauthia, S. K.; Sundaram, R. M.; Thukkaiyannan, P.; Viraktamath, B. C. Isolation and Characterization of *Ustilaginoidea Virens* and Survey of False Smut Disease of Rice in India. *Phytoparasitica* **2012,** *40* (2), 171–176.

Ladhalakshmi, D.; Laha, G. S.; Singh, R.; Karthikeyan, A.; Mangrauthia, S. K.; Sundaram, R. M. Isolation and Characterization of *Ustilaginoidea Virens* and Survey of False Smut Disease of Rice in India. *Phytoparasitica* **2012,** *40*, 171–176.

Lee, F. N.; Gunnell, P. S. False Smut. In *Compendium of Rice Diseases*; Webster, R. K., Gunnell, P. S., Ed.; APS Press: St. Paul, Minnesota, 1992; pp 5 and 28.

Li, Y.-S.; Huang, S.-D.; Yang, J.; Wang, C.-L. Analysis of Quantitativc Trait Loci for Resistance to Rice False Smut. *Acta Agron. Sin.* **2011,** 37, 778–783.

Li, Y. G.; Kang, B. J.; Zhang, B. D.; Zeng, H. Z.; Xie, K. X.; Lan, Y. T.; Ma, H.; Li, T. F. A Preliminary Study on Rice False Smut. *Guangdong Agricul. Sci.* **1986,** *4,* 45–47.

Li, W. L.; Li, L. Y.; Feng, A. Q.; Zhu, X. Y.; Li, J. X. Rice False Smut Fungus, Inhibits Pollen Germination and Degrades the Integuments of Rice Ovule. *Am. J. Plant Sci*. **2013,** *4* (12), 2295–2304.

Li, Y.; Shi, J.; Xiang, Z.; Hu, R.; Shi, S.; Peng, T.; Liu, D.; Huang, T. Analysis of the Resistance to Rice Blast and False Smut of 18 Varieties of Hybrid Rice in Sichuan Province, China. *Int. J. Agric. Biol.* **2017,** *19,* 880–886.

Li, Y. S.; Yang, J.; Huang, S. D.; Wang, C. L. Analysis on Quantitative Trait Loci for Resistance to Rice False Smut Under Different Environmental Conditions. *Agric. Sci. Technol.* **2014,** *15* (3), 449–452.

Li, Y.-S.; Zhu, Z.; Zhang, Y.-D.; Zhao, L.; Wang, C.-L. Genetic Analysis of Rice False Smut Resistance Using Mixed Major Genes and Polygenes Inheritance Model. *Acta Agron. Sin.* **2008,** *34,* 1728–1733.

Li, Y. S.; Zhu, Z.; Zhang, Y. D.; Zhao, L.; Wang, C. L. Genetic Analysis of Rice False Smut Resistance Using Mixed Major Genes and Polygenes Inheritance Model. *Acta Agron. Sin.* **2008,** *34* (10), 1728–1733.

Liang, Y. F.; Han, Y.; Wang, C. F.; Jiang, C.; Xu, J. R. Targeted Deletion of the *USTA* and *UvSLT2* Genes Efficiently in *Ustilaginoidea Virens* With the CRISPR-Cas9 System. *Front Plant Sci*. **2018,** *9,* 699.

Liang, Y.; Zhang, X.; Li, D.; Huang, F.; Hu, P.; Peng, Y. Integrated Approach to Control False Smut in Hybrid Rice in Sichuan Province, China. *Rice Sci.* **2014,** *21,* 354–360.

Lin, X. Y.; Bian, Y. F.; Mou, R. X.; Cao, Z. Y.; Cao, Z. Z.; Zhu, Z. W.; Chen, M. X. Isolation, Identification, and Characterization of *Ustilaginoidea Virens* From Rice False Smut Balls With High Ustilotoxin Production Potential. *J Basic Microbiol*. **2018,** *58* (8), 670–678.

Liu, Y. L.; Chen, Z. Y.; Zhou, M. G.; Zhang, J.; Song, F. P.; Liu, Y. Analysis on Antifungal and Antibacterial Activities of *Bacillus Subtilis* Strain Bs- 916 and its Extract. *Chin. J. Pest Sci.* (In Chinese With English Abstract) **2007,** *9* (1), 92–95.

Lu, D. H.; Chen, Y. P.; Wang, J.; Wang, P.; Yao, L.; Ye, H. L.; Hu, R. P.; Chen, Y.; Mao, J. H. Genetic Diversity of *Ustilaginoidea Viren* From Rice in Sichuan Province. *Southwest China J. Agric. Sci.* **2013,** *26* (3), 994–1000. (In Chinese With English Abstract)

Lu, M. H.; Liu, W. C.; Zhu, F. Epidemic Low and Control Technique of Rice False Smut in Recent Years. *China Plant Protect.* **2018,** *38* (5), 44–47. (In Chinese)

Lu, S. Q.; Sun, W. B.; Meng, J. J.; Wang, A. L.; Wang, X. H.; Tian, J.; Fu, X. X.; Dai, J. G.; Liu, Y.; Lai, D. W.; Zhou, L. G. Bioactive Bis-naphtho-Gamma-Pyrones From Rice False Smut Pathogen *Ustilaginoidea Virens*. *J. Agric. Food Chem.* **2015,** *63* (13), 3501–3508.

Lu, D. H.; Yang, X. Q.; Mao, J. H.; Ye, H. L.; Wang, P.; Chen, Y. P.; He, Z. Q.; Chen, F. Characterising the Pathogenicity Diversity of Ustilaginoidea Virens in Hybrid Rice in China. *J. Plant Pathol.* **2009,** *91* (2), 443–451.

Meng, J. J.; Sun, W. B.; Mao, Z. L.; Xu, D.; Wang, X. H.; Lu, S. Q.; Lai, D. W.; Liu, Y.; Zhou, L. G.; Zhang, G. Z. Main Ustilaginoidins and Their Distribution in Rice False Smut Balls. *Toxins* **2015,** *7* (10), 4023–4034.

Mew, T. W.; Leung, H.; Savary, S.; Vera Cruz, C. M.; Leach, J. E. Looking Ahead in Rice Disease Research and Management. *Crit. Rev. Plant Sci.* **2004, 23,** 103–127.

Mohiddin, F. A.; Bhat, F. A.; Gupta, V.; Gupta, D.; Kalha, C. S. Integrated Disease Management of False Smut of Rice Caused by Ustilaginoidea Virens. *Trends Biosci.* **2012,** *5* (4), 301–302.

Muniraju, K. M.; Pramesh, D.; Mallesh, S. B.; Mallikarjun, K.; Guruprasad, G. S. Novel Fungicides for the Management of False Smut Disease of Rice Caused by *Ustilaginoidea Virens. Int. J. Curr. Microbiol. App. Sci.* **2017,** *6* (11), 2664–2669.

Nakamura, K.; Izumiyama, N.; Ohtsubo, K.; Koiso, Y.; Iwasaki, S.; et al. "Lupinosis"-Like Lesions in Mice Caused by Ustiloxin, Produced by *Ustilaginoidea Virens*: A Morphological Study. *Nat. Toxins* **1994,** *2,* 22–28.

Nessa, B.; Salam, M. U.; Haque, A. H. M. M.; Biswas, J. K.; Kabir, M. S.; Macleod, W. J.; Antuono, D. M.; Barman, H. N.; Latif, M. A.; Galloway, J. Spatial Pattern of Natural Spread of Rice False Smut (*Ustilaginoidea Virens*) Disease in Fields. *Am. J. Agric. Biol. Sci.* **2015,** *10* (2), 63–73.

Nessa, B.; Salam, M. U.; Haque, A. H. M. M.; Biswas, J. K.; Latif, M. A.; Ali, M. A.; Ansari, T. H.; Ahmed, M.; Parvin, N.; Baki, M. Z. I.; Salam, S.; Islam, M. S.; Galloway, J. Rice False Smut Disease at Different Times of Flowering. *Bangladesh Rice J.* **2015c,** *19,* 29–35.

Ou, S. H. Diseases of Rice. Commonwealth Mycological Institute: Kew, England, 1985; vol (2), pp 306 311.

Pan, Y. J.; Wang, S.; Yang, H.; Liu, X. Z.; Xie, X. W.; Zhou, Y. L. Effect of Rice Varieties on the Population Structure of *Ustilaginoidea Virens*. *Acta Phytopathol. Sin.* **2007,** *32* (2), 214–216. (In Chinese With English Abstract)

Pan, X. B.; Rush, M. C.; Sha, X. Y.; Xie, Q. J.; Linscombe, S. D.; Stetina, S. R.; Oard, J. H. Major Gene, Non-Allelic Sheath Blight Resistance From the Rice Cultivars Jasmine 85 and Tequing. *Crop Sci.* **1999,** *39,* 338–346.

Paramjith, S. B.; Sweety, K. Evaluation of Fungicides for Controlling False Smut (*Ustilaginoidea Virens)* of Rice. *Indian Phytopathol.* **2006,** *59* (1), 115–117.

Parsons, C. E.; Wilson, C.; Cartwright, R. D.; Sutton, E. A.; Carpenter, C.; Lee, F. N. Evaluation of Rice Varieties for Performance and Disease Reaction on Farms. In *B.R. Wells*

Rice Research Studies 2003; Norman, R. J., Meullenet, J. F., Moldenhauer, K. A. K., Eds.; Arkansas Agricultural Experiment Station Research Series, **2003,** *517,* 79–87.

Prakobsub, K.; Ashizawa, T. Intercellular Invasion of Rice Roots at the Seedling Stage by the Rice False Smut Pathogen, *Villosiclava Virens*. *J. Gen. Plant Pathol.* **2017,** *83,* 358–361.

Raji, P.; Sumiya, K. V.; Renjisha, K.; Dhanya, S.; Narayanankutty, M. C. Evaluation of Fungicides Against False Smut of Rice Caused by *Ustilaginoidea Virens. Int. J. Appl. Nat. Sci.* **2016,** *5* (2), 77–82.

Raychaudhuri, S. P. Mode of Infection of Rice by Ustilaginoidea Virens (CKE) Tak. *J. Indian Bot. Soc.* **1946,** *25,* 145–150.

Sanne, G. Studies on False Smut Disease Caused by *Ustilaginoidea Virens* on Paddy in Karnataka, India. *Int. Rice Res. Newsl.* **1980,** *5,* 4–5.

Shan, T. J.; Sun, W. B.; Liu, H.; Gao, S.; Lu, S. Q.; Wang, M. G.; Sun, W. X.; Chen, Z. Y.; Wang, S.; Zhou, L. G. Determination and Analysis of Ustiloxins A and B by LC-ESI-MS and HPLC in False Smutballs of Rice. *Int. J. Mol. Sci.* **2012,** *13* (9), 11275–11287.

Shivamurthy, P. Studies on False Smut of Rice Caused *Ustilaginoidea Virens* (Cke.) Tkh. M. Sc. Thesis, Univ. Agric. Sci., Raichur, Karnataka (India), 2017.

Singh, R. A.; Dubey, K. S. Sclerotial Germination and Ascospores Formation of Claviceps Oryzae-Sativae in India. *Indian Phytopathol.* **1984,** *37,* 168–170.

Singh, M.; Gangopadhyay, S. Artificial Culture of Ustilaginoidea Virens and Screening of Rice Varieties. *Trans. Brit. Mycol. Soc.* **1981,** *77,* 660–661.

Singh, R. N.; Khan, A. T. Field Resistance to False Smut (FS) and Narrow Brown Leaf Spot (NBLS) in Eastern Uttar Pradesh. *IRRI Newsl.* **1989,** *14* (4), 16–17.

Singh, G. P.; Singh, R. N.; Singh, A. Status of False Smut (FS) of Rice in Eastern Uttar Pradesh, India. *IRRI Newsl.* **1987,** *12* (2), 28.

Singh, A. K.; Singh, R. N. Screening for Resistance to False Smut (*Ustilaginoidea Virens* Takahashi) of Rice (*Oryza Sativa* L.). *Indian J. Genet.* **2005,** *65,* 49–50.

Song, J. H.; Wei, W.; Lv, B.; Lin, Y.; Yin, W. X.; Peng, Y. L.; Schnabel, G.; Huang, J. B.; Jiang, D. H.; Luo, C. X. Rice False Smut Fungus Hijacks the Rice Nutrients Supply by Blocking and Mimicking the Fertilization of Rice Ovary. *Environ. Microbiol.* **2016,** *18* (11), 3840–3849.

Sreeramulu, R.; Vittal, B. P. R. Periodicity in the Air of Air Borne Conidia of Rice False Smut Fungus, Ustilaginoidea Virens. *Trans. Brit. Mycol. Soc.* **1996,** *49,* 443–449.

Sugha, S. K.; Sharma, O. P.; Kaushik, R. P. Performance of Rice Genotypes Against False Smut Pathogen Under Rainfed Conditions. *Pl. Dis. Res.* **1992,** *8,* 76–77.

Sun, W. B.; Dong, X. J.; Xu, D.; Meng, J. J.; Fu, X. X.; Wang, X. H.; Lai, D. W.; Zhou, L. G.; Liu, Y. F. Preparative Separation of Main Ustilaginoidins From Rice False Smut Balls by High-Speed Counter-Current Chromatography. *Toxins* **2016,** *8* (1), 20.

Sun, W.; Fan, J.; Fang, A.; Li, Y.; Tariqjaveed, M.; Li, D.; Hu, D.; Wang, W.-M. *Ustilaginoidea Virens*: Insights Into an Emerging Rice Pathogen. *Annu. Rev. Phytopathol.* **2020,** *58,* 3.1–3.23

Sun, X. Y.; Shu, K.; Zhang, Y. J.; Tan, X. Q.; Yu, Y. F.; He, H. Y.; Zhang, X. Y.; Liu, Y. F.; Shu, W.; Sun, W. X.; Cai, L.; Li, S. J. Genetic Diversity and Population Structure of Rice Pathogen *Ustilaginoidea Virens* in China. *PLoS One* **2013,** *8* (9), e76879.

Sun, W. B.; Wang, A. L.; Xu, D.; Wang, W. X.; Meng, J. J.; Dai, J. G.; Liu, Y. F.; Lai, D. W.; Zhou, L. G. New Ustilaginoidins From Rice False Smut Balls Caused by *Villosiclava Virens* and Their Phytotoxic and Cytotoxic Activities. *J. Agric. Food Chem.* **2017,** *65* (25), 5151–5160.

Takahashi, Y. On Ustilago Virens Cooke and A New Species of Tilletia Parasitic on Rice Plants. *Botanical Magazine*, Tokyo, 1896; vol 10, pp 16–20.

Talbot, N. J. On the Trail of a Cereal Killer: Exploring the Biology of *Magnaporthe Grisea*. *Annu. Rev. Microbiol.* **2003,** *57,* 177–202.

Tanaka, E.; Ashizawa, T.; Sonoda, R.; Tanaka, C. *Villosiclava Virens* gen. nov., comb. nov., Teleomorph of *Ustilaginoidea Virens*, the Causal Agent of Rice False Smut. *Mycotaxon* **2008,** *106,* 491–501.

Tanaka, E.; Kumagawa, T.; Tanaka, C.; Koga, H. Simple Transformation of the Rice False Smut Fungus *Villosiclava Virens* by Electroporation of Intact Conidia. *Mycoscience* **2011,** *52* (5), 344–348.

Tang, S.; Gao, D.; Chen, Y.; Li, Y.; Zhang, S.; Zhao, Z. Inheritance Pattern of Twoline Hybrid Rice Resistance to False Smut. *Agric. Sci. Technol.* **2014,** *15,* 557–560.

Tang, Y. X.; Jin, J.; Hu, D. W.; Yong, M. L.; Xu, Y.; He, L. P. Elucidation of the Infection Process of *Ustilaginoidea Virens* (Teleomorph: *Villosiclava Virens*) in Rice Spikelets. *Plant Pathol.* **2013,** *62* (1), 1–8.

TeBeest, D. O.; Jecmen, A.; Ditmore, M. Infection of Rice by the False Smut Fungus, Ustilaginoidea Virens. In *B.R. Wells Rice Research Studies 2010*; Norman, R. J., Meullenet, J. F., Moldenhauer, K. A. K., Eds.; Arkansas Agricultural Experiment Station Research Series, 2011; vol 591, pp 70–81.

Tsuda, M.; Sasahara, M.; Ohara, T.; Kato, S. Optimal Application Timing of Simeconazole Granules for Control of Rice Kernel Smut and False Smut. *J. Gen. Plant Pathol.* **2006.** *72* (5),301–304.

Upadhyay, A. L.; Singh, R. V. Yield Loss Assessment in Rice Due to False Smut. Ann. Plant Soil Res. **2013,** *15* (2), 173–174.

Vakiti, D.; Bhale, U.; Teja, T. R.; Bhale, M. S.; Koutu, G. K. Utilization and Identification of Rice Genetic Stock Against False Smut Disease Under Agro Conditions of Jabalpur (M.P.). *Agric. Update* **2017,** *12* (1), 79–82.

Wang, X. H.; Fu, X. X.; Lin, F. K.; Sun, W. B.; Meng, J. J.; Wang, A. L.; Lai, D. W.; Zhou, L. G.; Liu, Y. The Contents of Ustiloxins A and B Along With Their Distribution in Rice False Smut Balls. *Toxins* **2016,** *8* (9), 262.

Wang, X. H.; Wang, J.; Lai, D. W.; Wang, W. X.; Dai, J. G.; Zhou, L. G.; Liu, Y. Ustiloxin G, A New Cyclopeptide Mycotoxin From Rice False Smut Balls. *Toxins* **2017,** *9* (2), 54.

Wang, F.; Zhang, S.; Liu, M. G.; Lin, X. S.; Liu, H. J.; Peng, Y. L.; Lin, Y.; Huang, J. B.; Luo, C. X. Genetic Diversity Analysis Reveals That Geographical Environment Plays a More Important Role Than Rice Cultivar in *Villosiclava Virens* Population Selection. *Appl. Environ. Microbiol.* **2014,** *80* (9), 2811–2820.

Wang, W. B.; Zhang, R. S.; Luo, C. P.; Yin, X. L.; Liu, Y. F.; Lu, F.; Chen, Z. Y. Biological Characteristics and Genetic Diversity of *Ustilaginoidea Virens* From Rice Regions in China. *Sci. Agric. Sin.* **2014b,** *47* (14), 2762–2773. (In Chinese With English Abstract)

Willits, D. A.; Sherwood, J. E. Polymerase Chain Reaction Detection of *Ustilago Hordei* in Leaves of Susceptible and Resistant Barley Varieties. *Phytopathology* **1999,** *89* (3), 212–217.

Wilson, C. E.; Cartwright, R. D.; Gibbons, J. W.; Richards, A. L.; Frizzell, D. L.; Branson, J. W.; Runsick, S.; Parsons, C. E. Evaluation of Rice Varieties for Performance and Disease Reaction on Farms. In *B. R. Wells Rice Research Studies 2004*; Norman, R. J., Muellenet, J. F., Moldenhauer, K. A. K., Eds.; University of Arkansas Agricultural Experiment Station Research Series, **2005**; *529,* 149–158.

Wilson, C. E.; Scott, B.; Smith, K.; Cartwright, R.; Moldenhauer, K.; Gibbons, J.; Bernhardt, J.; Norman, R.; Slaton, N.; Blocker, M.; Tolbert, A.; Branson, J. CLEARFIELD* Rice – A New Approach to Red Rice. Rice Information, University of Arkansas Division of Agriculture, U. S. Department of Agriculture and County Governments Cooperating. No. 2003; p 153.

Xu, J. L.; Xue, Q. Z.; Luo, L. J.; Li, Z. K. Preliminary Report on Quantitative Trait Loci Mapping of False Smut Resistance Using Near-Isogenic Introgression Lines in Rice. *Acta Agric. Zhejiangensis* **2002,** *1*,14–19.

Yan, L.; Zhang, X. M.; Li, D. Q.; Huang, F.; Hu, P. S.; Peng, Y. L. Integrated Approach to Control False Smut in Hybrid Rice in Sichuan Province, China. *Rice Sci.* **2014,** *21* (4), 354–360.

Yong, M. L.; Liu, Y. J.; Chen, T. Q.; Fan, L. L.; Wang, Z. Y.; Hu, D. W. Cytological Studies on the Infection of Rice Root by *Ustilaginoidea Virens*. *Microsc. Res. Techn.* **2018b,** *81* (4), 389–396.

Yoshoda, H.; Anahosur, K. H.; Srikant, K. Chemical Control of False Smut of Rice Caused by *Claviceps Oryzae- Sativae* Hashioka. *Karnataka J. Agri. Sci.* **2000,** *13* (3), 623–627.

Yu, M. N.; Chen, Z. Y.; Yu, J. J.; Hu, J. K.; Yin, X. L.; Nie, Y. F.; Liu, Y. F. Genetic Diversity and Pathogenicity of *Ustilaginoidea Virens* Isolated From Different Rice False Smut Balls of a Diseased Spike. *Acta Phytopathol. Sin.* 2013, *43* (6), 561–573. (In Chinese With English Abstract)

Zhang, Z.; Du, X. F.; Chai, R. Y.; Mao, X. Q.; Qiu, H. P.; Wang, Y. L.; … Sun, G. Agrobacterium Tumefaciens-Mediated Transformation of the Pathogen of *Ustilaginoidea Virens*. *Chinese J. Rice Sci.* **2006,** *20* (4), 440–442.

Zhang, K.; Li, Y. J.; Li, T. J.; Li, Z. G.; Hsiang, T.; Zhang, Z. D.; Sun, W. X. Pathogenicity Genes in *Ustilaginoidea Virens* Revealed by a Predicted Protein-Protein Interaction Network. *J. Prot. Res.* **2017,** *16* (3), 1193–1206.

Zhang, Y.; Zhang, K.; Fang, A. F.; Han, Y. Q.; Yang, J.; Xue, M. F.; Bao, J. D.; Hu, D. W.; Zhou, B.; Sun, X. Y.; Li, S. J.; Wen, M.; Yao, N.; Ma, L. J.; Liu, Y. F.; Zhang, M.; Huang, F.; Luo, C. X.; Zhou, L. G.; Li, J. Q.; Chen, Z. Y.; Miao, J. K.; Wang, S.; Lai, J. S.; Xu, J. R.; Hsiang, T.; Peng, Y. L.; Sun, W. Y. Specific Adaptation of *Ustilaginoidea Virens* in Occupying Host Florets Revealed by Comparative and Functional Genomics. *Nat. Commun.* **2014,** *5,* 3849.

Zheng, D. W.; He, L. P.; Meng, C. M.; Hu, D. W. In *The Cytological Evidence on Infection of Ustilaginoidea Virens at the Germination Stage*. Proceedings of the Annual Meeting of Chinese Society for Plant Pathology; China Agricultural Science and Technology Press: Beijing, China, 2009; p 118. (In Chinese)

Zhou, Y.; Fan, J.; Zeng, C.; Liu, X.; Wang, S.; Zhao, K. Preliminary Analysis of Genetic Diversity and Population Structure of *Ustilaginoidea Virens. Acta Phytopathol. Sinica* **2004,** *34* (5), 442–448.

Zhou, Y. L.; Izumitsu, K.; Sonoda, R.; Nakazaki, T.; Tanaka, E.; Tsuda, M.; Tanaka, C. PCR-Based Specific Detection of *Ustilaginoidea Virens* and *Ephelis Japonica. J. Phytopathol.* **2003,** *151* (9), 513–518.

Zhou, Y. L.; Izumitsu, K.; Sonoda, R.; Nakazaki, T.; Tanaka, E.; Tsuda, M.; Tanaka, C. PCR-Based Specific Detection of Ustilaginoidea Virens and Ephelis Japonica. *J. Phytopathol.* **2003,** *151,* 513–518.

Zhou, L.; Lu, S.; Shan, T.; Wang, P.; Sun, W.; et al. Chemistry and Biology of Mycotoxins From Rice False Smut Pathogen. In *Mycotoxins*: *Properties*, *Applications and Hazards*; Melborn, B. J., Greene, J. C., Ed.; Nova Sci. Publ.: New York, 2012; pp 109–30.

Zhou, Y. L.; Pan, Y. J.; Xie, X. W.; Zhu, L. H.; Wang, S.; Li, Z. K. Genetic Diversity of Rice False Smut Fungus, *Ustilaginoidea Virens* and its Pronounced Differentiation of Populations in North China. *J. Phytopath.* **2008,** *156* (9), 559–564.

Zhou, Y.; Xie, X.; Wang, S.; Pan, Y.; Liu, X.; Yang, H.; … Li, Z. Detection of *Ustilaginoidea Virens* From Rice Field and Plants by Nested PCR. *J. Agri. Biotechnol.* **2006,** *14* (4), 542–545.

Zhou, Y.-L.; Xie, X.-W.; Zhang, F.; Wang, S.; Liu, X.-Z.; Zhu, L.-H.; et al. Detection of Quantitative Resistance Loci Associated With Resistance to Rice False Smut (*Ustilaginoidea Virens*) Using Introgression Lines. *Plant Pathol.* **2014,** *63,* 365–372.

Zhou, Y. L.; Xie, X. W.; Zhang, F.; Wang, S.; Liu, X. Z.; Zhu, L. H.; Xua, J. L.; Gao, J. M.; Li, Z. K. Detection of Quantitative Resistance Loci Associated With Resistance to Rice False Smut (*Ustilaginoidea Virens*) Using Introgression Lines. *Plant Pathol.* **2014,** *63* (2), 365–372.

CHAPTER 16

Emerging Biotechnological Tools for Rice *Bakanae*/Foot Rot Disease Resistance

RAJVIR KAUR[1], YUVRAJ CHOPRA[2], and DINESH KUMAR SAINI[1]

[1]*Department of Plant Breeding and Genetics, Punjab Agricultural University, Ludhiana, Punjab, India*

[2]*College of Agriculture, Punjab Agricultural University, Ludhiana, Punjab, India*

ABSTRACT

Rice is a staple food crop consumed by almost half of the world's population. *Bakanae* or foot rot disease, caused by the fungal pathogen *Fusairum fujikuroi* [telomorph: *Gibberella fujikuroi* (Sawada) Wollenw], has become a serious worldwide threat, leading to high losses in yield. Understanding of the interaction between rice and *F. fujikuroi* may be helpful to find out new resistant or tolerant cultivars and to devise the alternative disease control strategies. In this chapter, we have discussed the different molecular and chemical patterns involved in this host–pathogen interaction. Host plant resistance is considered the most effective, sustainable, and practically feasible approach for disease management. In recent decades, using conventional breeding approaches, resistance genes have been incorporated from resistant lines to rice cultivars. The efficiency of these breeding approaches has further improved with the development of molecular markers. These marker-based or molecular breeding approaches mainly include marker-assisted backcrossing and marker-assisted gene pyramiding. Identification of genes/QTLs and/or marker-trait associations is necessary for the successful application

Fungal Diseases of Rice and Their Management. Deepti Srivastava, Md. Shamim, Malik Mobeen Ahmad, and R. S. Upadhyay (Eds.)

of these marker-based approaches. Various QTL mapping and genome-wide association studies have been conducted in rice to identify markers associated with foot rot resistance loci. One of the main aims of this chapter is to update and summarize findings on foot rot disease resistance breeding of rice published so far. Furthermore, we have compiled a user-friendly table listing disease resistance QTL data and marker-trait associations providing a valuable resource for further foot rot disease resistance research. This chapter also provides a panorama of the application of OMICS approaches including both genomics and postgenomic technologies such as transcriptomics and proteomics to the improvement of foot rot disease resistance in rice crop.

16.1 INTRODUCTION

Rice (*Oryza sativa* L.) is the major staple crop cultivated all over the world. Around 90% of global area is cultivated and devoured only in Asian countries (http://ricep edia.org/rice-around-the-world /Asia). India ranks first in terms of area under rice cultivation globally, and China has the highest production, followed by India. Among the major diseases of rice crop, "*Bakanae*" is one of the most severe and oldest diseases concerning rice production. It was originally found in 1828 in Japan (Ito, 1931) and presently spotted in Asia, North America, Africa, and Europe (Pra et al., 2010). *Bakanae* disease arose as a notably vital disease in current times throughout the world in all the rice-cultivating countries (Bal and Barun, 2018). There are significant yield losses that are triggered by the disease and varied from 10% to greater than 70% in different rice-growing countries (Puyam et al., 2019). Yield losses of 25% have been stated in Bangladesh (Hossain et al., 2007). About a 40% decline in crop production was reported in Nepal (Desjardins et al., 2000). In Japan, t*his* disease is liable for 25–50% reduction in crop yield whereas in Thailand, 3.7–14.7% reduction in yield has also been documented (Ito, 1931). In Pakistan, this disease is acknowledged to affect the total yield losses of 10–50% (Ghazanfar et al., 2013).

In India, it has been reported that the *Bakanae* disease has a severe and pronounced effect on Basmati rice cultivars (Bal and Barun, 2018). The reduction of 15–25% in crop yields has been noted from the various states in India (Sunder et al., 2014). Actually, the word *Bakanae* is from Japanese language, which signifies "bad" or "foolish" elongated seedlings symptoms caused by the disease. In different divisions of the world, this disease is recognized by distinct names such as "otoke nae" in Japan, while in China as "white stalk" or "Fusariosis," in British as "palay lalake," in French Equatorial as "foot rot" or "*Bakanae,*" and "foot rot" or "foolish plant" in India (Singh and Sunder, 1997). Sometimes, it is also called "root rot" and/or "white head disease" (Saremi et al., 2008). In recent years, *Bakanae* has become an important disease in basmati rice varieties (Bashyal et al., 2015). It might be due to novel approaches of seedling raising, intensification of basmati/scented rice cultivation, inadequate management and crop cultivation practices, and other additional common intensifying factors, viz., climatic modification, cropping arrangement, intensive cultivation of vulnerable cultivars favors extreme disease incidence (Bashyal et al., 2015).

In this chapter, we have discussed the different methods for the identification of *Fusairum* species, molecular, and chemical patterns involved in *F. fujikuroi* interaction, whole-genome sequencing of different strains of *Fusairum* species, etc. We have enlisted the different resistant genotypes, resistant QTLs in a user-friendly manner. We have also elaborated on the genetics of *Bakanae* resistance and discussed prospects of molecular breeding approaches for the *Bakanae* resistance breeding of rice.

16.2 BAKANAE DISEASE: CAUSAL ORGANISM AND EPIDEMIOLOGY

Initially, Fusairum heterosporum Nees (teleomorph *Lisea fujikuroi*) was identified as the causal organism of *this* disease (Hori, 1898; Sawada, 1917). For many years, the nomenclature of *Bakanae* fungus has been in conflict among plant pathologists. Wollenweber and Reinking (1935) described the imperfect stage of the *bakanae* fungus as *Fusairum moniliforme* (Sheldon) with *Gibberella fujikuroi* (Sawada) Wr. as the perfect stage. It signifies a complex of polytypic *Fusairum* species as an anamorphs in the section Liseola (Leslie and Summerell, 2008). Thirteen biological species have been identified within the *Gibberella fujikuroi* species complex (GFSC) (Lima et al., 2012). Among these, three species, *F. fujikuroi* Nirenberg, *F. proliferatum* (Mats.) Nirenberg, and *F. verticillioides* (Sacc.) Nirenberg, have been found to be associated with disease incidence (Desjardins et al., 2000). Principally, *F. fujikuroi* Nirenberg has been identified as a causal organism involved in triggering the foot rot disease of rice and it has been detected to be more contagious than any other species of *Bakanae* fungus. *Bakanae* of rice is a monocyclic disease and the fungus produces conidia on infected plants. Conidia are readily spread by water and wind (Bal and Barun, 2018). Conidia infect the seeds with their high production on infected culms when its incidence coincides with the flowering and ripening of rice. Diseased kernels develop reddish shade symptoms as a result of conidial infection, and when there is a severe infection the entire seed gets discolored.

The fungus can be isolated from asymptomatic seeds that are collected from severely infected rice field. At the flowering stage of the crop, airborne ascospores have also been described as a source of infection (Sun, 1975). At early stages of seedling development, the fungus can infect seedlings as it becomes systemic in the plant without colonization of the floral organs. For the occurrence of the disease, the first 72 h after germination of seed are crucial as large amounts of sugars and amino acids are produced from germinating seeds. The optimum range of temperature for the growth of *F. fujikuroi* is 27°C to 30°C, and higher concentrations of nitrogen in the soil also promote the development of the pathogen. Further, excessive humidity leads to elongated culms, while low humidity produces stunted plants. The vascular bundles, larger vessels, and xylem gaps can also be infected due to the development of micro-conidia and mycelia of the pathogen, while the phloem and parenchyma remain uninfected by fungus. The main source of inoculum for following season is the infected seeds.

16.3 TYPICAL SYMPTOMS OF BAKANAE DISEASE

Bakanae disease produces different types of symptoms varying from seedlings rotting to seedling blight, rotting of the root, crown rot, stunting of plants, etiolation, hypertrophy, sterility, infection of grains at ripening, discoloration of grains, and bare panicle formation under diverse climatic conditions all around the world (Desjardins et al., 2000). The incidence of the disease is best-known to be in both the main crop and ratoon crop. The fungal growth can be seen at the coleoptiles immediately at the upper surface of soil and at the joint of palea and lemma of affected grains (Bashyal, 2018). In wet nurseries, the seedlings start dying 1 week after the sowing of the crop. The diseased plants produce brown discoloration. During initial phases, the brown discoloration is restricted to lower nodes and at the later stages, the upper nodes also give the same appearance of brown discoloration. The hollow internodes are known to bear mycelial masses of conidia. When there is a severe infection of the disease, the infection emerges on the surface of the nodes and plant surface changes to pink or white in color (Surek and Gumustekin, 1994). The lesions are also formed on the leaf surface. The flag leaves of diseased plants become yellow. The plants at maturity stage may yield lanky and thin tillers.

16.4 IDENTIFICATION OF DIFFERENT FUSARIUM SPECIES

The cultural, morphological, and pathological descriptions of the isolates of *F. fujikuroi* obtained from diverse rice-growing areas in the world show that the pathogen has a very high level of polymorphism. These isolates generally differ in radial growth, the presence of macro-conidia, and the degree of virulence. The knowledge of variability in pathogen can be helpful in defining effective strategies for the management of foot rot of rice. Greater genetic diversity increases the probability of a higher rate of sexual recombination, thus augmenting the rate of evolution of novel virulent pathogen species. The genetic variation in fungi may be caused by several factors including selection, reproduction system, mutation, population size, random genetic drift, and gene flow (McDonald and Linde, 2002). Interaction among these factors may result in the change of variability among and within the populations of these pathogens. Combined reproduction or mating system is considered a major cause of pathogen variability. Mixed coordination causes the maximum risk of evolution since it comprises recombination, where pressure exerted by selection (sexual stage) creates new genotypes with high fitness and maintenance of the fittest genotypes is through clones, spatial, and temporal distribution (asexual stage) (McDonald and Linde, 2002). There is a big impact of the mutation on pathogen's variability although the rate of mutation is still unknown.

For categorization and classification of different *Fusairum* species and strains, various parameters such as pigmentation of culture, their growth rate, and absence or presence of micro-conidial chain (Zainudin et al., 2008a), and gibberellic acid production (Sunder, 1995) have been used in different studies. However, identification based on such parameters only is not reliable; hence, a correct recognition of different *Fusairum spp.* requires the evaluation of their molecular, physiological, and biochemical characteristics.

16.4.1 *CULTURAL AND MORPHOLOGICAL CHARACTERIZATION*

Morphological characterization of *Fusairum* species is basically accomplished on the basis of structure and formation of macroconidia and microconidia, presence or absence of chlamydospores, the origin of conidiophores, the morphology and pigmentation of colony, colony growth diameter on potato dextrose agar, and its color (whitish pink with active sporulation) (Yadav et al., 2014; Baharuddin, 2017). Various studies regarding morphological characterizations of different species have been reviewed elsewhere (Puyam et al., 2018). Identification of the *G. fujikuroi* species complex based on the morphological traits has quite a few drawbacks as some species possess very comparable morphology, for example, *F. sacchari* and *F. subglutinans*, *F. proliferatum* and *F. fujikuroi*, and *F. verticillioides* and *F. andiyazi*. Moreover, identification based on morphology is not consistent as visual remarks in differentiating some closely related species as *F. moniliforme* and *F. proliferatum* may possibly sometimes be complicating. Irrespective of the drawbacks related to morphological characters, these traits however play a significant role in the categorization of isolates into minor groups before the application of other methods of identification. As mentioned in the previous section, identification on the basis of these morphological and cultural traits only is not reliable; therefore, the more sensitive and reliable biochemical and/or molecular-based methods are recommended for the accurate identification of the species.

16.4.2 BIOCHEMICAL CHARACTERIZATION

Numerous chemotaxonomic studies in past 20 years have shown that each *Fusairum* species produces distinctive profile of secondary metabolites (Zainudin et al., 2008b). The variation in pathogenicity of various strains of a pathogen is also attributed to these secondary metabolites (Puyam et al., 2017). Secondary metabolites, like mycotoxins, causing health-related problems to animals and humans are known to be produced by *F. fujikuroi* (Desjardins, 2006) and moreover, 45 gene clusters encoding putative secondary metabolite have also been studied. Fumonisins and gibberellins have been analyzed exhaustively and their production is known to be varied among strains. Researchers have outlined the relevance of chemotaxonomic criteria in *Fusairum* species by studying the specific secondary metabolite's profiles and identification of these on the basis of bio-chemical properties (Malonick et al., 2005; Zainudin et al., 2008b). Gibberellin, an eminent plant hormone, was initially recognized as a virulence factor in rice infestation by *F. fujikuroi*. Gene disruption studies showed the role of gibberellin in the initiation of fungal colonization in host roots of rice. Fumonisin was first discovered in *F. verticillioides*, a relative of *F. fujikuroi*. The fumonisin (FB1) is a carcinogenic mycotoxin and hinders the activity of enzyme ceramide synthase, involved in sphingolipid biosynthesis in animals and plants. This causes alteration in the cell function and also promotes cell autolysis (Gutierrez-Najera et al., 2005). Members of GFSC (*F. fujikuroi, F. proliferatum, and F. verticillioides*) which were originated from rice, have been studied to determine their abilities to produce fumonisin (FB1 and FB2) (Matić et al., 2013).

In summary, there are a number of secondary metabolites (SMs) produced by *Fusairum* species belonging to GFSC. These secondary metabolites include fumonisin (B1 and B2) produced by *F. verticillioides* and *F. proliferatum,* fusaric acid by *F. moniliforme,* and beauvericin and fusaproliferin by *Gibberella fujikuroi* complex isolated from maize, moniliformine by *Fusairum spp* isolated from rice and gibberellins by *Fusairum* species linked with *Bakanae* disease (Yabuta et al., 1937; Bacon and Hinton, 1996; Desjardins et al., 2000; Reynoso et al., 2004; Zainudin et al., 2008b; Puyam et al., 2017a). For instance, Puyam et al. (2017a) isolated 38 isolates of *Fusairum moniliforme* from basmati plants infected from foot rot which were collected from various agro-ecological zones and quantified GA and FA production from these isolates. All the isolates showed varying production of FA and GA. A total of thirty isolates out of 38 produced fusaric acid and all the 38 isolates produced GA; it showed that GA has maximum involvement in malfunctioning induced by this disease. These conclusions are supported by the results of Kaur et al. (2011) in which they studied and reported that the fungal pathogen causing *Bakanae* disease in rice produces GA. However, they also reported that not all the isolates produced FA. Various *Fusairum spp* have been grouped into three different categories such as low gibberellin producers, moderate gibberellin producers, and high gibberellin producers (Bhalla et al., 2010).

Seventy-eight distinct isolates of *Fusairum spp.* were analyzed by Bacon et al. (1996) and reported that different strains of *Fusairum moniliforme* show high variability for GA production which might be because of the differences in gibberellin production pathways. The fusaric acid gene cluster in *F. fujikuroi* which was identified and characterized by Niehaus et al. (2014), consists of *PKS* encoding gene and four co-regulated genes (*FUB1–FUB5)*. They described that *FUB1* and *FUB4* play important role in secondary metabolite's biosynthesis in rice. The existence or the absence of the fusaric acid gene cluster may regulate the production of FA by different strains, although it needs further analyses. Conclusively, to date, *F. proliferatum, F. fujikuroni, and F. verticilloides* of GFSC have been known to produce GA and related to foot rot disease. Niehaus et al. (2017) reported that two pathotypes of *F. fujikuroi* which produce secondary metabolites, a *Bakanae* type (elongated stem), producing gibberellin but modest or no fumonisin, and a stunting type, which produces fumonisin but not gibberellin, are phylogenetically different. These biochemical properties can be used to supplement the morphological or molecular characterization used in recognizing *Fusairum* species under GFSC, causing *Bakanae* disease of rice.

16.4.3 MATING POPULATIONS AND MOLECULAR CHARACTERIZATION

Morphological and biochemical analysis of *Fusairum* species is problematical and tiresome as diverse species under GFSC have comparable morphological pattern and biochemical pathways. These above-mentioned taxonomic difficulties can be overcome by the criterion of Mating Populations (MPs) which can effectively distinguish these species. The pattern of mating is regulated by two alleles at a single mating-type locus and heterothallism. There is a sexual compatibility between the members of the same MPs but not with the members from any other MPs. The *Fusairum fujikuroi* Nirenberg (heterothallic ascomycete) is a member of the *Fusairum fujikuroi* species complex (FFC), a monophyletic lineage that comprises a minimum of 11 sexually infertile mating populations (MPs A-K) and several different anamorphic species. These mating populations are known to be linked with *Bakanae diseases* from diverse geographical areas (reviewed by Puyam et al., 2018). In *Gibberella fujikuroi*, type of mating can be recognized on the basis of MAT-1 and MAT-2 region-based universal primers. Recently in 2020, Bashyal et al. recognized the population's mating type on the basis of these primers for *Gibberella fujikuroi*. Among the 63 isolates, around 18 percent were identified as MAT-1 and around 71% as MAT-2. Hsuan et al. (2011) conducted mating studies on 78 strains of *Fusairum* species section Liseola isolated from major crops including rice. From the crosses between tester strains of *Gibberella fujikuroi* species complex and tester strains of mating populations A to E, 64.1% were fertile. The results of the Mating studies showed that 19 out of 50 strains belonged to mating population A (*Gibberella moniliformis*), 18 belonged to MP-B (*Gibberella sacchari*), four to MP-E (*Gibberella subglutinans*), six to MP-D (*Gibberella intermedia*), and three to MP-C (*G. fujikuroi*). The availability of the two mating types in *F. fujikuroi* implies the chances of sexual reproduction in the population.

Different molecular-based techniques such as Simple Sequence Repeats (SSR) primers (Puyam, 2017), Universal Rice Primers (URP) (Bashyal et al., 2020), RAPD (Gherbawy et al., 2002), RFLP of IGS-rDNA region (Hsuan et al., 2010), SNP (Bashyal et al., 2017), and Universally Primed-Polymerase Chain Reaction (UP-PCR) (Cumagun et al., 2011), conventional and real-time PCR (Amatulli et al., 2012), CHEF gel karyotypes (Xu et al., 1995) have been extensively used to identify the members of GFSC either alternatively or complementary to methods based upon morphological and pathological characters. Presently, for the characterization of the *Fusairum* species at the species level, some of the frequently used techniques are sequencing of translation elongation factor 1-a gene internally transcribed spacer regions in the ribosomal repeat region and ß-tubulin (Bashyal and Aggarwal, 2013; Petrovic et al., 2013; Pakshir et al., 2020).

The translation elongation factor 1-a gene (EF-1a gene) encodes an essential part of the protein translation machinery and it is considered an excellent single-locus detection tool in *Fusairum* species. All *F. fujikuroi* strains from *Bakanae* infected rice fields are fumonisin nonproducers, and single nucleotide polymorphism (SNP) detected in the translation elongation factor 1 α *(TEF1α)* gene is sufficient to discriminate between fumonisin producers and nonproducers. Screening based on translation elongation factor 1-a gene has been the most comprehensively relevant and welcoming across the genus. It might be due to its inability to recognize the nonorthologous copies of the gene and its highly conservative and informative nature at the species level (Amatulli et al., 2012; Suga et al., 2014; Pakshir et al., 2020). In GenBank, the sequences for this gene are available with specific *FUSAIRUM*-ID *(http://isolate.fusariumdb.org/blast.php).* The *FUSAIRUM*-ID database comprises sequences that are concomitant with a strain, which is available publicly through either USDA NRRL collection or the collection maintained at the Pennsylvania State University. For instance, Amatullia et al. (2010) reported variable levels of aggressiveness of diverse *Fusairum spp* related to *Bakanae* disease of rice in Italy and separated them using EF-1a sequences. Rahjoo et al. (2008) also recognized the *Fusairum* isolates on the basis of morphology, species-specific primers (VER1/2 and PRO1/2), and *TEF* 1-a gene region.

Besides, nuclear rDNA can also be considered an ideal target for specific PCR primers, as the ribosomal DNA (rDNA) of fungi comprises both variable and conserved regions. One such region is the intergenic spacer (IGS) which splits the repeated ribosomal units and seems to be the most swiftly changing spacer region (Kim et al., 2001). PCR amplification of IGS region followed by RFLP analysis has been proved to be a valuable tool for the analysis of intraspecific and taxonomic variation of *Fusairum* isolates (Miyagawa et al., 2002; Kim and Min, 2004; Hsuan et al., 2010; Ghazanfar et al., 2016; Nosratabadi et al., 2018). Ghazanfar et al. (2016) sequenced ITS regions of rRNA gene complex of local isolates of *Fusairum* and compared them to determine the similarity and diversity of these isolates at a genetic level. Recently in 2018, Nosratabadi et al. analyzed (β)-tubulin region of 107 strains of *Fusairum* and differentiated them in different groups.

More recently in 2019, a user-friendly, quick, and efficient amplification method, named as "One-step loop-mediated isothermal amplification (LAMP)," was proposed for the quick and sensitive detection of *Fusairum fujikuroi* in *Bakanae* disease. In this method, a gene designated as *NRPS31* is targeted for designing the LAMP primers. As this gene plays a significant role in GA biosynthesis and only present in *F. fujikuroi*, and is absent from any other sequenced fungal genome, it can be used for more sensitive and faster discrimination of *F. fujikuroi* among the other *Fusairum* species present on rice seeds and seedlings (Zhang et al., 2019). This LAMP assay is less sensitive to inhibitors as compared to PCR and it allows the rapid detection of as minute as 1 femtogram of pure genomic *Fusairum* species DNA within 60 min. Moreover, the contrasting genomic regions might allow discrimination of closely related *Fusairum* strains that are positioned in the same group such as some mitochondrial DNA regions. Molecular studies based on mitochondrial small subunit (mtSSU), translation elongation factor (EF-1), rDNA, introns, and exons of *Fusairum verticilloides* were conducted in Japan by Hirata et al. (2001). The results obtained from these molecular methods are distinctive and highly sensitive as they are based on genotypic differences. Moreover, these studies allow the efficient detection of the target DNA molecules in the presence of complex mixtures, even when the viability of mycelia is lost.

16.4.4 GENOME SEQUENCING OF FUSARIUM SPECIES: REVEALING VARIATIONS AT GENOMIC LEVEL

The accessibility of the whole genome sequence of multiple *Fusairum* species has improved the understanding of host–pathogen interaction including virulence/defense mechanisms and the associated pathways (Piombo et al., 2020). Moreover, the whole-genome sequencing (WGS) of different strains of *Fusairum* species has helped in understanding the biology and evolution of the *Fusairum* as well as *G. fujikuroi* species complex. To our knowledge, so far eighteen genomes of *F. fujikuroi* have been reported by different researchers from various countries (Wiemann et al., 2013; Jeong et al., 2013; Chiara et al., 2015; Bashyal et al., 2017; Niehaus et al., 2017; Radwan et al., 2018; Urbaniak et al., 2018; Piomno et al., 2020) as reported in Table 16.1. The genome size of various strains varies from 42.4 (strain F250) Mbp to 49.1 Mbp (strain V64-1) divided among 12 chromosomes. The repetitive elements constituted only 1% of the total genome. The gene content of these strains varies from 13,000 to 15,000 genes, constituting approximately 1200 genes encoding for secreted proteins.

Earliest in 2013, Jeong et al. sequenced a *Fusairum fujikuroi* Korean strain, B14. They identified all previously categorized conserved gene clusters responsible for secondary metabolite production in the B14 genome including those for gibberellin, fumonisin and so on. Additionally, they confirmed *F. fujikuroi* (B14), as the *MAT*1-2 mating type strain, carrying 20 *PKS* (polyketide synthase) genes, including four *PKS-NRPS* (non-ribosomal peptide synthetase) genes and two non-reducing *PKS* hybrid genes. In the same year, Wiemann et al. (2013) reported a high-quality genome sequence of *F. fujikuroi* isolate (IMI58289). On the basis of the analysis of genome sequences, they showed that the cluster contains a polyketide synthase gene (*PKS19*) and the cluster that contains a non-ribosomal peptide synthetase gene (*NRPS31*) are distinctive to *F. fujikuroi*. Chiara et al. (2015) compared draft genome sequences of three geographically distinct isolates of *Fusairum fujikuroi* and observed that the diverse profiles of secondary metabolites were present within *F. fujikuroi*. They also provided additional evidences that duplication, gene divergence, and differential loss play vital roles in *F. fujikuroi's* genome evolution.

Later in 2017, Bashyal et al. conducted comparative genome analysis of *Fusairum fujikuroi* isolates and identified a total of 1194 secretory proteins in the genome of "F250" isolate of *F. fujikuroi*. They identified 356 genes that encode carbohydrate active enzymes capable of degradation of complex polysaccharides. In addition, they identified 585 genes necessary for the pathogen and host interactions. In the same year, 2017, Niehaus et al. sequenced genomes of eight additional *F. fujikuroi* isolates collected from distant geographic areas. These isolates differ in the size of chromosomes, type of asexual spores, and the number and expression of secondary metabolite (SM) gene clusters. Also, they analyzed that B14 isolate produces no GAs, but it produces high amounts of fumonisins on rice during infection period. It differs from other previously studied isolates due to the presence of three extra polyketide synthase genes (*PKS40, PKS43,*

PKS51) and the absence of the *F. fujikuroi*-specific apicidin F (*NRPS31*) gene cluster.

TABLE 16.1 Whole Genomes of Different Strains of *F. Fujikuroi* Sequenced by Various Researchers.

S. no.	Country of origin	Host	Strain	Genome size	Reference
1.	China	Infected rice	IMI 58289	43.9 Mbp	Wiemann et al., 2013
2.	South Korea	Infected rice	B14	44 Mbp	Jeong et al., 2013
3.	Thailand	Infected rice	KSU3368	43.1 Mbp	Chiara et al., 2015
4.	USA	Prairie grass	KSUX-10626	43.1 Mbp	
5.	Taiwan	Infected rice	FGSC8932	43 Mbp	
6.	India	Infected rice	F250	42.4 Mbp	Bashyal et al., 2017
7.	Japan	Infected rice	m567	44 Mbp	Niehaus et al., 2017
8.	Germany	Maize	FSU48	46.1 Mbp	
9.	Taiwan	Infected rice	C1995	45.8 Mbp	
10.	Philippines	Infected rice	MRC2276	45 Mbp	
11.	Italy	Infected rice	E282	46.1 Mbp	
12.	South Korea	Infected rice	B20	44.3 Mbp	
13.	India	Infected rice	NCIM 1100	45.3 Mbp	
14.	USA	Fuel environment	FUS01	49 Mbp	Radwan et al., 2018
15.	USA	Clinical isolate	COH1152	48 Mbp	Urbaniak et al., 2018
16.	Northern Italy	Infected rice	Augusto2	43.7 Mbp	Piombo et al., 2020
17.	Northern Italy	Infected rice	CSV1	45.6 Mbp	
18.	Northern Italy	Infected rice	I1.3	45.6 Mbp	

Further, they confirmed the strong correlation between the ability to produce either GAs or fumonisins and the pathotype (*Bakanae* or stunting/withering). Most recently, in 2020, Piombo et al. sequenced three *F. fujikuroi* strains, namely, Augusto2, CSV1, and I1.3. They performed SNP calling, gene prediction, and structural variant analysis with a reference genome and showed that all three strains have almost similar genomes (extremely limited polymorphisms in genes) but still exhibit strong differences in phenotype. Furthermore, they showed how change in temperature may result in the development of diverse symptoms in rice plants inoculated with a single strain *F. fujikuroi*. Conclusively, the whole genome data of different strains offers an improved understanding of the *F. fujikuroi* genome and likely to further magnify our knowledge on *Fusairum fujikuroi* and rice interactions.

16.5 *F. FUJIKUROI*-RICE INTERACTION: MOLECULAR AND CHEMICAL PATTERNS/RESPONSES

Understanding of the interaction between rice and *Fusairum fujikuroi* may be helpful in identifying novel resistant cultivars and to devise the alternative disease management strategies and also develop the more accurate and rapid molecular diagnostic tools and/or techniques. However, the underlying molecular mechanisms of defense responses against pathogen so far have not been completely explored in rice crop. With an aim to explain the factors that are involved in rice resistance against *Bakanae* disease, Matic et al. (2016) conducted an experiment to study RNA-seq comparative transcriptome on infected seedlings of twelve rice accessions at 1 and 3 weeks of post germination. They suggested 3 weeks post germination infection stage to study the resistance mechanisms as they observed appreciable transcriptional changes at this particular stage. They observed 3119 differentially expressed genes (DEGs) in Selenio (most resistant genotype) and 5095 in Dorella (most susceptible genotype). Moreover, they observed upregulation of germin-like protein, glycoside hydrolases, PR1, MAP kinases, and WRKY TF (transcriptional factors) in the resistant genotypes upon infection with pathogen. In the susceptible genotypes, they observed downregulation of WRKY transcriptional factors and MAP kinases and upregulation of chitinases. Their gene ontology (GO) enrichment analysis detected different GO terms for resistant and susceptible genotypes in response to *F. fujikuroi*. GO terms "jasmonic acid biosynthetic process," "response to chitin," and "plant type hypersensitive response" were detected in resistant genotype, while "response to salicylic acid stimulus" and "gibberellin metabolic process" were detected in susceptible genotype. GO terms in susceptible genotypes were in agreement with the production of GA3, which is essential for disease development. This was the first study, where researchers provided information regarding the cellular and molecular processes occurring in rice during infection period (Matic et al., 2016).

In the same year, a most important transcriptomic study was published (Ji et al., 2016). For transcriptomic analysis, they used a moderately resistant genotype, c.v 93-11 and a susceptible genotype, Nipponbare. They observed differential expression of 1152 and 1052 transcripts between the control (no infection) and the treatments (infection with *F. fujokuroi*) for 93-11 and Nipponbare, respectively. They observed specific defense-related GO terms in 93-11 and revealed that some specific WAK, WRKYs, and MAP3Ks were responsible for the *bakanae* disease resistance in resistant genotype, 93-11. They further suggested that these defense-related genes on chromosome one may play an important role in the rice–*F. fujikuroi* interaction. These resistance genes can be further categorized and underlying candidate genes can be identified. Once a candidate gene is identified, gene-specific marker can be developed which can be employed for rice *Bakanae* resistance breeding programs (Ji et al., 2016).

To better understand the molecular mechanisms of disease resistance, a six-plex tandem mass tag approach was further employed by researchers for relative quantitative proteomic comparison of treated and control rice seedlings (Ji et al., 2019). They observed differential accumulation of proteins (DEPs) in resistant and susceptible genotypes. In total 123 DEPs (of which 77.2% were upregulated and 22.8% were downregulated) and 91 (of which 94.5% were upregulated, and 5.5% were downregulated) were accumulated in resistant and susceptible genotypes, respectively. Contrasting protein regulation patterns were shown by cluster analysis of the two genotypes. It clearly indicated different interaction mechanisms of the "host" and the "pathogen" between resistant and susceptible genotypes. More importantly, they observed sharp upregulation with a fold change of 109.2 of an aquaporin, PIP2–2, in resistant genotype, which may be related to pathogen defense and the implementation of resistance to *Bakanae* disease (Ji et al., 2019).

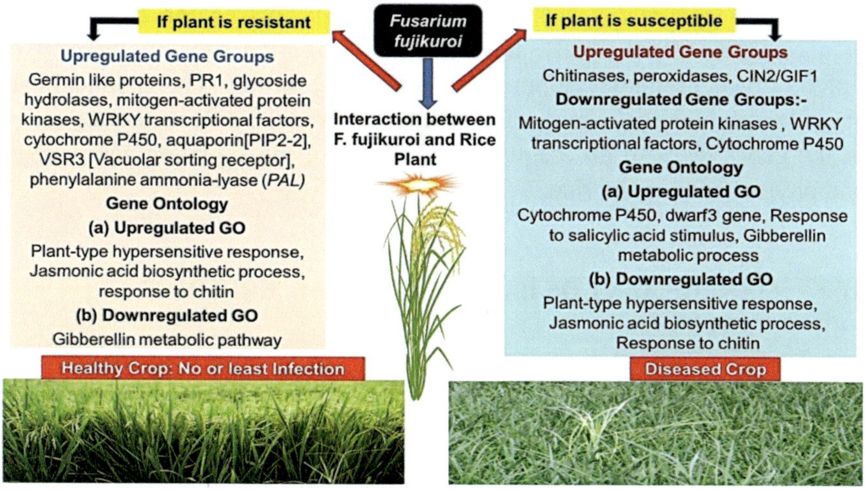

FIGURE 16.1 Diagrammatic representation of major molecular patterns involved in *F. fujikuroi*-rice -interaction.

Overall, these transcriptomic and proteomics studies revealed the molecular defense responses of the rice plant against *F. fujikuroi* infection. Major molecular patterns involved in *F. fujikuroi*-rice interaction are shown in Figure 16.1. These patterns based on molecular studies may be further compared with the chemical patterns or responses included in the rice- *F. fujikuroi* interaction as described in the paragraph below.

The differential expression patterns of IAA, GAs, and ABA have been observed in resistant (BR3) and susceptible (MR 211) varieties that were artificially inoculated with *F. proliferatum* at different time intervals. This differential expression was found to be linked with the response of rice plants to *Fusairum* species (Elshafey et al., 2018). Later, to get a better understanding of the basis of rice resistance, a quantitative method, HPLC-MS/MS, was developed to concurrently detect phytoalexins and phytohormones (Siciliano et al., 2015). Using this method, they reported induced production of phytoalexins, mainly sakuranetin in the resistant genotype and induced production of abscisic acid and gibberellin and inhibited phytoalexins and jasmonic acid in susceptible genotype upon application of pathogen inoculum. Their results suggested the involvement of the vast categories of secondary metabolites in plant defense against pathogens which could be essential factors for breeding disease-resistant rice (Siciliano et al., 2015). Studies have also shown that dwarf and semi-dwarf rice materials prove to be better resources for *Bakanae* resistance in rice as the materials carrying dwarf genes such as *d29, sd6, or sdq(t)* show resistance against *Bakanae* (Ma et al., 2008). Conclusively, these studies suggested the involvement of different SMs in the development of this disease in rice. The chemical data are in harmony with the molecular studies data and the differential activation of the genes leads to differential biosynthesis of the signaling components and phytoalexins in the rice.

16.6 MANAGEMENT OF BAKANAE DISEASE: BREEDING PERSPECTIVES

Various strategies such as thermal seed treatment or hot water immersion or fungicides have been devised for the management of *Bakanae*. This thermal seed treatment becomes ineffective when applied to critically infected rice seeds as the thermal effect is not proficiently transmitted to the pericarp layers (Hayasaka et al., 2001). Conversely, dressing of the seeds with fungicides has limited efficacy in destroying or killing the spores of the fungal pathogen, due to resistance against many generally used fungicides (Iqbal et al., 2011; Lee et al., 2011). Moreover, there has been an increasing occurrence of this disease, causing serious alarms among rice-producing areas globally and there is an urgent need to look into alternative disease management measures; identification of resistant genotypes and development of disease-resistant cultivars are some of them.

16.6.1 RESISTANT SOURCES

Since the first report by Ito et al. in 1931 on *Bakanae* disease-resistant genotypes, several studies have been published where authors have reported different resistant genotypes originated from different countries (Table 16.2). Unfortunately, only a few accessions were found to be highly resistant against *Bakanae*. An extensive screening on more than 400 rice accessions carried out by Li et al. in 1993 resulted in the identification of only a single highly resistant and 12 moderate resistant genotypes (Li et al., 1993). Screening has also shown that aromatic cultivars and germplasms are more susceptible to foot rot disease as compared to nonaromatic rice cultivar and germplasm (Fiyaz et al., 2014). Various susceptible genotypes have also been identified which are also important for genetic studies. In India, the high disease incidence and severity have been observed in cultivars, viz., Pusa Basmati 1509, Pusa Basmati 1121 while moderate incidence and severity

TABLE 16.2 Different Resistant Entries/Genotypes/Varieties Identified in Different Countries.

Entry/genotype/variety	Origin	Reference
Shiroka, Akage No. 3, Kairyo-Mochi No. 1	Japan	Ito et al., 1931
Wateribune (R), Aryan (R), and GEB 24 (R)	India	Thomas, 1933
Macan Bino, Apostal, Macan Binan, Guinongang	Philippines	Reyes, 1939
Strain 1		
HR 98 (R), HR 105 (R), Agami (R), and Monthakab (R)	India	Rajagopalan, 1961
Norin 18 (R) and Norin 20 (R)	Japan	Rajagopalan, 1961
Dharial, Surjamukhi, Charnock, Panbira, IR8	Bangladesh	Miah et al., 1973
RP 5-3 (R), RP 5-32 (R), RP 5-48 (R), RP 176-6 (R), CR 4-1048 (R), and CR 83-23 (R)	India	Amin and Reddy, 1976
MR 7, MN 62M	Thailand	Sukapanpotharam et al., 1980
IR 20 (R), IR 26 (R), IR 32 (R), IR 38 (R), IR 44 (R), and IR 45 (R)	Philippines	Gangopadhyay, 1983
PK 3303-15-1	Pakistan	Gill et al., 1993
Co 18 (R), Co 22 (R), Adt 8 (R), and PTB 7 (R)	India	Singh and Sunder, 1997
DM 107-4	Pakistan	Cheema et al., 1998
C 4-64 (R), Karjat x 13-21 (R), BR 4363-8-11-4-9 (R), BR 1067- 84-1-3-2-1 (MR), IR 58109-109-1-1-3 (R), BR 1257-31-1-1 (R), HKR 96-561 (MR), HKR 96-565 (R), HKR 07-40 (R), HKR 07-53 (R), HKR 08-13 (R), HKR 08-22 (R), HKR 08-21 (MR), PAU 3456-46-6-1-1 (R), MAUB 2009- 1 (R), PNR 600 (R), and RDN 01-2-10-9 (R)	India and Philippines	Sunder and Singh, 1998
IR 6, IR 8, IR 9, Shadab, KS 282, PK 1339-12-1-1-0-6, PK 3717-9	Pakistan, Philippines	Khan et al., 1999
PAU 2343-60-2-3, PAU 2383-2-3-1-1, IR 67418-204-4-4-2, IR 67423-234-3-3-3, IR 67423-208-2-1-1, IR 67418-131-2-3-3, IR 58755-158-1-3-2, IR64668-54-3-1, IR 66229-45-3-2, IR 67409-98-1-3-3-1, Double haploids 1, Double haploids 7, Double haploids 19	India, Philippines	Bagga et al., 2000

TABLE 16.2 *(Continued)*

Entry/genotype/variety	Origin	Reference
Nan-jing No.6 (R), Bei-ken-zi (R), Xue-he-ai-zao (R), Xue-he-ai-zao improvement (R),B5580A (R), CBR34 (R), CBR34 improvement (R), Xin-qian-ai (R), Xin-qian-ai 2 (R),Xin-gui-ai (R), KL803-29 (R), and KL803-29 improvement	China	Ma et al., 2008
Binam (HR), Kadous (MR), Shafagh Fajr (MR), Nemat (MR), Neda (MR), and Khazar (MR)	Iran	Saremi et al., 2008
IR 6 (R) and IR-8 (R)	Philippines	Iqbal et al., 2011
KSK-133 (R), Bas-Kernel (MR), Bas-198 (MR), and Bas-370 (MR)	Pakistan	Iqbal et al., 2011
Heukjinju (MR), Hwadong (MR), Inwol (R), Joan (MR), Nongho (MR), Samgwang (MR), Sampyeong (MR), and Seokjeong (MR)	Republic of Korea	Lee et al., 2011
GSL-5 (HR), GSL-9 (HR), GSL-12 (HR), GSL-44 (HR), GSL-60 (HR), GSL-66 (HR), GSL-67 (HR), GSL-36 (HR), GSL-68 (HR), GSL-4 (R), GSL-6 (R), GSL-8 (R), GSL-10 (R), GSL-13 (R), GSL-11 (R), GSL-23 (R), GSL-28 (R), GSL-31 (R), GSL-34 (R), GSL-37 (R), GSL-42 (R), GSL-43 (R), GSL-59 (R),GSL-69 (R), GSL-75 (R), GSL-82 (R), and GSL-296 (R)	India	Ahangar et al., 2012
KS-133 (MR) and IR-6 (MR)	Pakistan and Philippines	Ghazanfar et al., 2013
Pusa 1342 (HR), Apo (MR), ASD 18 (MR), Athad apunnu (HR), BPT 5204 (R), Bungpat subdmei (MR), C101A51 (HR), Chandana (HR), CO 39 (MR), CSR 30 (MR), Domsiah (MR), FL 478 (MR), Haryana Basmati 1 (MR), Himju (HR), Jhulhat (MR), Kala Fultas (MR), Kaw kaver (MR), Lalat (MR), Malviya dhan (MR), Mazari dhan (MR), Mirzag (MR), MTU 1010 (MR), Nagina 22 (MR), Panchami (HR), Pathara (MR), PAU 201 (HR), Peeli badam (R), Pokkali (MR), Punjab mehak (MR), Pusa 6B (MR), Pusa Basmati 1 (MR), Improved Pusa Basmati 1 (MR), Pusa 1602 (MR), Ramjawayan (MR), Rangoli (MR), Safed luchai (MR), Sona Mahsuri (MR), Suhagmani (MR), Suphala (R), Vandana (MR), and Varun Dhan (HR)	India	Fiyaz et al., 2014
IR 24 (MR), IR 50 (MR), IR 58025 B (HR), and IR 72 (MR)	Philippines	Fiyaz et al., 2014
Weonseadaesoo (HR), Hawn (R), Nep Huong (R), Hanareum 2 (MR), Nampyeong (R), and Goamibyeo (MR)	Republic of Korea	Kim et al., 2014

TABLE 16.2 *(Continued)*

Entry/genotype/variety	Origin	Reference
Giza 178 (HR), GZ 6296 (HR), and GZ 10101-5-1-1-1 (HR)	Egypt	El-shafey et al., 2016
IR 66298 (R), IR 65600 (R), and IR 66159 (R)	Philippines	Elshafey et al., 2016
HR 5824 (R), GZ 5310 (R), and BY 6-20 (R)	Egypt	Elshafey et al., 2016
C-4-63-G (MR),BPT 5204(R), IR 74712-115-3-3-3 (MR), Kasturi (MR), Vallabh Basmati 22 (MR), Improved Pusa Basmati (MR), Improved Pusa Basmati 1 (MR), IET 22289 (MR), Pusa Basmati (MR), RNR 2465 (MR), GR 101 (MR), GR 104 (MR), Narmada (MR), Sharbati (MR), Mugadh Sugandh (MR), Basmati 122 (MR), Rajendra Swetha (MR), UPRI 93-63-2 (MR), IR 74724-82-2-2-3 (MR), and IET 15832 (MR), Punjab Mehak 1(MR),	India	Jain et al., 2016
CSR-30(MR)		
GREPPI, ARSENAL, KING, GRALDO, L205, ADAIR, BENGAL, ALAN, MAYBELLE, A201, IAC32-52, MAIORAL	Italy, USA, Brazil, Portugal	Volante et al., 2017
GZ 10101-5-1-1-1	Giza	EL-Shafey et al., 2018

have been observed in other cultivars such as CSR 30, Pusa 2511, Pakistani basmati, Pusa Basmati 1401, and Dehradun basmati under natural conditions (Bashyal et al., 2017). Different resistant entries/genotypes/varieties identified in different countries are given in Table 16.2. Resistant levels of the genotypes are given in parenthesis (where HR, R, and MR refer to highly resistant, resistant, and moderately resistant, respectively). These genotypes can be used as donors for transferring the disease resistance into cultivated germplasm.

16.6.2 CONVENTIONAL BREEDING APPROACHES: POTENTIAL AND LIMITATIONS

Significant genetic gains have been achieved in numerous crops including rice by the use of conventional breeding strategies. Search for better lines and wild-type cultivars have been persistent in the field of crop improvement. The rice genus, *Oryza*, comprises 23 species and nine acknowledged genome types, which constitute a vast gene pool needed for genetic improvement of rice cultivars. Out of 23 species in the genus *Oryza*, two (diploids) are cultivated, namely, *O. sativa and O. glabberima*. The three major races of cultivated rice are Indica, Japonica, and Javanica. In rice, inter-specific crosses are extensively used in rice-breeding programs (Bah et al., 2018). Extensive screening for resistant varieties along with economic management strategies are required as the application of fungicides is not so effective.

Large-scale screenings for *Bakanae*-resistant rice germplasm have identified several resistant genotypes (Table 16.2). Initially, resistance to *Bakanae* disease was reported as monogenic as it was found to be dominant in cv. KS 282 and recessive in cv. IR 6 (Khan et al., 1999), but it is not the actual case. The resistance is believed to be imparted by multiple genes with a cumulative effect, some lines have only some of the genes from this cluster while the other lines might have other or a few or none; thus, their disease response varies greatly in response to the resistance to the pathogen. There is an extensive diversity of variation and varietal resistance to *Bakanae* and only a few numbers of varieties have been recorded as highly resistant to the *Bakanae* of rice (as shown in Table 16.2). So, classical breeding of transferring a resistance gene to the cultivar fails here. Moreover, monogenic resistance can be easily broken by new virulent strains. Therefore, it is very imperative to enhance the diverse genetic resources against *Bakanae* disease in rice-breeding programs, thus limiting the scope of classical breeding in the development of resistant cultivars. Hence, it becomes even more imperative for the need to have better tools to meet the expectations. Cultivating resistant cultivars represents the most economical and sustainable method for diseases management.

16.6.3 MOLECULAR BREEDING APPROACHES

16.6.3.1 FROM THE ADVENT OF MOLECULAR MARKER TO HIGH-THROUGHPUT GENOTYPING

In the present era of genomics, a typical shift in rice breeding for disease resistance can only be imagined with the advances in molecular marker technology and genome sequencing. Rice is a perfect choice among cereals for starting genomic studies as of its limited genomic size and diploid nature. Genomes of two domestic rice subspecies, that is, *japonica* (cultivar Nipponbare) and *indica* (cultivar 93-11), have been sequenced (Project IRGS, 2005; Yu et al., 2002). Recognizing the recent advancement of genomic sciences and research, and the completion of the rice genome sequence, now with a high level of precision we can identify and map several genes through linkage-based mapping. As of today, there is a little knowledge about mapped loci regulating resistance to *Bakanae* disease, although hundreds of loci have been mapped for other traits (Sharma et al., 2020). For the contemporary plant breeders, the tools for mapping desirable gene(s), mining superior alleles, and identifying recombinants possessing the desirable allelic combinations have been opened up through advancements in genotyping technologies along with the molecular markers like RAPD, AFLP SSR, SNP, InDel markers, etc., which have also been widely used to study the host and pathogen variability in rice for the *Bakanae* diseases.

This made the way for genomics-assisted breeding or molecular breeding, which in turn assures a very efficient and precise selection enabled by markers linked to the trait or the genes governing these traits (Muthu et al., 2020). Thousands of single nucleotide polymorphisms (SNPs) and insertion/deletions (InDels) between highly homologous genomes can be identified via re-sequencing. Moreover, high-throughput platforms of molecular markers have been developed for quantitative trait loci (QTL) mapping, gene identification, and molecular breeding (Kaur et al., 2020). Using linked markers, these QTLs or the candidate genes can be employed in the precision breeding programs such as the "Marker Associated Breeding (MAB) programs" for the development of resistant cultivars against *Bakanae* diseases of rice.

16.6.3.2 HIGH-THROUGHPUT PHENOTYPING: A LIKELY INTEGRAL PART OF THE FUTURE RICE-BREEDING PROGRAMS

Owing to the epistatic and dominance effects of genes regulating disease resistance and the drawbacks associated with year round screening, it is very difficult and often impossible to pyramid genes through conventional phenotype assays in breeding crop plants (Muthu et al., 2020). One of the major challenges is the need of a robust high-throughput phenotyping system that can help in phenotyping a great number of recombinants in a breeding programme, possibly in a nondestructive manner and deal with genotype x environment interactions. Recently, a RT-PCR kit was developed which uses SSR markers for the selection of a rice variety containing *bakanae* resistant gene *BK1*. Using this method, it is now possible to effectively and efficiently differentiate resistant and susceptible individuals (Park et al., 2013). Moreover, appropriate prediction models are further required to help breeders in identifying potential genotypes for the specific target environment. Various biological and conventional methodologies have been proposed for screening the diseased plants. Moreover, *F. fujikuroi* can be characterized by sequencing the genes encoding translation elongation factor (*TEF*), *histone H3, and b-tubulin* (Jeon et al., 2013) and many other methods of rapid identification of pathogen have also been discussed above. Various morphological and color abnormalities can be observed in plants affected by *F. fujikuroi*, such as stunting, elongation, a large angle between leaf and stem, and yellowish-green leaves.

A more recent and advanced technology, machine vision, can provide an automated inspection by the combined application of image analysis and machine learning techniques. Machine vision is a rapid and nondestructive method of phenotyping. Various techniques have already been used in grading the level of leaf spot diseases; distinguishing brown spot, blast, and narrow brown spot diseases; screening sheath blight, bacterial leaf blight, and blast diseases in rice (Sanyal et al., 2007; Shen et al., 2008; Kurniawati et al., 2009; Yao et al., 2009). Flatbed scanners have been used to collect the images of infected and healthy/control seedlings, their morphological and color traits have also been quantified. The genetic algorithm has been used for selecting essential traits and optimal model parameters for the support-vector machine (SVM) classifiers, which can be used for distinguishing the infected and healthy seedlings. This method has already shown the great accuracy (87.9%) and prediction value (91.8%) for differentiating the healthy and infected seedlings. This method is less subjective and tiresome in comparison to conventional examination techniques including clustering methods. Therefore, it can be used as a possible alternative in detecting disease infection and determining rice susceptibility and resistance in disease resistance breeding programs (Chung et al., 2016). High-throughput phenotyping and more efficient machine learning techniques are likely to be an integral part of future breeding programs.

16.6.3.3 UNDERSTANDING THE GENETICS OF BAKANAE DISEASE RESISTANCE IN RICE

16.6.3.3.1 Identifying Resistant Qtl/Genes and Linked/Flanking Markers: Linkage Based QTL Mapping

Information about mapped loci providing resistance to *Bakanae* in rice is very limited. To our knowledge, till now, only seven QTL mapping studies have been conducted which have identified a total of 26 resistant QTLs using different populations such as RILs (Recombinant Inbred Lines), NILs (Near Isogenic Lines), DH (Doubled Haploid), and F2:3 (F2 derived F3), and so on (Table 16.3). Out of these 26 QTLs, 12 QTLs on rice chromosome 1, two QTLs on chromosomes 3, four QTLs on each chromosome 6 and 10, and one QTL on each chromosome 8, 9, and 11 have been mapped. Resistance to *Bakanae* disease was measured using different parameters such as differences in seedling growth (Yang et al., 2006; mortality rate (Fiyaz et al., 2016; Ji et al., 2018; Cheon et al., 2019; Kang et al., 2019), the proportion of healthy plants (Hur et al., 2015; Lee et al., 2018), disease severity and colonization rate (Chen et al., 2019). For the first time in 2006, Yang et al. reported 2 QTLs, one, qB1, on chromosome 1 and another, qB10 on chromosome 10 using a DH population derived from a cross between Chunjiang 06 and TN1. Both of these parents were susceptible to disease but there was extensive segregation for the disease resistance in the generated DH population. Both of the mapped QTLs together explained greater than 26 % of the total phenotypic variation.

Later in 2015, another QTL mapping study was conducted using NIL (BC_6F_4) population derived from the cross between resistant variety "Shingwang"(*indica*) and the susceptible variety "Ilpum"(japonica). One major QTL on chromosome 1 was identified and mapped. It explained 65% of the total phenotypic variation (Hur et al., 2015). This major QTL was present in a 520 kb region between two flanking markers (RM8144 and RM 11295). This QTL interval was further delimited to only 35 Kbp by InDel 18 and InDel 19–14 markers. Detailed sequence analysis of this delimited region uncovered four putative candidate genes, *LOC_Os01g41770*,

LOC_Os01g41780, *LOC_Os01g41790*, and *LOC_Os01g41800* (Lee et al., 2019). Up-regulation of three candidate genes, *LOC_Os01g41770*, *LOC_Os01g41780*, and *LOC_Os01g41790* in susceptible cultivar, Ilpum; and of *LOC_Os01g41800* in resistant cultivar, Shingwang was detected after inoculation of plants with *G. fujikuroi*. Till now, this is the only report on fine mapping for identifying the gene present in any *Bakanae* resistant QTL.

In 2016, a RILs population derived from the cross between Pusa 1342, a highly resistant variety, and Pusa Basmati 1121, a highly susceptible variety was used for QTL mapping and four resistant QTLs were mapped on chromosome 1. These four QTLs explained 4.76, 24.74, 6.49, and 9.1% of phenotypic variation, respectively (Fiyaz et al., 2016). One of these QTL was mapped in a 0.26 Mb region, where a total of 55 genes were identified (Fiyaz et al., 2016). Later in 2018, a major QTL, qFfR1, was mapped on chromosome 1 using the F_3 population derived from the cross between Nampyeong variety and a susceptible Korean japonica line, DongjinAD. A total of 179 putative candidate genes were mapped in the QTL region, out of which 15 genes were associated with defense mechanisms against *Bakanae* disease (Ji et al., 2018). For QTL mapping, both parents were re-sequenced and identified DNA polymorphisms were used for the development of CAPS markers, which were then employed for QTL mapping.

In the same year, 2018, another study was published, which reported the identification of a major QTL, *qBK1WD*, on the chromosome 1 using the RIL population derived from the cross between Wonseadaesoo (R) and Junam (S). On a physical map, this QTL was located between chr01_13542347 (13.54 Mb) and chr01_15132528 (15.13 Mb) (Lee et al., 2018). Most recently in 2019, three QTL mapping studies have been published, one study reported the mapping of the two major QTLs, *qFfR1-1* and *qFfR6*, on chromosome 1 and 6 using two $F_{2:3}$ population derived from the cross between Junam and Nampyeong, and Saenuri and Nampyeong, respectively (Cheon et al., 2019). In the second study, a total of 14 QTLs including 1 mapped for both parameters, on rice chromosomes 1, 3, 4, 6, 8, 10, and 11 using a RIL population derived from the cross between IR64 and Nipponbare. In the third study, a major QTL, *qFfR9*, was mapped at 30.1 centimorgans (cM) on chromosome nine with an LOD score of 60.3 using the $F_{2:3}$ population obtained from the cross between Junam and Samgwang. Putative candidate genes present in the QTL interval were also reported. For QTL mapping, both parents were re-sequenced and identified DNA polymorphisms were used for the development of CAPS markers, which were then employed for QTL mapping (Kang et al., 2019). Further, there is a need for more QTL mapping studies using different mapping

TABLE 16.3 Different Resistant QTLs Reported in Various Published Studies.

QTL	Chr	Flanking markers/bp positions	PVE (%)	Add. effect	LOD score	Cross	Population type and size	References
qB1	1	RM7180-RM486	13.4	1.477	2.32	Chunjiang 06(S)/TN1(S)	DH	Yang et al., 2006
qB10	10	RM1108-RM304	13.3	1.450	2.5			
qBK1	1	RM8144-RM11295	65	31.74	33.21	Shingwang(R)/ Ilpum(S)	NILs (168)	Hur et al., 2015
qBK1.1	1	RM9-RM11282	6.49	−7.03	3.86	Pusa1121(R)/ Pusa1342(S)	RILs (168)	Fiyaz et al., 2016
qBK1.2	1	RM10153-RM5336	24.74	−13.7	12.07	Pusa1121(R)/ Pusa1342(S)	RILs (168)	Fiyaz et al., 2016
qBK1.3	1	RM10271-RM53	4.76	−6.08	2.73	Pusa1121(R)/	RILs (168)	Fiyaz et al., 2016
qBK3.1	3	RM411-RM3698	9.1	−8.34	3.31	Pusa1342(S)		
qFfR1	1	JND01005-JNC010809	40.1	21.71	22.7	Nampyeong(R)/DongjinAD (S)	$F_{2:3}$ (180)	Ji et al., 2018
qBK1WD	1	chr01_10336087-chr01_26628298	20.2	−9.53	8.29	Wonseadaesoo (R)/Junam(S)	RILs (200)	Lee et al., 2018
qFfR1-1	1	KJ01_075	49.9	19.45	21.4	Saenuri (S)/	$F_{2:3}$ (188)	Cheon et al., 2019
qFfR6	6	KJ06_055-KJ06_059	12.4	9.44	5.99	Nampyeong (R)		
qBK1.4	1	401,383-426,527bp	27					
qBK1.5	1	2,254,725-2,329,243bp	18					
qBK1.6	1	22,088,636-22,248,843bp	28					
qBK1.7	1	23,631,992-23,640,231bp	25					
qBK3.2	3	27,480,288-27,635,650bp	19			IR64/		
qBK4.1	4	22,371,038-22,428,157bp	33			Nipponbare	RILs (132)	Chen et al., 2019

TABLE 16.3 *(Continued)*

QTL	Chr	Flanking markers/bp positions	PVE (%)	Add. effect	LOD score	Cross	Population type and size	References
qBK6.1	6	3,276,254-3,639,339bp	22					
qBK6.2	6	4,866,345-5,059,806bp	28					
qBK6.3	6	25,298,288-25,638,876bp	20					
qBK8.1	8	6,142,736-6,239,144bp	28					
qBK10.1	10	5,678,051-6,024,705bp	25					
qBK10.2	10	6,849,663-6,864,693bp	21					
qBK10.3	10	9,090,969-9,337,961	26					
qBK11.1	11	22,576,995-22,582,906bp	13					
qFfR9	9	KJ09_024, 9FC30	59	35.15	60.3	Samgwang(R)/ Junam(S)	$F_{2:3}$ (188)	Kang et al., 2019

populations, high-throughput genotyping, and high-throughput phenotyping techniques, which will result in the identification of more resistant QTLs, which can be targeted for rice resistance breeding programs.

16.6.3.3.2 Identifying Resistant QTL/Genes and Linked/Flanking Markers: Genome Wide Association Studies (GWAS)

Various studies on strains collected from Asia, Africa, and Europe have revealed great genetic diversity among the pathogen population (Wulff et al., 2010; Jeon et al., 2013; Valente et al., 2017), which shows the importance of identification of additional loci conferring rice resistance to *Bakanae*. Genome-wide association studies (GWAS) are known to have better resolution than QTL mapping, thus reducing the confidence interval of QTLs and, eventually, the number of putative genes underlying individual QTLs (Volante et al., 2017; Chen et al., 2019). For the first time in 2017, a GWAS study was conducted for identifying the candidate genes providing resistance against *Bakanae* disease (Volante et al., 2017). For this purpose, first rice germplasm was collected from different countries, mainly from Italy, USA, Portugal, and Spain and then extensively screened for *Bakanae* disease resistance. This screening allowed the identification of different accessions with moderate to high resistance to disease. Then, GWAS was conducted which revealed the significant association of two genomic regions with the observed phenotypic variation for the response to pathogen infection on the chromosome arm 1S (named as *qBK1_628091*) and on the chromosome arm 4L (named as *qBK4_31750955*). A total of 156 and 4 significant SNPs were identified on chromosome arms 1S and 4L, respectively. A fine comparison of both genomic positions with the already published QTLs suggested the novelty of these genomic regions associated with *F. fujikuroi* resistance. An investigation for candidate genes identified 129 genes for *qBK1_628091*, of which 45 were functionally annotated. Among these annotated genes, 33 encode for protein kinase, mostly for receptor-like kinases.

In the same way, a total of 56 functionally annotated genes were identified for the second region, *qBK4_31750955*. Out of these 56 genes, several genes, including genes for NB-LRR resistance protein and SHR5-receptor-like kinases, and so on were identified which are known to have an association with resistance response to disease. These identified genes in both regions can be further characterized and employed for breeding disease-resistant rice (Volante et al., 2017). Recently in 2019, another GWAS was conducted using 231 diverse accessions from Rice Diversity Panel 1. They mapped 14 QTLs using the bi-parental population and then using the GWAS approach they uncovered 206 putative candidate genes within these 14 QTLs. Many of these identified genes are known to encode leucine-rich repeat (LRR)-containing and NB-ARC proteins, transcription factors, ubiquitination-related proteins, and oxidase/oxidoreductase proteins. Moreover, one QTL, *qBK1.7,* was delineated to a region of 8239 bp, and coding regions of two LRR-containing genes *Os01g0601625* and *Os01g0601675* were uncovered in this particular region which may be associated with *Bakanae* resistance (Chen et al., 2019). The information on the resistance performance of hundreds of rice accessions evaluated in the above-discussed studies and identified candidate genes will aid in resistant rice breeding. In the future, there is a need for more GWAS studies using rice accessions collected from different geographical locations which will enable the identification of more resistant accessions and marker-trait associations.

18.6.3.4 PROSPECTS OF MARKER-ASSISTED BREEDING FOR BAKANAE RESISTANCE RICE BREEDING

As discussed above, many major resistant QTLs and potential candidate genes have been identified in rice and the information on linked or flanking markers is available; these major resistant QTLs and candidate genes can be transferred from the donor species into the high yielding but susceptible varieties using marker-assisted breeding approaches. Transfer of QTLs/genes using molecular markers will enable the reduction in the amount of time and labor-intensive assays required for screening the disease-resistant plants. As far as disease resistance is concerned, the pyramiding of different QTLs/genes (having different underlying molecular mechanisms for providing disease resistance) in one cultivar is considered better over the transfer of single QTL/gene (Pradhan et al., 2015). In this way, a single plant is supposed to deliver higher and/or durable resistance against diseases. This is not a new concept; the effects of pyramiding different genes have been investigated for various plant-microbe interactions. For instance, pyramiding of three bacterial blight resistant genes revealed a high level of durable resistance (Pradhan et al., 2015). In the same way, marker-assisted pyramiding of two different *Bakanae* resistant QTLs (*qBKWD* and *qBK1*) provided increased resistance to *Bakanae* disease in rice (Lee et al., 2018), which showed the promising effect of this marker strategy in the development of *Bakanae* disease-resistant rice varieties. Unfortunately, this is the only report of marker-assisted gene/QTL pyramiding for *Bakanae* resistant in rice so far. In the future, we may expect enhanced development of *Bakanae* disease-resistant rice varieties using this strategy as we now have more sensitive and efficient marker systems and high- throughput phenotyping platforms that will further accelerate the breeding programs.

16.7 CONCLUSION AND FUTURE PERSPECTIVES

Foot rot/*Bakanae* disease is arising as a potential threat to rice cultivation globally which causes drastic yield losses. As shown by various studies, *F. fujikuroi* shows great level of variability in nature; therefore, identifying novel resistance sources against them is of much importance. Different cultural, morphological, biochemical, and molecular methods have been developed for the correct identification of *Fusairum* species. The knowledge of variability in the pathogen, its way of survival is needed to devise efficient diseases management strategy (Matic et al., 2017). There is a lack of information on the research of biochemical properties and their role in *Bakanae* disease. A total of eighteen strains of *F. fujikuroi* have been sequenced which have revealed extensive DNA polymorphisms among the strains (Table 16.1). Other strains can also be sequenced and comparative analysis can be conducted in the near future which may enable the identification of the key candidate genes required for pathogenicity of the pathogen. Different chemical and molecular patterns involved in the interaction between *F. fujikuroi* and rice have been uncovered in various studies. Understanding of the interaction between rice and *F. fujikuroi* may be helpful to find out novel resistant cultivars and to devise the alternative disease control strategies as well as to develop the more sensitive and rapid molecular diagnostic techniques. A great number of resistant genotypes have been identifying which can be employed for breeding programs. To overcome the limitations associated with conventional breeding, biotechnological tools or molecular breeding approaches, described elsewhere (Saini et al., 2020), can be employed for rice *bakanae* disease resistant breeding programs. Several QTL mapping and association mapping studies have uncovered the various QTLs and candidate genes associated with the resistance mechanism (Table 16.3). These QTLs/candidate genes can be pyramided in cultivars for durable and effective resistance.

New biotechnological techniques like genome editing including CRISPR/Cas9 and TALENs and genetic transformation are yet to be explored in rice with respect to *Bakanae* disease resistance. These techniques have been very successful in rice for resistance to bacterial blight, for instance,

Oliva et al. (2019) introduced mutation into three *SWEET* gene promoters using CRISPR–Cas9-mediated genome editing and produced rice lines with robust, broad-spectrum resistance to bacterial blight. Hopefully, these genome editing techniques will be used for *Bakanae* disease resistance in near future.

KEYWORDS

- ***F. fujikuroi*-rice interaction**
- **fungal foot rot**
- **marker-assisted breeding**
- **marker-trait association**
- **rice**

REFERENCES

Ahangar, M. A.; Najeeb, S.; Rather, A. G.; Bhat, Z. A.; Parray, G. A.; Sanghara, G. S.; Kashap, S. C.; Ahangar, F. A.; Ahmad, H. Evaluation of Fungicides and Rice Genotypes for the Management of *Bakanae*. *ORYZA-An Int. J. Rice* **2012,** *49*, 121–126.

Amatulli, M. T.; Spadaro, D.; Gullino, M. L.; Garibaldi, A. Molecular Identification of *Fusairumspp.* Associated With *Bakanae* Disease of Rice in Italy and Assessment of Their Pathogenicity. *Plant Pathol.* **2010**, *59*, 839–844.

Amatulli, M. T.; Spadaro, D.; Gullino, M. L.; Garibaldi, A. Corrigendum: Molecular Identification of *Fusairumspp.* Associated With *Bakanae* Disease of Rice in Italy and Assessment of Their Pathogenicity. *Plant Pathol.* **2012,** *61*, 820–820.

Amin, K. S.; Reddy, T. B. Seedling and Adult Stage Methods for Screening Rice to *Bakanae* [India]. *Indian Phytopathol.* **1977**, *29*, 46–49.

Application no. KR20130124934A, Inventors: Soo, P. D.; Hwang, U. H.; Park, S.-K.; Shin, D.-J.; Lee, J.-H.; Yoon, Y.-N.; Joon, H.; Han, S.-I.; Choi H.-W.; Song, Y.-C.; Kwak, D.-Y.; Yeoun, S.; Nam, M.-H. Composition Comprising DNA Marker for Selecting Rice Variety Containing Bakanae Disease-Resistant BK1 Gene and Method of Selecting Rice Variety Resistant to Bakanae Disease Using the DNA Marker.

Bacon, C. W.; Hinton, D. M. Symptomless Endophytic Colonization of Maize by *Fusairummoniliforme*. *Can. J. Bot.* **1996,** *74*, 1195–1202.

Bagga, P. S.; Vineet, K. Resistance to *Bakanae* or Foot-Rot Disease in Basmati Rice. *Indian Phytopathol.* **2000,** *53*, 321–322.

Bah, S.; Labuschagne, M.; van der Merwe, R. Genetic Diversity of Improved Varieties of Intraspecific (O. Sativa and O. Glaberrima) and Interspecific (O. Sativa × O. Glaberrima) Rice. *Genet. Resour. Crop. Ev.* **2018,** *65*, 797–809.

Baharuddin, S. Morphological Characteristics and Mating Populations of Fusarium Species in Gibberella Fujikuroi Species Complex (Gfsc) Associated With Stalk Rot Disease of Maize in Indonesia, Malaysia and Thailand. *Plant Pathol.* **2017,** *16*, 33–40.

Bal, R. S.; Biswas, B. Epidemiology and Management of Foot Rot in Basmati Rice. *J. Krishi Vigyan* **2018**, *6*, 87–94.

Bashyal, B. M.; Aggarwal, R. Molecular Identification of *Fusairum* Species Associated With *Bakanae* Disease of Rice (Oryza sativa) in India. *Indian J. Agric. Sci.* **2013,** *83*.

Bashyal, B. M. Etiology of an Emerging Disease: Bakanae of Rice. *Indian Phytopathol.* **2018,** *71*, 485–494.

Bashyal, B. M.; Aggarwal, R.; Rawat, K.; Sharma, S.; Gupta, A. K.; Choudhary, R.; Bhat, J.; Krishnan, S. G.; Singh, A. K. Genetic Diversity and Population Structure of *Fusairum Fujikuroi* Causing *Bakanae*, an Emerging Disease of Rice in India. **2020,** *58*, 45–52.

Bashyal, B. M.; Aggarwal, R.; Sharma, S.; Gupta, S.; Rawat, K.; Singh, D.; Singh, A. K.; Krishnan, S. G. Occurrence, Identification and Pathogenicity of *Fusairum* Species Associated With *Bakanae* Disease of Basmati Rice in India. *Eur. J. Plant Pathol.* **2016,** *144*, 457–466.

Bashyal, B. M.; Rawat, K.; Sharma, S.; Kulshreshtha, D.; Gopala Krishnan, S.; Singh, A. K.; Dubey, H.; Solanke, A. U.; Sharma, T. R.; Aggarwal, R. Whole Genome Sequencing of *Fusairum Fujikuroi* Provides Insight Into the Role of Secretory Proteins and Cell Wall Degrading Enzymes in Causing *Bakanae* Disease of Rice. *Front. Plant Sci.* **2017,** *8*, 2013.

Bhalla, K.; Singh, S. B.; Agarwal, R. Quantitative Determination of Gibberellins by High Performance Liquid Chromatography From Various Gibberellins Producing *Fusairum* Strains. *Environ. Monit. Assess.* **2010,** *167,* 515–520.

Cheema, A. A.; Awan, M. A.; Ali, Y. Screening of Basmati Rice Mutants Against Prevalent Diseases in the Punjab Province. *Pak. J. Phytopathol.* **1998,** *10*, 39–41.

Chen, S. Y.; Lai, M. H.; Tung, C. W.; Wu, D. H.; Chang, F. Y.; Lin, T. C.; Chung, C. L. Genome-Wide Association Mapping of Gene Loci Affecting Disease Resistance in the Rice-*Fusairum Fujikuroi* Pathosystem. *Rice* **2019,** *12*, 85.

Cheon, K. S.; Jeong, Y. M.; Lee, Y. Y.; Oh, J.; Kang, D. Y.; Oh, H.; Kim, S. L.; Kim, N.; Lee, E.; Baek, J.; Choi, I. Kompetitive Allele-Specific PCR Marker Development and Quantitative Trait Locus Mapping for *Bakanae* Disease Resistance in Korean Japonica Rice Varieties. *Plant Breed. Biotechnol.* **2019,** *7*, 208–219.

Chiara, M.; Fanelli, F.; Mule, G.; Logrieco, A. F.; Pesole, G.; Leslie, J. F.; Horner, D. S.; Toomajian, C.; Genome Sequencing of Multiple Isolates Highlights Subtelomeric Genomic Diversity Within *Fusairum Fujikuroi*. *Genome Biol. Evol.* **2015,** *7*, 3062–3069.

Chung, C. L.; Huang, K. J.; Chen, S. Y.; Lai, M. H.; Chen, Y. C.; Kuo, Y. F. Detecting *Bakanae* Disease in Rice Seedlings by Machine Vision. *Comput. Electron. Agric.* **2016,** *121*, 404–411.

Cumagun, C. J. R.; Arcillas, E.; Gergon, E. UP-PCR Analysis of the Seedborne Pathogen *Fusairum Fujikuroi* Causing *Bakanae* Disease in Rice. *Int. J. Agric. Biol*. **2011,** *13,* 1814–9596.

Desjardins, A. E. *Fusairum Mycotoxins: Chemistry, Genetics, and Biology*. American Phytopathological Society (APS Press), 2006.

Desjardins, A. E; Manandhar, H. K.; Plattner, R. D.; Manandhar, G. G.; Poling, S. M.; Maragos, C. M. *Fusairum* Species From Nepalese Rice and Production of Mycotoxins and Gibberellic Acid by Selected Species. *Appl. Environ. Microbiol.* **2000,** *66*, 1020–1025.

EL-Shafey, R.; Anis, G. B.; Elmoghazy, A. M. Evaluation of some developed high yielding rice genotypes resistant to blast, *Bakanae* and white tip nematode. *J. Plant Prod.* **2016**, *7*, 317–329.

Elshafey, R. A. S.; Tahoon, A. M.; El-Emary, F. A. Analysis of Varietal Response to *Bakanae* Infection *Fusairum Fujikuroi* and Gibberellic Acid Through Morphological, Anatomical and Hormonal Changes in Three Rice Varieties. *J. Phytopathol. Pest Manag.* **2018**, *22*, 63–87.

Fiyaz, R. A.; Krishnan, S. G.; Rajashekara, H.; Yadav, A. K.; Bashyal, B. M.; Bhowmick, P. K.; Singh, N. K.; Prabhu, K. V.; Singh, A. K. Development of High Throughput Screening Protocol and Identification of Novel Sources of Resistance Against *Bakanae* Disease in Rice (Oryza Sativa L.). *Indian J. Genet.* **2014,** *74*, 414–422.

Fiyaz, R. A.; Yadav, A. K.; Krishnan, S. G.; Ellur, R. K.; Bashyal, B. M.; Grover, N.; Bhowmick, P. K.; Nagarajan, M.; Vinod, K. K.; Singh, N. K.; Prabhu, K. V. Mapping Quantitative Trait Loci Responsible for Resistance to *Bakanae* Disease in Rice. *Rice* **2016,** *9*, 45.

Fiyaz, R. A.; Yadav, A. K.; Krishnan, S. G.; Ellur, R. K.; Bashyal, B. M.; Grover, N.; Bhowmick, P. K.; Nagarajan, M.; Vinod, K. K.; Singh, N. K.; Prabhu, K. V. Mapping Quantitative Trait Loci Responsible for Resistance to *Bakanae* Disease in Rice. *Rice* **2016,** *9*, 45.

Gangopadhyay, S. Current Concepts on Fungal Diseases of Rice. In *Current Concepts on Fungal Diseases of Rice*, 1983.

Ghazanfar, M. U.; Nazir, J.; Waqas, W.; Iqbal, M. Screening of Some Fine and Coarse Rice Varieties Against *Bakanae* Disease. *J. Agric. Res.* **2013,** *51*, 41–49.

Gherbawy, Y. A.; Adler, A.; Prillinger, H. Genotypic Identification of *Fusairum* Subglutinans, F. Proliferatum and F. Verticillioides Strains Isolated From Maize in Austria. *Mycobiology* **2002,** *30*, 139–145.

Gill, M. A.; Yasin, S. I.; Khan, T. Z.; Ahsanullah, Q.; Butt, M. A. Evaluation of Resistance to *Bakanae* and Foot Rot Disease. *Int. Rice Res. Notes* **1993,** *18,* 14–15.

Gutierrez-Najera, N.; Munoz-Clares, R. A.; Palacios-Bahena, S.; Ramírez, J.; Sanchez-Nieto, S.; Plasencia, J.; Gavilanes-Ruíz, M. Fumonisin B 1, a sphingoid Toxin, is a Potent Inhibitor of the Plasma Membrane H+-ATPase. *Planta* **2005,** *221*, 589–596.

Hayasaka, T.; Ishiguro, K.; Shibutani, K.; Namai, T. Seed Disinfection Using Hot Water Immersion to Control Several Seed-Borne Diseases of Rice plants. *Japanese J. Phytopathol.* **2001**, *67*, 26–32.

Hirata, T.; Kimishima, E.; Aoki, T.; Nirenberg, H. I.; O'Donnell, K. Morphological and Molecular Characterization of *Fusairum* Verticillioides From Rotten Banana Imported Into Japan. *Mycoscience* **2001,** *42*, 155–166.

Hori, S. Some Observations on '*Bakanae*'Disease of the Rice Plant. *Mem. Agric. Res. Sta.* **1898,** *12*, 110–119.

Hsuan, H. M.; Salleh, B.; Zakaria, L. Molecular Identification of *Fusairum* Species in *Gibberella Fujikuroi* Species Complex From Rice, Sugarcane and Maize From Peninsular Malaysia. *Int. J. Mol. Sci.* **2011,** *12*, 6722–6732.

Hsuan, H.; Zakaria, L.; Salleh, B. Characterization of *Fusairum* Isolates From Rice, Sugarcane and Maize Using RFLP-IGS. *J. Plant Prot. Res.* **2010**, *50*, 409.

Hur, Y. J.; Lee, S. B.; Kim, T. H.; Kwon, T.; Lee, J. H.; Shin, D. J.; Park, S. K.; Hwang, U. H.; Cho, J. H.; Yoon, Y. N.; Yeo, U. S. Mapping of qBK1, A Major QTL for *Bakanae* Disease Resistance in Rice. *Mol. Breed.* **2015,** *35*, 78.

Hwang, I. S.; Kang, W. R.; Hwang, D. J.; Bae, S. C.; Yun, S. H.; Ahn, I. P. Evaluation of *Bakanae* Disease Progression Caused by *Fusairum Fujikuroi* in Oryza Sativa L. *J. Microbiol.* **2013,** *51*, 858–865.

Iqbal, M.; Javed, N.; Sahi, S. T.; Cheema, N. M. Genetic Management of *Bakanae* Disease of Rice and Evaluation of Various Fungicides Against *Fusairummoniliforme* In Vitro. *J. Phytopathol.* **2011,** *23*, 103–107.

Ito, S. Studies on the *Bakanae* Disease of the Rice Plant. *Rep. Hokkaido Agric. Exp. Stn.* **1931,** *27*, 1–95.

Jain, J.; Sidhu, N.; Lore, J. S.; Gill, R. S. Evaluation of Aromatic Rice Genotypes for Resistance Against Foot Rot Disease. *Plant Dis. Res.* **2016,** *31*, 150–153.

Jeon, Y. A.; Yu, S. H.; Lee, Y. Y.; Park, H. J.; Lee, S.; Sung, J. S.; Kim, Y. G.; Lee, H. S. Incidence, Molecular Characteristics and Pathogenicity of *Gibberella Fujikuroi* Species Complex Associated With Rice Seeds From Asian Countries. *Mycobiology* **2013,** *41*, 225–233.

Jeong, H.; Lee, S.; Choi, G. J.; Lee, T.; Yun, S. H. Draft Genome Sequence of *Fusairum Fujikuroi* B14, The Causal Agent of the *Bakanae* Disease of Rice. *Genome Announc.* **2013**, *1*.

Ji, H.; Kim, T. H.; Lee, G. S.; Kang, H. J.; Lee, S. B.; Suh, S. C.; Kim, S. L.; Choi, I.; Baek, J.; Kim, K. H. Mapping of a Major Quantitative Trait Locus for *Bakanae* Disease Resistance in Rice by Genome Resequencing. *Mol. Genet. Genomic Med.* **2018,** *293*, 579–586.

Ji, Z.; Zeng, Y.; Liang, Y.; Qian, Q.; Yang, C. Transcriptomic Dissection of the Rice–*Fusairum Fujikuroi* Interaction by RNA-Seq. *Euphytica* **2016,** *211*, 123–137.

Ji, Z.; Zeng, Y.; Liang, Y.; Qian, Q.; Yang, C. Proteomic Dissection of the Rice-*Fusairum Fujikuroi* Interaction and the Correlation Between the Proteome and Transcriptome Under Disease Stress. *BMC Genom.* **2019,** *20*, 91.

Kang, D. Y.; Cheon, K. S.; Oh, J.; Oh, H.; Kim, S. L.; Kim, N.; Lee, E.; Choi, I.; Baek, J.; Kim, K. H.; Chung, N. J. Rice Genome Resequencing Reveals a Major Quantitative Trait Locus for Resistance to *Bakanae* Disease Caused by *Fusairum Fujikuroi*. *Int. J. Mol. Sci.* **2019,** *20*, 2598.

Kaur, B.; Mavi, G. S.; Gill, M. S.; Saini, D. K. Utilization of KASP Technology for Wheat Improvement. *Cereal Res. Commun.* **2020,** 1–13.

Kaur, J.; Pannu, P. P. S.; Sharma, S.; Lakhanpal, G. Morphological, Biochemical and Molecular Characterization of Different Isolates of *Fusairum* From Foot Rot Infected Rice Plants. *Plant Dis. Res.* **2011,** *26*, 108.

Khan, T. Z.; Gill, M. A.; Bajwa, K. M.; Khan, M. G. Screening of Rice Varieties/Lines Against *Bakanae* Disease. *Pak. J. Phytopathol.* **1999,** *11*, 166–168.

Khan, T. Z.; Gill, M. A.; Khan, M. G. Inheritance Studies for *Bakanae* Disease of Rice. *Pak. J. Phytopathol.* **1999,** *11*, 181–182.

Khokhar, L. K.; Jaffrey, A. H. Identification of Sources of Resistance Against *Bakanae* and Foot Rot Disease in Rice. *Pak. J. Agric. Sci.* **2002,** *17*, 176–177.

Kim, H. J.; Choi, Y. K.; Min, B. R. Variation of the Intergenic Spacer (IGS) Region of Ribosomal DNA Among *Fusairum* Oxysporum Formae Speciales. *J. Microbiol.* **2001**, *39*, 265–272.

Kim, M. H.; Hur, Y. J.; Lee, S. B.; Kwon, T.; Hwang, U. H.; Park, S. K.; Yoon, Y. N.; Lee, J. H.; Cho, J. H.; Shin, D.; Kim, T. H. Large-Scale Screening of Rice Accessions to Evaluate Resistance to *Bakanae* Disease. *J. Gen. Plant Pathol.* **2014,** *80*, 408–414.

Kim, H. J.; Min, B. R. Nucleotide Divergence Analysis of IGS Region in *Fusairum* Oxysporum and its Formae Speciales Based on the Sequence. *Mycobiology* **2004,** *32*, 119–122.

Kurniawati, N. N.; Abdullah, S. N. H. S.; Abdullah, S.; Abdullah, S. In *Investigation on Image Processing Techniques for Diagnosing Paddy Diseases.* 2009 International Conference of Soft Computing and Pattern Recognition, 2009; pp 272–277.

Lee, S. B.; Hur, Y. J.; Cho, J. H.; Lee, J. H.; Kim, T. H.; Cho, S. M.; Song, Y. C.; Seo, Y. S.; Lee, J.; Kim, T. S.; Park, Y. J. Molecular Mapping of qBK1 WD, A Major QTL for *Bakanae* Disease Resistance in Rice. *Rice* **2018,** *11*, 1–8.

Lee, Y. H.; Lee, M. J.; Choi, H. W.; Kim, S. T.; Park, J. W.; Myung, I. S.; Park, K. S.; Lee, S. W. Development of In Vitro Seedling Screening Method for Selection of Resistant Rice Against *Bakanae* Disease. *Res. Plant Dis.* **2011**, *17*, 288–294.

Leslie, J. F.; Summerell, B. A. *The Fusairum Laboratory Manual.* John Wiley & Sons, 2008.

Li, D. J. Studies on Resistance of Rice Materials to *Bakanae* Disease and Pathogenicity of Pathogen (*Fusairummoniliforme*). *Acta. Phytopathol. Sinica* **1993,** *23*, 315–319.

Lima, C. S.; Pfenning, L. H.; Costa, S. S.; Abreu, L. M.; Leslie, J. F. *Fusairum* Tupiense sp. nov., A Member of the *Gibberella Fujikuroi* Complex That Causes Mango Malformation in Brazil. *Mycologia* **2012,** *104*, 1408–1419.

Ma, L.; Ji, Z.; Bao, J.; Zhu, X.; Li, X.; Zhuang, J.; Yang, C.; Xia, Y. Responses of Rice Genotypes Carrying Different Dwarf Genes to *Fusairummoniliforme* and Gibberellic Acid. *Plant Prod. Sci.* **2008,** *11*, 134–138.

Malonek, S.; Bömke, C.; Bornberg-Bauer, E.; Rojas, M. C.; Hedden, P.; Hopkins, P.; Tudzynski, B. Distribution of Gibberellin Biosynthetic Genes and Gibberellin Production in the *Gibberella Fujikuroi* Species Complex. *Phytochemistry* **2005,** *66*, 1296–1311.

Matic, S.; Bagnaresi, P.; Biselli, C.; Carneiro, G. A.; Siciliano, I.; Valé, G.; Gullino, M. L.; Spadaro, D. Comparative Transcriptome Profiling of Resistant and Susceptible Rice Genotypes in Response to the Seedborne Pathogen *Fusairum Fujikuroi*. *BMC Genom*, **2016,** *17*, 608.

Matic, S.; Spadaro, D.; Prelle, A.; Gullino, M. L.; Garibaldi, A.; Light Affects Fumonisin Production in Strains of *Fusairum Fujikuroi*, *Fusairum* Proliferatum, and *Fusairum* Verticillioides Isolated From Rice. *Int. J. Food Microbiol.* **2013,** *166*, 515–523.

McDonald, B. A.; Linde, C. The Population Genetics of Plant Pathogens and Breeding Strategies for Durable Resistance. *Euphytica* **2002**, *124*, 163–180.

Miah, S. A.; Zaman, M. Q. Study on Reaction of High Yielding Dwarf Rice to Foot-Rot and *Bakanae* in Bangladesh. *Bangladesh J. Bot*. **1973**, *2*, 73–81.

Miyagawa, E.; Sato, M.; Asano, K.; Okamoto, E. Genetic Variation in *Fusairum* Oxysporum Isolated From Wilt-Infested Alfalfa in Hokkaido. *J. Nat. Sci.* **2002**, *27*, 69–77.

Muthu, V.; Abbai, R.; Nallathambi, J.; Rahman, H.; Ramasamy, S.; Kambale, R.; Thulasinathan, T.; Ayyenar, B.; Muthurajan, R.; Pyramiding QTLs Controlling Tolerance Against Drought, Salinity, and Submergence in Rice Through Marker Assisted Breeding. *PLoS One* **2020,** *15*, 0227421.

Niehaus, E. M.; Janevska, S.; von Bargen, K. W.; Sieber, C. M.; Harrer, H.; Humpf, H. U.; Tudzynski, B. Apicidin F: Characterization and Genetic Manipulation of a New Secondary Metabolite Gene Cluster in the Rice Pathogen *Fusairum Fujikuroi*. *PLoS One* **2014,** *9*, 103336.

Niehaus, E. M.; Kim, H. K.; Münsterkötter, M.; Janevska, S.; Arndt, B.; Kalinina, S. A.; Houterman, P. M.; Ahn, I.P.; Alberti, I.; Tonti, S.; Kim, D. W. Comparative Genomics of Geographically Distant *Fusairum Fujikuroi* Isolates Revealed Two Distinct Pathotypes Correlating With Secondary Metabolite Profiles. *PLoS Pathog.* **2017**, *13*, 1006670.

Nosratabadi, M.; Kachuei, R.; Rezaie, S.; Harchegani, A. B. Beta-Tubulin Gene in the Differentiation of *Fusairum* Species by PCR-RFLP Analysis. *System. Field Crop Diseases*, No. Ed. 3, 1999; Nyvall, R. F.; Iowa State University Press, 2018, vol *3*.

Oliva, R.; Ji, C.; Atienza-Grande, G.; Huguet-Tapia, J. C.; Perez-Quintero, A.; Li, T.; Eom, J. S.; Li, C.; Nguyen, H.; Liu, B.; Auguy, F. Broad-Spectrum Resistance to Bacterial Blight in Rice Using Genome Editing. *Nat. Biotechnol.* **2019,** *37*, 1344–1350.

Pakshir, K.; Farazmand, F.; Ghasemi, F.; Mirhendi, H.; Zomorodian, K.; Kharazi, M.; Pour, R. A.; Golestani, H.; Motamedi, M. Translation Elongation Factor 1-Alpha Gene as a Marker for Diagnosing of Candidal Onychomycosis. *Curr. Med. Mycol.* **2020,** *6*, 15.

Petrovic, T.; Burgess, L. W.; Cowie, I.; Warren, R. A.; Harvey, P. R. Diversity and Fertility of *Fusairum* Sacchari From Wild Rice (Oryza Australiensis) in Northern Australia, and Pathogenicity Tests With Wild Rice, Rice, Sorghum and Maize. *Eur. J. Plant Pathol.* **2013,** *136*, 773–788.

Piombo, E.; Bosio, P.; Acquadro, A.; Abbruscato, P.; Spadaro, D.; Different Phenotypes, Similar Genomes: Three Newly Sequenced *Fusairum Fujikuroi* Strains Induce Different Symptoms in Rice Depending on Temperature. *Phytopathology* **2020,** *110*, 656–665.

Prà, M. D.; Tonti, S.; Pancaldi, D.; Nipoti, P.; Alberti, I. First Report of *Fusairum* Andiyazi Associated With Rice *Bakanae* in Italy. *Plant Dis*. **2010,** *94,* 1070–1070.

Pradhan, S. K.; Nayak, D. K.; Mohanty, S.; Behera, L.; Barik, S. R.; Pandit, E.; Lenka, S.; Anandan, A. Pyramiding of Three Bacterial Blight Resistance Genes for Broad-Spectrum Resistance in Deepwater Rice Variety, Jalmagna. *Rice* **2015,** *8*, 19.

Project, I. R. G. S. The Map-Based Sequence of the Rice Genome. *Nature* **2005,** *436*, 793–800.

Puyam, A. Investigations on Foot Rot (Fusairum Moniliforme Sheld.) of Basmati Rice With Reference to Survival and Variability of Pathogen and Host-Resistance. Doctoral Dissertation, Punjab Agricultural University, Ludhiana, 2017.

Puyam, A.; Pannu, P. P. S.; Kaur, J.; Sethi, S.; Variability in Production of Gibberellic Acid and Fusaric Acid by *FusairumMoniliforme* and Their Relationship. *J. Plant Pathol.* **2017,** *99*, 103–108.

Puyam, A.; Pannu, P. P. S.; Kaur, J.; Sethi, S.; Kaur, K. Understanding *Bakanae*: A Major Threat and an Emerging Disease of Basmati Rice. *Indian Phytopathol.* **2019,** *72*, 599–605.

Puyam, A.; Pannu, P. P. S.; Kaur, J.; Shikha, S. Cultural, Morphological and Molecular Variability of Sheld Causing Foot Rot Disease of Basmati Rice in Punjab. *J. Mycol. Plant Pathol.* **2017,** *47*, 369–381.

Radwan, O.; Gunasekera, T. S.; Ruiz, O. N. Draft Genome Sequence of *Fusairum Fujikuroi*, A Fungus Adapted to the Fuel Environment. *Genome Announc.* **2018**, *6*, 3.

Rahjoo, V.; Zad, J.; Javan-Nikkhah, M.; Gohari, A. M.; Okhovvat, S. M.; Bihamta, M. R.; Razzaghian, J.; Klemsdal, S. S.; Morphological and Molecular Identification of *Fusairum* Isolated From Maize Ears in Iran. *J. Plant Pathol.* **2008,** 463–468.

Rajagopalan, K. Screening of Rice Varieties for Resistance to Foot-Rot Disease. *Curr. Sci.* **1961**, *30*, 145–147.

Reyes, G. M. Rice Diseases and Methods of Control. *Philipp. J. Agric.* **1939,** *4*, 419–436.

Reynoso, M. M.; Torres, A. M.; Chulze, S. N. Fusaproliferin, Beauvericin and Fumonisin Production by Different Mating Populations Among the *Gibberella Fujikuroi* Complex Isolated From Maize. *Mycol. Res.* **2004**, *108*, 154–160.

Saini, D. K.; Devi, P.; Kaushik, P. Advances in Genomic Interventions for Wheat Biofortification: A Review. *Agronomy* **2020**, *10*, 62.

Sanyal, P.; Bhattacharya, U.; Parui, S. K.; Bandyopadhyay, S. K.; Patel, S. In *Color Texture Analysis of Rice Leaves Diagnosing Deficiency in the Balance of Mineral Levels Towards Improvement of Crop Productivity*. 10th International Conference on Information Technology (ICIT 2007), 2007; pp 85–90.

Saremi, H.; Ammarellou, A.; Marefat, A.; Okhovvat, S. M. Binam a Rice Cultivar, Resistant for Root Rot Disease on Rice Caused by *Fusairummoniliforme* in Northwest, Iran. *Int. J. Bot.* **2008**.

Sawada, K. Beitrage über Formosas-Pilze no. 14. *Trans. Nat. Hist. Soc. Formosa* **1917,** *31*, 31–133.

Sharma, V.; Saini, D. K.; Kumar, A.; Kesh, H.; Kaushik, P. Breeding for Biofortification Traits in Rice: Means to Eradicate Hidden Hunger. In *Agronomy*; Intech Open, 2020.

Siciliano, I.; Amaral Carneiro, G.; Spadaro, D.; Garibaldi, A.; Gullino, M. L. Jasmonic Acid, Abscisic Acid, and Salicylic Acid are Involved in the Phytoalexin Responses of Rice to *Fusairum fujikuroi*, A High Gibberellin Producer Pathogen. *J. Agr. Food Chem.* **2015,** *63*, 8134–8142.

Singh, A. K.; Gopalakrishnan, S.; Singh, V. P.; Prabhu, K. V.; Mohapatra, T.; Singh, N. K.; Sharma, T. R.; Nagarajan, M.; Vinod, K. K.; Singh, D.; Singh, U. D. Marker Assisted Selection: A Paradigm Shift in Basmati Breeding. *Indian J. Genet. Plant Breed.* **2011**, *71*, 120.

Singh, R.; Sunder, S. Foot Rot and *Bakanae* of Rice: Retrospects and Prospects. *Int. J. Trop. Plant Dis.* **1997,** *15*, 153–176.

Suga, H.; Kitajima, M.; Nagumo, R.; Tsukiboshi, T.; Uegaki, R.; Nakajima, T.; Kushiro, M.; Nakagawa, H.; Shimizu, M.; Kageyama, K.; Hyakumachi, M. A Single Nucleotide Polymorphism in the Translation Elongation Factor 1α Gene Correlates With the Ability to Produce Fumonisin in Japanese *Fusairum Fujikuroi*. *Fungal Boil.* **2014,** *118*, 402–412.

Sukapanpotharam, S.; Chantarasnit, A. A. A Screening Method for *Bakanae* Resistance. *Int. Rice Res.* **1980,** *5*, 6.

Sun, S. K. The Disease Cycle of Rice *Bakanae* Disease in Taiwan. In *Proc. Natl. Sci. Counc. Repub. China* **1975,** *8*, 245–255.

Sunder, S. Survival and Variability in Fusairum Moniliforme Sheld. and Management of Bakanae Disease of Ricen. Doctoral Dissertation, College of Agriculture Chaudhary Charan Singh Haryana Agricultural University Hisar, 1995.

Sunder, S.; Singh, R. Vegetative Compatibility, Biosynthesis of GA3 and Virulence of *Fusairummoniliforme* Isolates From *Bakanae* Disease of Rice. *Plant Pathol.* **1998**, *47*, 767–772.

Sunder, S.; Singh, R.; Dodan, D. S. Management of *Bakanae* Disease of Rice Caused by *Fusairummoniliforme*. *Indian J. Agric. Sci.* **2014,** 84, 224–228.

Surek, H.; Gumustekin, H. Research Activities on Controlling Rice *Bakanae* and Foot Rot Diseases (*Fusairummoniliforme*) in Turkey. *Cah. Options Mediterr.* **1994,** *8*.

Thomas, K. M. The 'Foot Rot' of Paddy and its Control. *Madras Agric. J.* **1933**, *21*, 263–272.

Urbaniak, C.; Dadwal, S.; Bagramyan, K.; Venkateswaran, K. Draft Genome Sequence of a Clinical Isolate of *Fusairum Fujikuroi* Isolated From a Male Patient With Acute Myeloid Leukemia. *Genome Announc.* **2018**, *6*.

Volante, A.; Tondelli, A.; Aragona, M.; Valente, M. T .; Biselli, C.; Desiderio, F.; Bagnaresi, P.; Matic, S.; Gullino, M. L.; Infantino, A.; Spadaro, D. Identification of *Bakanae* Disease Resistance Loci in Japonica Rice Through Genome Wide Association Study. *Rice* **2017,** *10*, 29.

Weizheng, S.; Yachun, W.; Zhanliang, C.; Hongda, W. In *Grading Method of Leaf Spot Disease Based on Image Processing*, 2008 International Conference on Computer Science and Software Engineering. 2008; vol 6, pp 491–494.

Wiemann, P.; Sieber, C. M.; Von Bargen, K. W.; Studt, L.; Niehaus, E. M.; Espino, J. J.; Huß, K.; Michielse, C. B.; Albermann, S.; Wagner, D.; Bergner, S. V. Deciphering the Cryptic Genome: Genome-Wide Analyses of the Rice Pathogen *Fusairum Fujikuroi* Reveal Complex Regulation of Secondary Metabolism and Novel Metabolites. *PLoS Pathog.* **2013,** *9*, 1003475.

Wollenweber, H. W.; Reinking, O. A. *Die Fusarien, ihre Beschreibung. Schadwirkung und Bekämpfung*, 1935; pp 1–355.

Wulff, E. G.; Sørensen, J. L.; Lübeck, M.; Nielsen, K. F.; Thrane, U.; Torp, J. *Fusairumspp.* Associated With Rice *Bakanae*: Ecology, Genetic Diversity, Pathogenicity and Toxigenicity. *Environ. Microbiol.* **2010,** *12*, 649–657.

Xu, J. R.; Yan, K.; Dickman, M. B.; Leslie, J. F. Electrophoretic Karyotypes Distinguish the Biological Species of *Gibberella Fujikuroi* (*Fusairum* Section Liseola). *Mol. Plant Microbe In.* **1995,** *8*, 74–84.

Yabuta, T.; Kobe, K.; Hayashi, T. Biochemical Studies of the'*Bakanae*'Fungus of Rice. I. Fusarinic Acid, A New Product of the'*Bakanae*'Fungus. *Zeitschrift fur Pflanzenkrankheiten (Pflanzenpathologie) und Pflanzenschutz* **1934,** *10*, 1059–1068.

Yadav, R. S.; Tyagi, S.; Javeria, S.; Gangwar, R. K. Effect of Different Cultural Condition on the Growth of *Fusairummoniliforme* Causing *Bakanae* Disease. *Eur. J. Mol. Biotechnol.* **2014,** *2,* 95–100.

Yang, C. D.; Guo, L. B.; Li, X. M.; Ji, Z. J.; Ma, L. Y.; Qian, Q. Analysis of QTLs for Resistance to Rice *Bakanae* Disease. *Chin. J. Rice Sci*. **2006,** *6*, 657–659.

Yao, Q.; Guan, Z.; Zhou, Y.; Tang, J.; Hu, Y.; Yang, B. Application of Support Vector Machine for Detecting Rice Diseases Using Shape and Color Texture Features. In 2009 International Conference on Engineering Computation, 2009; pp 79–83.

Yu, J.; Hu, S.; Wang, J.; Wong, G. K. S.; Li, S.; Liu, B.; Deng, Y.; Dai, L.; Zhou, Y.; Zhang, X.; Cao, M. A Draft Sequence of the Rice Genome (Oryza Sativa L. ssp. indica). *Science* **2002,** *296*, 79–92.

Zainudin, N. A. I. M.; Razak, A. A.; Salleh, B. *Bakanae* Disease of Rice in Malaysia and Indonesia: Etiology of the Causal Agent Based on Morphological, Physiological and Pathogenicity Characteristics. *J. Plant Prot. Res.* **2008a**, *48*, 475–485.

Zainudin, N. A. I. M.; Razak, A. A.; Salleh, B. Secondary Metabolite Profiles and Mating Populations of *Fusairum* Species in Section Liseola Associated With *Bakanae* Disease of Rice. *J. Microbiol.* **2008b**, *4*, 6–13.

Zhang, S. Y.; Dai, D. J.; Wang, H. D.; Zhang, C. Q. One-Step Loop-Mediated Isothermal Amplification (LAMP) for the Rapid and Sensitive Detection of *Fusairum Fujikuroi* in *Bakanae* Disease Through NRPS31, an Important Gene in the Gibberellic Acid Bio-Synthesis. *Sci. Rep.* **2019,** *9*, 1–9.

CHAPTER 17

Potential of Bio-Control Agents for the Sustainable Management of Rice Fungal Diseases

SARBANI DAS[1], BHOLANATH SAHA[2], SUSHANTA SAHA[3], NANDINI ROY[4], PRASENJIT BARMAN[4], and RAJENDRA BAIRWA[5]

[1]*Department of Plant Pathology, Bidhan Chandra Krishi Viswavidyalaya, Mohanpur, Nadia, West Bengal, India*

[2]*Dr. Kalam Agricultural College, Bihar Agricultural University, Kishanganj, Bihar, India*

[3]*Bidhan Chandra Krishi Viswavidyalaya, Kalyani, Nadia, West Bengal, India*

[4]*Department of Plant Pathology, Uttar Banga Krishi Viswavidyalaya, Pundibari, Coochbehar, West Bengal, India*

[5]*Dr. Rajendra Prasad Central Agricultural University, Pusa, Samstipur, Bihar, India*

ABSTRACT

The quality of the agricultural produce is maintained by Crop Protection for years. The past century has been identified with remarkable use of chemical fertilizers and pesticides along with improved agronomical practices enabling growers to maintain quality crop productivity. However, the widespread use of chemicals in the last few decades for controlling pests and diseases led to adverse impacts on the environment, reducing the quality and

Fungal Diseases of Rice and Their Management. Deepti Srivastava, Md. Shamim, Malik Mobeen Ahmad, and R. S. Upadhyay (Eds.)

in turn affecting consumer health. In recent times, variable steps are being undertaken for the management of plant diseases. Biological control is one of the eco-friendly alternative approaches for disease management reducing the amount of human contact with harmful chemicals and their residues. Several biocontrol agents including fungi and bacteria have been recognized, but they require effective adoption and further development for appropriate efficacy on several grounds. Extensive study regarding bio-control through various mechanisms like direct antagonism, mixed-path antagonism by the production of allelochemicals and indirect antagonism needs to be done. This necessitates a better understanding of the interconnections among the pathogen, the host, and the environment that will lead to sustainable agriculture. Beyond the field assessment, analysis and extensive research on microbial communities together with culture-independent molecular techniques including sequencing technologies and genomic study commenced a new era of plant disease management.

17.1 INTRODUCTION

Rice has been the major important food crop for the developing world and is considered the staple food grain for greater than 60% of the population in India (C.R.R.I., Vision 2030. Central Rice Research Institute (ICAR), Cuttack, Odisha, India. (2011)). The growing population led to the necessity of increased quality production and yield. Green Revolution aided the country to fulfil the huge food demand. However, despite great success, the quality rice productivity declined due to several reasons. The major factor behind the declining quality and yield potential of the crop was "diseases" with special mention to rice fungal and bacterial diseases. Various fungus, bacteria, viruses, nematodes, and mycoplasma-like organisms were found infecting rice. Manandhar et al. (1998) recorded more than 70 diseases in rice caused by several pathogens; still, more has been added later. Notable diseases of rice include the blast disease caused by fungus *Pyricularia grisea*, sheath blight disease caused by fungus *Rhizoctonia solani*, sheath rot disease due to fungus *Sarocladium oryzae*, bacterial leaf blight of rice by *Xanthomonas oryzae pv.* oryzae, and rice tungro disease due to rice tungro bacilliform virus (RTBV) and rice tungro spherical virus (RTSV) leading to heavy losses in production and yield across the world (Gnanamanickam et al., 2002). Brown spot (*Bipolaris oryzae*) and stem rot (*Sclerotium oryzae*) are two other fungal diseases affecting rice. Chemical control has been quite effective to combat rice pathogens, but their increased user raised an alarm from the viewpoint of environmental hazards as well as economy. The possible emergence of resistant pathogens to these chemical fungicides was still another matter of concern (Gnanamanickam et al., 2002). Many fungicides were found to combat the pathogen under field and laboratory environment successfully, but they were reported to create a very low inhibitory effect on the spore production like the sclerotial germination in the case of *Rhizoctonia solani* (Chakraborti et al., 2018). All these considerations incited a hunt for biocontrol agents effective against rice pathogens. During the 1980s, preliminary research on biological control of rice diseases began and has been a huge area of exploration till now. Studies on bio-control agents against plant diseases are still based on identification, evaluation, and bio formulation for their exploitation on the field.

17.2 NATURAL CONTROL FOR PLANT DISEASE

All factors including the surrounding environment, moisture, temperature should be considered before opting for natural control of plant diseases. Plant diseases due to soil-borne pathogens are highly complex because plants are exploited by pathogens, predators, parasites, as well as decomposers in those cases. Several chemical control methods successfully inhibit the pathogen attack, but at the same time, they also prepare the plant to fall prey to other pathogens. The disease can be checked by taking natural means like the establishment of disease suppressive soil that includes proper drainage and good texture, optimum fertility level, types of soil organisms present, and also the presence of beneficial nutrients for plants, organic composting, and crop rotation. Pathogenic diseases are suppressed through induced resistance in plants (SAR and ISR), competition for nutrients, direct parasitism upon harmful pathogens, and secretion of antibiotics by beneficial soil microorganisms. Crop rotation plays an important role in the restriction of pathogen proliferation, but it has to be done with non-host crops for the existing pathogens. Also, crop rotation is effective for pathogens that survive in crop residue or soil, but not much applicable for airborne pathogens or those vectored by insects from elsewhere as they can survive for long in the absence of hosts (e.g., *Fusarium*). Adequate soil nutrient content and pH has a major role in disease management and plays a major role in impacting the pathogen's environment. Cover crop, compost application, manuring, mulching adds organic matter to the soil which plays a major role in the control of root rot diseases. In compost amended soil, the organic matter is already digested, while in other techniques decomposition occurs through aerobic and anaerobic means (Kamal et al., 2015).

17.3 ECO-FRIENDLY APPROACH

In soil environments, out of a variety of micro-organisms that exist, some of them are beneficial but most are pathogenic to plants. Application of organic amendments and other sound practices encourage the diversity of micro-organisms, which leads to disease suppression. World agriculture is leaning toward taking successful biological management of plant pests and diseases, thereby reducing the dependence on chemical inputs which is an important footstep in Integrated and Organic Pest Management systems (Kamal et al., 2015). Companies are also showing interest in manufacturing and marketing biocontrol agents (McSpadden Gardener and Fravel, 2002). Extensive research facilitated small essential commercial sectors producing BCAs most of which is dealt with by relatively small companies (Fravel, 2005). Though many biocontrol products are not yet registered, they are circulated and well accepted as plant growth promoters or strengtheners without any specific disease control claim by the distributors.

17.4 BIO-CONTROL: A SUSTAINABLE APPROACH FOR DISEASE MANAGEMENT

Disease management is very important for secured food supply for the population as well as for agriculture-based industries. This can be done either through chemical control or by undertaking biological means using predators or parasites. Extensive use of chemicals results in long-term adverse effects on the environment disturbing the ecological balance and contaminating the food chain through bioaccumulation of lethal residues (Sharma et al., 2016). G.F. Von Tabuef first used the term biological control. According to Kiewnick, 2007, the term "biocontrol agent" or "biopesticide," as a generic definition, has been used with a constricted focus on certain preparations including living microorganisms, to a broader definition taking into account botanical compounds and semichemicals (like pheromones).

17.5 BIOLOGICAL CONTROL AGENTS (BCAS)

In plant pathology and entomology the term "biological control" is commonly used. Biocontrol is the use of living agents like microbial pathogens, entomopathogenic nematodes, or predatory insects for suppression of harmful pathogens. Bio-formulation of these beneficial agents (biofertilizer or biopesticides) is used to control plant disease and restrict pathogen growth. *Streptomyces* isolated from the rhizosphere of *Araucaria* inhibits the growth of pine rot causing fungi, *Armillaria*, and *Fusarium* (Vasconcellos, 2009). The seed and seedling treatment of tomato with the talc-based *S. griseus* formulation reduces disease severity due to *F. oxysporum f. sp. lycopersici* (Anitha et al., 2009). As reported by Kanini et al. (2013) indigenous *Streptomyces* isolates inhibited *Rhizoctonia solani* fungus. *Streptomyces* have been used in controlling many pathogens like *Fusarium* wilt of cotton and *Pythium* damping-off in tomato by *S. ambofaciens, Rhizoctonia* root rot control in pea by *S. hygroscopius var. geldanus* due to the antibiotic production, geldanamycin, in vitro control of *R.solani* and *Pythium ultimum* by *Streptomyces lydicus* WYEC108 with the production of antifungal metabolites (Yuan and Crawford, 1995).

Localized resistance and induced systemic resistance to late blight pathogen of tomato are stimulated by VAM fungus, Glomus mosseae which has been demonstrated in split root experiments using mycorrhizal and non-mycorrhizal roots (Cordier et al., 1998). This occurs due to the phenolic compound accumulation which in turn induced defence response in the plant. The VAM-affected plant cells showed immunity against *P. parasitica* with callus development at the parasitic infection site. Also it characterized cell wall thickening of proteins and pectins in non-mycorrhizal root parts and callus formation at infection site as part of the systemic response to the resistance. No such observations were found in non-mycorrhizal but pathogen-infected plants.

17.5.1 BENEFITS OF BIO-CONTROL AGENT

- **Environment friendly:** Biocontrol agents are safer for the environment and persons using them and leave no toxic residues, unlike chemical disease control agents.
- **Easy and low manufacturing cost:** The manufacture of biocontrol agents is comparatively easier and often cheaper.
- **Pathogen sensitivity:** Bio-pesticides can effectively eradicate specific pathogens from the infection site and can be used together with bio-fertilizers.
- **Soil application:** Biocontrol agents are especially effective for many soil-borne pathogens where chemical fungicides cannot be applied as soil harbors several beneficial microorganisms as well which might be harmed due to chemical pesticides.

- **Host congenial:** Biocontrol agents besides disease control promote root and plant growth encouraging the accumulation of beneficial microflora in the rhizosphere which in turn enhances the crop yield potential. It also aids in the mobilization of nutrients making them available to the crop plant.
- **No resistance development:** Biocontrol agents do not only avoid the problem of resistance development which is often observed in the case of chemical pesticides but are also found to induce systemic resistance among various plant species.
- **Self-regulating:** Biocontrol agents do not involve any complicated management and help preserve the ecosystem.

17.5.2 CERTAIN DRAWBACKS OF BIO-CONTROL AGENT

- **Biocontrol agent as a detriment in new habitat:** An organism introduced from elsewhere to eradicate a pathogen in a new environment may itself turn predators into some existing beneficial organisms.
- **Delayed efficacy:** Unlike chemical pesticides, the immediate remedy in the case of biocontrol agents may not be observed as they work slowly and environmental conditions play a major role in their working efficacy.
- **Preventive tool:** Biocontrol agents function as a preventive tool against specific diseases and not much as a curative tool.
- **Short shelf life:** The shelf life of biocontrol agents like *Pseudomonas fluorescens* is only three months and that of *Trichoderma viridae* is four months. Periodical checking under the surveillance of skilled workers is required to maintain the effectiveness of biocontrol agents.
- **Skilled labor:** Skilled persons are required for uninterrupted multiplication and supply of biocontrol agents with no contamination.
- **Low production:** Biocontrol agents are obtainable only in very few places and production is very little.

17.6 SUSTAINABLE AGRICULTURE BY BCAS

Commercially biocontrol agents are marketed with a specific action that promotes sustainable agriculture

- Bio-fertilizers improve the potential of plants for nutrient uptake.

- Bio-protectants restrict several diseases associated with cultivable crops.
- Bio-stimulants induce the production of phytohormone. Many genera of bacteria like *Bacillus, Pseudomonas, Streptomyces, Paenibacillus, Burkholderia,* and *Agrobacterium* and also some fungal species show such properties. Many antagonistic fungi have been reported to control serious plant diseases; *Trichoderma* strains (*T. harzianum, T. viride, T.* virens) are the major source (Benitez et al., 2004).

Commercial mycofungicides are developed using a broad range of biocontrol agents (Benitez et al., 2004; Kim and Hwang, 2004; Fravel, 2005) which suppress plant diseases through ISR, MIR, antibiotics, siderophores production, or others. Biofertilizers increase crop nitrogen uptake from nitrogen-fixing bacteria like *Azospirillium*, and also iron uptake from siderophore-producing *Pseudomonas* bacteria. *Streptomyces* is also in use as a promising biocontrol tool for controlling several soil-borne bacteria. Phytohormones like IAA, gibberellins, cytokinins, and ethylene inhibitors are also produced by them. Such studies encouraged developing new biofertilizers using various antagonists (Harman et al., 2004). The cool temperature and wet soil promote the attack of seedlings by the fungus *Rhizoctonia solani. Rhizoctonia* is traced and attacked by a specific chemical released by the beneficial fungus *Trichoderma. Trichoderma* hyphae entangle the pathogen thereby releasing enzymes that dehydrate the cells of *Rhizoctonia* and ultimately killing them. *Trichoderma* cultures are marketed as biological seed treatment agents presently for the control of the damping-off disease.

17.7 BIOLOGICAL CONTROL AGENTS FOR RICE PATHOGENS

Diverse groups of bio-control agents like bacteria, viruses, and fungi are present. The easily manageable bacteria due to their rapidly growing and aggressive colonizing nature are considered the best antagonist for bio-control as compared to others. Bacteria have close interaction with host plants; this might be used as efficient biocontrol agents in sustainable agriculture. Soil serves as an important cache to isolate bacteria with biocontrol activities (Bloemberg and Lugtenberg, 2001). Bacterial antagonists, particularly *Bacillus* and *Pseudomonas*, are promising bio-control agents. Gram-positive and endospore-producing Bacilli bacteria are heat tolerant and also desiccation, making them ideal for field application. The rod-shaped gram-negative *Pseudomonas* have simple nutritional requisites and are exceptional colonizers and largely found in the rice rhizosphere. Various strains of fluorescent and nonfluorescent bacteria in the rice rhizosphere in upland and lowland have been found antagonistic to rice sheath blight fungus in vivo, in vitro, and also in the field condition. Nearly 23 bacterial antagonists under the genera *Pseudomonas, Erwinia, Serratia*, and *Bacillus* restricted the growth of *R. solani* mycelia, and also other harmful rice pathogens, *P. grisea* (blast), *Sarocladium oryzae* (sheath rot), *Sclerotium oryzae* (stem rot), *B. oryzae* (brown spot), and *Fusarium fuijkuroi* (bakanae). Almost 40 bacterial antagonists against rice sheath blight disease have been found that include *Chaetomium globosum, Trichoderma spp., Laerisaria arvalis, Penicillium,* and *Myrothecium verrucaria*. Nookongbut et al. (2020) studied the control of three important rice plant pathogens by the use of antifungal EPS of *R. palustris* KTSSR54 possibly applicable on the field. The EPS of *R. palustris* KTSSR54 acts as an efficient growth regulator for plant increasing grain yield and also to control serious fungal diseases all by itself. The combination of live KTSSR54 and its EPS indicated a remarkable increase in rice yield potential. The combination depicted satisfying results and proved promising to be used as a biocontrol agent in rice fields even under acidic soil conditions.

TABLE 17.1 List of Rice Fungal Diseases With Their Bio-Control Agents.

Disease	**Causal organism(s)**	**Biological control agents**
Bacterial leaf blight	*Xanthomonas oryzae pv. Oryzae*	*Bacillus sp.*
Paddy blast	*Pyricularia grisea*	*Pseudomonas fluorescens*
Brown spot	*Bipolaris oryzae*	*Pseudomonas sp.* *P. aeruginosa* *Bacillus sp.* *B. subtilis*
Sheath rot	*Sarocladium oryzae*	*Pseudomonas sp.* *P. fluorescens* *B. subtilis* *P. aeruginosa*
Stem rot	*Sclerotium oryzae*	*P. fluorescens* *P. aeruginosa* *B. subtilis* *B. pumilus*
Sheath blight	*Rhizoctonia solani*	*Bacillus sp.* *B. subtilis* *B. pumilus* *Serratia marcescens* *Pseudomonas sp.*

Adapted and modified from Vasudevan et al. (2002).

17.8 POTENTIAL OF BIO-CONTROL AGENTS

An appropriate application technique of biocontrol agent to restrict pathogen needs to be followed. Direct inoculation methods like spraying, seed treatment, application through a drip system or application through solid-phase inoculants can be uptaken.

1. Disease severity of several rice fungal pathogens can be reduced by 20–42% with prior application of *P. fluorescent* bacteria before pathogen inoculation both in vitro and on the field. Such bacteria also encourage an increase in rice plant height, tiller number, and grain yield by 3–160% (Kumar and Dangar, 2013).
2. Seed treatment with specific nonpathogenic bacteria reduces rice bakanae disease (*Fusarium fujikuroi*) by 72–96% (Kumar and Dangar, 2013). Sarwar et al. (2018) evaluated the biocontrol activity of surfactin-producing *Bacillus* (SPB) strains NH-100 and NH-217 and also purified surfactin A from these strains against rice bakanae disease on Super Basmati rice. The study revealed that surfactin A reduced bakanae disease up to 80%.
3. Prior treatment of rice with different *Bacillus* species like seed treatment before rice sowing, root dip of seedling before transplanting, or foliar sprays of antagonists before pathogen inoculation can restrict the rice sheath blight pathogen up to 59%, increasing the plant height and grain yield in turn (Kumar and Dangar, 2013).
4. Recent study also revealed that *P. fluorescens* imposes an insecticidal effect on rice tungro virus vector. Bacterial strains of PpV14i and Pf7-14 can lead to as high as 90% mortality if the vectors feed for 7 days on treated rice leaves (Kumar and Dangar, 2013).
5. In medium low land rainfed conditions blast disease in direct-seeded rice was targeted for management using bio-inoculants and *Trichoderma* spp. was selected as it proved to be a potential biological control agent against many economically important fungi. Seed treatment with *Trichoderma harzianum* improved seed germination and reduced disease intensity. Treated seeds showed 23.30–30.55% disease intensity compared to control seeds 40.50–48.09%, hence reducing 10–25% disease intensity approximately (Singh et al., 2012).
6. The efficiency of certain biocontrol agents and plant extracts was tested for reducing aflatoxin B1 (AFB1) in stored rice. The plant extract from *Syzigium aromatic* (5 g/kg) inhibited completely

Aspergillus flavus growth and AFB1 production. *Curcuma longa, Ocimum sanctum,* and *Allium sativum* also inhibited the A. flavus growth by 65–78% and AFB1 production by 72.2–85.7% (5 g/kg). Culture filtrate of Rhodococcus erythropolis used as biocontrol showed complete inhibition of AFB1 production (25 mL/kg). Other biocontrol agents, *Bacillus subtilis, Trichoderma virens,* and *Pseudomonas fluorescens* showed 68%, 80%, and 93% reduction of A. *flavus* growth and 58%, 72.2%, and 83.7% reduction of AFB1 (200 mL/kg), respectively (Reddy et al., 2009).

17.9 CHAETOMIUM SPP.

These are normally found in soil and organic composts and are efficient cellulase and laccase producers. Among various species, *C. cochlioides, C. globosum,* and *C. cupreum* are potential antagonists against various pathogens. *C. cupreum, C. globosum* are reported to be antagonistic to *P. oryzae* causing **rice** blast and *D. oryzae (Helminthosporium)* causing brown spot of rice. Soytong (2014) proved *Chaetomium cochliodes* to be a novel antagonist against rice brown spot var Pittsanulok 2 (causal organism: *Drechslera oryzae*) and inhibition of mycelial growth by 83.83 % and spore production bt 71.55% was observed. Different formulations were made out of *C. cochliodes* to test for the best control. Bio-powder formulation performed well in terms of leaf spot control and promoting plant growth (over 44%) followed by the application of a crude extract of the same, benlate and finally spore suspension of *C. cochliodes*. Bi-culture antagonistic test of rice blast pathogen infecting KoeKo 9 rice variety in Thailand and *Chaetomium lucknowense* showed significant efficacy of *Chaetomium* species in the inhibition of blast pathogen *P. oryzae*. Ethyl acetate and methanol crude extract of *Chaetomium* were found to be most effective in restricting the pathogen growth (Song et al., 2017).

17.10 TRICHODERMA SPP.

Trichoderma species are popular BCAs easily available, present everywhere, have high reproductive rate, can survive in unfavorable conditions, can modify rhizosphere environment, have great nutrient efficiency, promote plant growth, stimulate plant defence mechanism, and are highly aggressive against harmful fungal pathogens (Benitez et al., 2004). Although they fall mostly under fungal imperfect, some species shows morphological similarity to *Hypocrea* anamorph. The important species of *Trichoderma* used for bio control include *T. harzianum, T. viride, T hamatum, T. polysporum, T. koningii, T. pseudokoningii, T. piluliferum,* and *T. virens. Trichoderma* isolates belonging to *Trichoderma harzianum, T. virens*, and *T. atroviride* species were screened against rice brown spot disease due to *Bipolaris oryzae* in Iran. Results showed to significantly inhibit pathogen mycelial growth in vitro producing volatile and nonvolatile metabolites. In the glasshouse tests, remarkable disease control was observed for two strains of *T. harzianum* and increased seedling growth in the case of *T. atroviride* strain (Khalili et al., 2012). Swain et al. (2018) reported from Orissa, utilization of *Trichoderma* spp. from tree barks for higher productivity of rice. *Trichoderma erinaceum* controlled soil-borne pathogens *Rhizoctonia solani, Sclerotium oryzae, Sclerotium rolfsii*, and *Helminthosporium oryzae* and *R. solani* in field condition. *Trichoderma erinaceum* treatment showed the highest yield. *Trichoderma*-treated plants showed higher expression of stress-related enzymes that helped in better crop growth under stress conditions. The study thus encouraged workers to further incorporate *Trichoderma erinaceum* as a bio fertilizer and also as a bio control agent in integrated rice crop management. *Trichoderma asperellum* strains found positive for IAA release and phosphate solubilization were examined for their efficiency to suppress two rice pathogens, *R. solani* and *S. oryzae* in vivo and in vitro. Regression analysis study showed a positive correlation between the chlamydospores produced by *T. asperellum* and their antagonistic effect against the rice pathogens. TAIK-1 strain showed a strong antagonistic effect on fungal pathogens, besides promoting the growth and development of the rice plants (Chinnaswami et al., 2021).

17.11 MECHANISMS OF BIO-CONTROL AGENTS

17.11.1 DIRECT ANTAGONISM

17.11.1.1 PARASITISM

An interactive mechanism where two phylogenetically different organisms stay together for a long period. Here one organism acts as "host" and the other organism, "parasite," remains dependent on the host and is the one which gets benefitted. *Trichoderma* acts as mycoparasite-producing toxic metabolites and cell wall-degrading enzymes like chitinase and glucanase.

17.11.1.2 HYPERPARASITISM

Organisms parasitic to harmful plant pathogens are called hyperparasites. The mycoparasitic potential of *Arthrobotrys oligospora* was tested in vitro against *Rhizoctonia solani*. In *A. oligospora* hyphal coils developed around the hyphae of *R. solani*. Under field conditions, the biocontrol potential of *A. oligospora* against *R. solani* was also recorded. Application of *A. oligospora*, isolate VNS-1, in soil infested with *R. solani* reduced the disease incidence by 55.68–59.32% and lesion length by 54.91–66.66% under greenhouse and field conditions (Singh et al., 2012). "Hyperparasitism" is the direct antagonism to plant pathogens by their highly specific obligate parasites. Such plant pathogen parasites sometimes attack sclerotia (*Coniothyrium minitans*), or they may attack the hyphae (*Pythium oligandrum*). A harmful pathogen like that of powdery mildews' could be attacked by a number of hyper-parasites like *Gliocladium virens, Ampelomyces quisqualis, Cladosporium oxysporum,* and *Acremonium alternatum* (Kiss, 2003).

17.11.1.3 COMMENSALISM

The association of two unrelated species in which one organism gets benefitted but the other is not harmed is known as commensalism. Rhizosphere microbes control the soil-borne plant pathogens by competing with them for nutrients and also through antibiotic production (Kumar et al., 2016a). The microbes also increase the solubility of minerals and synthesize several other essential elements that are essential for plant growth. A study on Bacillus isolates from saline soils of Uttar Pradesh was done to check their biocontrol potential against *R. solani* for control of rice sheath blight disease. Dual-culture assay and microscopic studies showed significant antagonistic activity for the strain RH5 (84.41%). 16S rDNA sequencing analysis identified the biocontrol strain RH5 as *Bacillus subtilis*. The in vivo study confirmed that the strain significantly increased overall growth and activated resistance in rice plants producing defence-related antioxidant enzymes (Jamali et al., 2019).

17.11.2 MIXED-PATH ANTAGONISM BY SYNTHESIS OF ALLOCHEMICALS

17.11.2.1 SIDEROPHORES

Siderophores are low molecular weight ligand molecules with a high affinity to acquire iron from iron limiting microenvironment, thus preventing the growth of other microorganisms. Hydroxamate and catechols are two major classifications of siderophores based on their functional groups. A mix of these two groups has also been reported by Hider and Kong in 2010. *Streptomyces* strains, *S. lydicus* (Tokala et al., 2002), *S. pilosus* (Muller et al., 1984; Muller and Raymond, 1984), and *S. violaceusniger* (Buyer et al., 1989), produce siderophores. *Pseudomonas fluorescens* strains A1, B10, TL3B1, and BK1 biologically controlled Erwinia carotovora (Kloepper et al., 1980). Rhizobacteria produce siderophores that play a major role in suppressing the pathogenic fungus showing that a synthetic iron chelator can inhibit the fungus. There is evidence to support such a siderophore activity in the history of biocontrol like pseudomonads for wilt diseases by different *Fusarium* species and *Pythium* damping-off (Lim et al., 1999). In the study of Chaiharn et al. (2009) most isolates that showed major hostile activity were members of the genus *Bacillus* and *Pseudomonas* particularly the species *B. firmus* and *P. aureofaciens*. Investigations regarding the function enzymes, phytohormones, and siderophores produced by *Pseudomonads* in the rice rhizosphere and their biocontrol activities showed them to significant compounds inhibiting rice disease-causing pathogens (Roy et al., 2013).

17.11.2.2 ANTIBIOSIS

"Antibiosis" is derived from the word "antibiotics" which are organic substances affecting the metabolic activities and growth of other microbes by endolysis or breakdown of cell cytoplasm (Roshan et al., 2013). A variety of antibiotics have been identified, including compounds such as 2,4-diacetylphloroglucinol (DAPG), oligomycin A, oomycin A, amphisin, phenazine, hydrogen cyanide, pyoluteorin, pyrrolnitrin, tensin, tropolone, and cyclic lipopeptides produced by *pseudomonads*. Iturin suppresses fungal pathogens like *Rhizoctonia solani* and *Botrytis cinerea*, while phenazine carboxylic acid acts as an antagonist against rice pathogen *Rhizoctonia solani* (Padaria et al., 2016). As reported by many research workers, *Pseudomonas* produce various antibiotics that suppress the rice pathogens, for example, strains of *Pseudomonas fluorescens* (Pf7-14) produce antifungal antibiotics inhibiting germination of the blast pathogen conidia (Chatterjee et al., 1996).

17.11.2.3 VOLATILE SUBSTANCES

Volatile compounds are produced by some biocontrol agents besides antibiotics. The common substances include certain acids, ketones, aldehydes, sulfides, and hydrocyanic acid (HCN) (Bouizgarne, 2013). HCN production plays an important role in disease control (Wei et al., 1991), for example, Roy et al. (2013) reported HCN production by P. fluorescens strains that helped suppress rice plants infecting fungal pathogen. Reports of HCN production by beneficial microbes to minimize the harmful effect of pathogenic fungi are available (Gopalakrishnan et al., 2014). It has been reported by Chaiharn et al. (2020) rhizosphere actinomycetes successfully reduced disease symptoms in vivo and in vitro conditions. The study showed *Streptomyces palmae* PC 12 was potential as a biocontrol agent against rice blast fungus and also promoted the growth of rice variety RD6. The fungal inhibitory activity was possibly due to chitinase like cell-wall-degrading enzymes production and also HCN. Herbs like *Alpinia galangal, Zingiber officinale*, and *Curcuma longa* have been widely used for the treatment of many plant diseases as antifungal agents due to the presence of bioactive compounds. Nobila et al. (2021) studied extracts from these herbs that showed efficiency for rice blast pathogen, *Pyricularia oryzae*, in vitro. The best treatment was found to be *A. galangal* hexane crude extract that gave 52.9% growth inhibition of *P. oryzae*. However, field application needs to be done to assess the effectiveness on field condition and also chemical analysis to confirm the effective antifungal compound present in the plant crude extracts.

17.11.2.4 LYTIC ENZYME PRODUCTION

Microorganisms secrete lytic enzymes that can hydrolyse polymeric compounds like cellulose, hemicellulose, chitin, and proteins that plays an important role in pathogen suppression. Chitinase secreted by *Streptomyces sp., Serratia marcescens,* and *Paenibacillus sp.* was found inhibitory against *Botrytis cinerea, Sclerotium rolfsii*, and *Fusarium oxysporum f.sp. cucumerinum. Trichoderma* competes with the pathogenic fungi-producing antibiotics and lytic enzymes (glucanase and chitinase) in species like *T. harzianum* and *T. hamatum. Streptomyces corchorusii* UCR3-16 showed promising biocontrol potential for several rice fungal pathogens with the highest inhibition against *Rhizoctonia solani.* The strain produced fungal cell wall degrading enzymes like lipase, chitinase, protease, β-1,3-glucanase, and β-1,4-glucanase. The strain also enhanced several PGP traits. The strain was reported to enhance the growth of rice plants even under pathogen challenged conditions (Tamreihao et al., 2016).

17.11.2.5 UNREGULATED WASTE PRODUCTS

Such products are released by a few soil microbes, for example, ammonia, ethylene, hydrogen sulfide, methane, nitrite, carbondioxide, other volatile sulfur compounds, etc., and restrict the growth of harmful plant pathogenic bacteria. Ammonia-producing *Bacillus megaterium* has an inhibitory effect on the growth of Fusarium oxysporum (Shobha and Kumudini, 2012).

17.11.2.6 DETOXIFICATION AND DEGRADATION OF VIRULENCE FACTOR

Here, a specific protein of the biocontrol agent binds to the pathogen toxin detoxifying the pathogen virulence factors, either directly or indirectly, ultimately lowering the virulence of the pathogen toxin. Biocontrol strains like *Ralstonia solanacearum* and *B. cepacia* can hydrolyze phytotoxin fusaric acid produced by various *Fusarium* spp.

17.11.3 INDIRECT ANTAGONISM

Indirect antagonisms do not engage pathogen sensing by the biocontrol agents but include some enzymes, competition for root colonization, and induced resistance-mechanisms (Pal and McSpadden Gardener, 2006). Interaction between sedentary parasitic nematodes and mycorrhizal fungi ensures besides improving plant nutrition, resistance to soil pathogens is related to other issues as well. Studies showed a negative correlation for phosphorus nutrition and nematode population in mycorrhizal roots during the interaction between *Meloidogyne hapla* and VAM fungi in susceptible tomato plants (Cooper and Grandison, 1986). Also, nematodes were less in number and plants showed better growth in the case of mycorrhizal preinfected roots. Mycorrhizal fungi are thus found to confer tolerance, resistance, and other bio protection aids to host plants although the exact methods involved are still not clear.

17.11.3.1 COMPETITIVE ROOT COLONIZATION

Nutrient limitation is an important approach of antagonism of some biocontrol agents from a microbial point of view. Carbon plays an important role in the competition of root colonization for nutrients in species of *Trichoderma*

(Sivan and Chet, 1989). The main mechanism of *Fusarium* wilt suppression is the carbon competition between the pathogenic *Fusarium* and nonpathogenic *F. oxysporum* strains (Alabouvette et al., 2009). Biocontrol has been successfully applied in damping-off of seedlings due to competition between rhizospheric bacteria and *Pythium ultimum,* for the same carbon source and thereby suppression of the latter. Conidia germination of *Botrytis cinerea* is restricted by species of *Pseudomonas* species due to amino acid competition. However, biotrophs like rusts and powdery mildew cannot be controlled by competitive root colonization as no exogenous nutrients are required for the infection of hosts.

17.11.3.2 PLANT GROWTH PROMOTION THROUGH SAR AND ISR

Biocontrol agents like nonpathogenic plant growth-promoting fungi (PGPF) and nonpathogenic plant growth-promoting rhizobacteria (PGPR) produce chemical stimuli that induce local or systemic defence in plants or stimulate sustained change increasing the disease tolerance capacity of hosts against broad-spectrum pathogens. This is known as induced resistance and such induced resistance can be classified into induced systemic resistance (ISR) and systemic acquired resistance (SAR). ISR is mediated by jasmonic acid (JA) and/or ethylene in response to application of nonpathogenic rhizobacteria, while SAR is driven by salicylic acid (SA) frequently produced after infection of pathogen and thereby induces pathogenesis-related (PR) proteins expression. Colonization of *Glomus intraradices* on the roots of *Oryza sativa* conferred resistance through induction of defence-related genes (Campos-Soriano et al., 2012) chitinase Chit42 of *T. harzianum* expression increased resistance in potato and tobacco against the foliar pathogens, *B. cinerea, Alternaria solani,* and *A. alternata*, and soil-borne pathogen, *Rhizoctonia solani*. Several fluorescent pseudomonas and other PGPR develop ISR in rice in reaction to treatment with *P. fluorescens* strains like PpV14i and Pf7-14, which in turn suppresses rice blast disease. The level of SA shoots up increasing the ISR and disease suppression occurs up to 25%. Grain discoloration of rice is a complex symptom being associated with a number of saprophytic and pathogenic fungi infecting rice in both field and storage condition in turn reducing the production and productivity. These pathogens also produce mycotoxins harmful to human and animals. Seed biopriming showed enhanced plant growth and an increase in defense enzyme activities such as peroxidase, catalase, amylase, and superoxide dismutase.

The experiments showed efficient management of grain discoloration using bio-control agents showing plant growth promotion and induced systemic resistance (ISR) activities (Raghu et al., 2020). Plant growth-promoting rhizobacteria (PGPR) control plant diseases, induced systemic resistance (ISR) being an important defensive strategy manifested by plants especially infecting the aerial parts. The treatment with *Bacillus spp*. enhanced the antioxidant defence system in infected rice, thus alleviating *P. oryzae*-induced oxidative damage and suppressing blast disease (Rais et al., 2017).

17.12 MECHANISM OF PATHOGEN SUPPRESSION

Several factors are associated with rice disease control using biocontrol agents and understanding of the mechanism of pathogen suppression is necessary for successful exploitation of the biocontrol agent and the disease management thereby.

- For rice sheath blight disease bacteria as bio control agent is favored by acidic soil (pH 5.0) along with the presence of excess boron. The working efficiency is more in direct-seeded rice as compared to transplanted rice. Soil pH is a very important factor for the working potential of bio control agents. *Trichoderma* competes with the pathogenic fungi-producing antibiotics and lytic enzymes (glucanase and chitinase) in species like *T. harzianum* and *T. hamatum*. Chakraborti et al. in 2018 aimed at developing a sustainable management programme for sheath blight disease of rice that mainly reduces the crop yield by testing different combination biocontrol agents and they concluded that a combination of *Trichoderma viridae* + *Pseudomonas fluorescens* treatment gave maximum rice yield.
- For rice blast disease gelatinolytic bacteria from rice leaves and soil were screened that degraded the fungal extracellular matrix showing promising biological control by inhibiting host plant germling adhesion by airborne fungal pathogen like rice blast pathogen, *Magnaporthe oryzae* B. Coch. The 16S rDNA sequence-based selected bacteria were Acidovorax, Pseudomonas, Chryseobacterium, and Sphingomonas. The screened bacterial culture inhibited spore adhesion on plastic cover glass in vitro and also protected rice from disease development (Shimoi et al., 2012). Strain BC79, isolated from primaeval forest soil in China identified as *Bacillus methylotrophicus*, showed potential suppression of mycelial growth and conidia formation of several plant pathogenic fungi in

dual culture methods. Experiments in a greenhouse showed BC79 culture had 89.87% biocontrol efficiency to *Magnaporthe oryzae*, the causal organism of rice blast. Field trials indicated BC79 culture filtrate showed the highest efficiency (84.8%) against *M. oryzae*. The best periods to apply the BC79 control were seedling and post-transplant stages (Shan et al., 2013). Supratapa et al. (2012) in their study showed that three rhizobacteria species, *Seratia liquefaciens, Xanthomonas luminescens*, and *Enterobacter agglomerans*, apparently reduced the pathogen growth of *Pyricularia oryzae*, causing rice blast disease. They also studied that the potential of these biocontrol agents can be improved by developing the formulation and field application methods. Carvalho et al. (2015) suggested that *Waitea circinata* extract from a *Cerrado* orchid had inhibition potential for rice blast disease caused by *M. oryzae*. In vitro study revealed *W. circinata* to produce a wide inhibition zone reducing pathogen mycelial growth by 75%, appressoria formation by 100%, and AUDPC reduction by 25% against the control. Further analysis of the crude extract showed the presence of benzophenones. *Streptomyces* happens to be promising biocontrol for their ability to produce various bioactive compounds against the rice blast fungus under both laboratory and greenhouse conditions. In vitro studies have been shown to exhibit inhibitory activity against the pathogen, while in vivo studies showed 88.3% disease reduction (Law et al., 2017). Biocontrol strain BJ-1 was isolated from a contaminated plate of *Magnaporthe oryzae* identified as Bacillus subtilis based on morphological characters, 16S rDNA sequence analysis, and Biolog Microlog identification. Dual culture test showed BJ-1 to have a wide antimicrobial spectrum especially effective against *M. oryzae* (He et al., 2019).

- For bakane disease of rice Andargie et al. (2017) reported for the first time *A. placitae* as a biocontrol agent against false smut disease (*Ustilagonoidea virens*) on rice plants, both in vivo and in vitro. Varying degrees of inhibition in the growth of *U. virens* were observed in all cases, isolate E337 depicted major antagonistic activity against the pathogen which was identified as *Antennariella placitae* by molecular and morphological data analysis including 18S rDNA sequence study. The experiments confirmed conidia application of *A. placitae* protected rice plants improving yield and decreasing the disease severity of false smut disease on susceptible rice plants. In 2016,

Hossain et al. studied that endophytic bacterium *Bacillus oryzicola* YC7007 strain depicted good biocontrol potential for rice bakanae disease by direct inhibition and also induced systemic resistance against the plant pathogen activating the jasmonic acid pathway. Very recently, Saito et al. (2021) developed a novel method to control rice Bakanae disease, caused by *Fusarium fujikuroi*, using spore suspension of nonpathogenic *Fusarium commune* W5 as biocontrol agents onto rice flowers. Interaction between the nonpathogenic *Fusarium* and *Fusarium fujikuroi* inhibited hyphal extension of *F. fujikuroi* possibly by competing with the pathogen and their survival as nonpathogenic *Fusarium* probably mimics the disease cycle of the bakanae pathogen and thereby suppress the disease.

Host rhizosphere harbors different microorganisms directly/indirectly influencing plant growth. Both beneficial bacteria like plant growth-promoting rhizobacteria (PGPR) and harmful plant pathogens are present simultaneously. Soil-borne pathogens may lead to disease complex in plants. Competition among the beneficial and nonbeneficial soil microbes in the rhizosphere leads to biocontrol of pathogens. The diseased plants by soil-borne pathogens exhibit different blights, wilts, root decay, crown gall, leaf spot, tissue discolouration, etc. Arbuscular mycorrhizal fungi help their plant hosts in several ways including improved nutrient uptake, biotic and abiotic resistance. Saprophytes and symbionts mainly interact with AM fungi in mycorrhizosphere (Barea et al., 2002; Johansson et al., 2004). Sanchez et al. (2004) showed that AM fungus (*G. mosseae*) and a fluorescent pseudomonad both induced specific plant genes supporting the assumption that they may share some plant cell programmes during root colonization. In addition to the mycorrhizal fungi on nodules, the role of contribution to plant protection by PGPR such as *Azospirillum, Pseudomonas, Agrobacteria, Bacillus*, etc., was also recognized. Etesamy et al. (2016) researched that endorhiza and rhizosphere of rice plants harbor gram-positive bacteria and the isolates studied had the potential to be used against rice fungal pathogens as biocontrol agents. *B. cereus* REN3 and *B. mojavensis* REN4 are broad-spectrum antifungal strains with desirable PGP traits with the innate potential to restrict the growth of all fungal rice pathogens under in vitro conditions.

17.13 SUMMARY

Huge losses are incurred every year in agriculture across the world due to plant diseases. Since the ever-growing population needs to be fed continuously, there lies the necessity to manage the diseases for steady supply of fresh marketable produce. To control plant diseases, one must be very acquainted with plant disease etiology. Because of the increasing pressure to fulfil the demand for food, there grew the dependence on chemical amendments for disease control leading to an adverse effect on the soil, air, and water environment which in turn resulted in the augmentation of resistant pathogens. To save the environment and meet the need of time researchers went on to find alternative approaches like the development of resistant crop cultivars and also biological control means. Varietal development turns out to be a complex process that involves intricate research work and may gradually lose its resistant nature in the long run. Biological control on the other hand as an alternative approach to disease management pretty much has the potential in suppressing the disease and stimulate plant growth as well. Biological control of plant pathogens is done by manipulating indigenous microbes in the soil through different mechanisms, thereby enhancing the natural disease control or by introducing allochthonous microbes in the soil environment of the plant harboring the pathogen. Bio-control is the purposeful exploitation of living organisms either residing or introduced in the periphery of the pathogen infecting the host plant to suppress the harmful activity of the pathogen and encourage proper growth and development of the plant.

17.14 FUTURE PROSPECTS FOR BIO-CONTROL

The recent interest in the use of biological control methods for plant disease suppression has opened up new opportunities for microbial-based products to replace the chemical amendments in the worldwide market. The widespread commercial application of biocontrol agents is yet to take a pace because of their variable performance under different on-field environmental trials. To gear up the process, more field-oriented research works must be uptaken. Research on lesser exploited fields like the effect of environmental factors on biocontrol agents, development of new formulations in which the shelf life of bioagent expands, mass production of biocontrol agents in a loop for adequate supply at the time of need and undertaking advanced techniques in nanotechnology and biotechnology fields for improvement of biocontrol strategies. Increased crop production and high crop yield, maintaining soil fertility at the same time, are the need of today's world. Such an eco-friendly approach can be met using the novel concept of rhizoengineering. Future studies in rhizosphere biology depending on the development of biotechnological and molecular approaches might show some achievement in the field of integrated management of soil microbial ecosystems (Gupta et al., 2015). Much work has been done on phosphate solubilization and nitrogen fixation by PGPR, but quite little information is available on potassium solubilization, despite being the third major vital macronutrient for plant growth. Research in this field will surely open new frontiers for the betterment of crop health (Gupta et al., 2015). Special emphasis should be given to the next-gen formulations of antimicrobial secondary metabolites with longer shelf life and higher suppression efficiency against soil-borne plant pathogens like bacteria (Keswani et al., 2016). Formulation of a group of antagonistic metabolites against specific pathogens occurs regardless of the geographic location where the particular disease incidence is high. Such an approach would particularly target a specific pathogen while the remaining soil microorganisms would remain unaltered (Keswani et al., 2016). There is a significant increase in global registrations of microbial biocontrol agents. The modifications in the legislation undertaking new policies at the country level, reduction in the use of chemical control agents are expanding the market for biocontrol agents. Till now the quantity of bio-control agents produced worldwide are still not efficient in meeting the huge demand as compared to chemical pesticides, but they have a promising future. New technologies like proteomics, functional genomics might counter the ecological constraints for the development of biocontrol agent determining the expression of target genes present in the bio-control agents during their formulation, mass production, application, and storage. Necessity of efficient microorganisms, facilitating healthy plant cultivation adapted to extreme conditions in a plant environment with high biodiversity. The new microbial findings should not only rely on plant–microbe interactions but also exploit the microbe–microbe interactions like phage effects and microbial transmission. The shift of focus is from plant rhizosphere to plant holobiont. Keeping into account plant microbiome engineering, a large number of microbial inoculants should be merged in the biocontrol and biofertilizer products. Most studies regarding the discovery of biocontrol antagonists are restricted to laboratory conditions. A thorough analysis of the inoculants needs to be done in a stressful environment to check the potential of the bioagents in disease suppression and plant growth (Nian et al., 2021; Mendoza-Arroyo et al., 2020; Pirttila et al., 2021).

KEYWORDS

- **biocontrol agents**
- **sustainibility**
- **antibiosis**
- **direct antagonism**
- **volatile substances**
- **virulence factor**

REFERENCE

Alabouvette, C.; Olivain, C.; Migheli, Q.; Steinberg, C. Microbiological Control of Soil-Borne Phytopathogenic Fungi With Special Emphasis on Wilt-Inducing Fusarium Oxysporum. *New Phytol.* **2009,** *184,* 529–544.

Andargie, M.; Congyi, Z.; Yun, Y.; Li, J. Identification and Evaluation of Potential Bio-Control Fungal Endophytes Against Ustilagonoidea Virens on Rice Plants *World J. Microbiol. Biotechnol.* **2017,** *33,* 120.

Anitha, A.; Rabeeth, M. *Afr. J. Basic Appl. Sci.* **2009,** *1* (1–2), 9–14.

Barea, J. M.; Azcon, R.; Azcon-Aguilar, C. Mycorrhizosphere Interactions to Improve Plant Fitness and Soil Quality. *Anton. Leeuw.* **2002,** *81,* 343–351.

Benítez, T.; Rincón, M. A.; Limón, M. C.; Codón, C. A. Bio Control Mechanisms of Trichoderma Strains. *Int. Microbiol.* **2004,** *7,* 249–260.

Bloemberg, G. V.; Lugtenberg, B. J. J. Molecular Basis of Plant Growth Promotion and Bio Control by Rhizobacteria. *Curr. Opin. Plant Biol.* **2001,** *4,* 343–350.

Bouizgarne, B. Bacteria for Plant Growth Promotion and Disease Management. In *Bacteria in Agrobiology: Disease Management*; Maheshwari, D. K., Ed.; Springer: Berlin, 2013; pp 15–47.

Buyer, J. S.; Sikora, L. J.; Chaney, R. L. A New Growth Medium for the Study of Siderophore Mediated Interactions. *Biol. Fertil. Soils* **1989,** *8,* 97–101.

Campos-Soriano, L.; Garcia-Martinez, J.; Segundo, B. S. The Arbuscular Mycorrhizal Symbiosis Promotes the Systemic Induction of Regulatory Defence-Related Genes in Rice Leaves and Confers Resistance to Pathogen Infection. *Mol. Plant Pathol.* **2012,** *13,* 579–592.

Carvalho, J. C. B.; Sousa, K. C. I.; Brito, D. C.; Chaibub, A. A.; Luzini, A. P.; Cortes, M. V. C. B.; Filippi, M. C. C.; Kato, L.; Vaz, B. G.; Costa, H. B.; Romao, W.; Araujo, L. G. Bio Control Potential of Waitea Circinata, an Orchid Mycorrhizal Fungus, Against the Rice Blast Fungus. *Trop. Plant Pathol.* **2015,** *40,* 151–159.

Chaiharn, M.; Chunhaleuchanon, S.; Lumyong, S. Screening Siderophore Producing Bacteria as Potential Biological Control Agent for Fungal Rice Pathogens in Thailand. *World J. Microbiol. Biotechnol.* **2009,** *25,* 1919–1928.

Chaiharn, M.; Theantana, T.; Pathom-aree, W. Evaluation of Bio Control Activities of Streptomyces Spp. Against Rice Blast Disease Fungi. *Pathogens* **2020,** *9,* 126.

Chakrabarti, R. R.; Majumdar, N.; Nath, R. Sustainable Management of Rice Sheath Blight Disease Through Antagonists, Fungicide and Soil Amendments, *Int. J. Pure App. Biosci.* **2018,** *6* (5), 1005–1009.

Chatterjee, A.; Valasubramanian, R.; Ma, W. L.; Vachhani, A. K.; Gnanamanickam, S. S.; Chatterjee, A. K. Bio Control of Rice Diseases With *Pseudomonas Fluorescens* Strain Pf7–14: Isolation of Ant Mutants Altered in Antifungal Antibiotic Production, Cloning of Ant$^-$ DNA and Evaluation of a Role of Antibiotic Production in the Control of Blast and Sheath Blight. *Biol. Con.* **1996,** *7, 185–195.*

Chinnaswami, K.; Mishra, K. D.; Miriyala, A.; Vellaichamy, P.; Kurubar, B.; Gompa, J.; Madamsetty, S. P.; Raman, M. S. Native Isolates of Trichoderma as Bio Suppressants Against Sheath Blight and Stem Rot Pathogens of Rice. *Egypt. J. Biol. Pest Control* **2021,** *31,* 12.

Cooper, K. M.; Grandison, G. S. Interactions of VAM Fungi and Root Knot Nematode on Cultivars of Tomato and White Clover Susceptible to Meloidogyne Hapla. *Ann. Appl. Biol.* **1986,** *108,* 555–565.

Cordier, C.; et al. Cell Defense Responses Associated With Localized and Systemic Resistance to Phytophthora Induced in Tomato by an Arbuscular Mycorrhizal Fungus. *Mol. Plant Microbe Interact.* **1998,** *11,* 1017– 1028.

C.R.R.I., Vision 2030. Central Rice Research Institute (ICAR), Cuttack, Odisha, India, 2011.

De Vasconcellos, R. L. F.; Cardoso, E. J. B. N. Bio Control **2009,** *54* (6), 807–816.

Etesami, H.; Alikhani, H. A. Evaluation of Gram-Positive Rhizosphere and Endophytic Bacteria for Biological Control of Fungal Rice (Oryzia Sativa L.) *Pathogens. Eur. J. Plant Pathol.* **2016.**

Fravel, D. R. Commercialization and Implementation of Bio Control. *Ann. Rev. Phytopathol.* **2005,** *43,* 337–359.

Frey-Klett, P.; et al. The Mycorrhiza Helper Bacteria Revisited. *New Phytol.* **2007,** *176,* 22–36.

Gnanamanickam, S. S. Ed.; *Biological Control of Crop Diseases*; Marcel Dekker Inc.: New York, 2002, p 468.

Gopalakrishnan, S.; Srinivas, V.; Prakash, B.; Vijayabharathi, R.; Rupela, O. Evaluation of Streptomyces Strains Isolated From Herbal Vermicompost for Their Plant Growth-Promotion Traits in Rice. Microbiol. Res. **2014,** *169,* 40–48.

Gupta, G.; Parihar, S. S.; Ahirwar, N. K.; Snehi, S. K.; Singh, V. Plant Growth Promoting Rhizobacteria (PGPR): Current and Future Prospects for Development of Sustainable Agriculture. *J. Microb. Biochem. Technol.* **2015,** *7,* 096–102.

Harman, G. E.; Howell, C. R.; Viterbo, A.; Chet, I.; Lorito, M. Trichoderma Species-Opportunistic, Avirulent Plant Symbionts. *Nat. Rev. Microbiol.* **2004,** *2,* 1–14.

He, Y.; Zhu, M.; Huang, J.; Hsiang, T.; Zheng, L. Bio Control Potential of a BacillusSubtilis Strain BJ-1 Against the Rice Blast Fungus Magnaporthe Oryzae. *Can. J. Plant Pathol.* **2019,** *41,* 1, 47–59.

Hider, R. C.; Kong, X. Chemistry and Biology of Siderophores. *Nat. Prod. Rep.* **2010,** *27,* 637–657.

Hossain, M. T.; Khan, A.; Chung, E. J.; Rashid, M. H.; Chung, Y. R. Biological Control of Rice Bakanae by an Endophytic Bacillus Oryzicola YC7007. *Plant Pathol. J.* **2016,** *32* (3), 228–241.

Jamali, H.; Sharma, A.; Roohi, Srivastava, A. K. Bio Control Potential of Bacillus Subtilis RH5 Against Sheath Blight of Rice Caused by Rhizoctonia Solani. *J. Basic Microbiol.* **2019,** 1–13.

Johansson, J. F.; Paul, L. R.; Finlay, R. D. Microbial Interactions in the Mycorrhizosphere and Their Significance for Sustainable Agriculture. *FEMS Microbiol. Ecol.* **2004,** *48,* 1–13.

Kamal, R.; Gusain, Y. S.; Kumar, V.; Sharma, A. K. Disease Management Through Biological Control Agents: An Eco-Friendly and Cost Effective Approach for Sustainable Agriculture-A Review. *Agri. Rev.* **2015,** *36* (1), 37–45.

Kanini, G. S.; Katsifas, E. A.; Savvides, A. L.; Hatzinikalaou, D. G.; et al. *J. Appl. Microbiol.* **2013,** *114* (5), 1468–1479.

Keswani, C.; Bisen, K.; Singh, V.; Sarma, B. K.; Singh, H. B. Formulation Technology of Bio Control Agents: Present Status and Future Prospects. In Bioformulations: for Sustainable Agriculture; Arora, N., Mehnaz, S., Balestrini, R., Eds.; Springer: New Delhi, 2016.

Khalili, E.; Sadravi, M.; Naeimi, S.; Khosravi, V. Biological Control of Rice Brown Spot With Native Isolates of Three Trichoderma Species. *Braz. J. Microbiol.* **2012,** 297–305.

Kiewnick, S. Practicalities of Developing and Registering Microbial Biological Control Agents, CAB Reviews: Perspectives in Agriculture, Veterinary Science, Nutrition and Natural Resources No. 013, CABI Publishing, 2007.

Kim, B. S.; Hwang, B. K. Biolofungicides. In *Fungal Biotechnology in Agricultural, Food, and Environmental Applications*; Dilip, K. A., Ed.; Basel, New York, 2004; pp 123–133.

Kiss, L. A Review of Fungal Antagonists of Powdery Mildews and Their Potential as Bio Control Agents. *Pest Manag. Sci.* **2003,** *59,* 475–483.

Kloepper, J. W.; Leong, J.; Teintze, M.; Schroth, M. N. Enhancing Plant Growth by Siderophores Produced by Plant Growth-Promoting Rhizobacteria. *Nature* **1980,** *286,* 885–886.

Kumar; Dangar. Functional Role of Plant Growth Promoting Endo-and Rhizobacteria in Major Cereal Crops. *Pop. Kheti* **2013,** *1* (1), 37–40.

Kumar, R.; Kumar, S.; Pankaj, V. P.; Tarafdar, A.; Biswas, K. Microbes in Plant Endosphere: Pathogenesis and Disease Antagonism. In *Plant Pathogen Interaction: Recent Trends*; Mitra, R., Barman, A., Eds.; Sharma Publications and Distributors: New Delhi, 2016a; pp 113–130.

Law, J. W. F.; Ser, H. L.; Khan, T. M.; Chuah, L. H.; Pusparajah, P.; Chan, K. G.; Goh, B. H.; Lee, L. H. The Potential of Streptomyces as Bio Control Agents Against the Rice Blast Fungus, Magnaporthe Oryzae (Pyricularia oryzae). *Front. Microbiol.* **2017,** *8,* 3.

Lim, H. S.; Lee, J. M.; Kim, S. D. Role of Siderophore in Biological Control of Fusarium Solani by Pseudomonas Fluorescens GL 20. *Bull Inst. Agric. Sci. Kimki Univ.* **1999,** *7,* 47–58.

Manandhar, H. K.; Lyngs Jorgensen, H. J.; Mathur, S. B.; Smedegaard-Peterson, V. Suppression of Rice Blast by Preinoculation With Avirulent Pyricularia Oryzae and the Nonrice Pathogen Bipolaris Sorokiniana. *Phytopathology* **1998,** *88,* 735–739.

McSpadden Gardener B.; Fravel, D. Biological Control of Plant Pathogens: Research Commercialization and Application in the USA. *Plant Heath Prog.* [Online] **2002.**

Mendoza-Arroyo, G. E.; Chan-Bacab, M. J.; Aguila-Ramírez, R. N.; Ortega-Morales, B. O.; Canche Solis, R. E.; Chab-Ruiz, A. O.; Cob-Rivera, K. I.; Dzib-Castillo, B.; Tun-Che, R. E.; Camacho-Chab, J. C. Inorganic Phosphate Solubilization by a Novel Isolated Bacterial

Strain Enterobacter sp. ITCB-09 and its Application Potential as Biofertilizer. *Agriculture* **2020,** *10,* 383.

Muller, G.; Matzank, B. F.; Raymond, K. N. Iron Transport in Streptomyces Pilosus Mediated by Ferrichrome Siderophores, Rhodotorulic Acid and Enantio- Rhodotorulic Acid. *J. Bacteriol.* **1984,** *160,* 313–318.

Muller, G.; Raymond, K. N. Specificity and Mechanism of Ferrioxamine Mediated Iron Transport in Streptomyces Pilosus. *J. Bacteriol.* **1984,** *160,* 304–312.

Nabila, N. H.; et al. Evaluation of Selected Herbs for Bio Control of Rice Blast Disease. *IOP Conf. Ser.: Earth Environ. Sci.* **2021,** *685,* 012026.

Nian, J.; Yu, M.; Bradley, C. A.; Zhao, Y. Lysobacter Enzymogenes Strain C3 Suppresses Mycelium Growth and Spore Germination of Eight Soybean Fungal and Oomycete Pathogens and Decreases Disease Incidences. *Biol. Control* **2021,** *152,* 104424.

Nookongbut, P.; Kantachote, D.; Khuong, N. Q.; Tantirungkij, M. The Bio Control Potential of Acid-Resistant Rhodopseudomonas Palustris KTSSR54 and its Exopolymeric Substances Against Rice Fungal Pathogens to Enhance Rice Growth and Yield. *Biol. Control* **2020,** *150,* 104354.

Padaria, J. C.; Tarafdar, A.; Raipuria, R.; Lone, S. A.; Gahlot, P.; Shakil, N. A.; Kumar, J. Identification of Phenazine-1-Carboxylic Acid Gene (phc CD) From Bacillus Pumilus MTCC7615 and its Role in Antagonism Against Rhizoctonia Solani. *J. Basic Microbiol.* **2016,** *56,* 1–10.

Pal, K. K.; Gardener, B. M. Biological Control of Plant Pathogens. Plant Health Instr. **2006**. DOI: 10.1094/PHI-A-2006-1117-02.

Pirttila, A. M.; Mohammad Parast Tabas, H.; Baruah, N.; Koskimaki, J. J. Biofertilizers and Biocontrol Agents for Agriculture: How to Identify and Develop New Potent Microbial Strains and Traits. *Microorganisms* **2021,** *9,* 817.

Raghu, S.; Baite, M. S.; Patil, N. B.; Sanghamitra, P.; Yadav, M. K.; Prabhukarthikeyan, S. R.; Keerthana, U.; Pandi, G. G.; Aravindan, S.; Rath, P. C. Grain Discoloration in Popular Rice Varieties (Oryza Sativa L) in Eastern India, Associated Mycoflora, Quality Losses and Management Using Selected Bio Control Agents. *J. Stored Prod. Res.* **2020,** *88,* 101682.

Rais, A.; Jabeen, Z.; Shair, F.; Hafeez, F. Y.; Hassan, M. N. Bacillus Spp., a Bio-Control Agent Enhances the Activity of Antioxidant Defense Enzymes in Rice Against Pyricularia Oryzae. *PLoS One* **2017,** *12* (11), e0187412.

Reddy, K. R. N.; Reddy, C. S.; Muralidharan, K. Potential of Botanicals and Bio Control Agents on Growth and Aflatoxin Production by Aspergillus Flavus Infecting Rice Grains. *Food Control* **2009,** *20,* 173–178.

Roshan, K.; Tarafdar, A.; Saurav, K.; Ali, S.; Lone, S. A.; Pattnaik, S.; Tyagi, A.; Biswas, K.; Mir, Z. A. Isolation and Screening of Bioactive Compound From Actinomycetes Isolated From Salt Pan of Marakanam District of the State Tamil Nadu, India. *Elixir BioTechnol.* **2013,** *61,* 16826–16831.

Roy, M.; Dutta, S. G.; Venkata, R. C. Pseudomonads: Potential Bio Control Agents of Rice Diseases. *Res. J. Agric. For. Sci.* **2013,** *1* (9), 19–25. ISSN 2320-6063.

Saito, H.; Sasaki, M.; Nonaka, Y.; Tanaka, J.; Tokunaga, T.; Kato, A.; Thuy, T. T. T.; Vang, L. V.; Tuong, L. M.; Kanematsu, S.; Suzuki, T.; Kurauchi, K.; Fujita, N.; Teraoka, T.; Komatsu, K.; Arie, T. Spray Application of Nonpathogenic Fusaria Onto Rice Flowers Controls Bakanae Disease (Caused by Fusarium Fujikuroi) in the Next Plant Generation. *Appl. Environ. Microbiol.* **2021,** *87,* e01959–20.

Sanchez, L.; Weidmann, S.; Brechenmacher, L.; Batoux, M.; van Tuinen, D.; Lemanceau, P.; et al. Common Gene Expression in Medicago Truncatula Roots in Response to Pseudomonas Fluorescens Colonization, Mycorrhiza Development and Nodulation. *New Phytol.* **2004,** *161*, 855–863.

Sarwar, A.; Hassan, M. N.; Imran, M.; Iqbal, M.; Majeed, S.; Brader, G.; Sessitsch, A.; Hafeez, F. Y. Bio Control Activity of Surfactin A Purified From Bacillus NH-100 and NH-217 Against Rice Bakanae Disease. *Microbiol. Res.* **2018,** *209*, 1–13.

Shan, H.; Zhao, M.; Chen, D.; Cheng, J.; Li, J.; Feng, Z.; Ma, Z.; An, D Bio Control of Rice Blast by the Phenaminomethylacetic Acid Producer of Bacillus Methylotrophicus Strain BC79. *Crop Protect.* **2013,** *44,* 29–37.

Sharma, M.; Tarafdar, A.; Ghosh, R. Use of Genomic Approaches in Understanding the Role of Actinomycetes as PGP in Grain Legumes. In *Plant growth-promoting actinomycetes: a new avenue for enhancing the productivity and soil fertility of grain legumes*; Gopalkrisnan, A., Sathya, A., VijayaBharthi, R., Eds.; Springer: New York, 2016; pp 249–262.

Shimoi, S.; Inoue, K.; Kitagawa, H.; Yamasaki, M.; Tsushima, S.; Park, P.; Ikeda, K. Biological Control for Rice Blast Disease by Employing Detachment Action With Gelatinolytic Bacteria. *Biol. Control* **2010,** *55,* 85–91.

Shobha, G.; Kumudini, B. S. Antagonistic Effect of the Newly Isolated PGPR Bacillus spp. On Fusarium Oxysporum. *Int. J. Appl. Sci. Eng. Res.* **2012,** *1,* 463–474.

Singh, U. B.; Sahu, A.; Singh, R. K.; Singh, D. P.; Meena, K. K.; Srivastava, J.; Renu; Manna, M. C. Evaluation of Bio Control Potential of Arthrobotrys Oligospora Against Meloidogyne Graminicola and Rhizoctonia Solani in Rice (*Oryza Sativa L.*). *Biol. Control* **2012,** *60,* 262–270.

Singh, P. K.; Singh, A. K.; Singh, H. B.; Dhakad, B. K. Biological Control of Rice Blast Disease With Trichoderma harzianum in Direct Seeded Rice Under Medium Low Land Rainfed Conditions. *Environ. Ecol.* **2012,** *30* (3B), 834—837.

Sivan, A.; Chet, I. Degradation of Fungal Cell Walls by Lytic Enzymes of Trichoderma Harzianum. *Microbiology* **1989,** *135,* 675–682.

Song, J.; Soytong, K.; Kanokmedhakul, S. Fungal Metabolites of Chaetomium Lucknowense for Inhibition of a Rice Blast Pathogen, Pyricularia Oryzae. *Int. J. Agric. Technol.* **2017,** *13* (7.2), 1621–1626.

Soytong, K. Bio-Formulation of Chaetomium Cochliodes for Controlling Brown Leaf Spot of Rice. *Int. J. Agric. Technol.* **2014,** *10* (2), 321–337.

Suprapta, D. N. Potential of Microbial Antagonists as Bio Control Agents Against Plant Fungal Pathogens. *J. ISSAAS* **2012,** *18* (2), 1–8.

Swain, H.; Adak, T.; Mukherjee, A. K.; Mukherjee, P. K.; Bhattacharyya, P.; Behera, S.; Bagchi, T. B.; Patro, R.; Shasmita; Khandual, A., Bag, M. K.; Dangar, T. K.; Lenka, S.; Jena, M. Novel Trichoderma Strains Isolated From Tree Barks as Potential Bio Control Agents and Biofertilizers for Direct Seeded Rice. *Microbiol. Res.* **2018,** *214,* 83–90.

Tamreihao, K.; Ningthoujam, D. S.; Nimaichand, S.; Singh, E. S.; Reena, P.; Singh, S. H.; Nongthomba, U. Biocontrol and Plant Growth Promoting Activities of a Streptomyces Corchorusii Strain UCR3-16 and Preparation of Powder Formulation for Application as Biofertilizer Agents for Rice Plant. *Microbiol. Res.* **2016,** *192,* 260–270.

Tokala, R. K.; Strap, J. L.; Jung, C. M.; Crawford, D. L.; Salove, M. H.; Deobald, L. A.; Bailey, J. F.; Morra, M. J. Novel Plant-Microbe Rhizosphere Interaction Involving Streptomyces Lydicus WYEC108 and the Pea Plant (Pisum Sativum). *Appl. Environ. Microbiol.* **2002,** *68,* 2161–2171.

Vasudevan, P.; Kavitha, S.; Priyadarisini, V. B.; Babujee, L.; Gnanamanickam, S. S. Biological Control of Rice Diseases. In *Biological Control of Crop Diseases*; Gnanamanickam, S. S., Ed.; Marcel Dekker Inc.: New York, 2002, p 468, pp. 11–32.

Wei, G.; Kloepper, J. W.; Sadik, T. Induction of Systemic Resistance of Cucumber to Colletotrichum Orbiculare by Select Strains of Plant Growth-Promoting Rhizobacteria. *Phytopathology* **1991,** *81,* 1508–1512.

Yuan, W. M.; Crawford, D. L. Characterization of Streptomyces Lydicus WYEC108 as a Potential Bio Control Agent Against Fungal Root and Seed Rots. *Appl. Environ. Microbiol.* **1995,** *61,* 3119–3123.

CHAPTER 18

New Insights from the Bioinformatics Tools for Rice Fungal Disease Resistance

JAGMOHAN SINGH[1,2], KARANSHER SINGH SANDHU[3], and BALWINDER KAUR[4]

[1]*Division of Plant Pathology, Indian Agricultural Research Institute, New Delhi, India*

[2]*Institute of Molecular Plant Science, University of Edinburgh, Edinburgh, UK*

[3]*Department of Crop and Soil Sciences, Washington State University, Pullman, WA, USA*

[4]*Department of Entomology and Nematology, University of Florida, Gainesville, FL, USA*

ABSTRACT

In these last two decades, rapid advancements in fungal omics approaches have generated a huge amount of biological data to understand the basic biology of rice fungal pathogens. But, to make interpretations from this big data, we require sophisticated computational analyses. Bioinformatics is referred to as the application of computer information and technology to manage and interpret the biological data to generate meaningful information. Bioinformatics provides computational algorithms and software tools for mining the huge data to draw meaningful conclusions. It develops computational algorithms, software, and databases for management, visualization, analysis, and modeling data from biological experiments. Mining of fungal

Fungal Diseases of Rice and Their Management. Deepti Srivastava, Md. Shamim, Malik Mobeen Ahmad, and R. S. Upadhyay (Eds.)

omics data by using bioinformatics tools helps in the understanding of the complex fungal biological processes from genotype to phenotype, microbial interaction in a niche and pathogen–host interactions, and much more. *Magnaporthe oryzae* causing rice blast is the first fungal plant pathogen with the available whole-genome sequence. Since a decade several studies are conducted in *Magnaporthe oryzae* with omics approaches using bioinformatics databases and tools. A good amount of literature is available on the use of bioinformatics in the research studies of most of the important rice fungal pathogens. In this book chapter, we will discuss in detail the basic concepts, databases, and tools of bioinformatics used in rice fungal disease management.

18.1 INTRODUCTION

In the last few decades, molecular biology has advanced with the development of numerous genomic techniques, with which researcher has generated a huge amount of biological data (Jayaram and Dhingra, 2010; Singh et al., 2011). The challenge to interpret this huge amount of biological data has given origin to a new branch of science and technology named bioinformatics (Vassilev et al., 2006). Bioinformatics is referred to as the application of computer information and technology to manage and interpret the biological data to generate meaningful information (Edwards and Batley, 2004). Bioinformatics is interface between biology and informatics merged by computer technology. However, it is multidisciplinary science covering concepts from statistics, computer science, informatics, mathematics, biology, biochemistry, chemistry, physics, biotechnology, and linguistics (Singh et al., 2011). It practically comprises development of computational algorithms, software, and databases and their application to data from biological experiments to better understand of the living processes with aim to boost agriculture.

Rice is a staple food for most of the countries in the world; loss incurred by diseases is a serious issue (Lilly and Subramanian, 2019). Rice crop is usually infected by a number of fungal diseases from its vegetative stage to grain maturity. Most of fungal diseases of rice are brown spot, blast, sheath blight, sheath rot, sheath blotch, collar rot, leaf smut, leaf scald, bakane, root rot, eye spot, stackburn, kernel smut, false smut, grain discoloration, Glume blight, ear blight, black kernel, scab, and udbatta disease (https://www.apsnet.org/edcenter/resources/commonnames/Pages/Rice.aspx) (Table 18.1). The 25–40% losses in rice production worldwide are caused by pathogens and pests, among which fungal diseases alone reduce rice production by 14%.

Most of the fungal diseases are managed by chemical approaches, which have hazardous impact on our environment. The development and use of resistance cultivars are the most appropriate approach for disease management, which is least expensive with no adverse impact on environment and human health. Earlier plant breeding techniques to develop resistant varieties were time consuming which take years to develop a cultivar resistance against a prevailing race or strain of pathogen. Moreover, developed variety becomes susceptible as per boom and bust cycle when new race or pathogen emerges due to selection pressure exerted by mono cropping of resistant variety. The breeding process is accelerated with the intervention of molecular biology, biotechnology, and genomics technologies which are highly assisted by bioinformatics tools. The development and use of computational algorithms, databases, tools, and software help to handle a large amount of data and interpret it in a better way. Rice being an important crop in most of the regions of the world, a large amount of research fund is diverted for this crop to develop biotic stress challenging cultivars. Rice researchers are using available cutting-edge genomics and genome manipulating technologies to address challenges faced by the crop. As mentioned earlier, genomics has provided researcher with ample amount of data, which is difficult to handle manually and much more challenging to conclude it for implications. In this chapter, we are discussing the bioinformatics tools used in rice research for addressing fungal disease problems.

TABLE 18.1 Worldwide Spread Fungal Diseases of Rice and Their Causal Organisms.

Fungal disease of rice	Causal organism
Aggregate sheath of rice	*Rhizoctonia oryzae-sativae*
Black horse riding of rice	*Curvularia lunata* *Cochliobolus lunatus* [teleomorph]
Rice blast	*Pyricularia oryzae* *Magnaporthe grisea* [teleomorph]
Brown spot of rice	*Cochliobolus miyabeanus* *Bipolaris oryzae* [anamorph]
Crown sheath rot of paddy	*Gaeumannomyces graminis*
Downy mildew of rice	*Sclerophthora macrospora*
Eyespot of rice	*Drechslera gigantean*
False smut of rice	*Ustilaginoidea virens*
Kernel smut of rice	*Neovossia horrida*
Leaf smut of paddy	*Entyloma oryzae*
Leaf scald of paddy	*Rhynchosporium oryzae*

TABLE 18.1 *(Continued)*

Fungal disease of rice	Causal organism
Pecky rice disease	Complex of fungus genus such as *Cochliobolus miyabeanus* *Microdochium oryzae,* *Sarocladium oryzae, Curvularia* spp. *Fusarium spp.*
Root rot of rice	*Fusarium spp.* *Pythium spp.* *Pythium dissotocum* *Pythium spinosum*
Seedling blight of paddy	*Cochliobolus miyabeanus* *Curvularia spp.* *Fusarium spp.* *Rhizoctonia solani* *Sclerotium rolfsii* *Athelia rolfsii* [teleomorph] and other pathogenic fungi
Sheath blight of rice	*Thanatephorus cucumeris* *Rhizoctonia solani* [anamorph]
Sheath rot of paddy	*Sarocladium oryzae*
Sheath spot of rice	*Rhizoctonia oryzae*
Stackburn of paddy	*Alternaria padwickii*
Stem rot of rice	*Sclerotium oryzae*
Water-mold disease	*Achlya conspicua* *Achlya klebsiana* *Fusarium spp.* *Pythium spp.* *Pythium dissotocum* *Pythium spinosum*

Source: https://en.wikipedia.org/wiki/List_of_rice_diseases

18.2 BIOINFORMATICS IN PLANT DISEASE RESISTANCE

With advancements of sequencing and genomic techniques agricultural crops and a number of agriculturally important organisms are being sequenced, the bioinformatics tools are enhancing agricultural genomics globally. This breakthrough in research has led to the development of a better understanding of plant-pathogen interactions and resistance mechanisms in crop plants (Koltai and Volpin, 2003). Bioinformatics plays a vital role in plant disease management and is extensively used in rice fungal disease management. Bioinformatics is used in various aspects of plant disease managements that are mentioned below.

18.2.1 STUDY OF HOST–PATHOGEN INTERACTION

The disease development in a plant takes place due to an established infection by a biotic agent namely fungi, bacteria, phytoplasma, nematode, and virus. The interaction of plant and biotic agent is in a complex that involves multiple aspects. The study of host–pathogen interaction is very important for plant pathologist to develop disease management strategies. It was earlier done by classic approaches of plant physiology, microbiology, plant breeding, genetics, and histology, but it has reached to next level with advanced molecular biology, biochemistry, and genomics (Ojo and Maxwell, 2010). Pathogen has possessed specific molecular patterns that are recognized by the host for executing the first line of defense. These patterns are named PAMPs (pathogen-associated molecular patterns) which are recognized by surface molecules of the plant known as PRRs (Pattern Recognition Receptors), which activate basal defense against pathogen by PTI (PAMP Triggered Immunity) (Kamoun, 2006; Nemri et al., 2014). Successful plant pathogens produce molecules that suppress the PTI to cause infection; these molecules are called effectors. Plant also evolves the mechanism of active defense by acquiring surveillance proteins (R proteins) that directly or indirectly monitor the presence of effector proteins of pathogen and active downstream signaling to restrict the pathogen. This mechanism is known as effector triggered immunity (ETI) (Rao et al., 2008). The molecular basis of PTI and ETI is now much better understood, as a result of the available genomic data and tools (Koltai and Volpin, 2003).

Among cereals, rice is the first crop whose whole genome was sequenced, well mapped, and well characterized with estimated sequenced size of 430 Mb (Eckardt, 2000). Similarly, *Magnaporthe oryzae* is the first plant pathogenic fungus to be genome sequenced. The genomic data of rice and its fungal pathogens is used through bioinformatics tools to identify R-proteins and effector molecules.

A number of databases are developed and used by researchers to study host–pathogen interaction in rice and its fungal pathogen such as PHI (pathogen-host interaction) database, DFVF, FSRD, Secretool, FSD, and

FunSecKB database (Table 18.2) (Singh et al., 2020). High-throughput sequencing techniques have provided genome and numerous transcriptomes of fungal pathogens of rice, which provides the opportunities to identify effector proteins of pathogens. The characterization of several Avr genes has provided the criteria to develop computational algorithms to efficiently predict effector proteins. At present, many pipelines and servers are available which use various bioinformatics tools to predict putative effectors from genomic data on the basis of conserved motifs, structural features, and signature sequences of the proteins. Effector proteins are small secreted proteins that possess signal peptide of 20–30 amino acids at N-terminal followed by a hydrophobic region that guides them to translocate across the membrane. There are various bioinformatics tools based on computational algorithms such as neural networks, Hidden Markov model (HMM), and machine learning systems that predict signal peptide with accuracy (Table 18.3). The transmembrane (TM) proteins also contain hydrophobic segments as secretory proteins, so it is difficult to distinguish them (Sonah et al., 2016). The presence of transmembrane in secretory protein anchored them in endoplasmic reticulum (Singh et al., 2019). Therefore, secretory protein to be characterized as effector molecules must avoid the TM domain. There are numerous online tools available for TM-domains prediction (Table 18.4).

TABLE 18.2 Databases for the Prediction of Pathogenicity, Virulence, and Effector Proteins of Fungal Pathogens.

Database	Important features of database	Reference/website
PHI-base (pathogen–host interaction database)	It is database of genes related to pathogenicity and virulence, also effector proteins of fungal and bacterial pathogens derived from literature.	Winnenburg et al., 2008 http://www.phi-base.org/
Secretool	Provides the bioinformatic tools for the prediction of Secretome.	Cortázar et al., 2014 http://genomics.cicbiogune.es/SECRETOOL
DFVF (database of fungal virulence factors)	It consists of 2058 genes related to virulence of fungal pathogens from 85 genera.	Lu et al., 2012 http://sysbio.unl.edu/DFVF/
FSRD (fungal stress response database)	It comprises fungal stress response-related proteins with their validated physiological function in scientific literature.	Karányi et al., 2013 http://internal.med.unideb.hu/fsrd
FunSecKB (the fungal secretome knowledge base)	It includes all the secretary genes derived from openly accessible fungal protein databases such as NCBI RefSeq.	Lum and Min, 2011 http://proteomics.ysu.edu/secretomes/fungi.php

TABLE 18.2 *(Continued)*

Database	Important features of database	Reference/website
FSD (integrated platform for annotation of fungal secretome)	It consists of 2,08,883 putative secreted proteins from 158 fungal/oomycete species identified by nine prediction programs.	Choi et al., 2010 http://fsd.snu.ac.kr/
FunSecKB2 (fungal protein subcellular location knowledgebase)	It comprises updated database of fungal secretory proteins and subcellular location of protein.	Meinken et al., 2014 http://proteomics.ysu.edu

TABLE 18.3 Bioinformatics Tools for the Prediction of Secretory Proteins From Fungal Genome.

Bioinformatics tool	Important features of tool	Reference/website
SignalP server 5.0	Predicts the signal peptide domain in proteins from an organism using various artificial neural networks.	Nielsen et al., 2019 www.cbs.dtu.dk/services/SignalP
Philius server	Uses dynamic Bayesian networks to predict the signal peptides and topology of the protein.	Reynolds et al., 2008 http://www.yeastrc.org/philius
SPOCTOPUS server	It is appropriate for genome-scale studies to predict signal peptides and membrane domain in protein.	Viklund et al., 2008 http://octopus.cbr.su.se/
SPSCAN	It classifies on the basis of signal peptide score of sequence to predict secretory proteins in whole genome of an organism.	http://www.csd.hku.hk/bruhk/gcgdoc/spscan.html
Phobius web server	Uses the computational tools to predict transmembrane domain and signal peptide simultaneously and provide better results.	Käll et al., 2007 http://phobius.binf.ku.dk
ProtComp	It is AI based which uses neural networks to predict on the basis of comparison with existing homologous proteins. It predicts on the basis of five factors: signal peptides, GPI-anchors, signal-anchors, transmembrane sites, and transit peptides of mitochondria.	http://linux1.softberry.com
PrediSi (PREDIction of SIgnal peptides)	It uses a position weight matrix approach to develop this tool and it is trained on sequences derived from the SwissProt database.	Hiller et al., 2004 http://www.predisi.de/home.html
Signal-CF	It uses the computational methods that predict secretory proteins and later recognize signal peptide cleavage site.	Chou and Shen, 2007 http://chou.med.harvard.edu/bioinf/Signal-CF/

TABLE 18.3 *(Continued)*

Bioinformatics tool	Important features of tool	Reference/website
Signal-3L	Consists of OET-KNN and PseAA engines for prediction of secretory proteins; a subsite-coupled discrimination algorithm is used to predict cleavage sites of signal peptide.	Shen and Chou, 2007 http://chou.med.harvard.edu/bioinf/Signal-3L/
WoLF PSORT	It converts protein sequences into numerical localization features.	Horton et al., 2007 www.wolfpsort.org

TABLE 18.4 Bioinformatics Tools Available for the Prediction of Transmembrane (TM) Domain in Protein.

Bioinformatics tool	Important features of tool	Reference/website
TMHMM Server v. 2.0	Predicts trans-membrane (TM) domains.	www.cbs.dtu.dk/services/TMHMM/
TopPred 1.10	Used to predict topology of membrane proteins.	http://mobyle.pasteur.fr/cgi-bin/portal.py
PredictPrnotein	It is better for depicting secondary structure of protein.	Rost et al., 2004 www.predictprotein.org/
DAS-TMfilter server	False positive is major problem in most of bioinformatic tools used for the prediction of TM domain. It is a better tool to remove false positives.	http://mendel. imp.ac.at/sat/DAS/DAS.html
TMpred	The TMpred is a good tool to predict the TM region and orientation in protein.	Hofman, 1993 www.ch.embnet.org
TMbase	TMbase is an important tool for analyzing the properties of TM proteins.	Hofman, 1993 www.ch.embnet.org
CCTOP (consensus constrained TOPology prediction)	CCTOP predict TM topology of proteins.	Dobson et al., 2015 http://cctop.enzim.ttk.mta.hu/
HMMTOP	It is a webserver for predicting TM helices and topology of proteins.	Tusnady and Simon, 2001 www.enzim.hu/hmmtop/
PHDhtm	PHDhtm use artificial neural networks to predict helical TM segments in the membrane proteins.	https://npsa-prabi.ibcp.fr/cgi-bin/npsa_automat.pl?page=/NPSA/npsa_htm.html
MetaTM	Depicts the TM proteins via a consensus method.	Klammer et al., 2009 http://metatm.sbc.su.se/
ConPred	It predicts TM topology by combining various methods.	Arai et al., 2004 http://bioinfo.si.hirosaki-u.ac.jp/~ConPred2

TABLE 18.4 *(Continued)*

Bioinformatics tool	Important features of tool	Reference/website
SOSUI	It used to identify transmembrane protein and depicts its secondary structures.	Hirokawa et al., 1998 http://harrier.nagahama-i-bio.ac.jp/sosui/
MINNOU	It classifies the transmembrane proteins with and without using hydropathy profiles of protein and their alignments.	http://minnou.cchmc.org/
SCAMPI	Uses statistics-based topology predictors and position-specific amino acid impact on the membrane insertion free energy.	Bernsel et al., 2008 http://scampi.cbr.su.se/
PRED-TMR	It defines the hydrophobicity for the membrane protein.	Pasquier et al., 1999 http://athina.biol.uoa.gr/
TMBB-DB	It uses Freeman–Wimley algorithm for the TM protein predictions.	Freeman and Wimley, 2012 http://beta-barrel.tulane.edu/
SOMRuler	It is one of the commonly used TM helices predictor.	Yu et al., 2011 www.csbio.sjtu.edu.cn/bioinf/SOMRuler//
Phobius	It is a complete package with TM and signal peptide predictions.	Käll et al., 2007 http://phobius.sbc.su.se/poly.html

18.2.2 STUDY OF PATHOGEN AND POPULATION GENETICS

Pathogen genetics is important as genetic makeup of the pathogen determines the ability of pathogen to cause disease, that is, pathogenicity, survival of the pathogen, development of pesticide resistance, and virulence of pathogen (Croll et al., 2018). With advancements in molecular biology, early detection of pathogen is done by testing the presence of its DNA or molecule produce by pathogen or plant on its infection. Pathogen genetics is better understood with available NGS sequencing technologies, transcriptomics, and proteomics approaches. Bioinformatic tools have a crucial role in drawing interpretations from the omics technologies. Population genetics is very important as pathogen being small in size and large in number, it studied at population level, not at individual level. Population genetics is the study of origin, maintenance, and spatial and temporal distribution of genetic variations of the species governed by natural mutations, recombination, genetic drift, genetic flow, and selection (Zhan, 2016). Population genetics in rice fungal pathogens is studied well by using various molecular markers such as RAPD (Random amplified polymorphic DNA), AFLP (Amplified fragment length polymorphism), SNP (single-nucleotide polymorphism), SSR (simple sequence repeats), or genotyping by sequencing (GBS). Microbial typing is recently used for studying population genetics in pathogens with high genetic variability. Microbial typing uses various techniques such as Multilocus sequence typing (MLST) and single-locus sequence typing (SLST). Freely accessible online web server-based databases for MLST data analysis are enlisted in Table 18.5 (Saeb, 2018).

TABLE 18.5 Databases for MLST Data Analysis.

Sr. no.	Bioinformatic tool	*url*
1.	Multi-locus sequence typing	http://www.mlst.net
2.	pubMLST	http://www.pubmlst.org
3.	Institute pasteur MLST	http://www.pasteur.fr/mlst/
4.	European working group for legionella infections (EWGLI) Sequence-based typing database	http://www.hpabioinformatics.org.uk/legionella/legionella_sbt/php/sbt_homepage.php/
5.	Environmental research Institute, University College Cork	http://mlst.ucc.ie/

18.3 BIOINFORMATICS PLATFORM FOR R-GENE ANALYSIS

As discussed earlier in effector triggered immunity, resistance gene (R-gene) of the plant recognizes the specific effector molecules encoded by Avr genes of the pathogen (Ray et al., 2019). R-genes encode for some specific domains that are conserved across the species. These domains of R-protein interact with effector protein and with downstream defense response signaling pathways. R-genes are classified into five classes on the basis of the specific protein domains:

1. CNL class: Proteins contain a coiled-coil domain, a nucleotide-binding domain, and leucine-rich repeats [CC-NB-LRR].
2. TNL class: Protein with toll-interleukin receptor, a nucleotide-binding domain, and a leucine-rich repeat [TIR-NB-LRR].
3. RLP class: Receptor-like proteins include a receptor serine-threonine kinase-like domain and leucine-rich repeats [Ser/thr-LRR].
4. RLK class: R-proteins include a kinase domain and leucine-rich repeats [Kin-LRR].
5. Others class: it includes other R-genes that provide resistance through different mechanisms like asc-1 and mlo.

The most popular bioinformatic database available for the prediction of resistance gene in plants is PRGdb (Plant Resistance Gene Database).

This provides the plant science researcher with a comprehensive and timely updated database of R-genes from available literature and putative R-genes predicted through bioinformatics pipelines. The PRG data is deposited in a MySQL database openly available and freely accessible using a web interface http://www.prgdb.org. ESTScan software is used to translate unigene sequences from various plant species into potential protein sequences. The InterProScan software is used for domain analysis of selected gene sequences and to classify the R-genes into five classes as mentioned above (Sanseverino et al.,
2010) (Fig. 18.1).

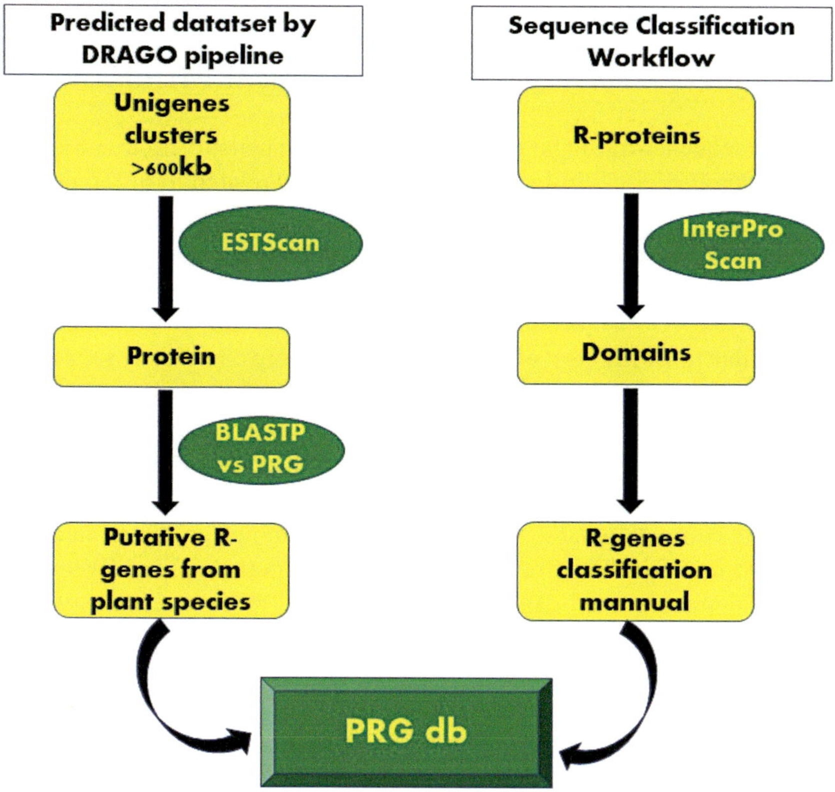

FIGURE 18.1 Diagrammatic representation of PRG database.

Disease Resistance Analysis and Gene Orthology (DRAGO 2) is the latest version of the bioinformatic pipeline used to annotate R-genes in plants. DRAGO 2 is compatible with either DNA or amino acid sequences in the FASTA format retrieved from transcriptome or proteome data. DRAGO2 pipeline uses HMMER v3 to detect LRR, NBS, Kinase, and TIR domains; COILS 2.2 for the detection of CC domain and TM domains is predicted using TMHMM 2.0c programs. PRGdb consist of 177072 putative PRGs analyzed with DRAGO 2 in 76 species, 153 reference PRGs in 35 plant species. PRGdb includes a total of 99 plant species with annotated PRGs. The PRGdb database contains information of 12 species of genus Oryza containing various R-genes-specific fungal pathogens. The noncultivated species of rice are also found as source of fungal disease resistance.

18.4 BIOINFORMATICS IN NGS OF RICE FUNGAL PATHOGENS

Genome sequence of fungal pathogen provides the genetic information from which pathogenicity, virulence factors, and genes related to different pathways involved in pathogenesis can be predicted. Genomic blueprint helps to understand the adaptability of pathogen in diverse environments and their evolution over a span of time (Aylward et al., 2017). With availability of a number of NGS sequencing technologies sequencing cost is drastically lowered due to which most of the plant pathogens are completely sequenced. *Magnaporthe oryzae* causing rice blast was first fungal pathogen to be completely sequenced in 2005 by Dean et al. Whole-genome sequencing of the important fungal pathogen of rice is available in public databases. Fungal genome sequences are available through online servers that utilize a number of computational tools to store the huge amount of data in databases that could be retrieved by researchers for further potential uses (Wee et al., 2018). Commonly used databases to submit and retrieve fungal genomic information are Genomes OnLine Database (GOLD;), NCBI Genome (www.ncbi.nlm.nih.gov/genome), MycoCosm (, 2013), Institute (www.broadinstitute.org). NGS sequencing platforms used for genome sequencing are Roche/454 (Margulies et al., 2005) Illumina/Solexa (Bennett et al., 2004; Shen et al., 2005), SOLiD, Ion torrent, etc. Now, third-generation sequencing (TGS) technologies are in market with better accuracy de novo assemblies including Pacific Biosciences (PacBio) single-molecule real-time (SMRT) sequencing, BioNano Genomics (BioNano) sequencing, and Oxford Nanopore Technologies (ONT) sequencing (YongKiat et al., 2018). All these sequencing technologies generate a number of long reads and short reads that are assembled with the help of various computational algorithms-based overlapping of reads, that is, OLC (overlap–layout–consensus) algorithms. The de novo assembles bioinformatics tools used for TGS sequencing technologies are presented in the from of decision tree for researchers in Figure 18.2. SMRT platform with long reads uses MinHash Alignment Process (MHAP), Hierarchical Genome-Assembly Process (HGAP), PBJelly, HINGE, and FALCON; BioNano platform uses RefAligner, a dynamic programming algorithm; the ONT platform makes use of PoreSeq and Nanocorr (YongKiat et al., 2018). Presently, scientists use a combination of NGS and TGS platforms to get better accuracy and high coverage. The bioinformatics tools used for the assembly in this hybrid strategy include DBG2OLC, NaS, OPERA-LG, Cerulean, and GMCloserT, etc., presented as a decision tree. The genome data analysis is nowhere possible to be manually carried by a human; it seeks assistance from computational tools (YongKiat et al., 2018). The genome-based alignment comprises various steps such as data processing and quality controls. The tools used for different purposes in TGS data processing are present in decision tree to assist the researcher to choose the available tool for their data analysis (Fig. 18.3).

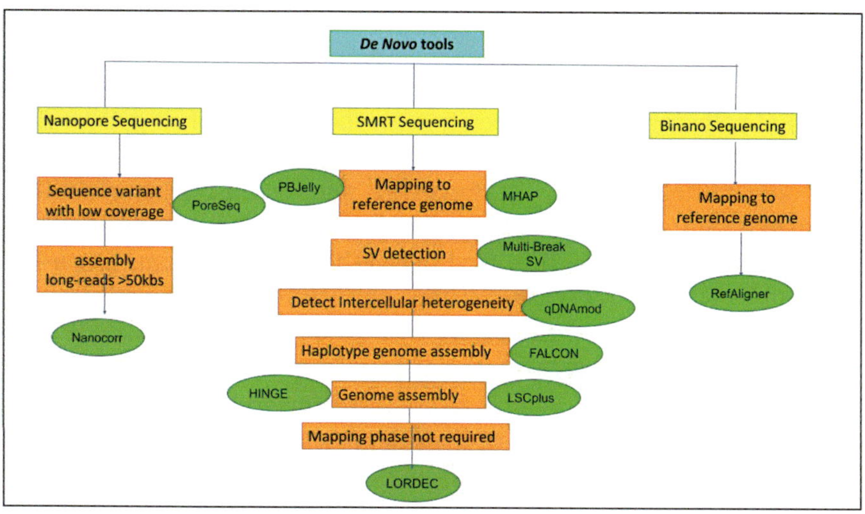

FIGURE 18.2 A decision tree for the selection of de novo assembles bioinformatics tools used for TGS sequencing technologies such as nanopore sequencing, SMRT sequencing, and binano sequencing.

Source: Adapted from Wee et al., 2019.

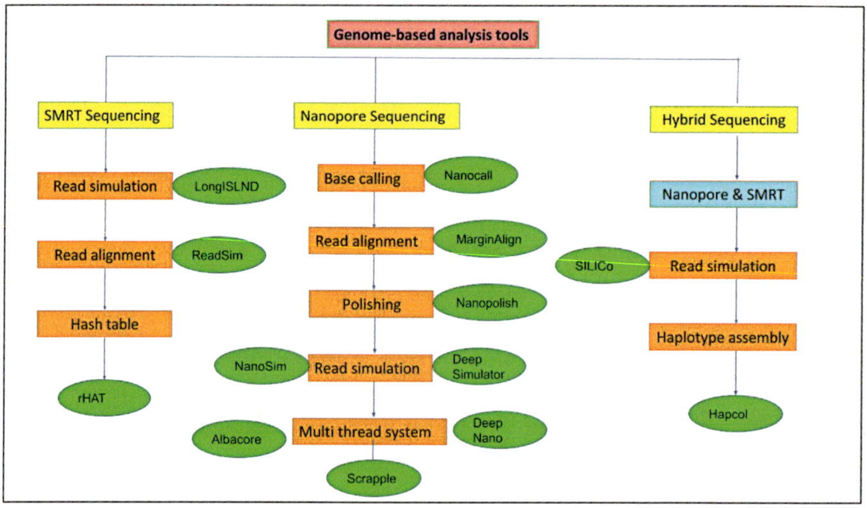

FIGURE 18.3 A decision tree for researchers to choose the suitable bioinformatics tools for sequencing analysis in TGS genome platforms such as nanopore, SMRT sequencing, and hybrid sequencing.

Source: Adapted from Wee et al., 2019.

18.5 CONCLUSIONS

Bioinformatics has become integral part of biological research at present and plant pathology is not expectation for this. Earlier plant pathological experiments were being carried out in laboratories, glass houses, and then move to fields, but now it starts with computer using computational tools (in silico) to analyze the available data and develop hypothesis for experiment planning. It is used in rice-breeding programs at various levels to develop the disease-resistant cultivars against fungal diseases of rice and successful varieties are available against rice blast. Bioinformatic tools have played a crucial role in understanding biology of *Magnaporthe oryzae*, *Bipolaris oryzae*, *Rhizoctonia solani*, and other pathogens. Various tools are used to study host pathogen interaction, to identify potential effector molecules of rice fungal pathogens and R-genes in rice plant. Molecular research in plant pathology cannot be thought of without intervention of bioinformatic tools and computational algorithms.

KEYWORDS

- **bioinformatics tools**
- **computational tools**
- **effector**
- **fungal diseases**
- **rice**
- **R-genes**
- **TGS sequencing**

REFERENCES

Arai, M.; Mitsuke, H.; Ikeda, M.; Xia, J. X.; Kikuchi, T.; Satake, M.; Shimizu, T. ConPred II: A Consensus Prediction Method for Obtaining Transmembrane Topology Models With High Reliability. *Nucleic Acids Res.* **2014**, *32* (suppl_2), W390–W393.

Armenteros, J. J. A.; Tsirigos, K. D.; Sønderby, C. K.; Petersen, T. N.; Winther, O.; Brunak, S.; von Heijne, G.; Nielsen, H. Signal ,P 5.0 Improves Signal Peptide Predictions Using Deep Neural Networks. *Nature Biotech.* **2019,** *37* (4), 420–423.

Aylward, J.; Steenkamp, E. T.; Dreyer, L. L.; Roets, F.; Wingfield, B. D.; Wingfield, M. J. A Plant Pathology Perspective of Fungal Genome Sequencing. *IMA Fungus* **2017,** *8* (1), 1–15. https://doi.org/10.5598/imafungus.

Bennett, S. Solexa Ltd. *Pharmacogenomics* **2004,** *5*, 433–8. 4.

Cao, B.; Porollo, A.; Adamczak, R.; Jarrell, M.; Meller, J. SP1. Multistage Protocol for Transmembrane Helix Prediction in MINNOU.

Choi, J.; Park, J.; Kim, D.; Jung, K.; Kang, S.; Lee, Y. H. Fungal Secretome Database: Integrated Platform for Annotation of Fungal Secretomes. *BMC Genom.* **2010,** *11* (1), 105.

Chou, K. C.; Shen, H. B. Signal-CF: A Subsite-Coupled and Window-Fusing Approach for Predicting Signal Peptides. *Biochem. Biophys. Res. Commun.* **2007,** *357* (3), 633–640.

Cortázar, A. R.; Aransay, A. M.; Alfaro, M.; Oguiza, J. A.; Lavín, J. L. Secretool: Integrated Secretome Analysis Tool for Fungi. *Amino Acids* **2014,** *46* (2), 471–473.

Croll, D.; McDonald, B. A. The Genetic Basis of Local Adaptation for Pathogenic Fungi in Agricultural Ecosystems. *Mol. Ecol.* **2017,** *26* (7), 2027–2040.

Dean, R. A.; Talbot, N. J.; Ebbole, D. J.; Farman, M. L.; Mitchell, T. K.; Orbach, M. J.; Thon, M.; Kulkarni, R.; Xu, J. R.; Pan, H.; Read, N. D. The Genome Sequence of the Rice Blast Fungus *Magnaporthe grisea*. *Nature* **2005,** *434* (7036), 980–986.

Dobson, L.; Reményi, I.; Tusnády, G. E. CCTOP: A Consensus Constrained TOPology Prediction Web Server. *Nucleic Acids Res.* **2015,** *43* (W1), W408–W412.

Eckardt, N. A. Sequencing the Rice Genome. *Plant Cell* **2000,** *12* (11). DOI: 10.1105/tpc.12.11.2011.

Edwards, D.; Batley, J. Plant Bioinformatics: From Genome to Phenome. *Biotechology* **2004,** *22* (5), 232–237.

Freeman, Jr, T. C.; Wimley, W. C. TMBB-DB: A Transmembrane β-barrel Proteome Database. *Bioinformatics* **2012,** *28* (19), 2425–2430.

Grigoriev, IV.; Nikitin, R.; Haridas, S.; Kuo, A.; Ohm, R.; Riley, R.; Salamov, A.; Zhao, X.; Korzeniewski, F.; Smirnova, T. MycoCosm Portal: Gearing up for 1000 Fungal Genomes. *Nucleic Acids Res.* **2013.** 10.1093/nar/gkt1183.

Grigoriev, IV.; Nordberg, H.; Shabalov, I.; Aerts, A.; Cantor. M.; Kuo, A.; Minovitsky, S.; Nikitin, R.; Ohm, R. A.; Otillar, R. The Genome Portal of the Department of Energy Joint Genome Institute. *Nucleic Acids Res.* **2012,** *40*, D26–D32.

Hiller, K.; Grote, A.; Scheer, M.; Münch, R.; Jahn, D. PrediSi: Prediction of Signal Peptides and Their Cleavage Positions. *Nucleic Acids Res.* **2004**, *32* (suppl_2), W375–W379.

Hirokawa, T.; Boon-Chieng, S.; Mitaku, S. SOSUI: Classification and Secondary Structure Prediction System for Membrane Proteins. *Bioinformatics (Oxford, England)* **1998,** *14* (4), 378–379.

Horton, P.; Park, K. J.; Obayashi, T.; Fujita, N.; Harada, H.; Adams-Collier, C. J.; Nakai, K. WoLF PSORT: Protein Localization Predictor. *Nucleic Acids Res.* **2007,** *35* (suppl_2), W585–W587.

Jayaram, B.; Dhingra, P. *Bioinformatics for a Better Tomorrow*; Indian Institute of Technology: New Delhi, India, 2010.

Käll, L.; Krogh, A.; Sonnhammer, E. L. Advantages of Combined Transmembrane Topology and Signal Peptide Prediction—The Phobius Web Server. *Nucleic Acids Res.* **2007,** *35* (suppl_2), W429–W432.

Kamoun, S. A Catalogue of the Effector Secretome of Plant Pathogenic Oomycetes. *Annu. Rev. Phytopathol.* **2006,** *44,* 41–60. DOI: 10.1146/annurev.phyto.44.070505.143436.

Karányi, Z.; Holb, I.; Hornok, L.; Pócsi, I.; Miskei, M. FSRD: Fungal Stress Response Database. *Database* [Online] **2013.** https://doi.org/10.1093/database/bat037

Koltai, H.; Volpin, H. Agricultural Genomics: An Approach to Plant Protection. *Eur. J. Plant Pathol.* **2003,** *109,* 101–108.

Lilly, J. J.; Subramanian, B. Gene Network Mediated by WRKY13 to Regulate Resistance Against Sheath Infecting Fungi in Rice (Oryza Sativa L.). *Plant Sci.* **2019,** *280*, 269–282.

Lu, T.; Yao, B.; Zhang, C. DFVF: Database of Fungal Virulence Factors. *Database* **2012**.

Lum, G.; Min, X. J. FunSecKB: The Fungal Secretome Knowledgebase. *Database* **2011**.

Margulies, M.; Egholm, M.; Altman, W. E.; Attiya, S.; Bader, J. S.; Bemben, L. A.; Berka, J.; Braverman, M. S.; Chen, Y. J.; Chen, Z.; Dewell, S. B. Genome Sequencing in Microfabricated High-Density Picolitre Reactors. *Nature* **2005,** *437* (7057), pp 376–380.

Meinken, J.; Asch, D. K.; Neizer-Ashun, K. A.; Chang, G. H.; Cooper JR, C. R.; Min, X. J. FunSecKB2: A Fungal Protein Subcellular Location Knowledgebase. *Comput. Mol. Biol.* **2014,** *4* (4).

Nemri, A.; Saunders, D. G.; Anderson, C.; Upadhyaya, N. M.; Win, J.; Lawrence, G. J.; Jones, D.; Kamoun, S.; Ellis, J.; Dodds, P. The Genome Sequence and Effector Complement of the Flax Rust Pathogen *Melampsora Lini*. *Front. Plant Sci.* **2014,** *5,* 98. DOI: 10.3389/fpls.2014.00098

Ojo, O. O.; Maxwell, O. Incorporating Bioinformatics Into Biological Science Education in Nigeria: Prospects and Challenges. *Infect. Genet. Evol.* **2010.** 10.1016/j.meegid2010.11.015

Pasquier, C.; Promponas, V. J.; Palaios, G. A.; Hamodrakas, J. S.; Hamodrakas, S. J. A Novel Method for Predicting Transmembrane Segments in Proteins Based on a Statistical

Analysis of the SwissProt Database: The PRED-TMR Algorithm. *Protein Eng.* **1999,** *12* (5), 381–385.

Ramadan, H. E.; Rossbach, C. J.; Porter, D. E.; Hofmann, O. S.; Bhandari, A.; Witchel, E. MetaTM/TxLinux: Transactional Memory for an Operating System. *ACM SIGARCH Comput. Archit. News* **2007,** *35* (2), 92–103.

Rao, V. S.; Das, S. K.; Rao, V. J.; Srinubabu, G. Recent Developments in Life Sciences Research: Role of Bioinformatics. *Afr. J. Biotechnol.* **2008,** *7 (5),* 495–503.

Ray, S. K.; Macoy, D. M.; Kim, W. Y.; Lee, S. Y.; Kim, M. G. Role of RIN4 in Regulating PAMP-Triggered Immunity and Effector-Triggered Immunity: Current Status and Future Perspectives. *Mol. Cells* **2019,** *42* (7), 503.

Reddy, T. B.; Thomas, A. D.; Stamatis, D.; Bertsch, J.; Isbandi, M.; Jansson, J.; Mallajosyula, J.; Pagani, I.; Lobos, E. A.; Kyrpides, N. C. The Genomes OnLine Database (GOLD) v. 5: A Metadata Management System Based on a Four Level (meta) Genome Project Classification. *Nucleic Acids Res.* **2015,** *43* (D1), D1099–D1106.

Reynolds, S. M.; Käll, L.; Riffle, M. E.; Bilmes, J. A.; Noble, W. S. Transmembrane Topology and Signal Peptide Prediction Using Dynamic Bayesian Networks. *PLoS Comput. Biol.* **2008,** *4* (11), e1000213.

Rost, B.; Yachdav, G.; Liu, J. The Predictprotein Server. *Nucleic Acids Res.* **2004,** *32* (suppl_2), W321–W326.

Saeb, A. T. M. Current Bioinformatics Resources in Combating Infectious Diseases. *Biomed. Inform.* **2018,** *14,* 31–35.

Sanseverino, W.; Roma, G.; Simone, M. D.; Faino, L.; Melito, S.; Stupka, E.; Frusciante, L.; Ercolano, M. R. PRGdb: A Bioinformatics Platform for Plant Resistance Gene Analysis. *Nucleic Acids Res.* **2010,** 38, 814–821. DOI: 10.1093/nar/gkp978

Shen, H. B.; Chou, K. C. Signal-3L: A 3-layer Approach for Predicting Signal Peptides. *Biochem. Biophys. Res. Commun.* **2007,** *363* (2), 297–303.

Shen, R.; Fan, J. B.; Campbell, D.; Chen, J.; Doucet, D.; Yeakley, J.; Bibikova, M.; Garcia, E. W.; McBride, C.; Steemers, F. High-Throughput SNP Genotyping on Universal Bead Arrays. *Mutat. Res. Fund. Mol. Mech. Mut.* **2005,** *573* (1–2), 70–82.

Singh, J.; Aggarwal, R.; Gurjar, M. S.; Sharma, S.; Jain, S.; Saharan, M. S. Identification and Expression Analysis of Pathogenicity-Related Genes in *Tilletia Indica* Inciting Karnal Bunt of Wheat. *Australas. Plant Pathol.* [Online] **2020.** https://doi.org/10.1007/s13313-020-00711-x.

Singh, J.; Aggarwal, R.; Gurjar, M. S.; Sharma, S.; Saharan, M. S. Identification of Carbohydrate Active Enzymes From Whole Genome Sequence of *Tilletia Indica* and Sporulation Analysis. *Indian J. Agric. Sci.* **2019,** *89* (6), 1023–1026.

Singh, V. K.; Singh, A. K.; Chand, R.; Kushwaha, C. Role of Bioinformatics in Agriculture and Sustainable Development. *Int. J. Bioinform. Res.* **2011,** *3* (2), 221–226

Sonah, H.; Deshmukh, R. K.; Bélanger, R. R. Computational Prediction of Effector Proteins in Fungi: Opportunities and Challenges. *Front. Plant Sci.* **2016,** *7*, 126.

Tusnady, G. E.; Simon, I. The HMMTOP Transmembrane Topology Prediction Server. *Bioinformatics* **2001,** *17* (9), 849–850.

Vassilev, D.; Nenov, A.; Atanassov, A.; Dimov, G.; Getov, L. Application of Bioinformatics in Fruit Plant Breeding. *J. Fruit Ornam. Plant Res.* **2006,** (14), 145–162.

Vencato, M.; Tian, F.; Alfano, J. R.; Buell, C. R.; Cartinhour, S.; DeClerck, G. A.; Guttman, D. S.; Stavrinides, J.; Joardar, V.; Lindeberg, M.; Bronstein, P. A.; Mansfield, J. W.; Myers, C. R.; Collmer, A.; Schneider, D. J. Bioinformatics-Enabled Identification of the HrpLRegulon and Type III Secretion System Effector Proteins of *Pseudomonas syringae* pv. 1448A. *Mol. Plant Microbe In.* **2006,** *19* (11), 1193–1206.

Viklund, H.; Bernsel, A.; Skwark, M.; Elofsson, A. SPOCTOPUS: A Combined Predictor of Signal Peptides and Membrane Protein Topology. *Bioinformatics* **2008,** *24* (24), 2928–2929.

Wee, Y.; Bhyan, S. B.; Liu, Y.; Lu, J.; Li, X.; Zhao, M. The Bioinformatics Tools for the Genome Assembly and Analysis Based on Third-Generation Sequencing. *Brief. Funct Genom.* **2018,** *18* (1), 1–12.

Winnenburg, R.; Urban, M.; Beacham, A.; Baldwin, T. K.; Holland, S.; Lindeberg, M.; Hansen, H.; Rawlings, C.; Hammond-Kosack, K. E.; Köhler, J. PHI-Base Update: Additions to the Pathogen–Host Interaction Database. *Nucleic Acids Res.* **2007,** *36* (suppl_1), D572–D576.

Yu, D.; Shen, H.; Yang, J. SOMRuler: A Novel Interpretable Transmembrane Helices Predictor. *IEEE Trans. Nanobiosci.* **2011,** *10* (2), 121–129.

Zhan, J. Population Genetics of Plant Pathogens. *eLS* **2016,** 1–7.

Index

C

N

O

P

Q

R